# Tactical Persistent Surveillance Radar with Applications

## Related titles on radar:

**Advances in Bistatic Radar** Willis and Griffiths
**Airborne Early Warning System Concepts, 3rd Edition** Long
**Bistatic Radar, 2nd Edition** Willis
**Design of Multi-Frequency CW Radars** Jankiraman
**Digital Techniques for Wideband Receivers, 2nd Edition** Tsui
**Electronic Warfare Pocket Guide** Adamy
**Foliage Penetration Radar: Detection and characterization of objects under trees** Davis
**Fundamentals of Ground Radar for ATC Engineers and Technicians** Bouwman
**Fundamentals of Systems Engineering and Defense Systems Applications** Jeffrey
**Introduction to Electronic Warfare Modeling and Simulation** Adamy
**Introduction to Electronic Defense Systems** Neri
**Introduction to Sensors for Ranging and Imaging** Brooker
**Microwave Passive Direction Finding** Lipsky
**Microwave Receivers with Electronic Warfare Applications** Tsui
**Phased-Array Radar Design: Application of radar fundamentals** Jeffrey
**Pocket Radar Guide: Key facts, equations, and data** Curry
**Principles of Modern Radar, Volume 1: Basic principles** Richards, Scheer and Holm
**Principles of Modern Radar, Volume 2: Advanced techniques** Melvin and Scheer
**Principles of Modern Radar, Volume 3: Applications** Scheer and Melvin
**Principles of Waveform Diversity and Design** Wicks *et al.*
**Pulse Doppler Radar** Alabaster
**Radar Cross Section Measurements** Knott
**Radar Cross Section, 2nd Edition** Knott *et al.*
**Radar Design Principles: Signal processing and the environment, 2nd Edition** Nathanson *et al.*
**Radar Detection** DiFranco and Ruby
**Radar Essentials: A concise handbook for radar design and performance** Curry
**Radar Foundations for Imaging and Advanced Concepts** Sullivan
**Radar Principles for the Non-Specialist, 3rd Edition** Toomay and Hannan
**Test and Evaluation of Aircraft Avionics and Weapons Systems** McShea
**Understanding Radar Systems** Kingsley and Quegan
**Understanding Synthetic Aperture Radar Images** Oliver and Quegan
**Radar and Electronic Warfare Principles for the Non-specialist, 4th Edition** Hannen
**Inverse Synthetic Aperture Radar Imaging: Principles, algorithms and applications** Chen and Marotella
**Stimson's Introduction to Airborne Radar, 3rd Edition** Baker, Griffiths and Adamy
**Test and Evaluation of Avionics and Weapon Systems, 2nd Edition** McShea
**Angle-of-Arrival Estimation Using Radar Interferometry: Methods and applications** Holder
**Biologically-Inspired Radar and Sonar: Lessons from nature lessons from nature** Balleri, Griffiths and Baker
**The Impact of Cognition on Radar Technology** Farina, De Maio and Haykin
**Novel Radar Techniques and Applications, Volume 1: Real aperture array radar, imaging radar, and passive and multistatic radar** Klemm, Nickel, Gierull, Lombardo, Griffiths and Koch
**Novel Radar Techniques and Applications, Volume 2: Waveform diversity and cognitive radar, and target tracking and data fusion** Klemm, Nickel, Gierull, Lombardo, Griffiths and Koch

# Tactical Persistent Surveillance Radar with Applications

David Lynch, Jr

SciTech Publishing
an imprint of the IET
theiet.org

Published by SciTech Publishing, an imprint of the Institution of Engineering and Technology, London, United Kingdom.

The Institution of Engineering and Technology is registered as a Charity in England & Wales (no. 211014) and Scotland (no. SC038698).



First published 2018



The Institution of Engineering and Technology
Michael Faraday House
Six Hills Way, Stevenage
Herts SG1 2AY, United Kingdom

www.theiet.org



**British Library Cataloguing in Publication Data**
A catalogue record for this product is available from the British Library

**ISBN 978-1-78561-650-1 (hardback)**
**ISBN 978-1-78561-651-8 (PDF)**

Typeset in India by MPS Limited
Printed in the UK by CPI Group (UK) Ltd, Croydon

# Contents

# List of figures

# List of tables

# Preface

The object of this book is to introduce technologists to the essential elements of persistent surveillance of tactical targets from both a hardware and software point of view, using simple Mathcad, Excel and Basic examples with real data. The Mathcad programs are contained in the Appendices for each chapter, along with an Adobe Acrobat file of the text of the programs as well as the plotted results of a program run. The real data for use in the enclosed or other programs is also included. Many of the nuts and bolts of practical real processing and detection are shown. More sophisticated methods are alluded to but not demonstrated. Everything from passive target detection and tracking, radar detection, antenna monopulse, active electronic scanned antennas (AESA), moving target tracking (MTT) with radar for cueing of EO and IR sensors leading to recognition and weapon delivery, motion compensation, tactical target spectral characteristics, moving target detection (MTI), space time adaptive processing (STAP), active surface target imaging (ISAR), synthetic aperture radar (SAR) imaging, change detection (CCD) and synthetic monopulse are introduced with examples. The Mathcad software should allow the curious to experiment with their own parameters and notions (play-with) to achieve a greater understanding of the underlying behaviors. The text introduces monopulse, which is the simplest multiple phase center antenna. It builds on this notion for both tracking and as a stepping stone to modern space-time adaptive arrays.

This book is a summary of the courses that my colleagues and your author taught between 1977 and 2014 on the general topic of surface tactical target surveillance from aircraft and spacecraft. Although the author has taught air-to-ground weapon delivery as well as weapon guidance, there is very little of that in the present text. When originally taught in the late 1970s, the 1980s and early 1990s, almost all of the airborne platforms were manned. Obviously the spacecraft weren't generally manned, but the spacecraft were not really surveilling tactical targets.

Since the mid-1990s things have changed, the focus has been on "conflicts-other-than-war" and irregular or guerrilla warfare. The author participated in a Rand study on this very topic in 1995–1996. In addition to technologists, we had retired military, active Navy SEALs and Army Delta Force members with real extensive combat experience in irregular warfare as well as very experienced active FBI and Los Angeles SWAT team members. As you might imagine, there were many ideas. One of the central themes was: how anticipate trouble before it starts or as it starts? What was required was persistent surveillance in all dimensions of information, space, time and wavelength. In separate programs some of us experimented with small crawlers, swimmers and fliers; detection of concealed weapons, tracing the source of small projectiles (bullets) in real time, as well as fixed covert surveillance means harvesting stray electric energy. We also explored nonlethal means of marking and disabling potential assailants. Like you might imagine much of this went nowhere, either for legal or practical reasons. One especially frustrating set of experiments was placing ultra-small TV cameras hosted on insects and small reptiles for use as covert surveillance in hostage and sniper situations. The idea was to use small electrical zapping pulses from the covert camera to make the host go where one wanted. It turns out that it is easier to kill a cockroach or reptile by zapping than it is to get them to go where you want!

Today, small to very large drone/unmanned autonomous vehicle (UAV) swimmers, crawlers and fliers are available that can carry acoustic, EO/IR cameras, passive RF sensors and radars (as well as Amazon packages!). Some drones now have endurance of days or more, which allows persistent surveillance. Spacecraft provide long-term surveillance of most of the earth at many wavelengths but at times that are *not* persistent or timely. Fortunately, with the advent of long-endurance drone platforms like Predator, Reaper, Scan Eagle, as well as many others, persistent surveillance is possible in many areas of concern around the world. There are many other world regions where persistent surveillance can only be accomplished with new sensors and platforms.

Your author made many contributions to these technologies, *but* most methods and notions in this book are the creation of others including Stan Aks, Pete Bogdanovic, Gary Graham, Joel Mellema, fred harris, Ralph Hudson, Eddie Phillips, Jack Pearson, Sam Blackman, Nate Greenblatt, Ralph Gifford, Jeff Hoffner, Jack Stein, Dan Rivers, Rudy Marloth, Ivan Bottlik, Tom Kennedy, Kan Jew, Howie Nussbaum, John Pearson, Hugh Washburn, Lee Tower, Steve Iglehart, Fred Williams, Don Stuart, Milt Radant, Dave Kramer, Charlie Smith, Mark Landau, Gene Gregory, Atul Jain, Bill Milroy, Don Parker, David Whelan, Chuck Krumm, Ken Perko, Hank McCord and many others. Although some of the ideas in this book are commonly attributed to people on the US East Coast and Canada, most of the work done at Hughes Aircraft and later General Motors Hughes Electronics predated by 5 years or more the material disclosed or patented by these other individuals and companies. The author witnessed a real-time 16 phase center STAP demonstration at Hughes in 1970. The first real-time digital FFT-based airborne SAR maps of 20 foot resolution were demonstrated in 1970, 11 ft in 1971, 7 ft in 1973 and 1 ft in 1977 over very large patch sizes (e.g., all of California from the Mexican border to just north of Oakland) at Hughes Aircraft. As often happens when conditions are right, smart people invent the same thing without the knowledge that others are also doing it. The author doesn't claim anything original in this book, but it will be new to many.

I also want to thank my reviewers, especially Dr. Joe Guerci and Dr. Eric Jensen; they are valued colleagues.

David Lynch, Jr.
IEEE Life Fellow, Senior Member AIAA,
Stealth Pioneer, Life Member Sigma Xi
2017

# About the author

The author is a technology weenie who found himself managing thousands of people. He was neither equipped by education, aptitude or experience to do so. Sometimes, to realize a vision of the future, one must take on jobs that are necessary but not in your life plan. Luckily, the author was surrounded by brilliant, hard-working people who could succeed often enough that together we shaped the future. First, in high-voltage machinery, then in space programs, in digital communications, in digital signal processing, in radar, in stealth, in tactical surveillance and in active antenna arrays. The author has been in on the beginning of many world-changing technologies. More by impatience with the status quo, than by vision of the future, the author has been at the right place at the right time in all these areas.

The author was an inventor, leader or contributor to many world firsts including manned spaceflight, telecommunications, digital signal processing, synthetic aperture radar and stealth. The author was involved in stealth programs including Have Blue, Tacit Blue, F-117, Sea Shadow, Advanced Cruise Missile, B-2, F-22 and others. The author was an engineer on the Gemini and Mercury spacecraft, test equipment for the Saturn V booster, T-1 through T-4 PCM communications systems, helicopter radars, engineer and leader in radar signal processing and displays; in real-time digital SAR systems, space object imaging systems, programmable signal processors, air defense, fighter radars, stealth systems, and leader in major weapon systems, lean manufacturing, engineering quality, electric automobiles, microwave integrated circuits, active array antenna technology; as well as management/technical teaching at companies, universities, government and foreign agencies.

The author is a Pioneer of Stealth and a former president of Pioneers of Stealth. He is a life member of research honorary Sigma Xi, a Senior Member of American Institute of Aeronautics and Astronautics and a Life Fellow of the Institute of Electrical and Electronic Engineers. The author was a company officer of General Motors Hughes Electronics. He is currently president of DL Sciences, Inc. He has written many papers, has eight patents, several technical books, as well as contributions to several other technical books.

The author was an avid mountain climber, ski mountaineer, hiker, runner, swimmer, biker and triathlete. Ultimately, age overtakes these activities and one finds themselves just walking or biking on level ground. Oh well, nobody is perfect.

*Chapter 1*

# Persistent tactical surveillance elements

## 1.1 Introduction

More and more we live in an environment of tactical persistent surveillance. Tactical persistent surveillance is almost continuous observation of a person, place or region primarily by electromagnetic means. If your cellphone is on, its location is likely known within 5 miles or less. If your cellphone or car Onstar™ is GPS enabled, then your location is known within tens of feet. Legal strictures prevent wide use of this information, primarily to protect politicians in developed countries. Every time you use a credit/debit card, your location is known to the accuracy of the service provider's location. Every day there are so many transaction reports for several billion people that even though a complete track of individuals could be maintained, it usually isn't. Your primary privacy protection is your ordinariness.

Downtown London has widespread use of video surveillance. Even with automatic image screening, this surveillance is labor intensive. There are traffic light cameras in many parts of the United States that record the driver, license number and time if they run a red light. These systems are primarily automatic. They are unquestionably income producing for the traffic control agency.

You can easily find a satellite picture of your house on Google Earth™. If you are willing to pay for it, you can purchase a high-resolution (1–2 ft) spacecraft image of your house. The author has a 0.3-m-resolution image (Russian) of his house at Big Bear Lake, CA. These images are usually taken at a specific time of day when the sky is clear. They could be months or years old. The images are hardly persistent surveillance but they do suggest what can be done with significant resources. There are numerous imaging spacecraft operated by the developed nations of the world.

Many countries have radar surveillance of aircraft by both military and civilian authorities. The Federal Aviation Agency (FAA) maintains enroute radar surveillance in the United States. As many US general aviation pilots know, there are many times when their aircraft is not visible even with active augmentation and thousands of radars. In many countries, automatic dependent surveillance broadcast (ADS-B) is used to provide persistent surveillance for all aircraft when radar coverage is limited. ADS-B and most FAA equipment are cooperative systems, that is, they depend on aircraft to aid in the determination of their own location. Another cooperative system is the Automatic Identification System (AIS), a VHF radio system primarily identifying ships at sea.

In undeveloped areas or when people specifically don't want their location or activities to be known, it is possible to greatly reduce or eliminate location accuracy and behavior tracking. Non-cooperative persistent tactical surveillance depends on surface, airborne and spaceborne assets. The last decade has seen substantial investment in persistent surveillance by many countries to control their borders, to provide counter insurgency support and to provide Indications and Warning (I&W) of hostile actions by neighboring countries. Airborne systems such as Wedgetail, AWACS, JSTARS, ASTOR, TR-1, Global Hawk, Predator, Hunter, Scan Eagle, and Reaper shown in Figure 1.1, are examples. Spaceborne systems such as LandSat, SPOT, RADARSAT, TerraSAR-X, and SEASAT also are examples. Many of these platforms carry high-resolution electro-optical (EO) or infrared (IR) imaging sensors. EO sensors are ones that use light near the visible part of the spectrum. That light could be quite dim but there are EO sensors, which are sensitive enough to see in only starlight. IR sensors depend on the radiation given off by the temperature of the target body. Since that body temperature may not be much different than the background, IR sensors often must be cooled to well below 0°F. Even at microwave frequencies there is radiation given off by warm bodies. In fact that is the basis for the microwave surveys of the edges of the early universe done by spacecraft.

The discipline of activity-based intelligence (ABI) is the use of collection and analysis focused on the activity and transactions associated with an item or entity, a population or an area of interest (AOI). It uses intelligence such as measurement and signature intelligence (MASINT), geospatial intelligence or information (GEOINT), signal intelligence, information or intercept (SIGINT), human information or intelligence (HUMINT), pictorial or image intelligence (IMINT), communications intelligence, information or intercept (COMINT) and electronic intelligence, information or intercept (ELINT) to place everything that is known about an item of interest in both space and time. All of these techniques ultimately are forensic, that is, like crime scene investigation (CSI).

EO and IR sensors near the earth's surface are typically limited to 10 km or less in range. The only sensors that can provide long-range (10s to 1000s of

*Figure 1.1 MQ-9 persistent surveillance aircraft [1, courtesy General Atomics]*

kilometers), all-weather, day–night operation are radars and SIGINT/ELINT/COMINT sensors. Radars can provide resolution of a few inches at ranges an order of magnitude beyond EO or IR sensors. EO and IR images are qualitatively different than those from radar returns and other VHF (30–300 MHz), UHF (300–1000 MHz) or microwave emissions (1000 MHz to 40 GHz), as well as COMINT, ELINT, SIGINT, synthetic aperture radar (SAR) and ground moving target indication (GMTI). One can lump all these latter emissions into the category of radio frequency (RF).

Most RF sensing is much more long range but often with much poorer resolution than optical sensors. Successful ABI requires the use of a hierarchy of sensing with the longest range lowest resolution sensors cueing progressively higher-resolution shorter-range sensors until everything desired to be known is observed as shown in Figure 1.2. The location and typical feature accuracy is stated for each information source. The process typically starts with what can be read on the internet or in the newspaper and ends with direct observation with human eyes. The typical RF sequence is SIGINT/ELINT/COMINT then GMTI radar then SAR. What follows next is EO/IR imaging and finally eyes. Each sensor cues the next higher-resolution and shorter-range sensor. The RF-based sensors can sometimes

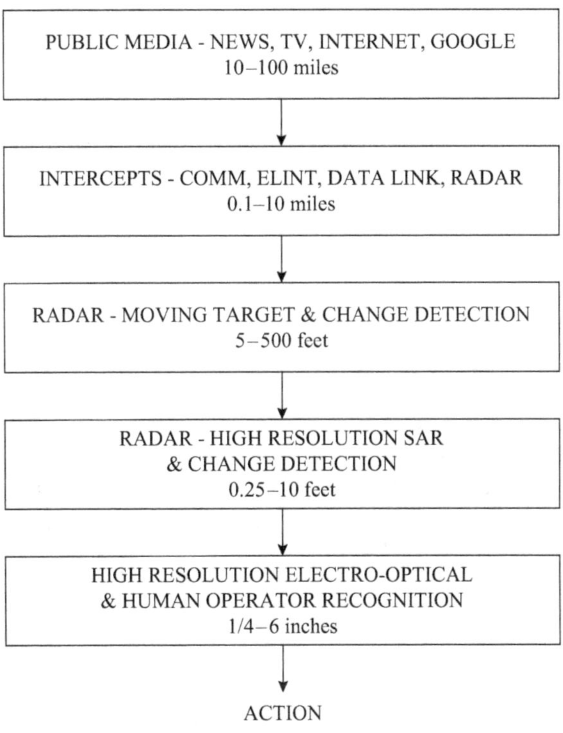


*Figure 1.2 ABI sequence*

provide recognition of a potential target. They cannot provide identification except in the case of voice prints of a cellphone or telephone user. In general, identification by a human using visible light is the only method accurate enough to allow a military or law enforcement action. Nowadays that identification is done with high-resolution video cameras and telescopic lenses.

This text is primarily about the use of radar for tactical ABI. However, the use of RF intercepts for radar cueing as well as techniques to enhance geolocation for the next sensor will be discussed in several chapters.

## 1.2 RF sees through clouds and haze and at night

VHF, UHF or microwave emissions can be detected in all weather with SIGINT, COMINT or ELINT sensors. Radars provide their own "light" to illuminate the surface or targets. At optical wavelengths there are only a few atmospheric windows not obscured by gases and dust. This limits visibility to tens of miles at best and a few miles under most conditions. Radars and intercept systems in most bands can "see" for 100s of miles. In the visible part of the spectrum, only 50% of the light reflected from an object arrives over a 5 mile path length in normal air as shown in Figure 1.3.

Figure 1.3 also shows that even in the IR wavelengths transmittance is only a little better. If there are intervening clouds, fog, rain and dust, the transmittance is much lower. However, at common radar and communications wavelengths, more than 99% of the energy reflected or emitted from an object arrives over a 5 mile path in the presence of intervening clouds, fog, rain and dust.

One should also note that there are parts of the so-called optical spectrum that are opaque! There are many places in the world where the air is so hazy for much of the year that, although the surroundings are very bright, one cannot tell where the

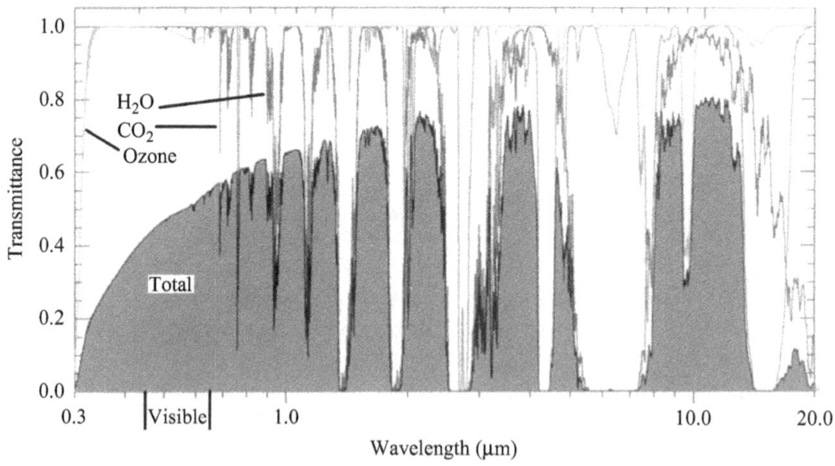


*Figure 1.3 Atmospheric transmittance over 5 mile path length; adapted from [2]*

sun is in the sky. The scattering from haze and dust is virtually invisible to radar and communications because the scattering particles are so small relative to the operating wavelengths. Furthermore, at the low end of radar and communications frequencies (VHF and UHF), these emissions easily penetrate foliage, again because the wavelengths are long relative to the feature size of trees and brush.

Unfortunately, we humans are used to seeing in the visible part of the optical spectrum. Radar images and RF intercepts don't look the same as our visual experiences. To compensate for that difference, these detections must be mapped into a context, which facilitates interpretation. These RF signals must be coupled with existing terrain maps or optical images to provide successful 24-h all-weather persistent surveillance for ABI.

## 1.3 Mapping of emitters onto terrain

There are millions of emitters in many parts of the planet. The only places where emitters are quite sparse are at the poles. The longer a stationary emitter is observed the more accurately the emitter can be geolocated. The accuracy is dependent on the cumulative ratio of signal strength to interference and noise ratio (SINR). For the case of moving emitters, SINR alone is not enough because the observation time is limited to a single measurement cell. If the movement is fast enough, and can be related to the surface, that is, surface vehicles, then the likelihood is that the vehicle is on a road. If the road is in a terrain database, then the detections can be "pushed" onto the closest road in the database, which may improve location accuracy. Similarly, if the terrain is very steep, anything over walking speed is improbable for a surface vehicle or person (commonly called a dismount). Fast over steep terrain must either be an aircraft, a fixed mover (example later) or a false target.

Usually there are so many intercepts that they can't be dealt with as individuals unless there is some other source of information cue focusing on a single one. Unique cellphone ID (SIM card), voice print and key word detection methods are often used to focus on the needle in the haystack. Also, these intercepts can be counted, and if enough of them appear in a spot, the intercepts can be plotted on topographic map to alert observers. Such a plot is shown for Poway High School in San Diego County, CA, in Figure 1.4. Poway High School has seven cellphone base stations on its campus. Accurate geolocation of individual cell users in the vicinity of the campus is trivial. Figure 1.5 shows a snapshot of a short period of time during a school day superimposed on a Google Earth™ map of the school and surroundings. Each black dot in the figure is an active user during the snapshot period.

Figure 1.6 shows a snapshot of the same area during Christmas break with very little activity. It wouldn't take a rocket scientist to figure out from the activity when school is in and out of session. Aggregate activity alone is often enough to see trends.

Space intercepts are usually not accurate enough by themselves to initiate action. Follow up with aircraft or terrestrial sensors is necessary to cue the next level of observation. Most intercepts are in angle space relative to the intercept platform. There are methods of providing distance/range estimates to fixed or slow-moving emitters from moving platforms.

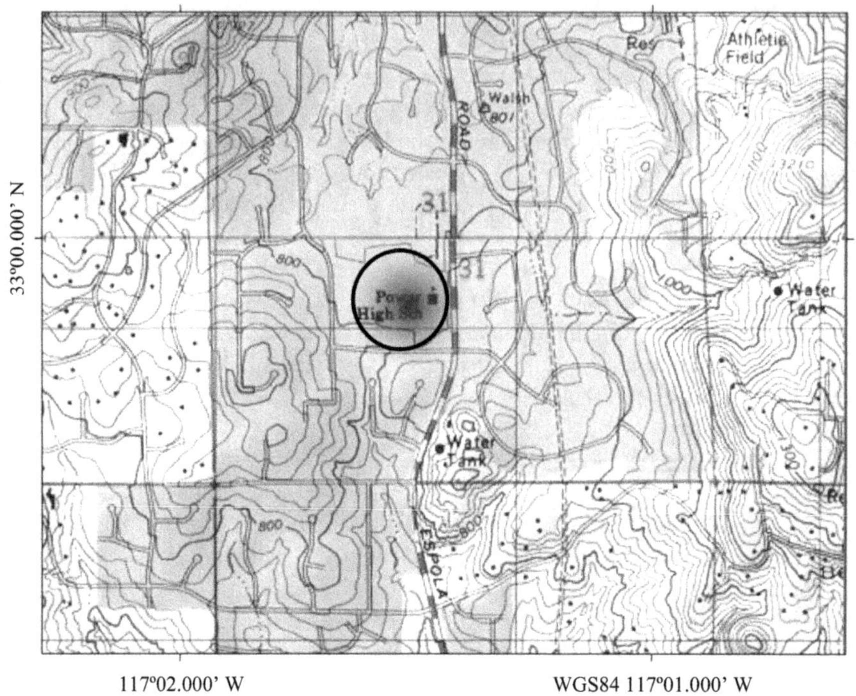


*Figure 1.4 High cellphone activity at Poway High School*

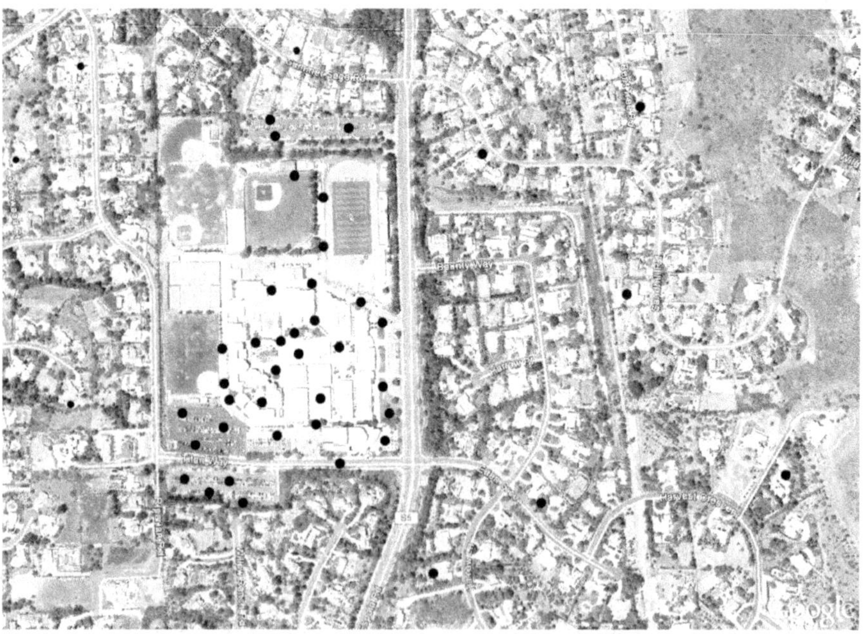

*Figure 1.5 Normal cellphone activity at Poway High School (author)*

*Figure 1.6 Lack of cellphone activity during holiday break (author)*

In all cases these estimates must be transformed from sensor coordinates to earth-stabilized coordinates. This could be latitude, longitude and altitude in a global grid or into what are commonly called "site" coordinates. Site coordinates are relative distances (e.g., Northings and Eastings) from an accurately known reference point, which could be map center or corner or some accurately known point such as a cell tower, bridge abutment, survey benchmark, mountain peak or building. The advantage of site coordinates is that the relative errors to a geolocated object can be smaller to allow handoff to another platform, which uses the same reference point. One reason for this is that, although the digital terrain elevation database (DTED) exists for our planet, there are major errors in absolute elevation between individual mapped areas and hence errors in the true three-dimensional (3D) location of a point. By referencing everything to a spot in the local area, the absolute errors washout.

It is also common to place a beacon transponder in the theater whose geolocation can be accurately established by GPS or survey. Similarly, ground fixed pseudo-satellites, which transmit a GPS-like signal, can dramatically improve the location accuracy of the interceptor platform as well as the accuracy of observed emitters.

In the last few years radio-controlled commercial drones are now able to provide surveillance and small weapon delivery over distances of a few miles. There have been numerous instances where drones have interfered with civilian aircraft and emergency services. The news media have reported drones in use by terrorist groups in Syria. These commercial drones are relatively inexpensive. The author has witnessed operators flying drones much like model airplanes using both smart phone and video game-style joystick controllers. They relay back video images to the controller or smart phone. The drones are large enough to carry a few pounds of explosive.

In addition to the cellular bands to be discussed in Section 1.5, commercial radio-controlled drones also use more power-limited short-range device (SRD) and industrial, scientific and medical (ISM) radio bands, which are legally limited to 100 mW power. Unfortunately, it is easy to modify these radio emitters up to 3 W radiated power for longer range. (You can by kits over the internet.) These are the same bands used for garage and car door openers, Wi-Fi and Bluetooth. Garage door openers have been modified to set off improvised explosive devices (IEDs). The frequencies are typically 433 MHz, 2.4 GHz and 5.8 GHz. Intercept sensors searching for potential hostile activity must now also search and locate both drones and their control stations (a man with a drone controller) [3,4].

## 1.4 Long-range detection of activity

There is not much happening on most of the earth's surface most of the time. Vast expanses of the seas are empty of human activity. Large areas of the poles are empty of much activity. Most of Siberia has more reindeer than people. The empty quarter of Saudi Arabia is *really* empty! Some manner of alerting and cueing of activity is necessary. The most common and ubiquitous form of emission today is from some form of mobile or cellphone. Commercial ocean shipping uses the AIS at roughly 162 MHz and radio communication. Aircraft use multiple communication and navigation aids. Terrorists and paramilitary forces use tactical communication, run generators, build fires, drive vehicles, build shelters and cause changes to the surrounding environment. Such forces are aware that satellite surveillance can observe them from time to time. When spacecraft are overhead, they may stay hidden. But if they must stay hidden most of the time, then they have less time for mischief. The detection of such activities can be as simple as reading the newspaper but also it depends on movement, image change detection (CD), thermal and communications emissions intercept.

The most common very long-range detection means is intercept of electromagnetic emissions (SIGINT/ELINT/COMINT). The elements contributing to the intercept signal to noise ratio ($SNR_I$) of an intercept from an emitter are shown in Figure 1.7. Obviously the $SNR_I$ of a RF emission from a transmitter depends on the transmitted power, $P_T$, the emitter antenna gain, $G$, as well as the decrease of the power density with range assuming spherical propagation, $1/(4\pi R^2)$. The signal propagation also experiences atmospheric losses as well as grazing angle scattering. The whole reason to use RF emission intercept for ABI is the potential for very long-range cueing. This inevitably causes the signal to propagate very close to the limb of the earth for a part of its path. Those low-grazing angles cause signal scattering losses. The intercept receiver captures some fraction of the emission

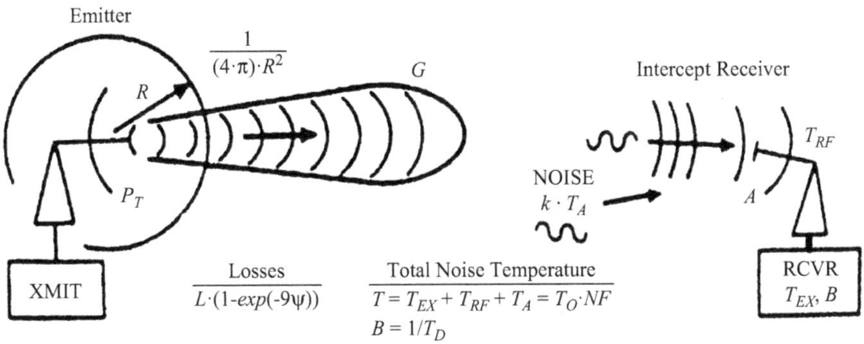


$$SNR_I = \frac{P_T \cdot G \cdot A \cdot T_D \cdot L \cdot (1\text{-}exp(\text{-}9\psi))}{4 \cdot \pi \cdot R^2 \cdot k \cdot T_O \cdot NF}$$

*Figure 1.7 Intercept equation elements*

based on its antenna area, $A$, the signal must compete with RF thermal noise entering from space and atmosphere, $T_A$, as well as the self-noise in the RF chain, $T_{RF}$, and the exciter, $T_{EX}$. Thermal noise is very broadband and so the noise power is proportional to the receiver bandwidth, $B$, the equivalent temperatures, $T$, and Boltzmann's constant, $k_B$. Normally, the search dwell time, $T_D$, at each intercept angle is matched to the reciprocal of the receiver equivalent bandwidth. Sometimes, the total equivalent noise temperature is represented by a standard temperature, $T_0$, multiplied by a noise factor, $NF$, representing all the noise contributors.

Intercept can be accomplished from space, aircraft or surface sensors. Usually these intercept detections especially from space or aircraft are at long range and so propagation path length grazing angle must be taken into account. Intercepts can provide positional cues to higher-resolution sensors for radar detection or EO/IR imaging. Equation (1.1) gives one version of the intercept equation for the interceptor signal to noise ratio ($SNR_I$) from an emitter [5].

$$SNR_I = \frac{P_T \cdot A_{eR} \cdot T_D \cdot G_T \cdot L \cdot \left(1 - \exp\left(-9 \cdot \psi\right)\right)}{4 \cdot \pi \cdot R^2 \cdot k_B \cdot T_0 \cdot NF}$$

$P_T$ = transmit power (watts), $L$ = all losses (number $\leq 1$),

$A_{eR}$ = receive antenna effective area $\left(\text{meters}^2\right)$,

$G_T$ = transmit antenna power gain including aperture efficiency,

$\psi$ = grazing angle $\left(\text{radians}\right)$, $T_0$ = standard temperature (290K),

$T_D$ = time interceptor dwells on emitter $\left(\text{seconds}\right)$,

$R$ = range from interceptor to emitter $\left(\text{meters}\right)$, $NF$ = noise factor,

$k_B$ = Boltzmann's constant $1.38 \cdot 10^{-23} \left(\text{watts}/\left(\text{hertz} \cdot \text{kelvin}\right)\right)$. (1.1)

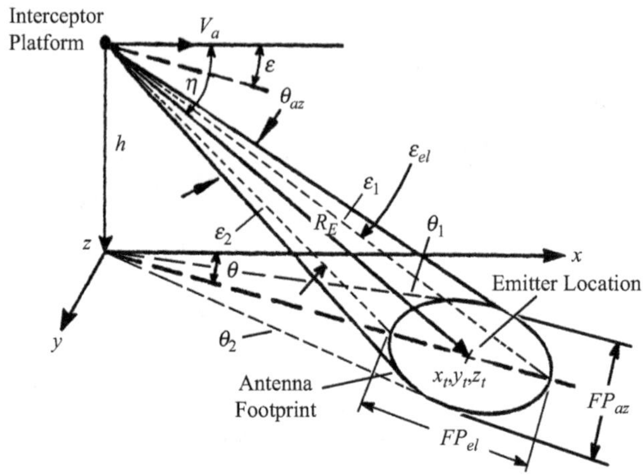


*Figure 1.8 Interceptor to emitter geometry*

## 1.5 Intercept direction finding [6,7]

In general intercept direction finding only provides angle of arrival (AOA) with little or no distance information. The basic geometry for an intercept from an airborne or spaceborne platform is given in Figure 1.8. The platform has some velocity, $V_a$, arbitrarily assumed along the $x$ axis. There is some bearing angle, $\eta$, between the velocity vector and the emitter, $x_t$, $y_t$, $z_t$. The interceptor is at some altitude, $h$, above the emitter. The interceptor antenna(s) have some azimuth beamwidth, $\theta_{az}$, and some elevation beamwidth, $\varepsilon_{el}$, which give rise to a footprint, $FP_{el}$ and $FP_{az}$, on the earth's surface. The composite bearing angle, $\eta$, is made up of a platform centered azimuth, $\theta$, and elevation, $\varepsilon$. In some cases the platform antenna(s) are small and their beams are almost omnidirectional. The distance to the emitter must be estimated by scaling power received to range, by triangulation or both.

To find accurate distance, one must use some form of triangulation and/or make an assumption about the altitude of the emitter since elevation angle may provide a large uncertainty in distance. Triangulation can be provided by using two AOA measurements along a known baseline as shown in Figure 1.9. The location where the two angles cross is the range to the emitter. If the emitter is moving, those measurements must be nearly simultaneous. The composite angles, $\eta$, have a component in azimuth and elevation since the emitter is very seldom at the same altitude as the interceptor [4].

An estimate of the range, $R_E$, to an emitter using triangulation is provided in (1.2):

$$R_E = \frac{D_I \cdot \sin(\eta_1) \cdot \sin(\eta_2)}{\sin(\eta_2 - \eta_1) \cdot \sin\left(\frac{\eta_2 + \eta_1}{2}\right)} \tag{1.2}$$

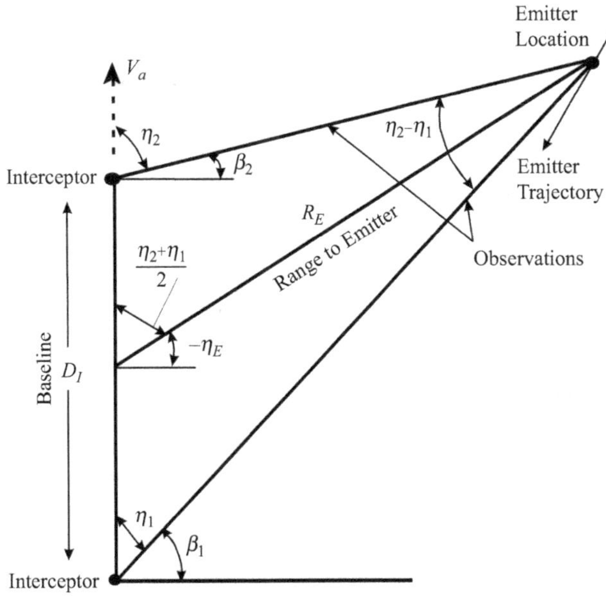


*Figure 1.9 Direction finding by triangulation [5]*

Both antennas must be able to detect the same emitter since initially the angle to the emitter is unknown. This implies a wide beamwidth for each antenna. On an airborne or spaceborne platform, a single antenna might be used to "fly out" the baseline. As long as the emitter is moving slowly relative to the velocity of the interceptor platform, a single antenna might be adequate to perform triangulation as long as the antenna continues to illuminate the emitter. If the intercept antenna(s) have low gain and a wide beamwidth, then an estimate of the AOA may be coarse for each individual measurement and the location accuracy for each set of angles can be poor.

The location estimate can be improved by calculating the partial derivatives in x and y for each angle measurement (sometimes also in z) and iteratively estimating the change in x and y for each new measurement as suggested in Figure 1.10. This particular form of position location is called a maximum likelihood estimate (MLE). The matrix, G, is commonly called a gradient matrix. Partial derivatives that are changing over the observation time are required in this matrix because angle and square root functions are nonlinear. The angle Δs are usually close enough together that approximations can be used. The iterative method requires an initial guess of position based on AOA and emitter received power in order to estimate range (±20% of range typical). Iteration continues until a local minimum is achieved, which may be different in x and y. This method bears some similarity to an extended Kalman filter (EKF), which will be explained in more detail in Chapter 4. The MLE method dramatically improves the location accuracy assuming if there are enough angle measurements [6]. Obviously, the

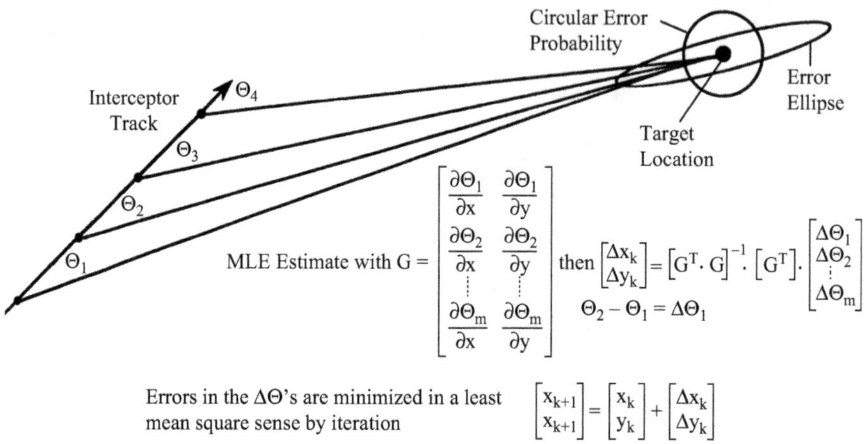


*Figure 1.10 Triangulation error reduction*

accuracy also depends on the SNR of each intercept.

Measured angles with noise: $\theta_i = \theta_{true} + \partial\theta_i$

Estimated target location vector: $\widehat{\mathbf{X}} = \begin{bmatrix} \widehat{x_t} \\ \widehat{y_t} \end{bmatrix}$

Sensor position changes: $\Delta x_i = \widehat{x_t} - xs_i\ ,\ \Delta y_i = \widehat{y_t} - ys_i\ ;\ r_i = \Delta x_i^{\ 2} + \Delta y_i^{\ 2}$

Angle error estimate: $\partial\Theta\left(\widehat{\mathbf{X}_m}\right) = \partial\theta_i = \theta_i - g_i\left(\widehat{\mathbf{X}}\right)\ ;\ g_i\left(\widehat{\mathbf{X}}\right) = \arctan\left(\frac{\Delta y_i}{\Delta x_i}\right)$

Partial derivative matrix angle to position at emitter location:

$$\mathbf{Gx} = gx_i\left(\widehat{\mathbf{X}}\right) = \begin{bmatrix} \frac{-\Delta y_0}{r_0} & \frac{-\Delta y_1}{r_1} & \cdots & \frac{-\Delta y_{i_{max}}}{r_{i_{max}}} \\ \frac{\Delta x_0}{r_0} & \frac{\Delta x_1}{r_1} & \cdots & \frac{\Delta x_{i_{max}}}{r_{i_{max}}} \end{bmatrix}$$

$$\widehat{\mathbf{X}}_{\mathbf{m-1}} = \widehat{\mathbf{X}}_{\mathbf{m}} + \left(\mathbf{Gx}^T \cdot \mathbf{S}^{-1} \cdot \mathbf{Gx}\right)^{-1} \cdot \mathbf{Gx}^T \cdot \mathbf{S}^{-1} \cdot \partial\Theta\left(\widehat{X_m}\right) \tag{1.3.1}$$

Where:

Diagonal covariance matrix: $\mathbf{S} = \begin{bmatrix} \theta_0^{\ 2} & 0 & \cdots & 0 \\ 0 & \theta_1^{\ 2} & \cdots & 0 \\ \vdots & \vdots & \ddots & \vdots \\ 0 & 0 & \cdots & \theta_{i_{max}}^2 \end{bmatrix}$

*m* is iteration index, bold capitals are vectors/matrices. (1.3.2)

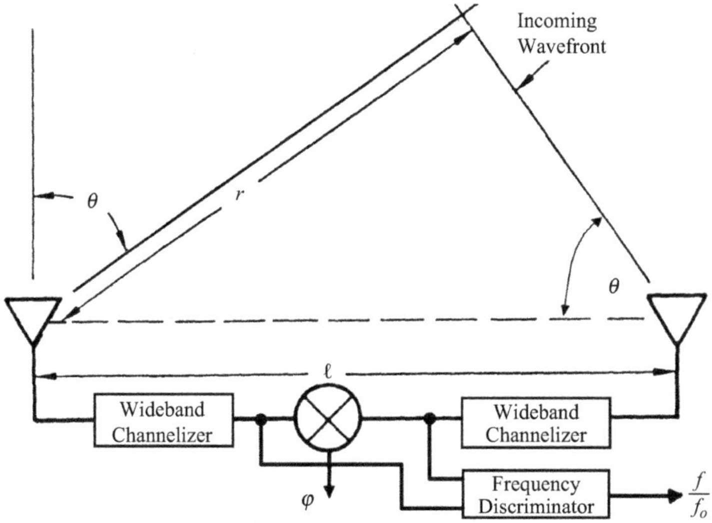


*Figure 1.11 Direction finding by phase measurement [5]*

More details for an MLE iterative method used for finding the two-dimensional (2D) location of an emitter on or near the surface is given in (1.3). The method works by using all the measurements compared to a possible target location at $x_t$ and $y_t$. Each iteration reduces the error until a "best" fit occurs. Since this is an iterative method, an initial guess for the x and y location is required to start. Bad guesses don't seem to affect the number of iterations very much. Usually only a small number (3–6) of iterations is required to arrive at the position estimate. Obviously, the best accuracy is achieved with the longest overall baseline and measurements spaced by at least 1% of emitter range. The number of samples inside the total baseline only helps until the SNR is high. Since the raw AOA measurements have a random component, the overall accuracy can vary significantly but is typically about 1%–2% of range.

There are at least three ways to estimate AOA. One is to use the phase (or time) difference between two or more closely spaced antennas to estimate the AOA. The first method is shown in Figure 1.11. This method depends on an altitude assumption in most cases, which is usually the earth's surface. As long as the real elevation angle is under 0.2 radians, the error in ignoring elevation is quite small.

Phase difference AOA depends on measuring the difference of the same signal to a small fraction of a wavelength at two or more antennas by comparing the relative phases from each antenna in a phase discriminator (multiplier) as shown in Figure 1.11. An incoming wavefront from a distant point source will arrive as almost a plane wave. For a wavefront arriving at an angle of $\theta$ relative to the normal to the antenna baseline, then the distance, $r$, is $\ell\cdot\sin\theta$ which can be related to time by dividing by the velocity of light, $c$. For a single emitter frequency, the phase difference between the two antennas is the product of the distance and the angular frequency divided by the speed of light as shown in (1.4).

$$\theta = \operatorname{asin}\left(\frac{r}{\ell}\right) \text{ and } \varphi = \frac{2\cdot\pi\cdot f\cdot\ell}{\mathrm{c}}\sin(\theta)$$

$$\text{Since: } c = f_0\cdot\lambda_0 \text{ then: } \varphi = 2\cdot\pi\cdot\left(\frac{f}{f_0}\right)\cdot\left(\frac{\ell}{\lambda_0}\right)\sin(\theta) \tag{1.4}$$

The frequency of the arriving signal must be measured accurately in order to correctly determine the AOA. Phase AOA is clearly a narrow band technique and filters before and after the phase detector are required. There is usually a frequency discriminator function in parallel with the phase detector to provide the other variable ($f/f_0$) required for AOA estimation. Also note that noise through an arcsine function has a nonlinear distortion, which gets progressively worse as emitter AOA approaches endfire causing some bias errors at low SNRs. These phase AOA bias errors are not nearly as severe as for amplitude AOA. Also there may be phase ambiguities that must be resolved.

Then, solving (1.4) for $\theta$ provides phase comparison AOA. For an emitter at $\theta_E$, the result is given in (1.5).

$$\theta_E = \operatorname{asin}\left(\frac{\lambda_0}{2\cdot\pi\cdot\ell}\cdot\frac{f_0}{f}\cdot\varphi\right) \tag{1.5}$$

where $\lambda_0$ is the wavelength at band center, $f_0/f$ is the reciprocal of the output from a frequency discriminator centered at band center, $f_0$, $\ell$ is the spacing between phase measuring antennas and $\varphi$ is the measured phase.

The overall phase AOA error has three components: the emitter multichannel SNR, the emitter frequency measurement error and the phase measurement error, which is given in (1.6) [5]. The frequency and phase measurements, in turn, have signal to noise dependence. The frequency measurement is the least sensitive to SNR and hardware limitations can easily be orders of magnitude lower than phase or antenna location errors. One can think of many other sources of error but these are the dominant contributors. The one sigma angle error is a function of SNR, antenna spacing, $\ell$, and calibration errors as given in (1.6). That error when coupled with multiple angle estimates results in an error ellipse of the type shown in Figure 1.10.

$$\sigma_\theta \approx \sqrt{\frac{\lambda_0^2}{2\cdot SNR}\cdot\left[\frac{1+0.5\cdot c_\varphi^2}{\pi^2\cdot\ell^2}+\frac{c_m^2\cdot B^2\cdot\sin^2\theta_E}{c^2}\right]+\frac{\lambda_0^2\cdot\sigma_{\varphi match}^2}{2\cdot\pi^2\cdot\ell^2}} \tag{1.6.1}$$

Where: $c_\varphi$ = phase detector angle coefficient, typically 1/2 to 2,

$c_m$ = scintillation coefficient, $c$ = velocity of light,

$\sigma_{\varphi\text{match}}$ = interchannel phase match, $B$ = frequency estimation bandwidth,

$\theta_E$ = emitter angle of arrival relative to baseline normal. (1.6.2)

Another AOA method uses a monopulse antenna with reasonably high gain. Monopulse will be discussed in Chapters 2 and 6 in more detail. A monopulse antenna is one in which a symmetrical sum pattern and an antisymmetrical difference pattern is synthesized in the antenna.

The approximate azimuth and elevation beamwidths: $\theta_{az} = \dfrac{1.5 \cdot \lambda}{D_x}$ and $\varepsilon_{el} = \dfrac{1.5 \cdot \lambda}{D_y}$

The approximate azimuth and elevation mainbeam footprints on the surface:

$$FP_{az} = 2 \cdot R \cdot \tan\left(\theta_{az}/2\right) \text{ and } FP_{el} = 2 \cdot R \cdot \frac{\tan\left(\varepsilon_{el}/2\right)}{\sin(\psi)}$$

Where: $\lambda$ = operating wavelength, $\psi$ = grazing angle, $R$ = slant range to beam center, $D_x$ = antenna horizontal length along the flight path, $D_y$ = antenna vertical height along the flight path. The elevation and azimuth location accuracies are approximately:

$$\varepsilon_{\text{acc}} \cong \frac{FP_{el}}{\sqrt{2 \cdot SNR_I}} \text{ and } \theta_{\text{acc}} \cong \frac{FP_{az}}{\sqrt{2 \cdot SNR_I}} \tag{1.7}$$

The ratio between the sum and difference pattern signals from an emitter is used to form a monopulse discriminant. The model presumes that the emitter is on the surface and the antenna beam footprint on the surface can be split with an accuracy determined by the SNR of the emitter intercept. Even though communication intercept power has an $R^2$ decrease with range, SNRs for "flea-powered" cellphones and push-to-talk radios are often quite high. For a single emitter in a narrow band, the discriminant amplitude is proportional to the AOA. Such patterns are shown in Figure 1.12. Anywhere that a program name is shown in parentheses

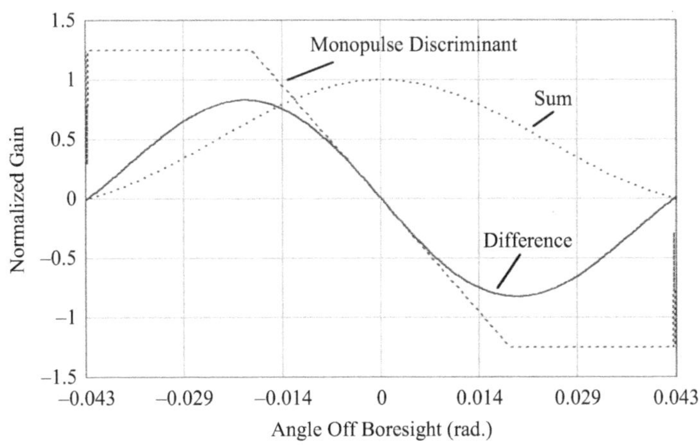


*Figure 1.12 Direction finding by monopulse beamsplitting (monopulse1)*

as in Figure 1.12, that is, (monopulse1), it refers to a Mathcad analysis program used to generate the graph or data. It is contained in that chapter's appendix in both Adobe Acrobat and Mathcad formats. Monopulse AOA estimates can be accurate to 1% of a beamwidth. This was successfully demonstrated on the Pave Mover program using a curve fit to the as-built monopulse pattern. The error model for monopulse location errors is given (1.7).

Another ranging scheme usable from aircraft is phase-rate-of-change (PRC) induced by the interceptor motion. PRC combines phase AOA with measurement to measurement phase derivatives. The emitter signal must be reasonably coherent to use PRC. This method is used primarily at higher frequencies. The two-phase AOA antennas must be widely spaced (100–500 wavelengths) and must have an SNR that is high enough to resolve ambiguities. For a given platform velocity, there is an ambiguous locus in x and y, which defines all the possible range and angle locations that could give rise to that PRC. That locus combined with an AOA defines an x and y emitter position. Again, the assumption is that the emitter is moving slowly relative to the interceptor. The idea is shown in Figure 1.13. The lower the emitter operating frequency and the slower the interceptor platform velocity, the longer it takes to achieve high (1% of range) location accuracy with this method. At L band and with a propeller-driven airplane platform, it may take more than 5 min to achieve 1% of range accuracy.

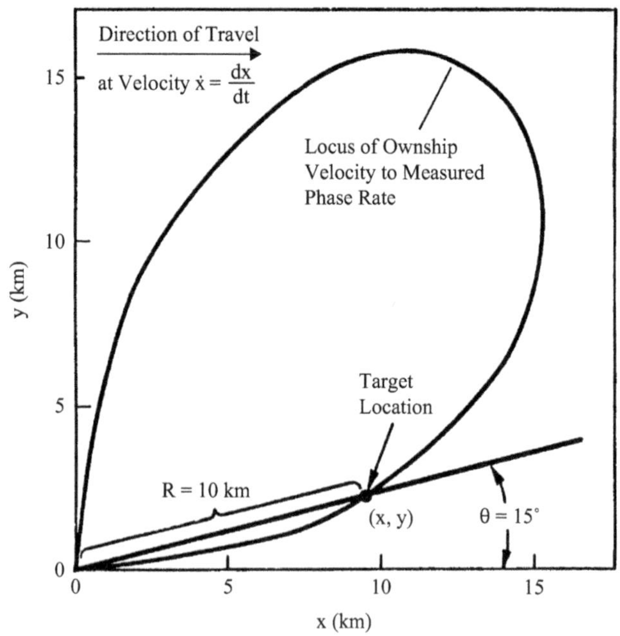


*Figure 1.13 Passive ranging by phase rate of change [5]*

The equation to be solved to find range with PRC is shown in (1.9). For a given ownship velocity assumed to be in the $x$ direction (without loss of generality) and an almost stationary emitter, the angle and PRC will be as given in (1.8).

$$\frac{d\theta}{dt} = -\frac{\sin(\theta)}{R}\cdot\dot{x} \text{ and } \dot{\varphi} = 2\cdot\pi\cdot\left(\frac{f}{f_0}\right)\cdot\left(\frac{\ell}{\lambda_0}\right)\cos(\theta)\cdot\frac{d\theta}{dt}$$

$$\text{Where: } \dot{x} = \frac{dx}{dt} \text{ and } \dot{\varphi} = \frac{d\varphi}{dt} \tag{1.8}$$

Solving for range yields (1.9).

$$R = -2\cdot\pi\cdot\left(\frac{f}{f_0}\right)\cdot\left(\frac{\ell}{\lambda_0}\right)\cos(\theta)\sin(\theta)\frac{\dot{x}}{\dot{\varphi}} \tag{1.9}$$

Since the rate of change of phase is an experimental derivative, it magnifies the errors, that is, SNRs must be >50 dB. The principal feature is that convergence times can be short if SNR is very high and wavelength is short.

Many other multilateration schemes have been tried such as those in the USAF/IBM ALS time of arrival location system and the USAF/Lockheed PLSS precision location systems of 30 years ago, which depend on time of arrival multistatic intercept rather than AOA. ALS was used as an adjunct range instrumentation system during the Pave Mover competition between Hughes and Grumman for JSTARS. They all suffer from different versions of the same problem, that is, noise and multipath does them in unless the SNR is very high.

---

**Example 1.1: Cellphone intercept from space**

Many spacecraft on orbit have a primary antenna mesh reflector aperture that could be adapted to locate as well as service cellular communications [8]. The primary aperture can be very broadband even if the feed aperture is narrowband. A broadband four quadrant direction-finding monopulse feed can be overlaid on the secondary feed aperture to localize individual emitters without compromising revenue-generating performance. Such a hypothetical system is shown in Figure 1.14. There will be significant Doppler shifts between the nominal cellular frequency channel and the signal intercepted by a space system that must be compensated for signal recognition [9,10]. Intercept from synchronous orbit is too inaccurate for tactical utility.

Assume for Example 1.1 that the parameters in Table 1.1 represent a possible system. Then using (1.1) and (1.7), the approximate cellphone location accuracy from space is given in Figure 1.15. The detailed calculations are in Mathcad and PDF files in Appendix A.1 over the internet or DVDR.

In the example, individual cellphones can be easily detected and located from space. Figure 1.15 shows the SNR limited location accuracy from a spacecraft in a 1000 km orbit with a 70 m$^2$ antenna effective aperture area. As one might expect the further away the emitter is from the orbital ground track, the poorer the

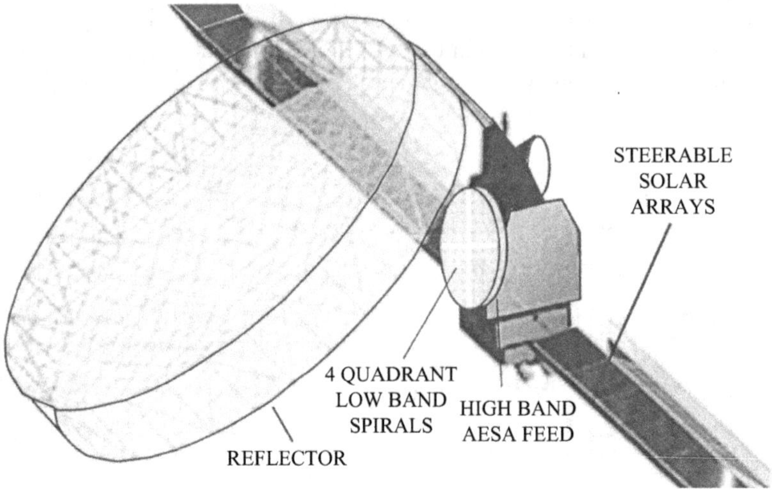


*Figure 1.14 Example space intercept system*

*Table 1.1 Example 1.1 Parameters*

| Parameter | Value |
|---|---|
| Cell power, $P_T$ | 0.3 W |
| Receive antenna area, $A_{eR}$ | 70 m$^2$ |
| Cellphone gain, $G_T$ | 1.26 |
| Dwell time, $T_D$ | 2 s |
| Grazing angle, $\psi$ | 0.105 rad |
| Maximum slant range, $R$ | 3108 km |
| Noise factor, $NF$ | 2.5 |
| Wavelength, $\lambda$ | 0.3 m |
| Azimuth aperture, $D_x$ | 10 m |
| Elevation aperture, $D_y$ | 10 m |
| Orbit | 1000 km |
| Maximum losses, $L$ | 0.83 |

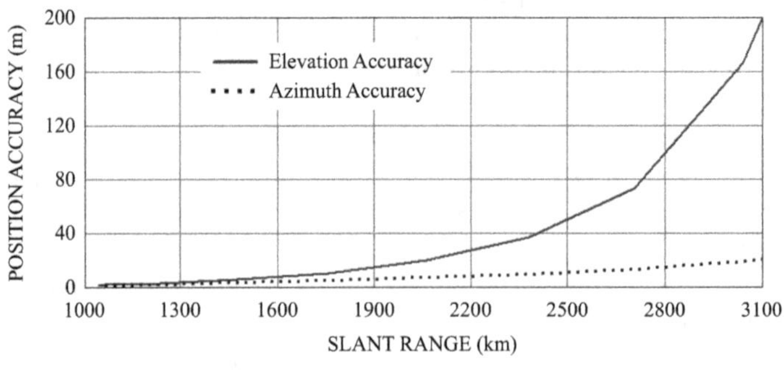


*Figure 1.15 Cellphone emitter location from space (cellphone13)*

*Figure 1.16 Example aircraft cellphone intercept; adapted from [11]*

accuracy. However, at 2800 km slant range, the one sigma location accuracy is still about 100 m. Usually, the next sensor to image a single cellphone location might be a 4 × 4 km radar or EO patch, so location to one sigma accuracies of ±20–200 m would be more than adequate. Although an EO image would be more desirable than a radar image, often much of the earth is covered by clouds and a radar image maybe all that is possible. Depending on potential vehicle speeds in the cellphone detection area and number of detections, an image area of 8 × 8 km or 20 × 20 km might be used.

Cellphone and drone control emissions can also be detected and localized from aircraft. Figure 1.16 shows the bottom of a light plane with aerodynamic blade (circled in white) and patch antennas in common use for communications and GPS reception. These antennas can be used for interception of cellphone and drone emissions and direction finding.

Cellphones are routinely interrogated by local cellular base stations to find a certain cellphone and complete calls. Both the cellphone and the cell base station can be detected and tracked. Two-way drone communications can also be tracked although the power may be much lower. Popular four-rotor drones don't move all that fast and so one can assume they are stationary for short periods of time. Direction finding can be accomplished by measuring the time delay or phase between antennas on a few pulses or alternatively by continuously measuring the change in time delay as the aircraft flies a long baseline.

Many modern intercept systems, cellular phones and cellular base stations are software defined radios (SDRs). Most of the transmitted waveforms and the receiver signal processing are performed in what is commonly called a digital radio transceiver (DRT). The original motivation for this approach was the proliferation of different communications waveforms and protocols. This forced new systems to handle legacy protocols as well as new more efficient methods in the same communication bands. All of the different cellular protocols in use have forced communications carriers and cellphone manufacturers to handle all of them in single hardware devices. These devices must handle base station protocols for Advanced Mobile Phone System (AMPS) first and second generations, Global System for

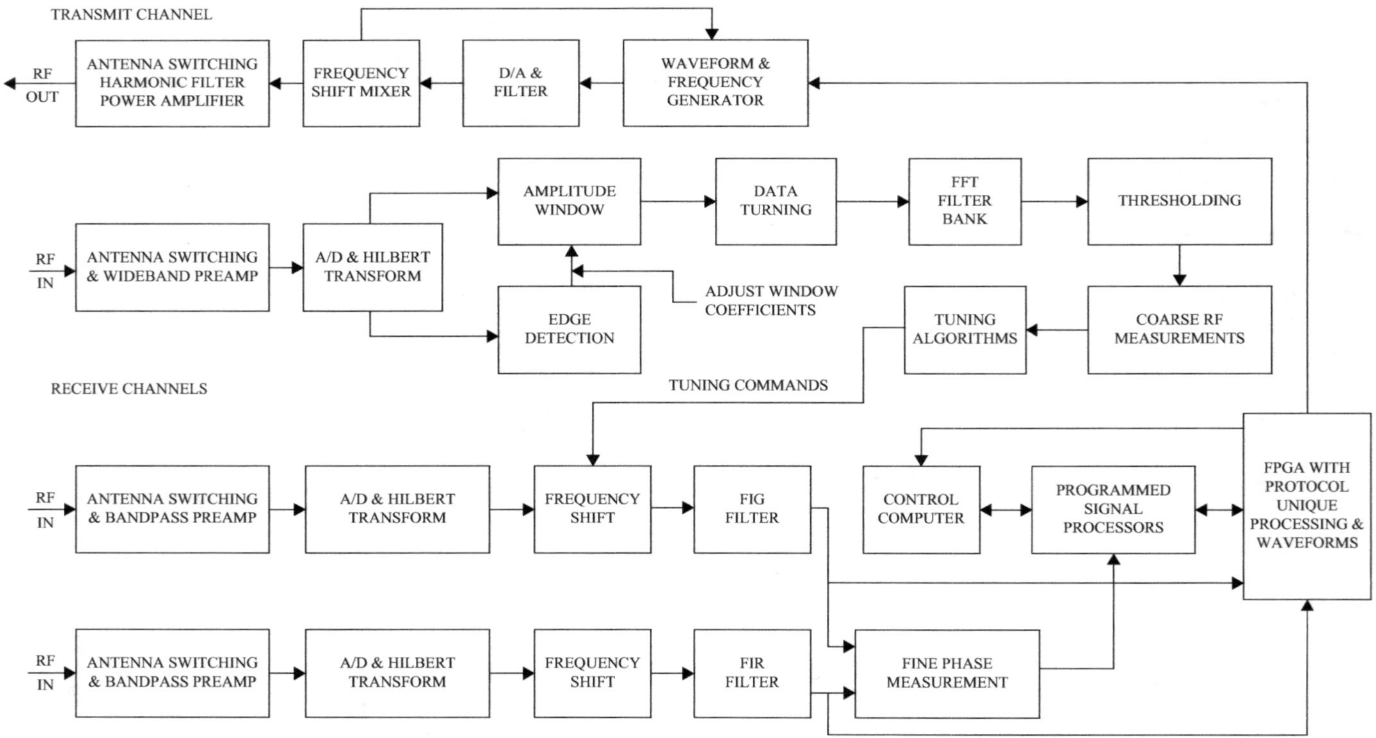


*Figure 1.17   Block diagram of typical digital radio transceiver; adapted from [12]*

Mobile communication (GSM) second and third generations, Universal Mobile Telecommunication System (UMTS) a third-generation upgrade to GSM that uses Wideband Code Division Multiple Access (WCDMA), High Speed Packet Access (HSPA), a third-generation upgrade for GSM with UMTS, Code Division Multiple Access (CDMA), Frequency Division Duplexing (FDD), a Long-Term Evolution (LTE) of third- to fourth-generation cellular networks, Time Division Multiple Access (TDMA), Frequency Division Multiple Access (FDMA), Orthogonal Frequency Division Multiple Access (OFDMA), which uses useful symbol time durations between 3.2 and 1000 μs, etc. These waveforms use phase and amplitude modulation in each band up to 64 states over a bandwidth of 400 MHz to 3 GHz. It is outside the scope of this book to explain each of these in detail. Suffice it to say that multiple companies have developed software for all of these protocols and can host them in a single DRT [12].

The ISM and SRD bands, which some drones use, can also be handled by SDR. Protocols such as IEEE 802.11 Wi-Fi, Bluetooth, HIPERLAN, IEEE 802.15.4 ZigBee and WDCT are well documented, so software should be available [4].

Figure 1.17 shows the block diagram of a typical DRT. Obviously, there might be more transmit or receive channels in a DRT but this is all that is required to stimulate a particular cellphone of interest SIM card and perform emitter location. A DRT such as Figure 1.17 should be able to detect and track nearby drones and the controller location.

---

**Example 1.2: Cellphone intercept from aircraft**

A possible set of parameters for an airborne cellphone intercept example using a transceiver of the type in Figure 1.17 is given in Table 1.2. Since blade antennas of the kind in Figure 1.16 have an equivalent intercept area of approximately $\lambda^2$, the AOA model will be like that of Figure 1.11 and (1.1), (1.5) and (1.6). The MLE iteration in (1.3) was used to determine location. The MLE iteration used 16 measurements spaced 562.5 m apart with a total baseline of 8.4375 km (Figure 1.18).

Figure 1.19 shows the SNR versus range for cellphone intercepts using the parameters of Table 1.2, which is more than adequate for accurate location. The MLE iteration requires an initial guess and was set at 18 km in y and 9 km in x. The iteration is not sensitive to the guess. Table 1.3 shows each iteration in x and y as well as the errors.

Since the measurements are from a random process, the location accuracy isn't always this good but with the parameters used 1% accuracy can be expected. The details of this example are in the internet or DVDR Appendix A.1 in MathCad and PDF files.

*Table 1.2 Example 1.2 parameters*

| Parameter | Value |
|---|---|
| Cell power, $P_T$ | 0.3 W |
| Receive antenna area, $A_{eR}$ | 0.1 m$^2$ |
| Cellphone gain, $G_T$ | 1.26 |
| Dwell time/measurement, $T_D$ | 2 s |
| Minimum grazing angle, $\psi$ | 0.105 rad |
| Maximum slant range, $R$ | 25.5 km |
| Noise factor, $NF$ | 2.5 |
| Wavelength, $\lambda$ | 0.3 m |
| Altitude, h | 5 km |
| Maximum losses, $L$ | 0.5 |
| Phase detector coeff., $c_\varphi$ | 2 |
| Scintillation coeff., $c_m$ | 2 |
| Velocity of light, $c$ | $2.9979 \times 10^8$ m/s |
| Phase center separation, $\ell$ | 0.3 m |
| Phase match, $\sigma_{\varphi\text{phasematch}}$ | 0.1 |
| Freq. est. band, $B$ | 200 kHz |
| Look angle, $\theta_E$ | $\pi/3$ |
| True target location, $x_T$ | 12,500 m |
| True target location, $y_T$ | 21,651 m |
| Sensor platform speed, $V_a$ | 125 m/s |
| Sensor platform heading | 10° |
| Time between measurements | 4.5 s |
| Number of measurements | 16 |
| Total intercept time | 67.5 s |

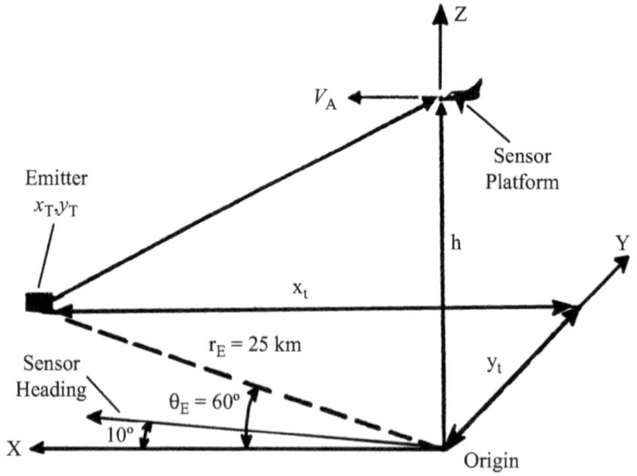


*Figure 1.18 Aircraft cellphone intercept example*

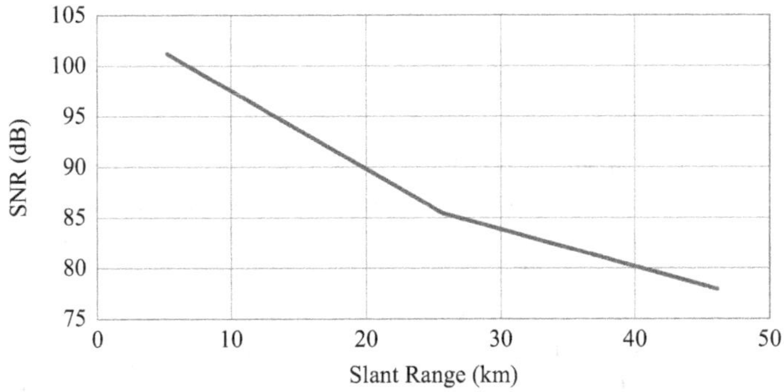


*Figure 1.19 Aircraft intercept SNR with range (Cellphone14d2)*

*Table 1.3 MLE location estimates (Cellphone14d2)*

| Iteration | X (km) | Y (km) | X error (m) | Y error (m) |
|---|---|---|---|---|
| 1 | 11.715 | 20.892 | −785.19 | −758.65 |
| 2 | 12.347 | 21.56 | −152.77 | −90.32 |
| 3 | 12.374 | 21.59 | −126.13 | −60.89 |
| 4 | 12.374 | 21.59 | −126.13 | −60.81 |
| 5 | 12.374 | 21.59 | −126.13 | −60.81 |

## 1.6 Navigation for geolocation

Navigation and communication have come a long way since the 1940s. There was a famous incident at the end of World War II that illustrates the point. On the day after cessation of hostilities in Europe, a British bomber force attacked a small town on Lake Como in Italy. Not only was the war over but the town was on the wrong lake – 100 miles from the true objective. Aids to navigation were so bad that many locations gave the same photo or radar picture. Furthermore, radio/radar-based velocity measurements for navigational dead reckoning in inclement weather were poor enough that errors of 100s of miles occurred.

Even today in the era of GPS, radar and radio aids to navigation are necessary to improve accuracy in the presence of jamming, own-ship motion, solar flares, altitude accuracy limitations and restricted coverage. Older navigation aids emit signals that can be intercepted by vehicles and used to determine location. Radio aids such as VOR, Decca, Loran, Tacan and DME are in use in various forms throughout the world but most are inaccurate at best. GPS satellites emit RF signals that contain time signals and ephemeris (orbit) data. There are usually enough (~12) satellites detectable by a GPS receiver that a location can be calculated for a moving or stationary platform. The GPS signal is quite weak (−130 to −140 dBm) and subject to inaccuracies caused by interference, multiple reflections and atmosphere. This often requires an additional RF sensor to improve navigation accuracy.

There basically are two general types of active ladar and radar aids to navigation: scene matching and velocity estimation by Doppler. Figure 1.20 shows an example of scene matching in which high-resolution images are matched to optical images or maps. The bridge abutment at the road interchange is surveyed and known within inches in longitude, latitude and altitude. Estimation of the abutment location relative to a radar sensor can correct for drift errors in inertial sensors as well as accuracy limits in navigation aids such as GPS. The radar operator designates the bridge abutment in each progressively higher-resolution image. Since each radar pixel in the image is the output from a specific filter matched to a terrain location relative to the sensor, designating the pixel designates the relative location of the platform. If that location is accurately known in absolute terms, then the difference between platform navigation position estimate and the relative location to an absolute point can be used to update the navigator. Nowadays this sort of realignment is usually performed in an adaptive Kalman filter.

Even during World War II, crude radar-to-map scene matching was performed for navigation and improved bombing accuracy. On a worldwide basis, thousands of significant terrain features and manmade structures change daily. This requires persistent updating for mapping and navigation.

Similarly, radar Doppler-based velocity estimation depends on measuring the mean line-of-sight velocity ($V_{LOS}$) in multiple angular directions to patches of the ground. This idea is shown in Figure 1.21. The Doppler centroid is estimated using multiple-range Doppler bins for each patch and converted through the operating

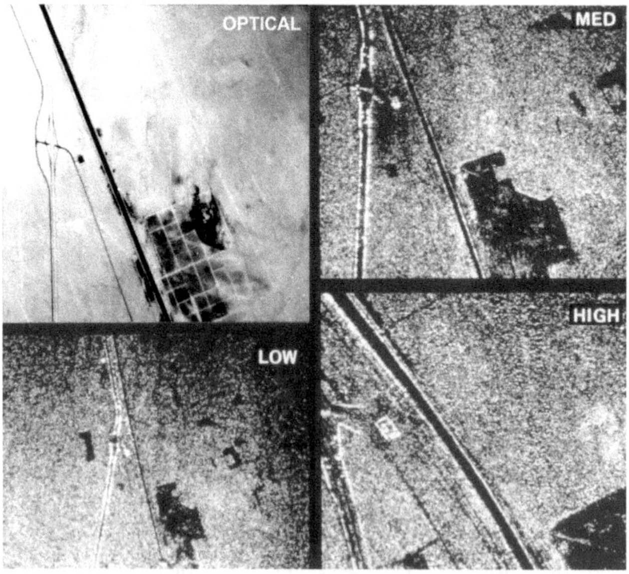


*Figure 1.20 Finding a road intersection through cloud cover [13]*

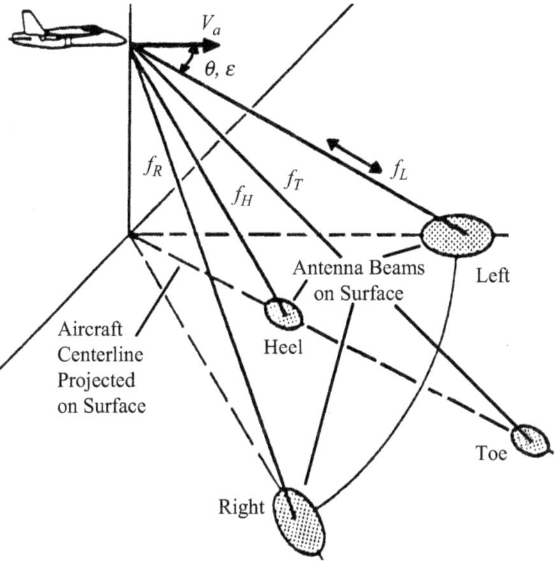


*Figure 1.21 Doppler velocity navigation*

wavelength to a velocity estimate. Although the antenna angles from the platform centerline are well known, the attitude of the platform relative to the earth may not be well known. Therefore, before true velocity can be estimated the roll, pitch and heading (yaw or drift) must be estimated [14]. Each velocity estimate is transformed through the look angles to an estimate of the platform velocity. Equations (1.10) provide one method of deriving platform velocity. At least four measurements are required to derive velocity as shown in Figure 1.21.

Platform roll: $\gamma_p = \operatorname{atan}\left[\dfrac{f_L + f_R}{f_L - f_R} \cdot \cot(\theta)\right]$

Platform pitch: $\beta_p = \operatorname{atan}\left[\dfrac{f_T - f_H}{f_T + f_H} \cdot \cot\left(\dfrac{\varepsilon_T - \varepsilon_H}{2}\right)\right] - \dfrac{\varepsilon_T + \varepsilon_H}{2}$

Platform drift (yaw): $\alpha_p = \operatorname{atan}\left[\left(\dfrac{f_L + f_R}{f_T + f_H}\right) \cdot \left(\dfrac{\cos\left(\dfrac{\varepsilon_T + \varepsilon_H + 2 \cdot \beta_p}{2}\right) \cdot \cos\left(\dfrac{\varepsilon_T - \varepsilon_H}{2}\right)}{\sin(\theta) \cdot \sin(\gamma_p)}\right)\right]$

Platform velocity: $V_a = \dfrac{\lambda \cdot f_T}{2} \cdot \sec(\varepsilon_T + \beta_p) \cdot \sec(\alpha_p)$ (1.10.1)

Where: $f_L$ = left beam Doppler, $f_R$ = right beam Doppler,

$\varepsilon_T$ = toe beam elevation angle, $\varepsilon_H$ = heel beam elevation angle

$f_T$ = toe beam Doppler, $f_H$ = heel beam Doppler,

$\theta$ = azimuth angle left & right, $\lambda$ = wavelength. (1.10.2)

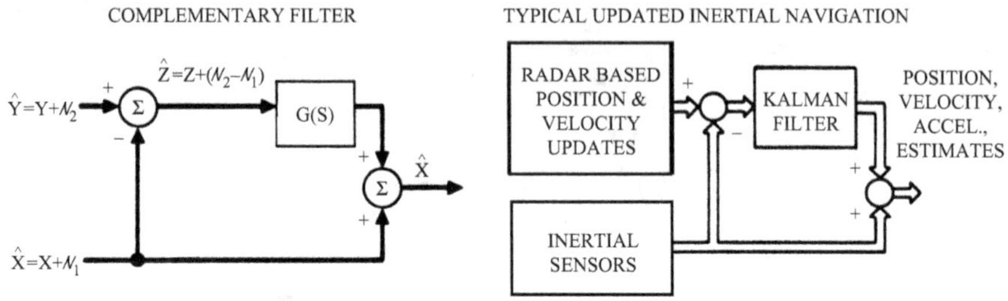


*Figure 1.22 Updated inertial navigation*

All of the estimates are combined again in an EKF for an estimate of platform true velocity ($V_a$). This method is often combined with scene matching and an inertial sensor to provide estimates of jerk, acceleration, velocity and position on a millisecond-by-millisecond basis for the radar and its carrying platform. These techniques can be used over water but they are much less accurate due to local water surface currents. Ocean surface currents have been mapped for hundreds of years, especially in the Atlantic. Surface current errors can be partially compensated but never completely eliminated.

Usually the way that estimates of position, velocity, etc., are combined with inertial sensors is in a complementary filter in which two sets of location information are subtracted from each other with the assumption that what is left is just some form of noise. The "noise" terms are filtered based on some knowledge of the characteristics of the error sources in the two independent forms of measurement. This idea is shown in Figure 1.22.

## 1.7 General mapping of terrain and movement

Although we have maps of terrain of most of the world, some of it is old or inaccurate. Also terrain is continually changing either by natural forces or by manmade development. Thus mapped areas must be continually updated by survey, optical or RF means. Figure 1.23 shows a typical Doppler beam sharpened (DBS) radar map of farmland. One can easily see roads, buildings, stream beds, tall trees, fenced farmland and residential areas.

In many countries, government presence in wilderness areas is very slight. They must depend on EO/IR and radar surveillance to assess crops, local development, illegal operations, smuggling, border penetration, guerrilla operations, etc. Ground moving target (GMT) surveys can provide I&W of mass movements of troops, material, rail and truck traffic, etc., as shown in Figure 1.24. There are many thousands of moving vehicles in Figure 1.24. Obviously, this requires enough GMT surveys of areas of interest to establish a baseline of "normal" traffic. To be successful the GMT baseline must be able to distinguish each general class of movement, for example, cars from trucks from tracked vehicles from people and so on. Similarly, SAR surveys must be able to detect changes to manmade and natural

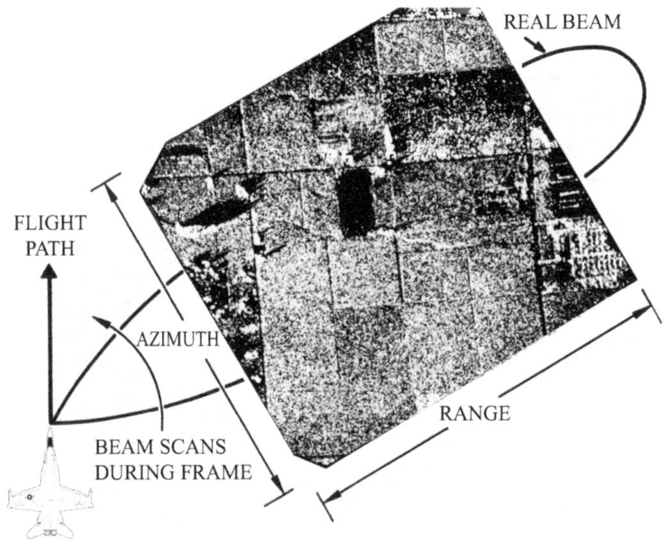


*Figure 1.23   DBS map: farm land and small town, San Joaquin Valley, CA [13]*

*Figure 1.24   Massed ground-moving targets; adapted from [15]*

features and categorize them as to type, that is, normal agricultural growth from illicit growth/harvesting from road building from mining from missile deployments, etc. Such surveys require persistent surveillance but at the same time most territory will have nothing of interest happening and must be screened automatically before presentation to an analyst.

There are software-based image exploitation systems that have been developed to detect changes and specific patterns of interest. Unfortunately, even with automatic target recognition (ATR) and CD, these systems are very labor intensive. This is at least partly because the very high resolution and accuracy necessary creates very large databases. These large databases, even if 90% of the data is automatically removed, still require trillions of picture elements (pixels) to be examined.

## 1.8 Tactical target detection

### *1.8.1 Stationary targets*

Often, tactical targets such as military missile sites have a characteristic shape established by the missile system design, military training and doctrine. The site may have a fenced perimeter, revetments, an access road, vehicle complements, semi-mobile shelters, electrical power generation or access to power lines in a somewhat standardized configuration. Figure 1.25 shows a series of progressively higher resolutions on a missile site, which contains all of the features cited earlier. If such sites start to appear near a country's border as they did in the run-up to 1973 Mid-East war, that would be a cause for alarm. Similarly, the appearance of intermediate-range ballistic missile sites (SS-4, -5) in Cuba in 1962 was a cause for a national crisis. In both of those cases, long-range radar sensors were not available and detection depended on good weather and permissive airspace.

Radar surface surveillance is now performed routinely from aircraft and spacecraft to prevent such surprises before they spin out of control. Just as missile sites were identified by the expert analyses of photographic means during the Cuban missile crisis, modern radar data can be used by experts to identify military deployments and large-scale guerilla activities. Large missiles, aircraft, radars,

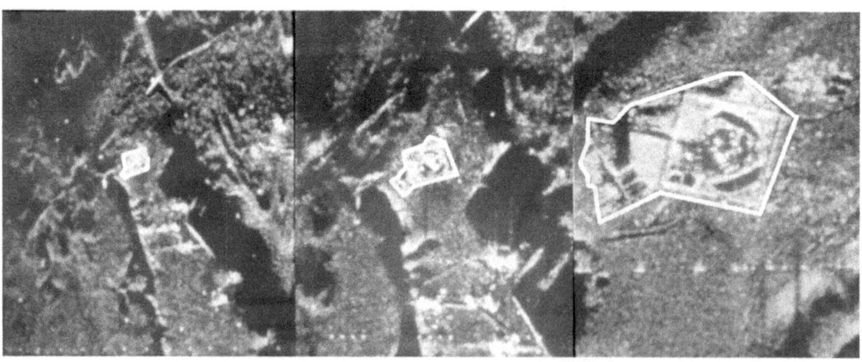

*Figure 1.25 Missile site (outlined in white) [13]*

military vehicles and guerilla training areas can be readily identified in relatively low-resolution (5′–20′) SAR and GMTI sensors. Analysts can easily identify the missile site in Figure 1.25 by the equipment deployment, roads, fencing and power sources in the image. The principal new advantage of radar is much longer-range observation under all weather and light conditions.

Military targets have recognizable SAR and GMTI signatures. For example, tanks like those in Figure 1.26 have a SAR radar signature. Figure 1.27 shows a SAR image of the two rows of the tanks pictured in Figure 1.26. As instructive as the radar reflection may be, the shadow cast by the tank is often more important because it shows the true shape better. One can also see the background brush as well. The radar image also shows that the tanks have been there long enough for the brush to grow up around them and cover the tracks they must have made getting to

*Figure 1.26 Tanks near USMC logistics base [1, courtesy General Atomics]*

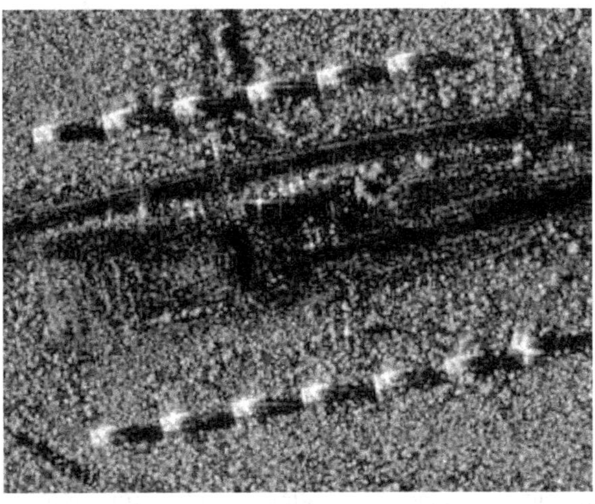

*Figure 1.27 Simulated tank company USMC logistics base [1, courtesy General Atomics]*

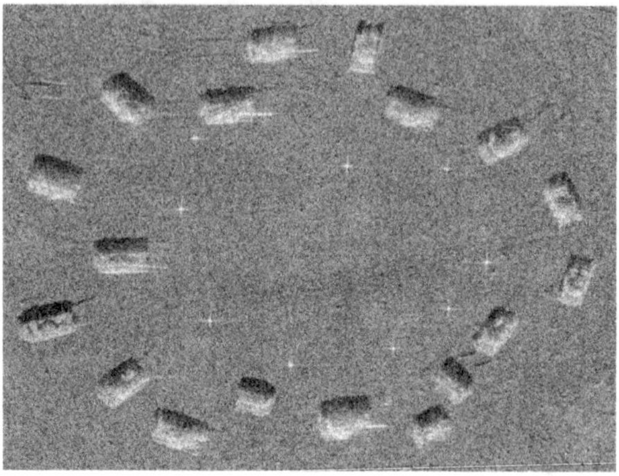

*Figure 1.28 High-resolution vehicle image allows recognition [1, courtesy General Atomics]*

their current location. Some other less obvious features are the power line and its associated maintenance "pole road" in the center of Figure 1.27.

Figure 1.28 shows a higher-resolution SAR image of tanks and other vehicles in a circle at various orientations. Inside the circle are some bright calibration reflectors that appear as crosses in the image. One will notice that in some orientations the tank gun barrel is very distinct and in others the barrel is faint but the shadow shows the barrels. One can also see the track marks of some of the vehicles as they were positioned in the field. There are also other track marks not directly associated with the current vehicle positions. Typically, vehicle track marks only need to be a few inches deep to be observable at X-band or above depending on grazing angle. Often there is as much intelligence to be gained from the track marks as the vehicles themselves since it may show where the vehicles came from or went to a hiding place. High-resolution images may also show where the ground or a road was disturbed, indicating an IED or tunneling. A common trick is to dig a trench large enough for a vehicle, cover the trench and vehicle with a tarp and throw a small amount of dirt on top of the tarp. This makes the vehicle virtually invisible to EO/IR but easily visible to radar. Many more SAR details will be provided in Chapter 7.

### *1.8.2 Moving targets*

In recent years there has been emphasis on slow-moving targets such as people walking sometimes called dismounts. The main difficulty for detection of interesting moving targets is that almost everything is moving. The foliage on trees and bushes is moving, fences move in the wind, insects, animals and birds move. Cable suspended stoplights, stoplight intersection intermittent stop and go, air conditioners and wind turbine ventilators all have moving parts. There are so many movers that the real challenge is separating the interesting movement from everything else. One major way to separate the interesting from the uninteresting is to

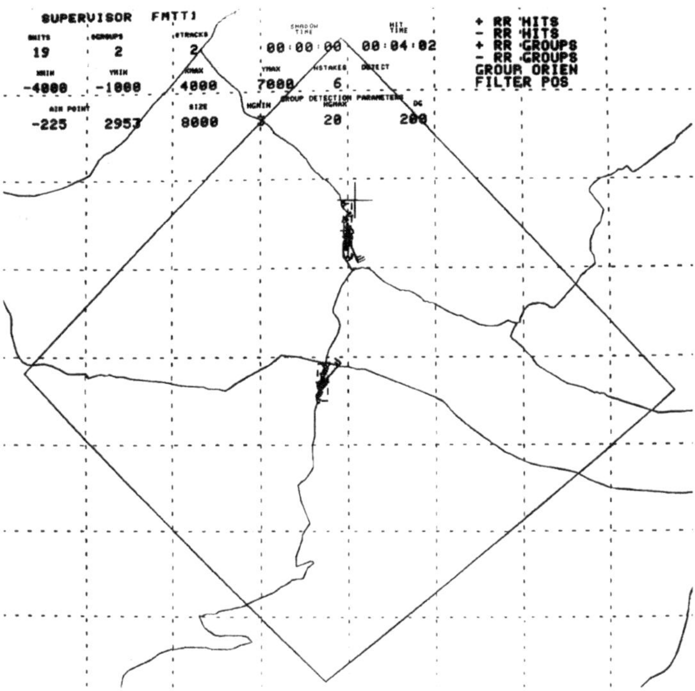


*Figure 1.29 GMTI tracking of two vehicle clusters on roads [16]*

track every mover in the field of view. The fences, ventilators, insects, etc., aren't going anywhere. They can be eliminated since the stay in roughly the same place. This may require the tracking of thousands of movers in the field of view. It may seem extreme but that is what is required to detect those movers that are moving with a directional purpose. Once the stationary movers are eliminated then other signatures can be used to discriminate between people/animals and vehicles. Many more GMTI details will be provided in Chapter 6.

As will be shown in Chapter 3, a moving tank has a narrower Doppler signature than a wheeled vehicle. It also usually travels slower and often is not on roads. That signature plus other cues such as order of movement and emissions allows operators to recognize tank companies. Tanks have a relatively large radar cross section (RCS) typically well in excess of 100 m$^2$ as will be shown in Chapter 3. Since they are easy to detect, the real issue is sorting them out from all the other vehicle traffic that may accompany them. In many types of terrain, roads and pathways are lined with buildings, hedgerows, power lines, fences and trees. These features tend to momentarily obscure individual vehicles in a tank company or leadership caravan. Thus, it is difficult to track individual vehicles in a cultural background but tracking clusters of associated targets is reasonably routine.

Figure 1.29 shows the tracking of moving target clusters in two groups (dashed rectangles) traveling in opposite directions on a road. The tracks are inside a 4 $\times$ 4 km "patch" (large diagonal box) with the local road network shown. There are a total of 19 movers in the two groups. The Beaufort symbols, circle with line and chevrons, in the cluster tracks are the combined speed and direction of the groups. The Beaufort symbol was originally used on weather maps to show the wind direction with chevrons showing speed. Often GMTI displays also contain road visibility probabilities relative to the radar as well as the road network. Clearly, ground mover down-range and cross-range accuracy must be good enough to place the vehicles on the correct road and in the correct cluster since they may pass side roads as well as clusters traveling in the opposite direction.

## 1.9 Sea surveillance

Another persistent surveillance application for GMTI/SAR radar is sea surveillance. Ships have generally large RCS and so are usually detectable from aircraft or space radars. The real problem is recognition. There are only a few thousand major naval combatants in the world and only a few hundred thousand large commercial vessels. Their characteristics are easily storable in less memory than the typical "thumb drive." Using a combination of their AIS transmissions, GMTI signature and SAR or inverse SAR (ISAR) imaging, ships at sea can be tracked and recognized.

Inverse SAR uses the motion of the ship to provide a high-resolution image such as those shown in the eight-frame sequence (movie) in Figure 1.30. These images easily show the outline and super structure of a ship as well as bright scatterers such as railings, antennas, masts and bridge. The bright streak in each of the images is an extremely bright scatterer, most likely a trihedral or rotating antenna. The ship length and width are easily estimated from a series of images. The roll, pitch and yaw of a ship at sea will cause scatterers on the ship to appear to

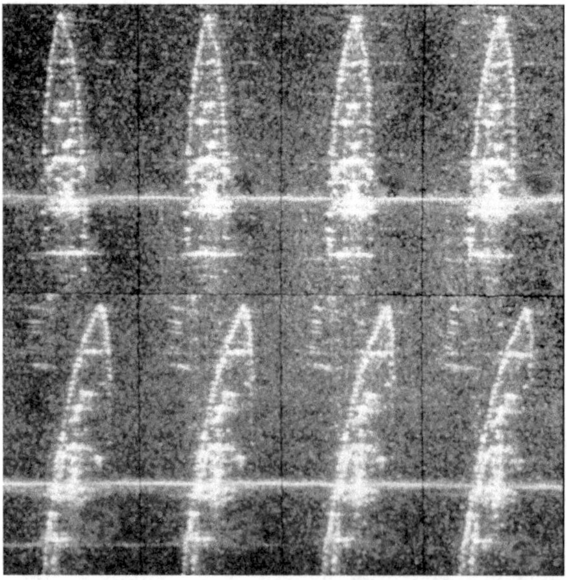

*Figure 1.30 Inverse synthetic aperture images of a ship at sea [17,18]*

be moving at different speeds for short periods of time. Since a ship is a quasi-rigid body, all of the motion will be along arcs of circles about the center of motion of the ship. Those circles will be projected as ellipses to a radar observer. Each will have a characteristic Doppler history that can be match filtered to resolve individual scatterers. Clearly, the entire ship must be tracked to keep the imaged patch centered on the ship. Depending on size, ship motions are usually measured in seconds and tracking should be sustained for 10–100 s. More details of ISAR will be provided in Chapter 7.

## 1.10 Object imaging

Another important aspect of persistent surveillance is recognition of small objects of all kinds. Unfortunately, radar bandwidths are usually small relative to feature sizes on many objects. A single frequency or angle look at an object will result in only a few "glints" from the object and it may be very difficult to tell what the object is. There are two ways to improve this situation. One way is to observe the object multiple times with different frequencies or look angles. Then to geometrically rectify each image and sum each corresponding pixel over all the images. This creates a more continuous tone image more nearly approximating visible light images. The other way, which requires very low noise, high dynamic range images, uses the shadow cast by the object.

Figure 1.31 shows a SAR image of a single-engine light aircraft in a parking area at a small airport. The reflected return from the airframe is small and very specular. It would be almost impossible to recognize the type of aircraft from the return. The shadow however clearly shows the platform of the aircraft from which the wingspan, wing area, fuselage length, horizontal stabilizer dimensions, etc., can be estimated. With this information, the aircraft can be recognized by referring to one of the many "Aircraft Observer's Recognition Handbooks."

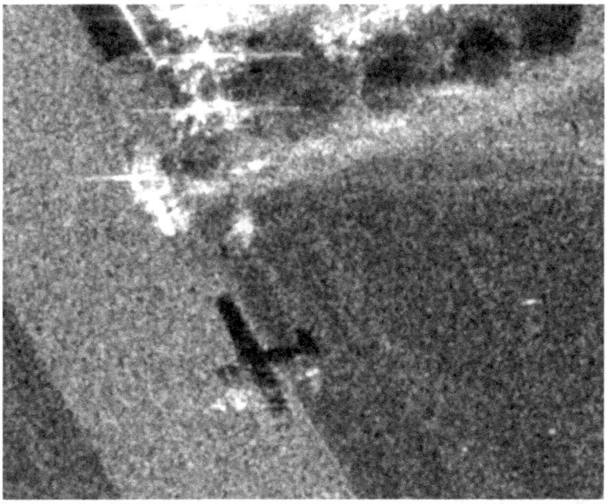

*Figure 1.31 Radar shadow of a light plane on taxiway [1, courtesy General Atomics]*

## 1.11 Radar patch size and minimum velocity requirements

In the late 1970s, the United States was searching for a non nuclear means to counter the Warsaw Pact armies' massive numerical superiority over NATO armies. Several study groups were empanelled with many famous US technical leaders to address this seeming disadvantage. Out of these groups came recommendations for both stealth aircraft and battlefield management similar to the airborne battle management that existed with AWACS. The author contributed to these study groups but is certainly not famous. Ultimately programs like the F-117, the B-2 and JointSTARS grew out of those recommendations. Several other programs were started as well. The focus was on a European battlefield against the Warsaw Pact. Fortunately, those systems were flexible enough to fulfill many other missions during the last 40 years.

There were 10–11 tanks in a Soviet tank company. The normal battle strategy is for tanks to space themselves far enough apart that a single weapon would not disable more than one tank, which is about 200–300 m. Depending on orientation of the tank formation, a patch must be 4 × 4 km to encompass a tank company. Similarly, a Warsaw Pact tank battalion was made up of 31 tanks, which requires an 8 × 8 km patch. For larger formations 24 × 24 km and 50 × 50 km were required. General theater surveillance requires up to 400 × 400 km coverage. The nation still uses these patch sizes in both potential and deployed systems for SAR maps as well as moving target detection. In spite of mechanization, armies move at about walking speeds. Therefore, a good lower bound for radial velocities is 1.5 m/s. This is a very slow walking speed, but because of grazing angle issues for some observations fast walking will appear to be only 1.5 m/s radial speed. Even at 1.5 m/s, humans and many animals have parts of their bodies moving at 2–3 times that speed allowing detection.

These patch size and speed requirements were established independently of patch range-azimuth resolution. A 4 km square patch with 0.1 m SAR resolution has $1.6 \times 10^9$ radar pixels but the achievable mover resolution is probably 3–20 m. This is at best $1.6 \times 10^6$ mover radar pixels. Figure 1.29 uses an 8 × 8 km patch with 2 m resolution.

## 1.12 Cueing of EO sensors

Once targets of interest are located within a few feet, the next step is use of EO/IR sensors to confirm the nature of the target and perhaps to launch a weapon with the aid of that sensor. These sensors are not only shorter range but also require an accurate angle to place the field of view on the potential target. This means that the radar gimbal viewing angle must be transformed to platform or site coordinates and through a second gimbal on the EO sensor.

Most of the time the cueing sensor and the EO sensor are at different locations on the platform or not even on the same platform. Radars measure range and Doppler very accurately but angle may be 100 times worse than the EO sensor angle accuracy (milliradians vs. microradians). The best that radars can do is about

0.15 milliradians standard deviation. Radars can improve cueing by using GPS and a local terrain database. The overall cumulative angular cueing error for radar to EO standard deviation must be less than 0.7 milliradians. Everything must be taken into account including atmospheric refraction to achieve this accuracy. Typical necessary coordinate transforms are covered in Chapter 2.

Figure 1.32 shows radar coverage from an aircraft on an aeronautical map including detection of a moving target (enclosed in black) with the corresponding EO image cued from the moving target detection. The slant range to the target is approximately 20,528 ft with a depression angle of 30.5°. Overall cueing accuracy was approximately 0.7 milliradians.

Radar having cued an EO sensor the next step is laser designation of the target for possible launch of a weapon. Figure 1.33 shows such a designation.

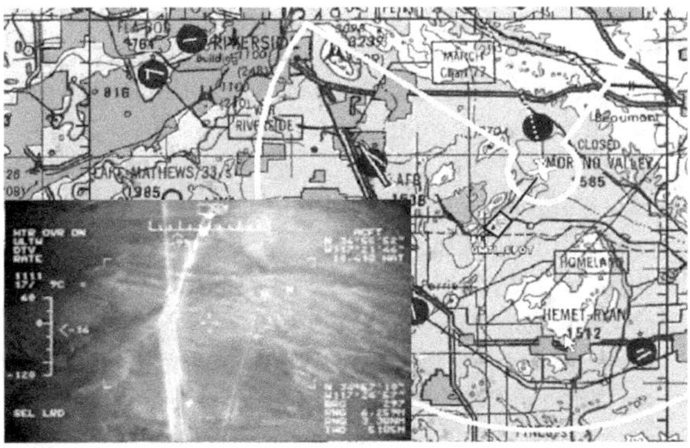

*Figure 1.32 Aeronautical map with GMTI cued EO sensor image [1, courtesy General Atomics]*

*Figure 1.33 Laser range finder on target with visible light sensor [1, courtesy General Atomics]*

## 1.13 Tactical target persistent surveillance from space [9,10]

Aircraft such as Hunter, Scan Eagle, Predator and Reaper can provide persistent surveillance over limited regions by dwelling in orbit above an area of 100 km$^2$ for up to 24 h. Longer-range radar sensors in aircraft such as JointSTARS, Global Hawk, U-2 and Astor have limited persistence for a variety of reasons. Spacecraft have been proposed and studied as a persistent surveillance asset but none has been deployed.

Suppose a constellation of spacecraft with hybrid arrays of the type depicted in Figure 1.14 were deployed. How well would they perform in a tactical persistent surveillance mission similar to JointSTARS or Predator? This particular question was the subject of a study by the author and colleagues. Since on orbit weight is a strong predictor of cost, the study used Beginning of Life on Orbit Mass (BOLOM) in a Monte Carlo simulation. The results are summarized in Table 1.4. There are 17 JointSTARS surveillance aircraft. The study chose an orbital altitude of 1,300 km so that 17 spacecraft would cover 90% of the earth (missing the poles where nothing is going on anyway). Thirteen hundred kilometers is the highest a low earth orbit (LEO) spacecraft can be without requiring extreme antiradiation measures necessary for the first Van Allen belt. One significant advantage from 17 spacecraft versus 17 aircraft is the operation and maintenance cost difference. The other advantage is prompt response since at most a spacecraft will be overhead in 15 min. The cost for such a system is approximately the same as JointSTARS ~20 B$. JointSTARS has distinguished itself in several Mid East conflicts but now is in need of replacement. The various scenarios depicted in the table are summarized in Figure 1.34.

The collection times range from 1 to 30 min, which is about all the time a specific target area might be in spacecraft view. With proper orbits another spacecraft will be over the target area within an average of 5 min. This would then provide persistent surveillance anywhere on the populated earth at any time without prior deployment to a theater or hotspot AOI.

Figure 1.35 shows more detail of the limitations that a large space radar will have from single beam, limited electronic scan coverage and finally rotation time

*Table 1.4 GMTI and SAR collection time comparisons (HA6GMTISAR, GMTI dwell, GMTI totals) [10]*

| Type/ BOLOM (kg) | Total time to collect 100 GMTI targets (s) | | | Total time to collect 20 SAR targets (s) | | |
|---|---|---|---|---|---|---|
| | 400 × 400 km AOI | One quadrant of FOR | Entire FOR | 400 × 400 km AOI | One quadrant of FOR | Entire FOR |
| HA/2000 | 220 | 240 | 420 | 1700 | 1720 | 1820 |
| HA/4000 | 120 | 140 | 320 | 1650 | 1670 | 1750 |
| HA/6000 | 90 | 110 | 210 | 1650 | 1670 | 1750 |
| HA/8000 | 80 | 100 | 200 | 1650 | 1670 | 1750 |
| HA/10000 | 65 | 80 | 180 | 1650 | 1670 | 1750 |
| HA/14000 | 65 | 80 | 180 | 1650 | 1670 | 1750 |

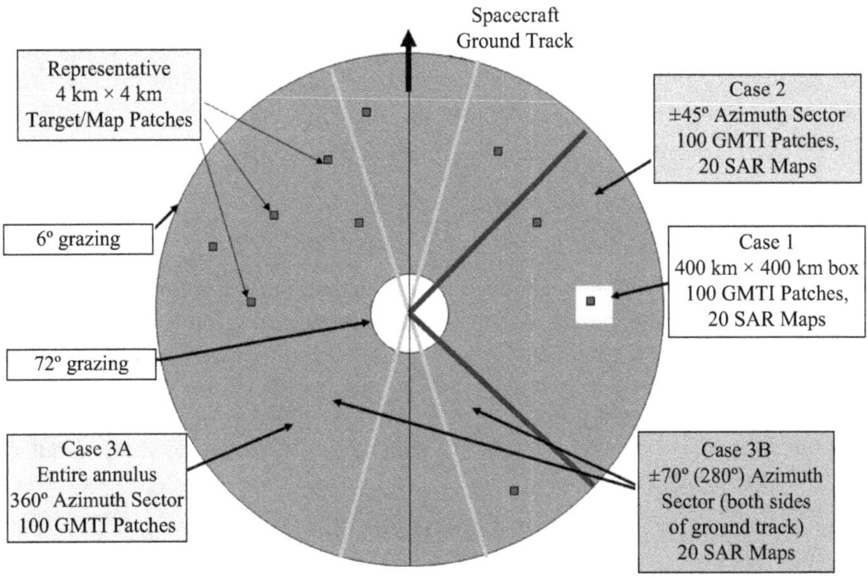


*Figure 1.34 Spacecraft Monte Carlo cases [10]*

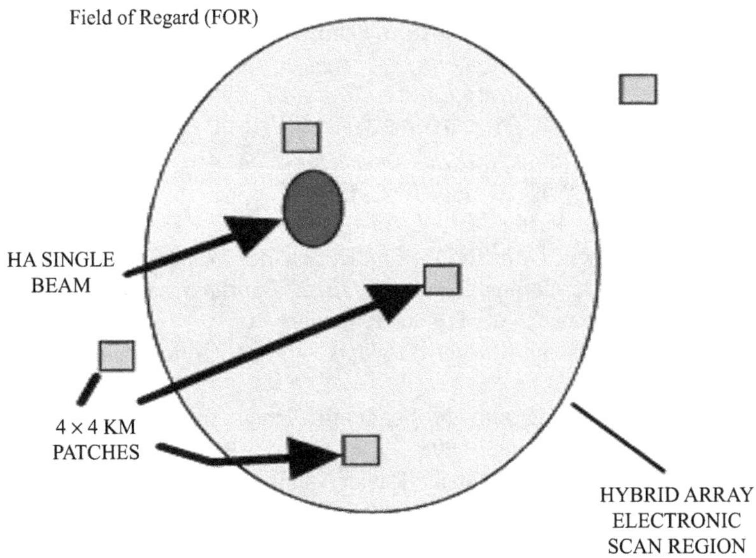


*Figure 1.35 Spacecraft field of regard [10]*

that the spacecraft requires to engage a new field of regard (FOR). Electronic scanning is of limited use because cross-range accuracy goes down as the square of the scan angle from the normal to the array. Typically 100–500 m cross-range accuracy is required to designate a target for higher-resolution observation no matter what the range resolution may be (1/2 a blob is still a blob).

## 1.14 Summary

Obviously the applications mentioned in this chapter are just a sampling of radar-based persistent surveillance. Also, the descriptions have left out many details necessary for understanding, analysis and design. Some of the basics will be discussed in the chapters that follow. Hopefully, the reader will have a good intuitive feel for all the important aspects of Intercept, GMTI and SAR by the end of this introductory book. The references and appendices will provide more detail on many of the topics in this book.

## References

[1] General Atomics Aeronautical Systems, Reconnaissance Systems Group, Interim Report CDROMs and Web Page, 2008.

[2] Olsen, R.C., *Remote Sensing from Air and Space*, SPIE Press, (2007), p. 65.

[3] Hindle, P. ed., "Drone Detection and Location Systems", *Aerospace and Defense Electronics*, June 2017, pp. 6–24.

[4] https://en.wikipedia.org/wiki/ISM_band.

[5] Lynch, D., *Introduction to RF Stealth*, SciTech Publishing, (2004), pp. 22–29, 67–97, 156–168.

[6] Gavish, M., Weiss, A., "Performance Analysis of Bearing-Only Target Location Algorithms", *IEEE Transaction on Aerospace and Electronic Systems*, Vol. 28, No. 3, July 1992, pp. 817–828.

[7] Socheleau, F., Pastor, D., Aissa-El-Bey, A., "Robust Statistics Based Noise Variance Estimation: Application to Wideband Interception of Noncooperative Communications", *IEEE Transactions on Aerospace and Electronic Systems*, Vol. 47, No. 1, January 2011.

[8] Martin, D., *Communication Satellites*, 4th ed., Aerospace Press, (2000), pp. 147–149.

[9] Lynch, D., "Limitations of Scanning Arrays for Space", *IEEE Radar Conference*, 2009, Paper 3005.

[10] Jensen, E., Lynch, D., "Space Radar Architecture Tradeoffs", *AIAA Space 2009 Conference*, Paper AIAA-2009-6787.

[11] Barrett, D., "Police Tactic Was Kept as Classified", *Wall Street Journal*, #/18/2016, P. A3.

[12] Octasic, Inc. Web Page: www.octasic.com, OCTBTS 3600 software defined radio.

[13] Hughes Aircraft Co., "Forward Looking Advanced Multimode Radar (FLAMR)", Report P78-09R, 1977, Declassified 12/31/87.

[14] Schetzen, M., *Airborne Doppler Radar: Applications Theory, and Philosophy*, AIAA, (2006), pp. 33–36.

[15] Northrop Grumman Inc., JSTARS marketing handout, 2006.

[16] Hughes Aircraft Co., "Pave Mover TAWDS Final Report", 1984.

[17] Aks, S., Lynch, D., Pearson, J., Kennedy, T., "Advanced Modern Radar", *Educational Technology Inc. Course Notes*, 1994.

[18] Lynch, D., "Surface Ship Stealth Update", *Educational Technology Inc. Course Notes*, 2007.

# Chapter 2
# What coherent radars do

## 2.1 Basic principles

Everything from your house electricity to cosmic rays involves electromagnetic energy. There is electromagnetic radiation bathing us all the time from your electric blanket to TV to cellphones to radioactive decay of the bricks in your house to cosmic rays. Each has its own characteristic oscillation frequency and corresponding wavelength related by the velocity of light. Figure 2.1 shows much of the electromagnetic spectrum including the general location of frequency and source of emissions in that band. In addition the energy per photon increases as the wavelength gets shorter. So even though it is easy to feel those big fat photons coming out of an open fire or oven, the much shorter wavelengths have far more energy to penetrate most materials. Nonetheless, all of these wavelengths can carry significant energy.

Your eye is an image sensor that has a lens (cornea) and a focal plane array of light sensors (rods and cones in the retina). It is limited by the same physics as all other electromagnetic sensors including radars. The resolution that a healthy, clean eye can achieve is limited by the operating wavelength and the diameter of the entrance pupil as shown in Figure 2.2. The typical human visual wavelength, $\lambda$, might be 0.56 μm (yellow light) and a typical eye pupil opening in bright light, $D_D$, might be 0.2 cm. Thus, you may be able to achieve a visual acuity of about 0.28 milliradian if your eyes are clean under ideal circumstances. That's the equivalent of being able to just resolve two golf balls side by side at 100 yards, something most can do.

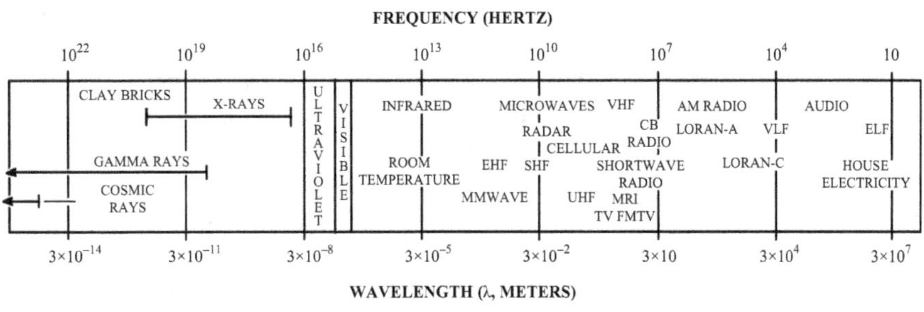


*Figure 2.1 Electromagnetic spectrum*

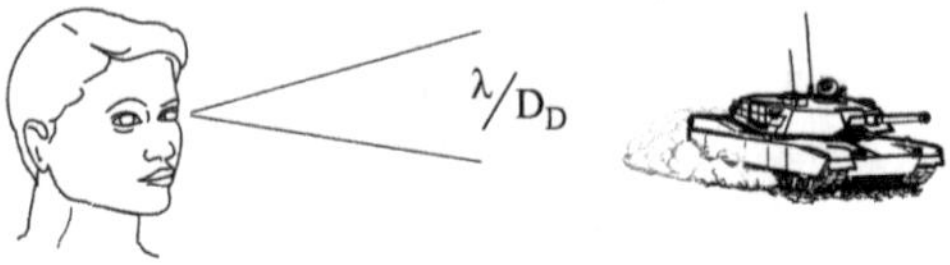


*Figure 2.2 Visual acuity*

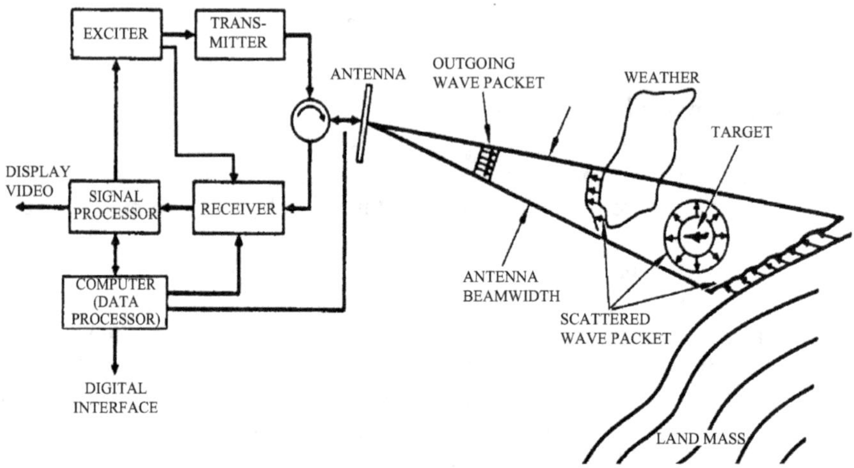


*Figure 2.3 Basic elements of a radar [1]*

Radars are limited in the same way as your eye but, with significant signal processing and high signal to noise and interference ratios (SNIR), dramatic improvements in accuracy and acuity can be achieved. This is enabled by use of coherent processing not available to the naked eye.

## 2.2 Basic radar

Figure 2.3 shows the basic elements of most radars. There is usually a very stable microwave reference and waveform generation circuitry in the exciter, which is sent to the transmitter circuitry. The transmitter increases the waveform power from the exciter and passes this on to one or more antenna elements where it is coupled to free space. The antenna usually has some means for steering the beam in a direction of interest much like one might point a flashlight beam. The pointing may be electronic, mechanical or both. The outgoing wave packet is mostly contained in the antenna main beamwidth. The packet propagates at the speed of light in the medium (usually air or space). The radar wave packet steadily expands as it propagates and so the energy per unit volume gets steadily smaller with the square of the distance traveled. The packet may encounter weather, targets, the earth's surface or animals. (In radar parlance, everything that you are actually looking for with the radar is a target and everything else is clutter.) Each object that the packet encounters reflects (scatters) some portion of the energy in the wave packet. Some of the energy that is reflected propagates back toward the radar – a return echo.

Part of the return echo is captured by the radar receiving antenna, which may be the same as the transmitting antenna. The energy captured by the antenna is sent to the receiver for amplification. Since the transmit energy is so much greater than the received energy, there usually is some protection and isolation circuitry between the transmitter and the receiver. The small size of the returned echo amplitude requires very large amplification and multiple filters to remove as much of the noise as possible. In addition, there are usually competing echoes from closer and much larger scatterers. Thus, the ratio between the smallest and largest echoes (dynamic range) can be billions to one.

Everything that has a temperature above absolute zero emits microwave thermal noise including the air. If the radar wave propagation path is entirely through the atmosphere, the received signal will be accompanied by thermal noise equivalent to the ambient temperature. In addition, the radar has internal noise, which also competes with the target return. Successive waveform packets are transmitted to obtain enough energy on the target so that it may be detected in the presence of noise.

After the returned echo amplitude has been increased enough so that it can be easily manipulated, it is sent to a signal processor. The signal processor applies successive filters and other processes, which maximizes the target signal and minimizes the accompanying noise and clutter. The signal processor records and sums all the wave packets to allow detection of targets. In modern parlance, these processes are called algorithms. The signal processor is a special purpose computer optimized for the required algorithms. In addition, there is usually a general purpose computer much like a desktop or laptop PC, which commands and controls the radar. It determines antenna pointing direction, frequency of operation, waveform shape, dwell time, what is to be recorded as well as what is to be provided to displays and other equipment. It may also continuously calculate the position of the radar in space and its relationship to other objects both clutter and targets.

Modern radars often have many different modes of operation and the computer data processor may optimize the radar behavior for each class of target [1]. For example, weather is clutter if you are looking for airplanes but airplanes are clutter if you are looking for weather. A weather radar must behave much differently than an air traffic control radar but they might be embodied in the same radar hardware.

## 2.3 The matched filter notion

When you look in a well-made mirror, you see an image, which you have come to believe is your image. You move to the left and the image moves to the left. You raise your hand to your face and the image does likewise. Mentally you have correlated your actions to those in the image. You have remembered the earlier actions and observed responses as well as the most current one and mentally integrated all those into a future expectation that the image will look as you look. This is called cross-correlation and it has a mathematical definition as given in (2.1).

$$R_{rs}(t) = \lim_{T \to \infty} \frac{1}{2T_{OT}} \int_{-T_{OT}}^{T_{OT}} r(\tau)\bar{s}(t-\tau)\,d\tau$$

where: $r(\tau)$ is the received signal, $\tau$ is a time delay variable,

$\bar{s}(t-\tau)$ is the conjugate of the desired signal,

$T_{OT}$ is the observation time, and $t$ is the time variable. (2.1)

Suppose you were to look at your reflection in a pond with wind-driven ripples. Part of your actions might be seen in the reflected image but many other parts of the image at any point in time would reflect things that have nothing to do with you. You mentally correlate your actions as well as your prior self-image to assess your current state of self (hair mussed, face dirty, etc.). You are "matched filtering" in the presence of noise.

A matched filter is a theoretical circuit that maximizes the output peak signal power relative to the mean noise power. In general, this circuit is not realizable but approximations to the matched filter can get within 10% of the optimum. One possible version is a correlation receiver that multiplies the received waveform point by point times a delayed complex conjugate replica of the transmitted waveform and integrates for the time extent of the transmitted waveform as shown in Figure 2.4. Since range is measured in a radar by time delay and returns can come from almost any range, the time delay must in fact be a whole range of possible time delays each integrated separately to create a range bin by range bin composite output. This operation is cross-correlation between the received signal, $r(t)$, and a delayed complex conjugate of the transmitted waveform, $\bar{s}(t-\tau)$. There may be thousands of different time delays involved at least one for every range bin.

Suppose you could find a filter circuit, $h(t)$, that could perform the same function for all possible time delays. It turns out that a circuit that has a filter time response (usually called impulse response) that is $s(t)$ run backwards in time plus some fixed time delay, $\tau$, is a good approximation of the matched filter, that is,

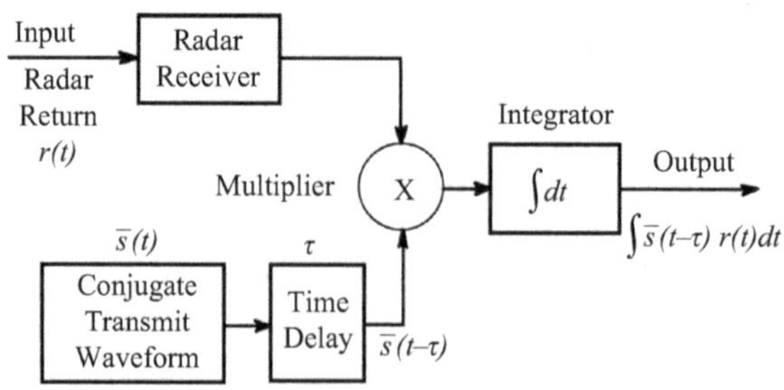


*Figure 2.4 Matched filter concept*

$h(t) = \bar{s}(\tau - t)$. The fixed time delay is required because you can't anticipate the future. When a filter has a time signal applied to it, it performs the mathematical function of convolution, which is given in (2.2). The time output of the filter is the convolution of the input signal $r(t)$ and the impulse response $h(t)$ of the filter. The fixed time delay, $\tau$, just adds a phase warp term to the product of $R$ and $H$. For many cases a simple 2 pole bandpass filter whose 3 dB bandwidth is equal to half the comparable bandwidth of the transmitted waveform gets within 1 dB of optimum [2]. The matched filter notion will be used repeatedly in the coming chapters.

$$r(t) * h(t) = \int_{-\infty}^{\infty} r(\tau) \cdot h(t-\tau)\, d\tau \leftrightarrow R(\omega) \cdot H(\omega)$$

where: $R(\omega)$ and $H(\omega)$ are the Fourier transforms of $r(t)$ and $h(t)$. (2.2)

## 2.4 Fourier transform

Your ear is a sound sensor. Your inner ear has an ever-narrowing spiral something like a snail shell. Each location along the spiral is sensitive to an ever-higher tone or frequency. A well-trained musician can easily tell these frequencies apart even when many tones are together in a chord. The ear is sensing each of these frequencies separately and integrating them to form knowledge of the harmonic content of the sound and ultimately interpretation of music or speech received. Your eye also senses color (frequency) intensity. The mathematical equivalent of that sensing process is called the Fourier transform (FT). The math asks the question how much of any frequency is contained in a signal.

A received radar time waveform will typically be of the form of (2.3).

$$r(t) = a(t) \cdot \sin\left(2\pi f_0 t + \phi(t)\right)$$

where: $a(t)$ is the amplitude modulation on the envelope of the sinewave, $f_0$ is the center frequency of the wave, $\phi(t)$ is the phase modulation of the sinewave. (2.3)

Once there is amplitude and/or phase modulation, the single sinewave carries many frequencies. FM radio encodes sound on to a radio transmission by phase (frequency) modulation. The FT must sense both the amplitude and phase of the signal that your ear and eye can't do. One way to sense both is by the use of two analyzing sinewaves that are 90° apart. A mathematical way to keep these two waveforms separate but carry them along through many calculations is by use of complex numbers, that is, numbers that contain $\sqrt{-1}$, which is sometimes represented as "j." Equation (2.4) shows one version of the FT in the top line with the two analyzing waves. A more compact way of stating the same thing mathematically uses the properties of the complex exponential function and is shown in the second line of (2.2).

$$R(2\pi f) = \int r(t)(\cos(2\pi ft) - \mathrm{j}\sin(2\pi ft))\,dt$$
$$R(2\pi f) = \int r(t)\exp(-\mathrm{j}2\pi ft)\,dt = \mathrm{FT}[r(t)]$$
where: exp is the exponential function and $\mathrm{j} = \sqrt{-1}$ (2.4)

A common convention is the use of upper case to represent the FT of a lowercase time waveform. The FT can be used not only for sound, electrical signals and electromagnetic waves but also to represent spatial frequencies in a TV image, heat conduction, sunspot cycles, etc. The integral of (2.4) multiplies the received waveform, $r(t)$, by successive sines and cosines spaced $df = 1/dt$ apart and accumulates the result for each incremental step to represent the frequency content at that step. The FT process is completely reversible with an inverse FT. Fortunately, there are books full of common (and not so common) FTs for waveforms. There are also simple rules for deriving new FTs from existing ones. In addition, there are computationally efficient methods of calculating FTs from raw data called discrete Fourier transforms (DFTs). Since FT outputs are orthogonal, they can be cascaded to increase speed or increase frequency resolution. The DFT is so easy nowadays that, if your cellphone takes or receives pictures, it uses DFTs in image compression and reconstruction.

## 2.5 Decibel notation

In the old-old days before desktop computers, many equations that involved significant numbers of multiplications, division and exponents were calculated using logarithms. A system of nomenclature was established to facilitate these calculations using Bel's and decibel's. The Bel was named after Alexander Graham Bell, the inventor of the telephone. It is used to represent power levels throughout a network including radar calculations. The common usage is decibels relative to some reference power level such as a watt. (A typical incandescent light bulb might be 60 W.) A power of 10 W relative to a watt would be as given in (2.5). Similarly the power gain through an amplifier would be the logarithm to the base 10 of the ratio between the output and input powers as shown in Figure 2.5. Voltage or

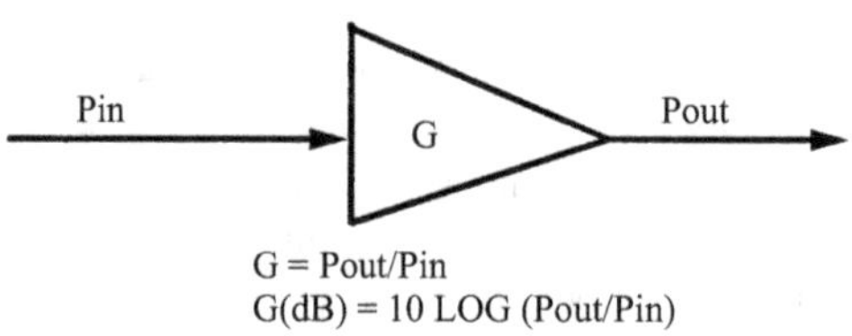


*Figure 2.5 The decibel notion*

current ratios do not represent power and are squared assuming a common impedance when carried along in equations. Referring to (2.5) again, a voltage of 10 V relative to a volt would be 20 dBv.

$$P_{dBW} = 10 \cdot \log_{10}\left(P_{watts}\right) = 10 \text{ dBW}$$
$$V_{dBv} = 10 \cdot \log_{10}\left(V_{volts}^2\right) = 20 \cdot \log_{10}\left(V_{volts}\right) = 20 \text{ dBv} \tag{2.5}$$

Since the magnitudes of powers and other parameters in radars can vary by billions, it is very convenient to express many parameters in decibels. Also, radar engineers begin to think in decibels and it is useful to have common decibel ratios on the tip of your tongue. Antenna gains, RCSs, sidelobes, losses, noise figures and even range are often quoted in dB. For example, a 60 W light bulb would be 17.8 dBw. Table 2.1 has most of the commonly used decibel ratios.

## 2.6 Antenna principles

It was observed hundreds of years ago that the resolving power of an optical system was limited by the area of collecting optics, sometimes called the entrance pupil. Just as suggested in Figure 2.2 for the human eye, the distant acuity, usually called beamwidth or angular resolution for radars is limited by the dimensions of the radar antenna (entrance pupil) and the operating wavelength. In the case of radars, the operating wavelengths are commonly between 2 ft and ¼ in. Therefore, to achieve the same acuity as the eye, the antenna aperture must be proportionately larger. For example, a radar operating at a wavelength of 0.03 m (X band) would have to be over 100 m in diameter to have the same angular resolution as your eye. Most travelers have observed large radar antennas around airports. Because of the operating wavelength, most of these radars don't have nearly the acuity of the human eye but they can "see" much farther. Even in rainy and cloudy weather,

*Table 2.1 Decibel ratios*

| Power Ratio | Decibel value | Power Ratio | Decibel value |
|---|---|---|---|
| 1 | 0 | 0.9 | −0.5 |
| 2 | 3 | 0.8 | −1 |
| 3 | 4.8 | 0.7 | −1.5 |
| 4 | 6 | 0.6 | −2.2 |
| 5 | 7 | 0.5 | −3 |
| 6 | 7.8 | 0.4 | −4 |
| 7 | 8.5 | 0.3 | −5.2 |
| 8 | 9 | 0.2 | −7 |
| 9 | 9.5 | 0.1 | −10 |
| 10 | 10 | 0.02 | −17 |
| 20 | 13 | 0.01 | −20 |
| 30 | 14.8 | 0.001 | −30 |

these radars may detect aircraft at several hundred miles. The antenna-limited resolution makes all these detections look like blobs. Amazingly, modern Doppler processing allows imaging aircraft at very long range, as we will see in later chapters.

First, the basic properties of antenna apertures will be described. Every antenna has a beamwidth that is usually compared to an isotropic radiator, that is, one that radiates uniformly in all directions. An isotropic radiator would spread its energy over a sphere of 4 $\pi$ steradians solid angle. Such a radiator is hard to make; so it is a fiction used only to standardize the definitions of antenna gain. The gain of an antenna is the ratio of its solid angular beamwidth, $\Omega$, in steradians to an isotropic radiator, 4 $\pi$ steradians. This is shown in the left of Figure 2.6. The far-field solid angle beamwidth is the half power area of the beam divided by the range squared at the point of measurement. This is usually achieved on an antenna far-field range by using a remote single-frequency transmitter. The far field is defined as the region in space beyond the distance $2D_D^2/\lambda$, where $D_D$ is the diameter of the antenna aperture. The antenna beam has a constant angular resolution in the far field. The near-field antenna beam is in the shape of a tube of approximately the same size as the antenna radiating area. The antenna patterns, which are calculated using simplifying assumptions, are close but not exactly correct.

The far-field distance between a transmitter and an antenna under test must be at least two times the square of the largest dimension of the antenna divided by the operating wavelength (i.e. $>2D^2/\lambda$ in Figure 2.6). The antenna under test is used in receive only and it is mounted on a gimbal that typically sweeps out 120° in elevation and 360° in azimuth. The relative half power extent is measured, divided by the known distance squared and the gain is calculated. This is the directive gain of the antenna. For an unweighted rectangular antenna, the directive gain can be calculated as shown in the right hand of Figure 2.6. The gain is approximately 4 $\pi$

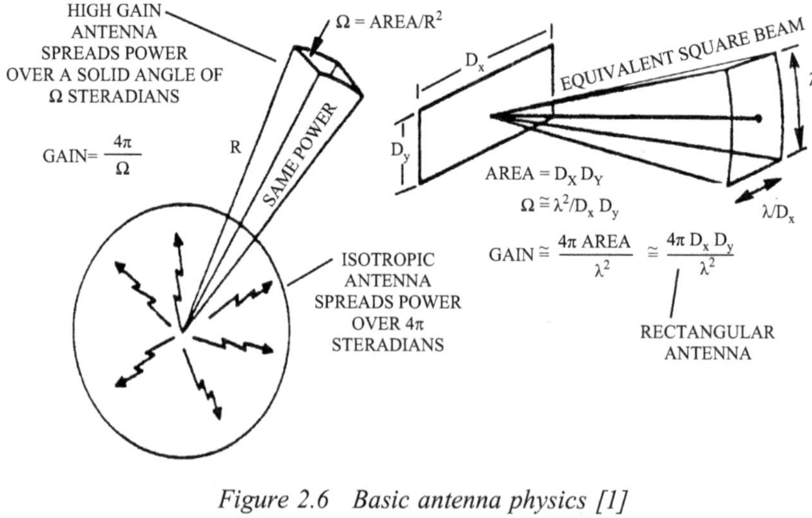


*Figure 2.6 Basic antenna physics [1]*

times the antenna area divided by $\lambda^2$. In electromagnetics, nothing ends precisely at a sharp edge due to diffraction and so the real antenna's gain is a little less than Figure 2.6. In addition, the directive gain does not take into account the inevitable losses in the antenna feed networks and the efficiency of coupling between the antenna elements and free space. Usually everything in the antenna itself has impedances between 25 and 300 Ω but free space is very close to 377 Ω. Imperfect impedance matches throughout the antenna lead to losses, so the net gain is always lower than the directive gain. A uniformly weighted antenna pattern has relatively poor sidelobes, so aperture weighting that reduces sidelobes also reduces the net gain. In optical terms this is called apodization (literally: cutting off the feet).

An antenna aperture whose dimensions are finite (they all are) has sidelobes. Sidelobes are responses to energy arriving from angles, which are well away from the main beam. Your eye, cameras and even rainbows have sidelobes. You may have taken a photo near the limb of the sun and found that the picture was washed out near the sunny side; that's a sidelobe. It turns out that the distribution of electromagnetic energy across a planar antenna aperture gives rise to a far-field pattern that can be approximated by the FT of the aperture distribution. The electric field at some point in the far field is approximated by (2.6). Equation (2.6) is easily recognized by the cognoscente as a spatial 2D FT of the aperture distribution to far-field angle space [3].

$$E(\theta,\phi) \cong \frac{\mathrm{j}\cdot\exp(-\mathrm{j}\cdot k\cdot R)}{\lambda\cdot R}\cdot\iint_{area}\left[F(x',y')\cdot\exp\left[\mathrm{j}\cdot 2\cdot\pi\cdot\left(\sin\theta\cdot(x'\cdot\cos\phi+y'\cdot\sin\phi)\right)\right]\right]\cdot dx'dy'$$

where: $\lambda$ = wavelength, $x' = x/\lambda$, $y' = y/\lambda$, $R$ = distance from aperture center to observer,

$\theta =$ angle from z axis to observer, $\phi =$ angle from x-z plane to observer, $\mathrm{j}=\sqrt{-1}$

$F(x',y') =$ phase and amplitude distribution on the aperture, $k=\dfrac{2\cdot\pi}{\lambda}$ the propagation constant.

Define: $u = \sin\theta\cdot\cos\phi$, $v = \sin\theta\cdot\sin\phi$, $w = \cos\theta$.

Then: $E(u,v) \cong \dfrac{\mathrm{j}\cdot\exp(-\mathrm{j}\cdot k\cdot R)}{\lambda\cdot R}\cdot\displaystyle\iint_{area}\left[F(x',y')\cdot\exp\left[\mathrm{j}\cdot 2\cdot\pi\cdot(x'\cdot u+y'\cdot v)\right]\right]\cdot dx'dy'$

which is a 2D Fourier transform to $u,v$ space multiplied by a complex constant. (2.6)

Not shown explicitly in (2.6) is that $F(x',y')$ has a polarization property and is often split into two illumination functions representing the polarization of the illumination. If the aperture distribution is uniform, then the far-field pattern will have a decaying sinewave shape. The first sidelobe of which is down to about 5% of the peak of the beam. Figure 2.7 suggests a single far-field plane cut in u,v space through the mainlobe and sidelobe structure similar to most antennas when plotted for an equivalent total of $\pm 180°$ about the main beam. In addition to the first sidelobe, there are many other sidelobes, which are usually characterized by their root mean square (RMS) level and their integrated sidelobe ratio (ISLR) relative to the peak of the beam. The sidelobe ratios are calculated using the power from the first null outward or an arbitrary level below the main beam peak. The half power beamwidth is usually used as the size of the mainlobe in radar calculations.

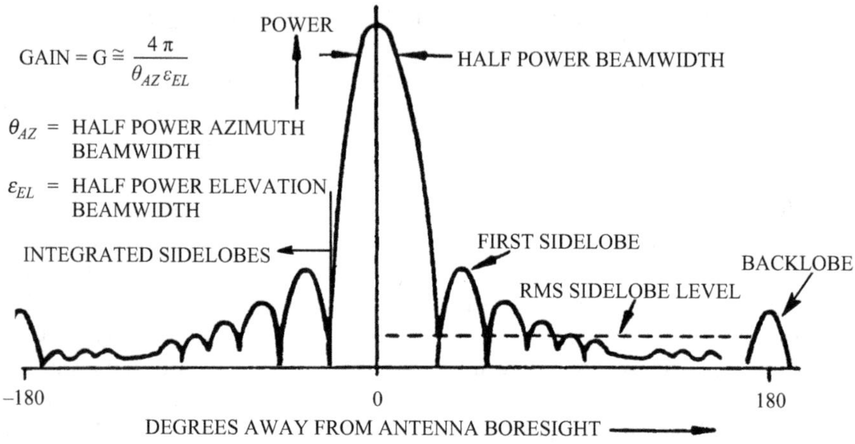


*Figure 2.7 Typical antenna properties [3]*

The gain of the antenna can be approximated by dividing 4 $\pi$ by the product of the half power beamwidths in radians in two orthogonal axes, say azimuth and elevation. Because there is leakage, diffraction, refraction and reflection around the antenna, there will usually also be a backlobe similar in size to the first sidelobe as shown in Figure 2.7.

Obviously not all antennas are flat as suggested by Figure 2.6. Often the aperture distribution is calculated on a surface (usually spherical) near but not on the physical antenna. Then, the distribution on that surface (x′,y′,z′) is transformed to a far-field pattern in (u,v,w) space as suggested in (2.6). Sometimes an antenna is measured on a spherical near-field range and its pattern is calculated using the 3D equivalent of (2.6).

A reflector antenna with some weighting will have a first sidelobe of 0.5% (−23 dB), an RMS sidelobe level of 0.03% (−35 dB) and an ISLR between 25% and 50% (−6 to −3 dB). A well-designed planar array with weighting and careful manufacture will have a first sidelobe of 0.03% (−35 dB), RMS level of 0.0003% (−55 dB), ISLR of 0.6% (−22 dB) and backlobe of 0.3% (−25 dB). Reflector antennas can have the highest bandwidth at the expense of poorer sidelobes.

$$G(\theta,\phi) \cong \frac{4\cdot\pi\cdot D_y\cdot D_x}{\lambda^2}\cdot\frac{\sin^2(U)}{U^2}\cdot\frac{\sin^2(V)}{V^2}$$

$$\text{where}: U = \frac{\pi\cdot D_x}{\lambda}\sin(\theta)\cdot\cos(\phi) \text{ and } V = \frac{\pi\cdot D_y}{\lambda}\sin(\theta)\cdot\sin(\phi)$$

$D_x$ = antenna horizontal length, $D_y$ = antenna vertical height, $\lambda$ = operating wavelength, $\theta$ and $\phi$ are as defined previously-not azimuth and elevation.

For better than -30 dB sidelobes typical 3 dB beamwidths: $\theta_{az} \cong \frac{1.3\cdot\lambda}{D_x}$ and $\varepsilon_{el} \cong \frac{1.3\cdot\lambda}{D_y}$

(2.7)

The two most common radar antenna geometries are rectangular and circular. An unweighted rectangular antenna geometry has a pattern given in (2.7). An unweighted circular antenna geometry has a pattern given in (2.8). Just as most rectangular antennas give rise to sin(x)/x shapes, circular or oval antennas give rise to cylindrical functions that are decaying Bessel function shapes.

$$G(\theta,\phi) \cong \frac{4\cdot\pi^2\cdot a^2}{\lambda^2}\cdot\frac{J_1^2(D_u)}{D_u^2}$$

For better than -30 dB sidelobes typical 3 dB beamwidth is:

$$BW_0 = \theta_{az} = \varepsilon_{el} \cong \frac{1.3\cdot\lambda}{2\cdot a}$$

where: $D_u = \dfrac{2\cdot\pi\cdot a}{\lambda}\sin(\theta)$, $a$ is the circular antenna radius,

$J_1(\ )$ is a Bessel function of the first kind. (2.8)

Real antennas are somewhat different. Nonetheless, these theoretical patterns suggest what most antennas are like, especially within 30° of the mainlobe. Other important parameters of antenna performance include amplitude weighting of the aperture to reduce sidelobes and the ISLR defined in (2.9), which is the power in the sidelobes divided by the power in the mainlobe. The integral limit of *2.3U3dB* is approximately the 10 dB down point on the mainlobe. As will be seen later the peak sidelobe ratio (PSLR) and ISLR often determine target detectability in high clutter and interference environments.

$$ISLR = \frac{\int_{-\infty}^{\infty} dv \int_{-\infty}^{\infty} |E(u,v)|^2\, du - \int_{-2.3V3dB}^{2.3V3dB} dv \int_{-2.3\cdot U3dB}^{2.3\cdot U3dB} |E(u,v)|^2\, du}{\int_{-V3dB}^{V3dB} dv \int_{-U3dB}^{U3dB} |E(u,v)|^2\, du} \quad (2.9)$$

Aperture weighting is used to reduce sidelobes but at the expense of mainlobe gain. Table 2.2 shows some common weighting functions and their effect on beamwidth, sidelobes, rolloff and ISLR. It expands the beamwidth, $BW_{EXP}$, and hence reduces the angle resolution. It also helps improve ISLR but an ISLR of roughly 160 to 1 or −22 dB is about the best that can be achieved. Similar weighting functions are used for filter weighting both for pulse compression and Doppler filtering [3].

It is very common for antennas to have more than one illumination function for the aperture and consequently far-field patterns, each may point in different directions, have different beamshapes and even separate operating wavelengths. This type of antenna has multiple "phase centers" that can be used to estimate direction of arrival of individual scatterers and interference. The simplest of these antennas has two phase centers and can form monopulse angle estimation in one dimension. There have been monopulse designs, which have had as many as

*Table 2.2 Antenna weighting functions*

| Name | $BW_{EXP}$ 3 dB | PSLR, dB | Sidelobe rolloff, dB/OCT. | Coherent gain | ISLR – 2.3U3dB, dB |
|---|---|---|---|---|---|
| Rectangular | 1.0 | −13.3 | −6 | 1.0 | −10 |
| Parabolic | 1.3 | −21.3 | −12 | 0.83 | −20.7 |
| Bartlet | 1.44 | −26.5 | −12 | 0.5 | −23.2 |
| Hanning | 1.63 | −31.5 | −18 | 0.5 | −21.9 |
| Hamming | 1.48 | −42.6 | −6 | 0.54 | −22.4 |
| Cosine3 | 1.81 | −39.3 | −24 | 0.42 | −21.9 |
| Blackman | 1.86 | −58.1 | −18 | 0.42 | −23.7 |
| Parzen | 2.06 | −53.1 | −24 | 0.38 | −22.6 |
| Zero Sonine | 1.24 | −24.7 | −9 | 0.59 | −24.6 |
| Mod. Taylor | 1.22 | −28 | −6 | 0.7 | −22.2 |
| Dolph-Chebyshev | 1.49 | −50 | 0 | 0.53 | −22 |

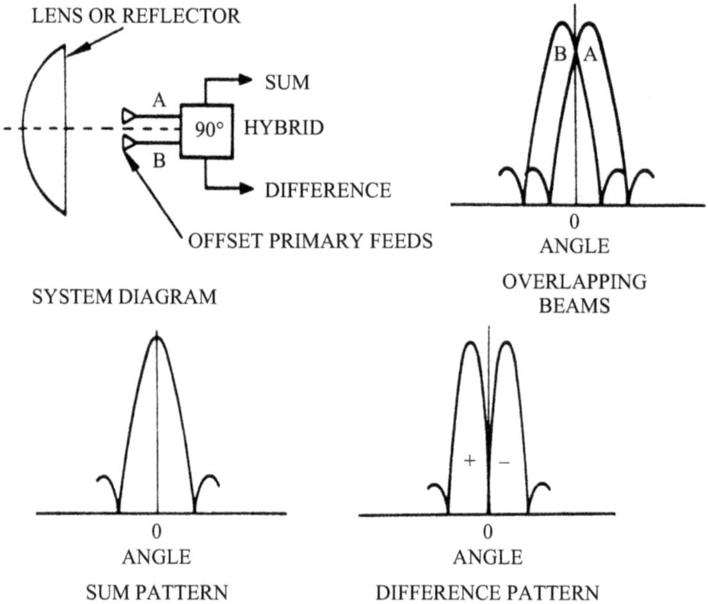


*Figure 2.8 Multiple phase centers provide monopulse [1]*

16 phase centers. Monopulse is used for target angular extent estimates, tracking, clutter and jammer cancellation as well as many other uses.

Figure 2.8 shows a monopulse example. Two phase centers are obtained using side-by-side feed horns near the focal point for a reflector. The two overlapping

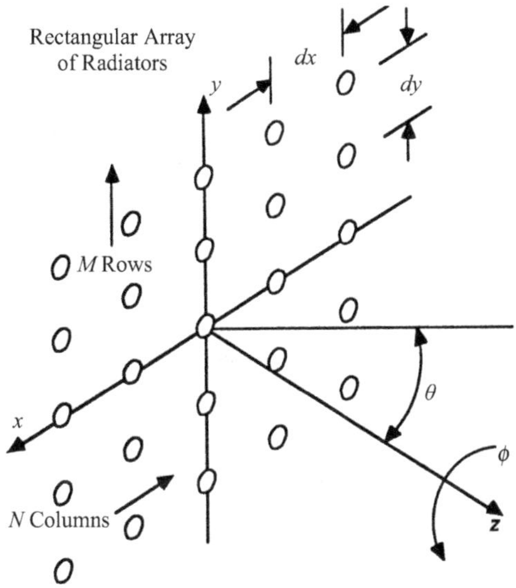


*Figure 2.9 Rectangular array of antenna radiators [3]*

beams can be combined to form sum and difference beams in a microwave hybrid. Obviously, such a pattern also can be established in the orthogonal dimension and then there would be four phase centers and two difference outputs. The antenna difference channel is normally 90° out of phase from the sum channel, which improves isolation and simplifies some processing. The sum and difference combination is often formed even when the ultimate processing will only use the original overlapped beams to improve dynamic range. In more complex antennas, there will usually be multiple beamforming networks (multiple coefficient sets for digital beamforming) for each phase center.

Many modern antennas are arrays of individual radiators, which may have active or passive microwave circuits for each radiator. This allows electronic scanning but also gives rise to some of the same sampling ambiguities mentioned later in Section 2.7. Figure 2.9 shows a notional array of individual radiators.

Such an antenna array can be scanned electronically by adjusting the phase from each radiator by an incremental amount $\beta$ as shown for a 1D case in Figure 2.10. The required value of $\beta$ is related to the look direction, element spacing and operating wavelength given by the last line of (2.10). Because the antenna is now a sampled system, it has ambiguities, which are called grating lobes, a term that comes from optics. The antenna beam from the individual radiators called the element pattern is usually adjusted so that the grating lobes don't degrade performance over the range of electronic scan angles (typically limited to less than ±60°). If cross-range accuracy is important, electronic scanning may be limited to ±20°.

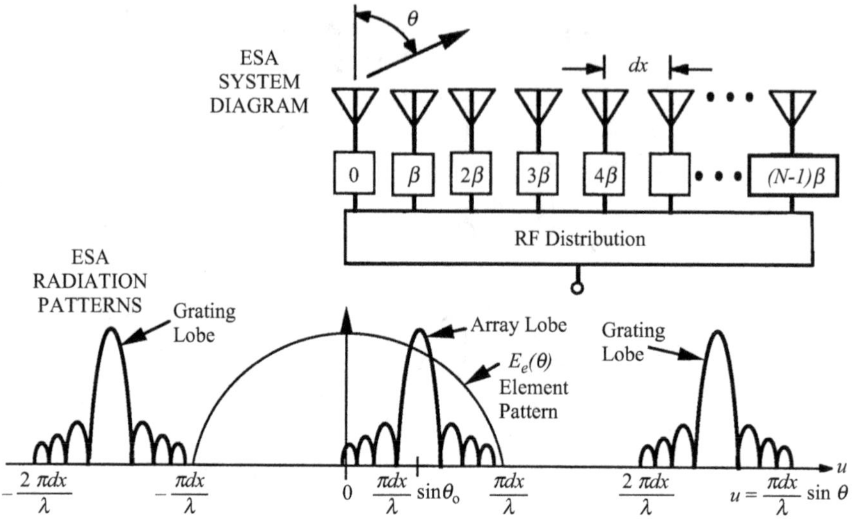


*Figure 2.10 Electronically scanned array [3]*

A uniformly weighted ($a_0$) 2D electronically scanned array of $N$ by $M$ elements as in Figure 2.9 has a far-field pattern as given in (2.10).

$$E(\theta,\phi) \cong N \cdot M \cdot a_0 \cdot E_e(\theta,\phi) \cdot \frac{\sin(N \cdot U_x - \beta_0/2)}{N \cdot \sin(U_x - \beta_0/2)} \cdot \frac{\sin(M \cdot V_y - \beta_1/2)}{M \cdot \sin(V_y - \beta_1/2)}$$

where: $\mathrm{U_x} = \pi \cdot dx \cdot u/\lambda$ and $\mathrm{V_y} = \pi \cdot dy \cdot v/\lambda$ and

$$\beta_0 = 2 \cdot \pi \cdot dx \cdot \sin(\theta_0) \cdot \cos(\phi_0)/\lambda \text{ and } \beta_1 = 2 \cdot \pi \cdot dy \cdot \sin(\theta_0) \cdot \sin(\phi_0)/\lambda. \tag{2.10}$$

Another important property of electronic scanning is that the beamwidth broadens as the commanded scan angle moves away from the normal (broadside) to the array. This is because the projected aperture in the "look" direction is smaller. Approximations for the gain and beamwidth as a function commanded electronic scan angle are given in (2.11). The variables in (2.11) are defined in Figure 2.11. As the commanded beam angle goes toward 90°, the array is in endfire and there is some maximum broadening represented by the constant, $c_{min}$, unique to the particular antenna details.

$$G(\theta,\phi) \cong \frac{4 \cdot \pi \cdot A \cdot (c_{\min} + \cos(\theta_0))}{\lambda^2 \cdot (1 + c_{\min})} \text{ and } \Omega(\theta,\phi) \cong \frac{BW_0 \cdot (1 + c_{\min})}{(c_{\min} + \cos(\theta_0))} \tag{2.11}$$

Surface moving target detection/indication (MTI) radar systems usually require a long antenna in the azimuth direction so that the Doppler spread of clutter across the mainbeam is as small as feasible. This allows the detection of targets that are moving slowly in the direction of the radar. As will be seen in Chapter 3, this

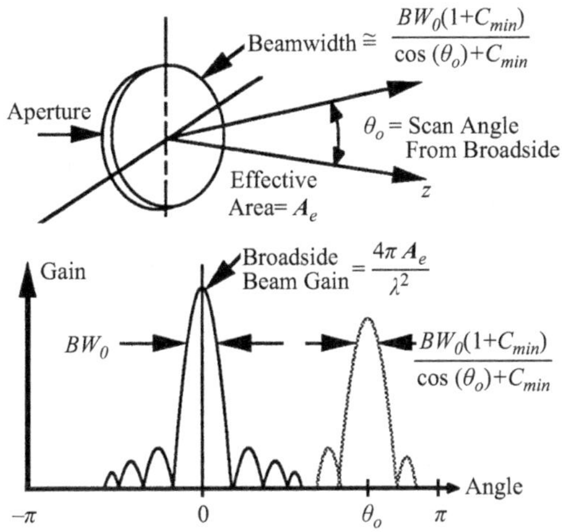


*Figure 2.11 Beam broadening with electronic scan angle [3]*

does not mean that the surface targets are actually slow but only that the component of velocity in the radar observer direction is small. A long dimension in one axis often requires the antenna to be aligned along the fuselage for an aircraft or along the orbital trajectory for a spacecraft, that is, it is scanning normal to the velocity vector. This is quite unfavorable for conventional ground MTI (GMTI) but great for synthetic aperture radar (SAR).

---

**Example 2.1** For example, consider the antenna in Figure 2.12. The active aperture was 29″ × 120″ and the overall pod deployed below the aircraft was 36″ in diameter and over 14′ long. It was aligned with the fuselage as were antennas on JointSTARS, ASTOR, VADER, Erieye, Phalcon, Wedgetail, etc. The TAWDS ESA antenna was scanned electronically in azimuth and was scanned in elevation by the gimbal motors at the ends of the array shown in Figure 2.12. It could look out both sides of the aircraft by rotating through 180°. It had azimuth monopulse for clutter cancellation and tracking. The monopulse was realized by utilization of

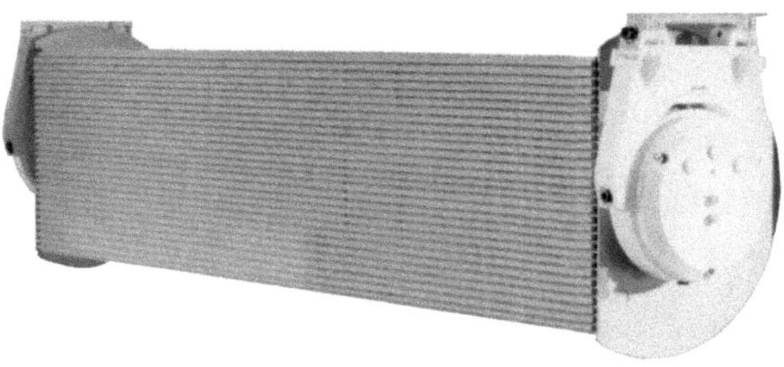

*Figure 2.12 TAWDS electronic scan antenna (ESA) (photo courtesy author)*

two independent antenna feeds, one for the sum channel and one for the difference channel. Figure 2.13 shows a simplified schematic of the 1D monopulse feed network.

The phase and amplitude distributions in the two feeds are given in (2.12). They consist of amplitude and phase distributions shown in Figure 2.14(a) and (b).

The phase distribution is for a steering angle of approximately 9° from broadside in Figure 2.14(b). That angle was chosen so that the phase plot would not be so busy that one could not see what is going on. The beam is steered by roughly one beamwidth for each additional $2\pi$ of phase taper. Notice that the difference pattern phase angle leads the sum pattern by 90° because the pattern is synthesized orthogonal to the sum. In the far field, because the difference pattern is antisymmetric and the sum is symmetric, they are both real.

These patterns are excellent for monopulse clutter cancellation using space time adaptive processing, adaptive displaced phase center or adaptive monopulse

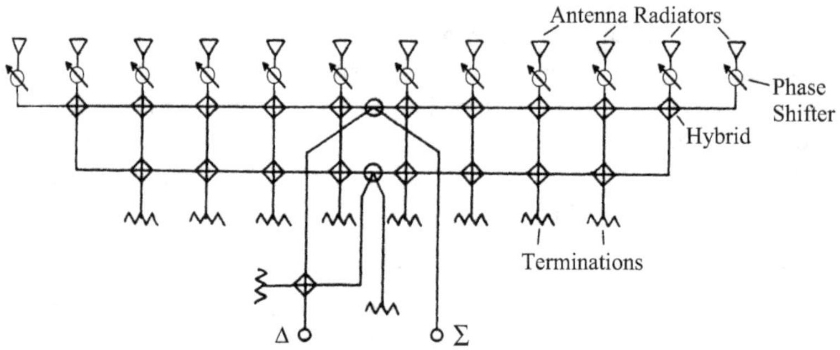


*Figure 2.13 Independent sum and difference feed*

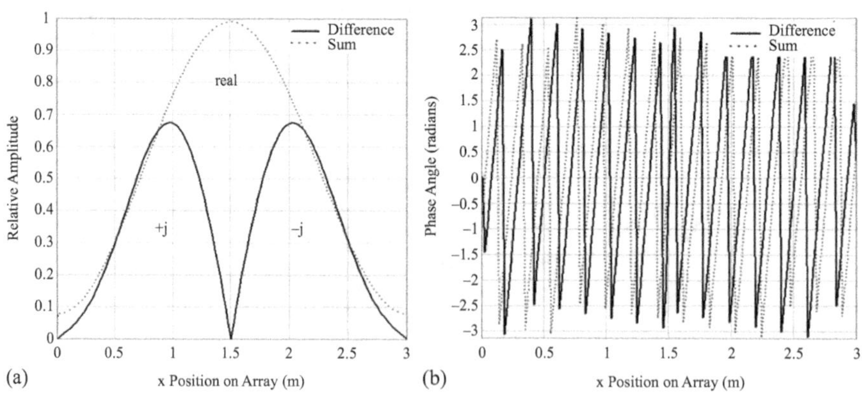


*Figure 2.14 Monopulse near-field illumination (Monopulse7): (a) amplitude, (b) phase*

concepts, which will be covered in Chapter 6. The equations of (2.12) are one of many monopulse possibilities and several others will be introduced later.

$$\Sigma(n) = C_1 \cdot \left( a_0 + (1 - a_0) \cdot \cos\left( \frac{2 \cdot \pi \cdot n}{\mathrm{N}} \right) \right) \cdot \exp(n \cdot \beta_0) \cdot \frac{dx}{\lambda}$$

$$\Delta(n) = C_2 \cdot \left( \left( a_1 + (1 - a_1) \cdot \cos\left( \frac{2 \cdot \pi \cdot n}{\mathrm{N}} \right) \right) \cdot \sin\left( \frac{2 \cdot \pi \cdot n}{\mathrm{N}} \right) \right) \cdot \exp(n \cdot \beta_0) \cdot \frac{dx}{\lambda}$$

where: $a_0$ typically $= 0.54$, $a_1$ typically $= 0.57$, $\beta_0 = k \cdot dx \cdot \cos(\theta_0)$,

$C_1$ and $C_2$ are normalizing constants, element spacing is $dx = \frac{D_x}{N}$, $k = \frac{2 \cdot \pi}{\lambda}$,

$\lambda$ is operating wavelength, $D_x$ is aperture length in x dimension,

$N$ is number of elements in x dimension, x dimension is assumed along velocity vector, $\theta$ measured from velocity vector. (2.12)

Figure 2.15 shows the far-field patterns of sum, difference and the sum to difference ratio for a steering angle of 45°. In general for angles off the normal, the phase of both the sum and difference channels change very rapidly and so the full amplitude and phase must be preserved through the monopulse processing. The term "u space" means that the dimension in the x direction is of the form $\sin(\theta)\cos(\phi)$ or in this case $\pi \cdot \sin(\theta) \cdot D_x/\lambda$ because $\phi$ is 0° (see (2.6)). Often in both phased and active array antennas, even the subarrays are combined to form sum and difference monopulse patterns that are combined to form the overall beamshapes. Although not obvious, forming "sum + j delta" and "sum − j delta" in two or more channels improves dynamic range and error compensation. The author is aware of ESA antennas, which contain dozens of subarrays each producing monopulse patterns.

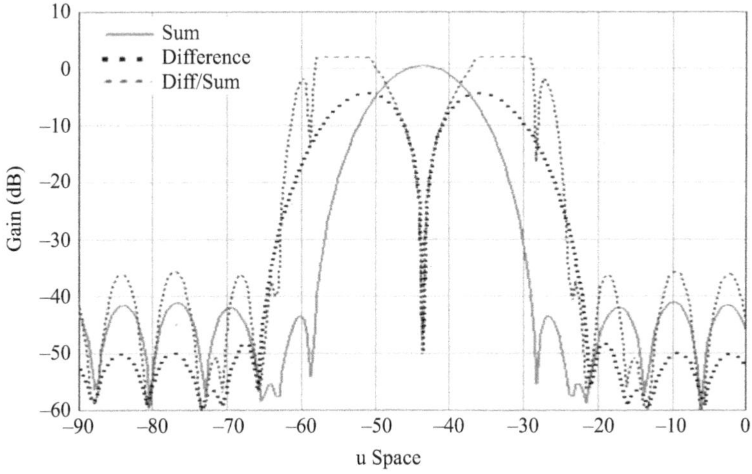


*Figure 2.15 Monopulse far-field patterns (Monopulse7)*

### 2.6.1 Active electronic scan antennas (AESA)

Modern intercept and radar systems often use active array antennas, which have a transmit and receive (T/R) or receive only channels at each radiating element. Figure 2.16 shows the block diagram of a typical AESA front end. Those channels are driven by a phase locked exciter chosen for the best short-term (<10 s) stability. In turn, the oscillator combined with synthesized offset frequencies produces a transmit oscillator that is used to drive the transmit side of the T/R channels through a signal distribution manifold. On receive, each channel may drive multiple manifolds, which are configured to provide desirable angle space properties. Those multiple manifold channels are converted to an intermediate frequency (IF) with the local oscillator allowing more gain control and frequency selectivity in the receiver block.

A typical T/R channel is shown in Figure 2.17. It contains power regulation, amplitude and phase control, a T/R switch, channel control logic, a driver and

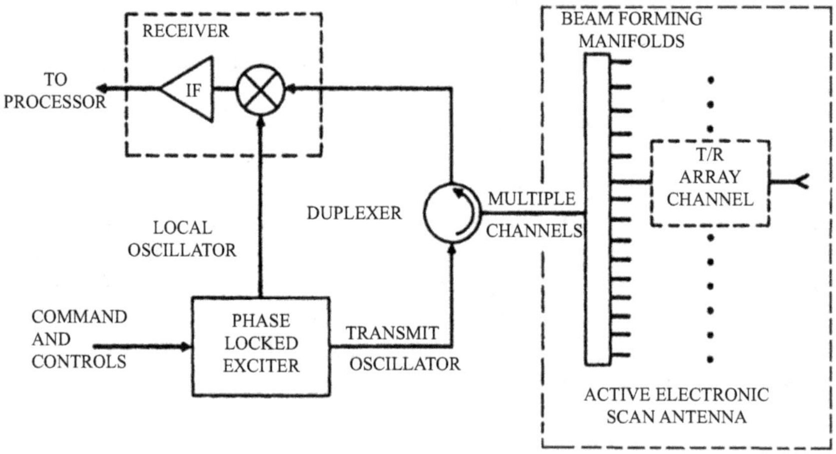


*Figure 2.16 Active electronic scan array (AESA) RF front end*

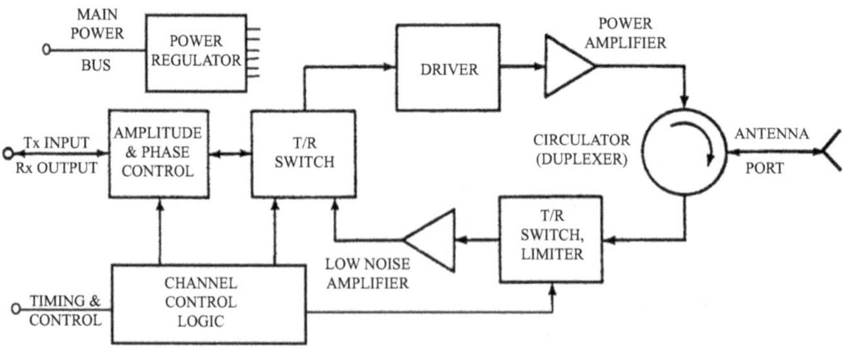


*Figure 2.17 T/R channel block diagram*

power amplifier, a circulator or duplexer switch, a low noise amplifier (LNA) and protection for the LNA in T/R switch and limiter. One output of the circulator drives an antenna radiating element. The channel control logic is embodied in a single logic chip, which is quite complex. Some of the functions contained in the channel control logic are shown in Figure 2.18. Part of the control logic function is adapting commands from the overall control computer in the sensor to the nonlinearities in each channel. The nonlinearities may occur with commanded frequency, phase and amplitude as well as array temperature. These calculations must usually be updated as often as each pulse, coherent dwell or second. Figure 2.19 is a photo of a single T/R channel. Although it is dated (mid-1990s), it is easier to see a one for one correspondence with Figure 2.17 functions. The LNA chip is rightmost in Figure 2.19 and the control logic chip is leftmost in the figure. The power amplifier chips are in the middle.

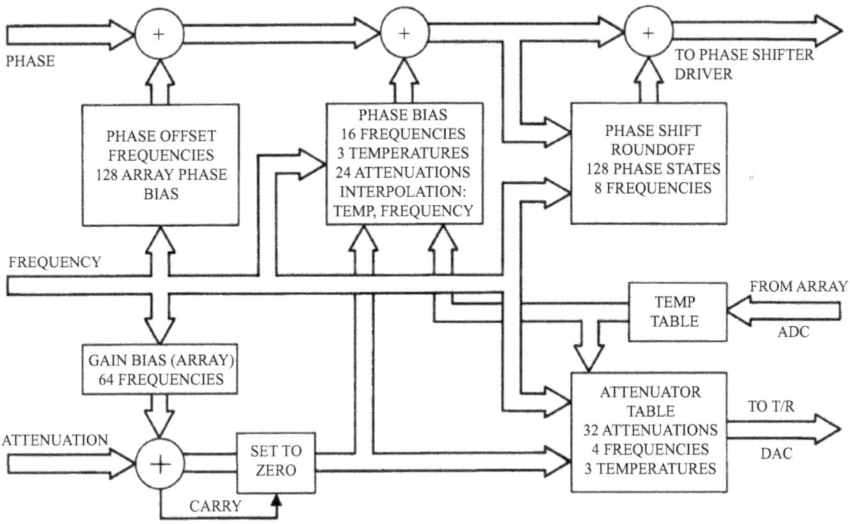


*Figure 2.18 T/R channel control logic*

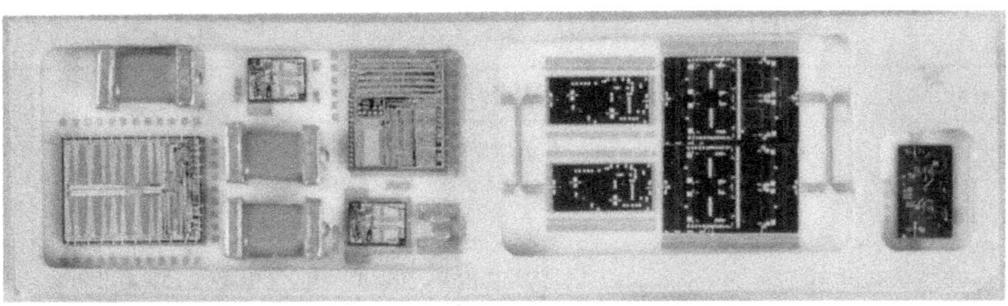

*Figure 2.19 Typical T/R channel (photo courtesy author)*

### 2.6.2 Antenna Example 2.2

Suppose that an AESA antenna for a GMTI radar had the parameters shown in Table 2.3. One feature of an AESA is that the transmit and receive patterns don't have to be the same. The receive patterns have already been calculated and are shown in Figures 2.13 and 2.14. The transmit pattern for an active array might use a cosine on a pedestal amplitude weighting. Such a weighting is very popular for active arrays because it provides more power in the mainbeam with easy amplitude weighting but its sidelobes aren't that good. The edge weighting might be ¼ of full-scale but it has a gain, which is 81% of an unweighted aperture. This provides more power on targets and when combined with the receive pattern has adequate performance. A modified Taylor difference pattern and a Hamming weighting on receive provides realizable sidelobes and almost linear sum to difference ratio.

The cosine on a pedestal amplitude function of Table 2.1 generates two lower-amplitude replicas of the unweighted $(\sin N x)/(N \sin x)$ pattern offset by 2/3 of a beamwidth. This provides some sidelobe cancellation and beam spread but significantly more on-target power than other low sidelobe weightings. The aperture pattern synthesis is given in (2.13). In this particular example, the weighting function is the same in both axes.

$$w_{AESA} = w_r \cdot w_y \cdot w_x$$

$$\text{where: } w_y = a + (1-a)\cdot\cos\left(\frac{4\cdot\pi\cdot m}{3\cdot M}\right)$$

$$w_x = a + (1-a)\cdot\cos\left(\frac{4\cdot\pi\cdot n}{3\cdot N}\right)$$

$$w_r = P_{rect}(n,m), \text{ a rectangular aperture,}$$

$$\text{where: } P_{rect}(n,m) = 1 \text{ if } n = -N/2 \text{ to } N/2 \text{ and}$$

$$\text{if } m = -M/2 \text{ to } M/2 \text{ and } P_{rect}(n,m) = 0 \text{ otherwise.} \tag{2.13}$$

*Table 2.3 Antenna Example 2.2*

| Parameter | Value |
|---|---|
| Type | Rectangular |
| Height | 1 m, $M = 42$ elements, y dimension |
| Width | 3 m, $N = 125$ elements, x dimension |
| Transmit weighting | Raised cosine weight, $a = 1/2$, $[(4\pi/3)/(N \text{ or } M)]$ |
| Receive weighting | Hamming and modified Taylor monopulse |
| Wavelength | 0.03 m |
| Element spacing | 0.024 m (0.8 $\lambda$) |

The illumination functions are separable in x and y. Thus when transforming to the angle space domain, the pattern is just the product of the x and y patterns as shown in (2.14).

The element pattern is assumed to be cos($\theta$) and element spacing is assumed to be 0.8 $\lambda$.

$$E(\theta,\phi) \cong N \cdot M \cdot E_e(\theta,\phi) \cdot \frac{\alpha x_n \cdot \sin(N \cdot U_x + Kx_n)}{N \cdot \sin(U_x + Kx_n)} \cdot \frac{\alpha y_n \cdot \sin(M \cdot V_y + Ky_n)}{M \cdot \sin(M \cdot V_y + Ky_n)}$$

where: $U_x = \pi \cdot dx \cdot \sin(\theta) \cdot \cos(\phi)/\lambda$ and $V_y = \pi \cdot dy \cdot \sin(\theta) \cdot \sin(\phi)/\lambda$,

$Kx_0 = 0,\ Kx_1 = \frac{2 \cdot \pi}{N},\ Kx_{-1} = \frac{-2 \cdot \pi}{N},\ Ky_0 = 0,\ Ky_1 = \frac{2 \cdot \pi}{M},\ Ky_{-1} = \frac{-2 \cdot \pi}{M},$

$\alpha x_0 = 1/2,\ \alpha x_1 = 1/2,\ \alpha x_{-1} = 1/2,\ \alpha y_0 = 1/2,\ \alpha y_1 = 1/2,\ \alpha y_{-1} = 1/2.$

(2.14)

Figure 2.20 shows the amplitude weighting on the left and the far-field pattern on the right.

If the elements in the x direction were exactly vertical row to row, then the antenna sidelobes would be oriented with the horizon. That would cause more sidelobe clutter power competing with targets. Often the elements are slightly skewed from horizontal and vertical so that many of the sidelobes are above and below the horizon, reducing sidelobe clutter and jammer power. As can be seen from Figure 2.20, the first sidelobes are about −23 dB relative to the mainlobe peak.

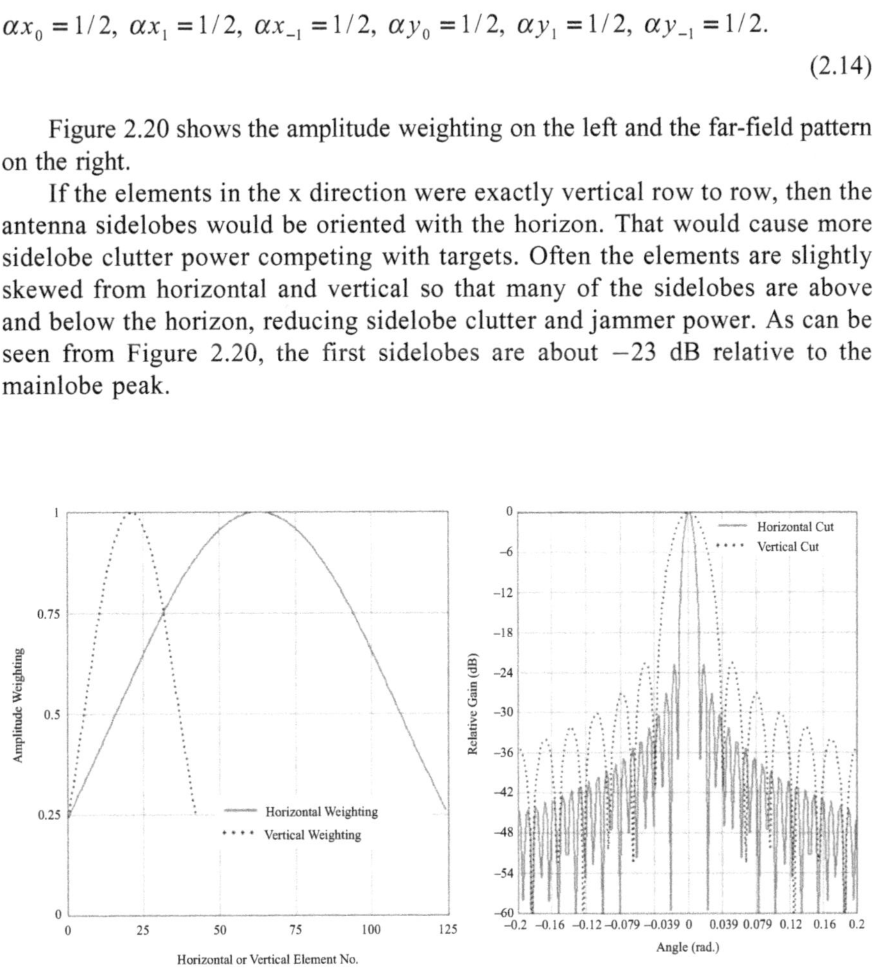


*Figure 2.20 Example 2.2 AESA transmit pattern (AESAWeighting002)*

## 2.7 Multichannel receivers

The typical modern radar receiver has four or more channels as suggested by Figure 2.21. These could be the traditional sum, elevation difference, azimuth difference and guard channels. They might also be from different antenna beamforming manifolds, which are switched to provide conventional monopulse, multiple phase centers in azimuth for GMTI adaptive processing, or a single mainbeam and multiple guard channels to cancel jamming. For ease of design, manufacture and operation, these channels are usually identical. An AESA front end may only have 30 dB of gain per T/R channel. In addition, channels at the edge of the array may be amplitude weighted to provide improved sidelobes but which result in receive gains as small as 4 dB. That still can create a very wide dynamic range at the output of the antenna manifolds for an AESA of a few thousand elements.

Focused signals could grow by thousands and random signals can grow by the square root of thousands, that is, 30–60. Clutter can vary by 10,000:1 in cross section. Signals at different ranges vary as the third or fourth power of range. Thus in aggregate the output from an AESA manifold could range over 200,000,000:1 or 83 dB. The thermal noise floor at the input to the LNA for a 250 MHz T/R channel bandwidth might be −118 dBw and bright clutter returns might be −8 dBw at minimum range. The LNA might be required to accommodate a 110 dB dynamic range. Unfortunately, −8 dBw is above the third-order intercept point for most LNAs.

If the transmitted pulse width is approximately 10 km (66 μs) in length, then the receiver is off during the transmitted "main bang" (pulse). When the receiver is turned on, the maximum clutter return at 10 km is down to −35 dBw for a 6-m-range resolution compressed pulse width and a 1.8 $m^2$ antenna area. Thus the LNA would see a dynamic range of only 83 dB with a peak of −35 dBw, which is well within the state of the art of a mostly distortion free LNA front end. This implies that no additional gain control is required in an AESA T/R channel. However, after the antenna beamforming manifolds, the dynamic range could grow by 15–30 dB. This will require some coarse automatic gain control (CAGC) before the first local oscillator mixer, which often has a third-order intercept point at approximately −30 dBw.

After the first LO mixer there is usually a fixed gain stage to move the signal and noise near the optimum operating point, which is then followed by another coarse gain control to adjust the signal as a function of range commonly called a sensitivity time control (STC). This is followed by a bandpass filter to limit noise. The output from the bandpass filter may go into a variable frequency oscillator (VFO) and mixer, which may dechirp clutter and center it in the desired frequency band for the A/D converter. Usually the VFO mixer output goes into a gain stage followed by a fine AGC and ultralinear A/D driver amplifier to present the optimum signal to the A/D. Modern practice is to use a single fast A/D followed by a digital Hilbert transform filter to derive in phase and quadrature vector components that are down shifted and subsampled to baseband for further processing. These digital outputs

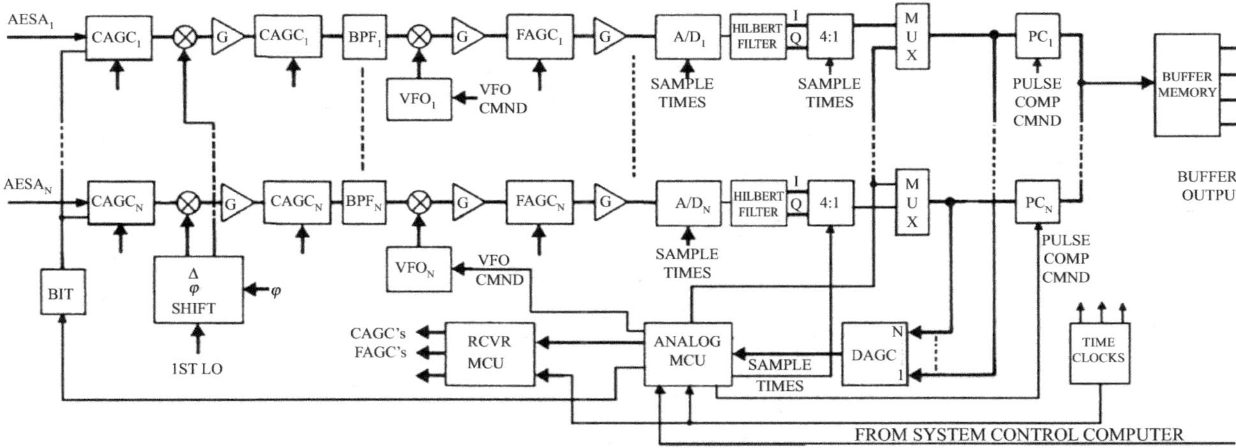


*Figure 2.21 Typical multichannel receiver*

are multiplexed and may be pulse compressed before being read into a large buffer memory. Later the digital signals are retrieved for the remaining processing. (More in a later chapter.)

## 2.8 Measuring the distance to a scatterer in wavelengths and the Doppler effect

A coherent radar can measure the instant time distance to a target scatterer to a fraction of a wavelength. This accuracy is necessary to make SAR or ISAR images as well as to detect GMTs. Furthermore, it can be done from both stationary and moving platforms. For X and Ku band, it means that measurements must be made to ¼ to 1/8 of an inch over distances of hundreds of kilometers. How is this possible? Hopefully the reader will see the answers by the completion of this text. For short ranges, a frequency modulated continuous wave (FMCW) radar can perform the necessary measurements. For example, many police speed-trap and missile seeker fuze radars work with such a waveform. For longer ranges, transmitter, receiver, reference oscillator noise and stability limit detection performance only to very large targets. The long-range small-target applications usually depend on pulse Doppler (PD) waveforms. A PD radar measures both the phase and amplitude of the echo from a target as suggested by Figure 2.22. A radar pulse is transmitted with starting phase, $\phi_0$, and amplitude, $A$. The pulse travels to the target and part of the energy is reflected back to the radar as shown in Figure 2.22. On return the phase and amplitude of the echo pulse are measured relative to a local oscillator reference. The echo signal amplitude may be a trillion times smaller than the transmitted pulse and there may be billions of wavelengths to the target and back. Somewhere inside the radar all these wavelengths are counted to make an estimate of the range and the reference-to-echo-signal phase comparison will be used to detect Doppler.

Each wave packet is transmitted, received at some later time and compared to a continuously running receiver local oscillator as shown in Figure 2.23. There will be some phase difference. If there is relative motion between the radar and target, then on each successive pulse return the measured phase will change. Assuming

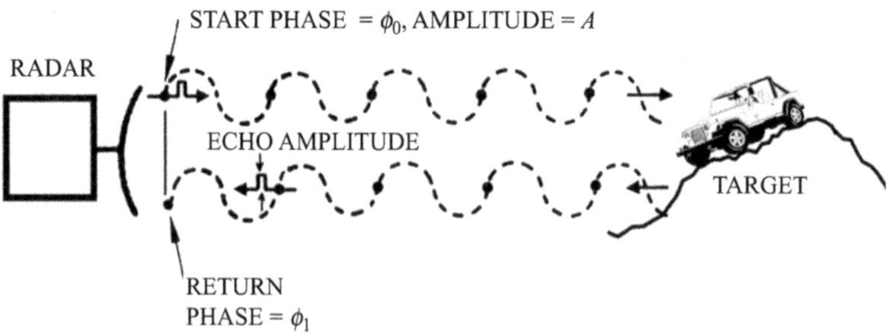


*Figure 2.22 Radar measures phase and amplitude echo from target*

that the sampling rate is high enough so as not to miss a rotation of 360° or more pulse to pulse, then those measurements can be used to estimate the change in true position as well as to estimate the Doppler frequency, since the time between pulses is accurately known (the rate of change of phase is frequency).

The Doppler shift is well known to most people as the tone shift from higher to lower one experiences when a high-speed vehicle passes by. A similar effect exists for radar waves reflected or emitted from an object with a relative velocity, v, to the radar observer. Visible light as well as radar microwaves is made up of multiple photons, which exhibit both a particle and a wave nature. Consider a single photon coming from a distant object that is traveling away from the radar observer as shown in Figure 2.24. The leading edge of that photon arrives at time $t_1$ and sometime later the trailing edge of the photon arrives at time $t_2$. When the photon arose on the object, it had a frequency of $f = 1/\tau$ and a wavelength of $\lambda = c \times \tau$,

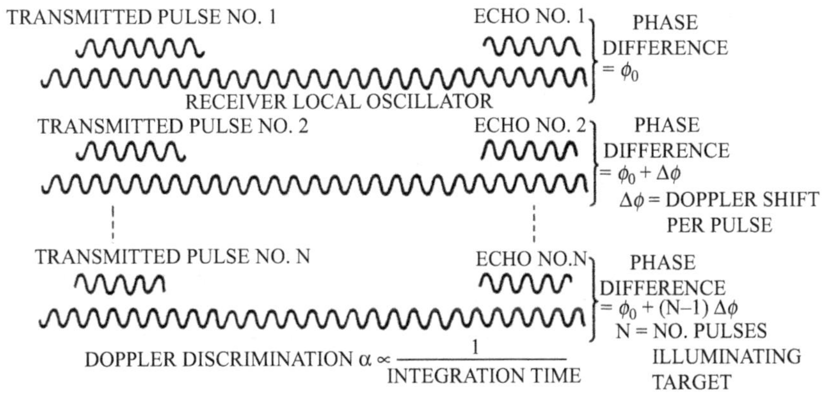


*Figure 2.23 Coherent radars measure phase to a fraction of a wavelength*

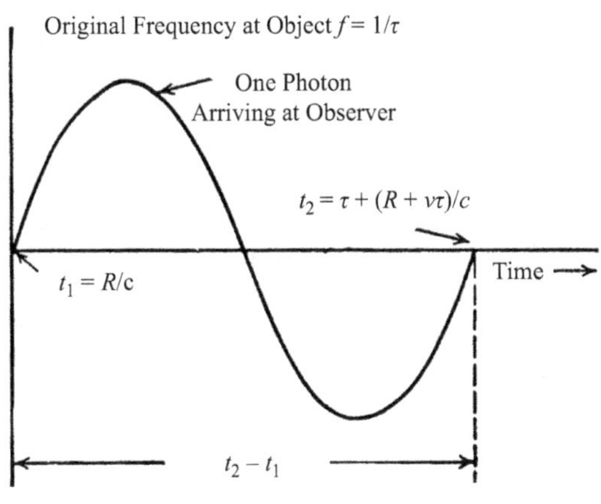


*Figure 2.24 Doppler notion for a single photon*

where c is the velocity of light and $\tau$ is the time extent of the photon. The photon traveled a distance, $R$, to get to the observer. The leading edge of the photon got here at $t_1 = R/c$. The trailing edge got here at $t_2$, a time slightly longer than $\tau$ because in the time length of the photon, the object moved away from the observer by a tiny amount. The velocity of light in a vacuum doesn't change but for the observer time appears to stretch.

That means that the received frequency is

$$f_{received} = \frac{1}{(t_2 - t_1)} = \frac{1}{\tau + \left(\frac{R + v \cdot \tau}{c}\right) - \frac{R}{c}} = \frac{1}{\tau\left(1 + \frac{v}{c}\right)} \tag{2.15}$$

Then making use of the power series expansion shown in (2.16).

$$\frac{1}{1+x} = 1 - x + x^2 - x^3 + \ldots\infty \text{ where } x < 1 \tag{2.16}$$

The approximate Doppler frequency will be

$$f_{received} \cong \frac{1}{\tau} \cdot \left(1 - \frac{v}{c}\right) = f \cdot \left(1 - \frac{v}{c}\right) \equiv f + f_{Doppler} \tag{2.17}$$

If the object is moving away from the observer, the received frequency is lower. If the object or energetic particles from an object are moving at relativistic velocities toward the observer, the frequency is shifted toward higher frequency. As you may know the electrons in vacuum cathode ray tubes and traveling wave tubes are traveling at relativistic velocities and so their wavelengths are shifted substantially. In the case of a monostatic radar, the effect is twice the radar to target relative velocity (out and back) for a factor of 2. As long as the relative velocity is small compared to the velocity of light, it is as shown in (2.18).

$$f_{Doppler} = \frac{2 \cdot V_r}{\lambda} = \frac{2 \cdot V \cdot \cos\eta}{\lambda} \tag{2.18}$$

where $V_r$ is the relative velocity along the line of sight (LOS), $\lambda$ is the operating wavelength, $V \cos\eta$ is the net vector velocity dot product for radar platform and target (see Figure 2.27). The Doppler resolution, $B_{filt}$, is inversely proportional to the pulse repetition frequency (PRF) times the number of pulses, $N$, as suggested in (2.19). Typically, those pulses are integrated in a matched filter, one for each possible Doppler with its corresponding pulse-to-pulse phase change (more on matched filters later in the text).

$$B_{filt} \propto \frac{1}{N \cdot PRF} \tag{2.19}$$

## 2.9 Pulse Doppler

Although the benefits of a PD waveform are substantial, once a repetitive waveform is used there will be sampling ambiguities (measurements that might represent more than one possible range or Doppler). PD radar systems are coherent sampled data systems. Figure 2.25 shows an actual oscilloscope image of a pulse train and its corresponding power spectrum.

Figure 2.26 shows an idealized pulse train spectrum that might be observed from the transmitter of a PD radar. There will be a train of pulses of width, $\tau$ and repetition period, T, which modulate a coherent sinusoidal carrier with angular frequency, $\omega_0$. The coherent sinusoidal carrier is phase locked to the receiver local oscillator reference. The PRF is the reciprocal of the period, T, and it gives rise to a sin x/x spectral envelope shown in the lower part of the figure. Under the envelope

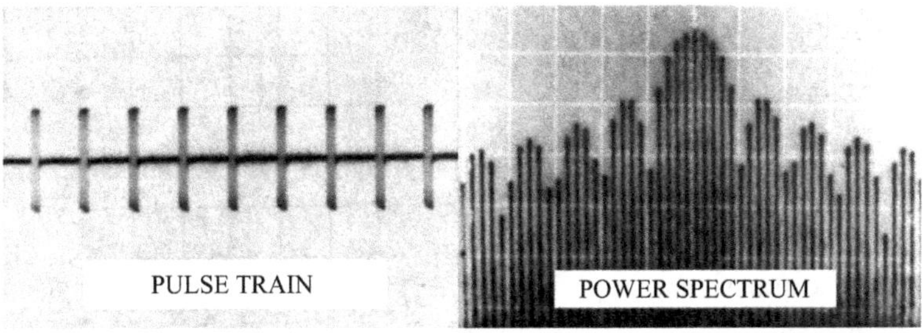


*Figure 2.25 Actual pulse Doppler spectrum (photo courtesy author)*

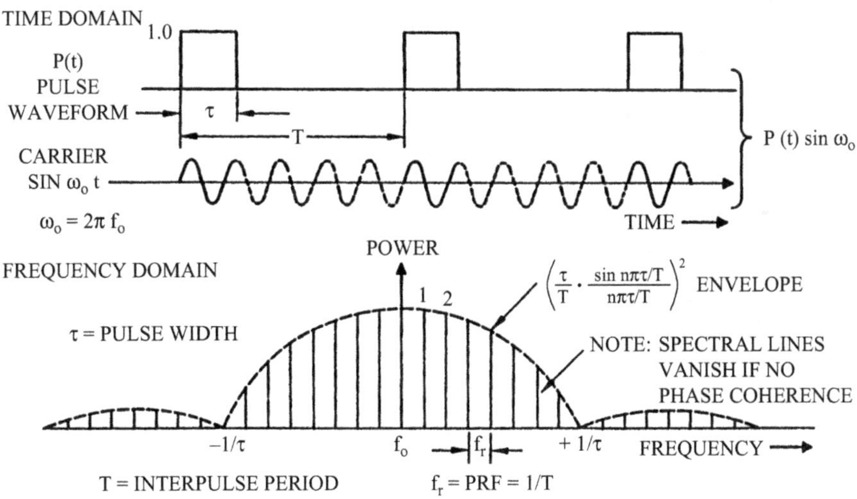


*Figure 2.26 Pulse train spectrum [1]*

are a series of spectral harmonics shown as 1, 2, etc. in the figure. The sampled nature of the waveform causes ambiguities or folding in the time and frequency domain. There are many manipulations of this basic waveform but the ambiguities are basic physics and can only be moved around – not eliminated. The true shape of the range and Doppler profile will be folded about each of the spectral harmonics. The more foldings the harder it is to resolve the ambiguities. If the PRF is too low, there may be too many Doppler foldings. If the PRF is too high, there may be too many range foldings.

Usually, some combination of antenna beamshape, PRFs and other waveform attributes allows the ambiguities to be resolved for a reasonable target or terrain subset. The antenna and its sidelobes are crucial in finding a waveform, which will provide adequate performance. Ideally, the antenna main beamwidth is small enough in both the range dimension and the Doppler dimension so that there are no ambiguities in the mainlobe and near sidelobes. A small antenna beam requires a large physical aperture relative to a wavelength (typically >30:1).

First, let us consider the geometric properties of a moving radar illuminating a moving ground target. Figure 2.27 shows the basic geometry and salient Doppler features of objects in the antenna mainbeam. The aircraft or spacecraft is moving with velocity vector of $V_a$ and with the center of the antenna mainbeam at an angle of $\eta_c$ relative to the velocity vector. Inside the antenna mainbeam is a moving target at some range, R, and angle, $\eta_t$, traveling at a velocity, $V_t$, at angle, $\zeta_t$ , with respect to the LOS angle. The target is traveling in the surface tangent plane (i.e. the plane defined by the grazing angle, $\psi$). The angle, $\eta_t$, is a compound angle made up of the radar platform azimuth, $\theta$, and elevation, $\varepsilon$, angles relative to the platform velocity

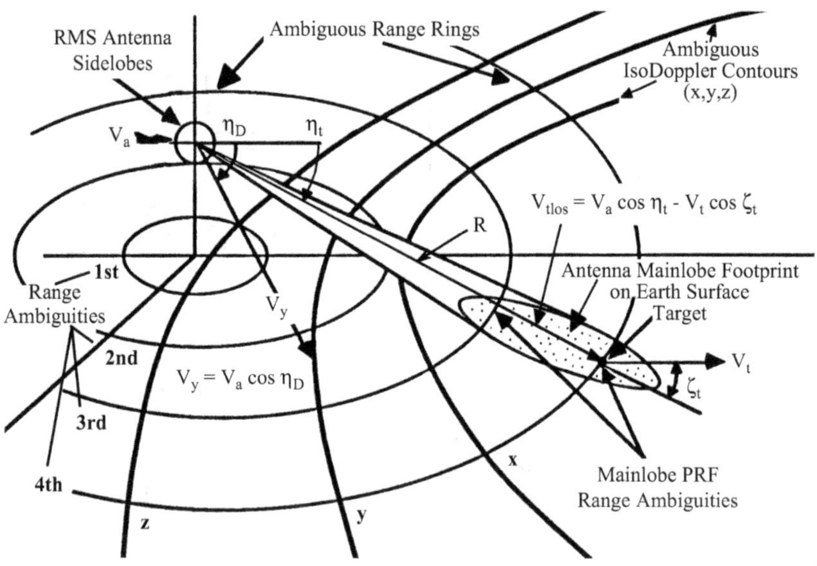


*Figure 2.27 Pulsed Doppler radar geometry [4]*

vector. The azimuth and elevation angles described earlier may not be aligned with the local horizontal or platform fiducials but are assumed so for simplicity. The compound angle, $\zeta_t$, is made up of the grazing angle, $\psi$, and the moving target heading in the surface tangent plane, $\alpha_t$.

Because the radar is pulsed at some PRF, it is a sampled data system that gives rise to sampling ambiguities. These sampling ambiguities arise both in the time or range domain (sometimes called fast time) as well as in the frequency or Doppler domain (sometimes called slow time). Figure 2.27 shows those ambiguities as range rings (1, 2, 3, 4) and constant Doppler (often called isoDoppler) contours (x,y,z) where they might intersect the earth's surface. At certain ranges and angles, these ambiguities may fall in the antenna mainbeam and always fall in the antenna sidelobes. In addition to a possible moving target, the antenna mainbeam illuminates some portion of the earth's surface commonly called the antenna footprint. That footprint may be used to create synthetic aperture maps or it may be considered clutter if the objective is to detect surface moving targets. For moving target detection, it is necessary to eliminate that competing surface return echo to find the movers. This geometry applies whether the observer radar platform is an aircraft or spacecraft.

Referring to Figure 2.27, one can see that the surface clutter that might compete with a moving target is approximately contained in the antenna mainlobe footprint area on the surface. It is made up of an azimuth footprint ($FP_{az}$) across the antenna beam and an elevation footprint ($FP_{el}$) along the beam. The azimuth footprint is the product of the antenna beamwidth and the range to the center of the footprint. When Doppler filtering is used, only that part of the clutter Doppler that falls in a single filter will compete with a target in that cell. The fraction of the azimuth beamwidth ($d_{az}$) that falls in a single filter can be estimated by calculating the total Doppler azimuth mainlobe clutter bandwidth across the beam ($B_{amlc}$) and then taking the ratio of a single Doppler filter bandwidth ($B_{filt}$) to the whole spread times the azimuth footprint. To a first order only the integrated sidelobes in the cardinal axes of range and Doppler will compete with a target. These notions are summarized in (2.20).

$$FP_{az} \cong R \cdot \theta_{az} = \frac{R \cdot 1.3 \cdot \lambda \cdot (1 + c_{\min})}{D_x \cdot (\cos(\theta_0) + c_{\min})}$$
$$B_{amlc} \cong \frac{2 \cdot V \cdot \theta_{az} \cdot \sin(\eta)}{\lambda}, \quad B_{filt} \cong \frac{1}{T_{OT} \cdot eff_n}$$
$$d_{az} \cong \frac{FP_{az} \cdot B_{filt}}{B_{amlc}} \tag{2.20}$$

For SAR mapping, the surface may not be flat and regular; therefore, some surface footprint segments will appear to be moving on a short-term basis (e.g. rising or falling terrain, wind-blown vegetation, etc.). In addition, the electrical center of the antenna beam on the surface cannot be known precisely due to atmospheric refraction, relative altitude uncertainties and other propagation anomalies. Therefore, the range and Doppler centroid of the mainbeam must be estimated in order to make properly illuminated and well-focused SAR maps. Even with centroid estimation, the edges of the map may not be in focus unless those apparent short-term surface motions are compensated. This compensation is often called autofocus and will be described later.

The ideal method for detecting moving targets in a clutter background is to form matched filters for each possible GMT direction of travel and speed. More commonly matched filters are formed on the quasi-stationary surface return, which is deleted from the signal. Then whatever is left is called a moving target. This almost works for many cases, but as the reader will see, much more effort is usually required.

## 2.10 Radar equation

The so-called "radar equation" is a method for calculating the most likely signal power and noise power sensed at a receiver based on parameters of the transmitter, of the target scatterer and of the receiver. The development of the radar equation is depicted in Figure 2.28.

The transmitter sends a waveform of average power, $P_T$, through the antenna with gain, $G$. The waveform propagates spherically in space spreading as $1/(4\pi R^2)$. Some portion of the transmitted waveform is scattered from the target echoing cross section, $\sigma$. The scattered signal waveform again propagates spherically in space spreading as $1/(4\pi R^2)$. Some portion of the scattered waveform is captured by the radar receiver antenna with area, $A$. This could be the same as the transmit antenna or a separate antenna. Some amount of thermal noise, $k_B T_A$, from the atmosphere and other sources also enters the receiving aperture. The receive antenna and associated plumbing also have an equivalent noise temperature, $T_{RF}$.

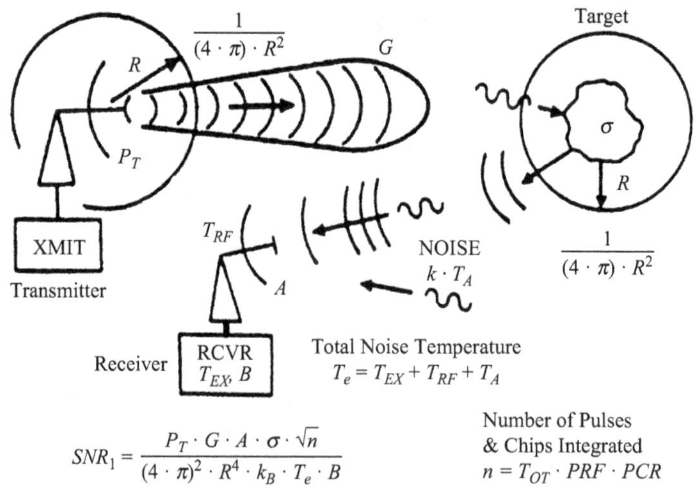


*Figure 2.28 Radar equation development*

Unfortunately, the receiver also exhibits some noise in the front-end amplifiers and mixers, $T_{EX}$, which is in excess of its actual temperature. The result is a total noise temperature, $T$, which when multiplied by Boltzmann's constant, $k_B$, and the final detection equivalent noise bandwidth, $B$, estimates the total noise power competing with the desired radar signal waveform. There usually are multiple pulses or repetitive waveforms, $n$, which are summed for final detection. $n$ is the time on target, $T_{OT}$, times the $PRF$ times the $PCR$. If the signal waveform is small but completely correlated pulse to pulse and the noise is uncorrelated, then the signal to noise ratio, $SNR_I$, improves by the square root of $n$. Combining all the terms results in an estimate of the final SNR given at the bottom of Figure 2.28.

A slightly more generalized version of the radar equation is given in (2.21.1) and (2.21.2). It includes losses, which may arise from a number of environmental sources as well as radar signal processing losses. This equation assumes noise is "white" thermal noise only. In general, there is additional interference caused by clutter, jamming and internal radar hardware and software. When added to thermal noise, SNR is sometimes designated as SNIR. It will be used interchangeably throughout the text.

Target signal to noise ratio:

$$SNR_I = \frac{P_T \cdot D_T \cdot G_T \cdot A_{eR} \cdot \sigma_t \cdot eff_n \cdot \sqrt{T_{OT} \cdot PRF \cdot PCR}}{(4 \cdot \pi)^2 \cdot R^4 \cdot k_B \cdot T_e \cdot B \cdot L}$$

Ground clutter signal to noise ratio:

$$SNR_{ICL} = \frac{P_T \cdot D_T \cdot G_T \cdot A_{eR} \cdot \sigma_o \cdot d_{az} \cdot d_r \cdot eff_n \cdot \sqrt{T_{OT} \cdot PRF \cdot PCR}}{(4 \cdot \pi)^2 \cdot R^4 \cdot k_B \cdot T_e \cdot B \cdot L} \tag{2.21.1}$$

$P_T$ = peak power during the pulse(watts), $PRF$= pulse repetition frequency (hertz),
$G_T$ = transmit antenna power gain including aperture efficiency,
$A_{eR}$ = receive antenna effective area (meters$^2$), $d_{az}$ = azimuth resolution (meters),
$\sigma_o$ = normalized ground clutter RCS (meters$^2$/meters$^2$), $d_r$ = range resolution (meters),
$\sigma_t$ = radar cross section of target (meters$^2$), $D_T$= transmitter duty ratio,
$T_{OT}$ = time radar illuminates target (seconds), $PCR$= pulse compression ratio,
$n = T_{OT} \cdot PRF \cdot PCR$, $eff_n$ = integration efficiency of $n$ during $T_{OT}$,
$R$ = radar to target range (meters), $B$ = equivalent noise bandwidth (hertz),
$k_B$ = Boltzmann's constant $1.38 \cdot 10^{-23}$ (watts/(hertz · kelvin)),
$T_e$ = total equivalent noise temperature (kelvin), $T_e$ sometimes replaced by $T_0 \cdot NF$,
$L$ = all other losses(a number larger than one).

(2.21.2)

Other versions of the radar equation will be introduced in later chapters specific to the target type and geometry.

## 2.11 Pulse compression

It often turns out that a single pulse of the right resolution in time (range) $\tau$ doesn't contain enough energy to provide a detection. One way around this problem is to transmit a longer pulse containing more energy and encode the pulse. Then by match filtering the encoded pulse on receive one can obtain better time resolution and the extra energy for detection. There have been dozens of ways invented to do this.

---

**Example 2.3** One simple way is shown in Figures 2.29 and 2.30. A binary phase code is applied to the transmitter frequency and amplified as shown in Figure 2.29.

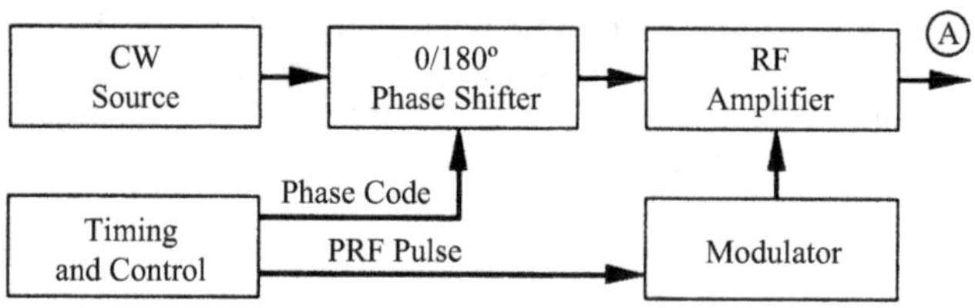


*Figure 2.29 Binary phase code modulation of the transmitter [3]*

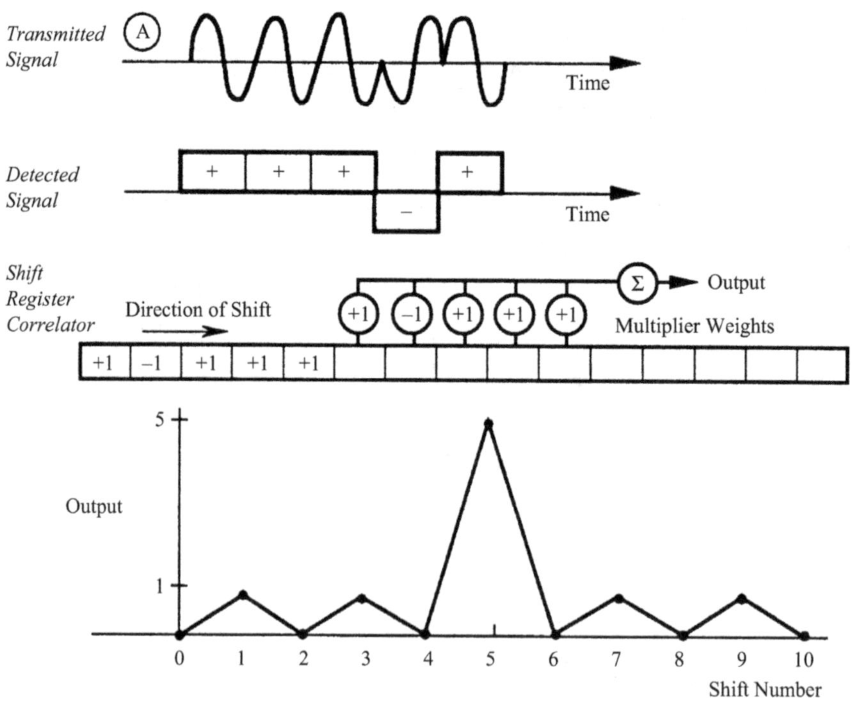


*Figure 2.30 Binary phase code pulse compression [3]*

Each code bit is applied in sequence into a time slot called a chip. The chip takes on a relative phase of either 0° or 180°. The output at point A in Figure 2.29 is shown in the first line of Figure 2.30. On receive the signal is synchronously demodulated as shown on the second line in Figure 2.30 and has a relative phase that is either in phase or 180° out of phase with the adjacent chips. That signal is applied to a matched filter correlator shown on the third line in Figure 2.30. The output is shown at the bottom of Figure 2.30. In the example in Figure 2.30, a five chip phase code is used in a transmitted pulse five times longer than the desired resolution. When match filtered, the output is five times higher than the transmitted amplitude. That is described as a pulse compression ratio (PCR) of 5:1. Note that the output does have some sidelobes just as an antenna might have.

In modern systems, PCRs of 20,000:1 are common. This places very stringent linearity and phase accuracy constraints on the waveforms. Otherwise the sidelobes can be quite poor and detection performance is compromised. This is partially mitigated by amplitude weighting on receive to suppress the known sidelobes. Adaptive processing is sometimes used in the range/pulse compression dimension to improve the range impulse response. Some of the weighting functions are unique to a specific waveform and some are the same weighting functions mentioned in the antenna section.

One of the most important properties of pulse compression waveforms used for moving target detection is the ISLR. The reason is that all the clutter that is distributed throughout the radar illuminated antenna footprint competes with the target. Even

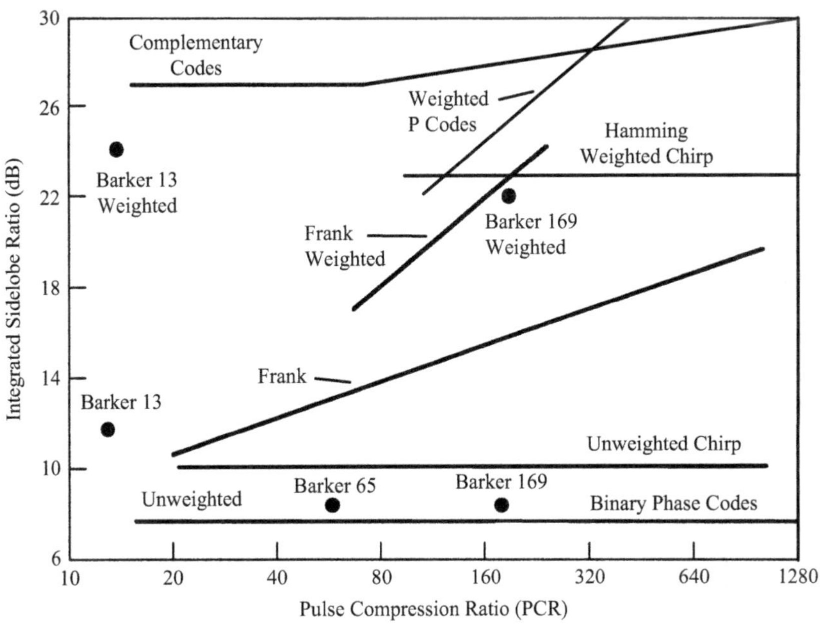


*Figure 2.31 ISLR for pulse compression waveforms; adapted from [3]*

though that clutter power in any one cell might be quite low, there could be many thousands of contributors. Furthermore, those integrated sidelobes are uncorrelated and appear noise-like and are not easily cancelled. Figure 2.31 shows a summary of typical pulse compression waveforms and their ISLR as a function of PCR.

### 2.11.1 Radar equation Example 2.4

There are three different radar equation elements that are important for a GMTI radar. First is the SNR at the output of the last filter before thresholding and detection. Obviously one must transmit enough power and do enough integration to detect a moving target. Second, the SNR on mean clutter in a single resolution cell provides information for threshold setting and space-time-adaptive-processing (STAP, more later). Third, the SNR on the uncorrelated sidelobe clutter represented by the ISLR is important because it may limit target detections more than thermal noise.

For example, suppose a GMTI radar had the parameters of Table 2.4 what would its performance be?

Then substituting into (2.11), (2.18), (2.19), (2.20) and (2.21) yields

$$
\begin{aligned}
G_T &\cong \frac{4\cdot\pi\cdot 1.8\cdot\left(0.05+\cos\left(0^o\right)\right)}{(0.03)^2\cdot(1+0.05)} = 25130 \\
SNR_I &= \frac{10^4\cdot 0.2\cdot 25130\cdot 1.8\cdot 10\cdot 0.88\cdot\sqrt{0.125\cdot 3000\cdot 1670}}{(4\cdot\pi)^2\cdot\left(150\cdot 10^3\right)^4\cdot 1.38\cdot 10^{-23}\cdot 290\cdot 1.6\cdot 25\cdot 10^6\cdot 2.5} = 19.69 \\
FP_{az} &= \frac{150\cdot 10^3\cdot 1.3\cdot 0.03\cdot(1+0.05)}{\left(0.05+\cos\left(0^o\right)\right)\cdot 6} = 975\ \text{m},\ FP_{el} = 150\cdot 10^3\ \text{m} \\
B_{amlc} &= \frac{2\cdot 200\cdot 1\cdot 0.0065}{0.03} = 86.67\ \text{Hz},\ B_{filt} = \frac{1}{0.125\cdot 0.88} = 9.09,\ \text{d}_{az} = \frac{975\cdot 9.09}{86.67} = 102.27\ \text{m} \\
SNR_{ICL} &= \frac{10^4\cdot 0.2\cdot 25130\cdot 1.8\cdot 6\cdot 102.27\cdot 10^{-3}\cdot 0.88\cdot\sqrt{0.125\cdot 3000\cdot 1670}}{(4\cdot\pi)^2\cdot\left(150\cdot 10^3\right)^4\cdot 1.38\cdot 10^{-23}\cdot 290\cdot 1.6\cdot 25\cdot 10^6\cdot 2.5} = 1.21
\end{aligned}
\qquad \text{(ex 2.4)}
$$

No gain other than that which naturally occurs in beamforming and integration is assumed for Example 2.4. In addition, the inevitable self-noise in the sensor system is not included. Furthermore, the square root of $N$ gain on noise may not be correct depending on processing.

An SNR of roughly twenty (13 dB) is normally required for reliable detection with a low false alarm rate. The clutter and target received power as a function of range are shown in Figure 2.32. Note that at the desired maximum range of 150 km both the radar equation computation and the plot of Figure 2.32 imply the required SNR. The duty ratio is 0.2 and so the 10 km around the main bang is blanked in the receiver. These parameters are similar to actual radars and the slow scan rate (0.125–0.250 s per dwell) necessary to see slow moving targets might require a minute or more to surveil a large area. The competing clutter is related to the

*Table 2.4 Radar equation Example 2.4*

| Parameter | Value | Parameter | Value |
|---|---|---|---|
| $P_T$, kW | 10 | $\lambda$, m | 0.03 |
| $D_T$ | 0.2 | $A_{eR}$, m$^2$ | 1.8 |
| $T_o$, K | 290 | $k_B$ | $1.38 \times 10^{-23}$ |
| $L$ | 2.5 | $NF$ | 1.6 |
| $\theta_o$, deg. | 0° | $\sigma_t$, m$^2$ | 10 |
| $G_T$ | Calculate | $T_{OT}$, s | 0.125 |
| $PRF$, kHz | 3 | $R$, km | 150 |
| $eff_n$ | 0.88 | $B$, Hz | $25 \times 10^6$ |
| $c_{min}$ | 0.05 | $T_{OT}$ x $PRF$ | 375 |
| $PCR$ | 1670 | $n$ | 626,250 |
| $\sigma_o$, m$^2$/ m$^2$ | 0.001 | Beamwidth: el x az, rad. | $0.13 \times 0.0065$ |
| $\eta$ , deg. | 90° | $ISLR$ | $5.75 \times 10^{-3}$ |
| $d_r$, m | 6 | $FP_{el}$, km | 150 |

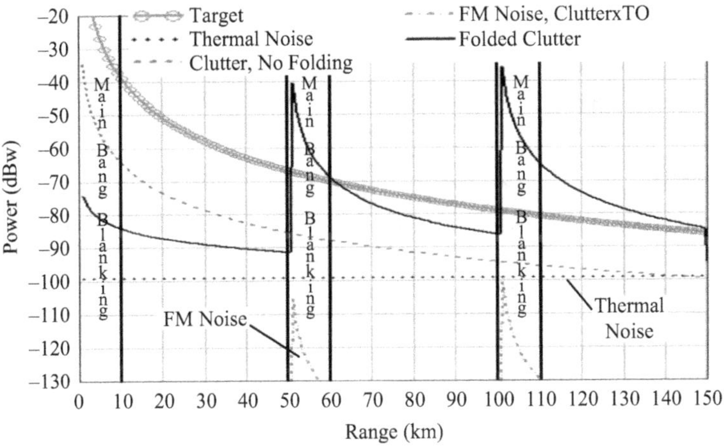


*Figure 2.32 SNR for target and clutter for Example 2.4 (GMTIPower006)*

azimuth and elevation antenna footprint on the surface. Because there are three range ambiguities out to 150 km, there are multiple clutter returns competing with the target. The ambiguous power is substantially above the power returned from ranges near the target range. Also the clutter area grows as the third power of range. The natural question to ask is why not go to a lower PRF with no range ambiguities. The problem is that lower PRFs provide a very small Doppler clear region as well as much smaller power on target for the same dwell time using practical hardware.

Figure 2.32 also includes the effect of the antenna pattern, which has been pointed to allow best performance at 150 km. One feature observable in Figure 2.32 is that in the absence of Doppler filtering the clutter will be above the target. Even though the RCS of a clutter patch might only be 0.001 m$^2$, there is so much competing area that the clutter signal is larger than the signal at the output of the processing for some ranges. This means that for detection as much as 13 + 13 dB of clutter rejection is required to detect a surface moving target in the region between 110 and 150 km. As long as the clutter is not in the same Doppler filter and angle bin as the target, this requirement is fully feasible. The clutter cancellation requires good filter sidelobes as well as angle rejection to a small fraction of a beamwidth. In this particular case, the transmitted FM noise modulated onto the clutter lies well below noise and is not a performance limitation.

## 2.12 Radar surface returns

The surface echo return competing with a ground mover comes from across the antenna mainbeam. Since each angle of arrival has a slightly different LOS, the equivalent Doppler velocity is spread out. Figure 2.33 qualitatively shows the three types of ground return to an airborne or spaceborne platform. The dominant elements are the altitude return at zero Doppler (unless you are crashing!), mainlobe ground return and sidelobe ground return. The relative amplitude of each changes depending on antenna performance, observer altitude and range of target observation. The relative surface return power in any one range and Doppler bin can be reduced by reducing resolution cell size, but only down to the size in range and Doppler of the desired target class. At the front end of the radar, the receiver sees all the return echo dynamic range and must be compensated to accept a range of signal amplitudes easily 10,000,000,000:1. This usually requires overload protection, range based gain control (sometimes called STC), and total ambient received power control to prevent saturation at the high-end and noise-level control at the low end (usually called AGC). Also, sometimes SAR or surface maps are made of the surrounding terrain about the vehicle and, since the terrain aspect changes relatively slowly, they are used to set receiver gain and other thresholds in the radar.

Similarly if the objective is to make radar maps in the antenna mainlobe, then the sidelobe and altitude return including GMTs are clutter and will limit the achievable dynamic range of the map. Although the surface backscatter may be low on the average, there will be some very bright returns called bright discretes, which may saturate the front end causing false returns in many radar pixels. Folding about the PRF can cause mirages in dark portions of a SAR image.

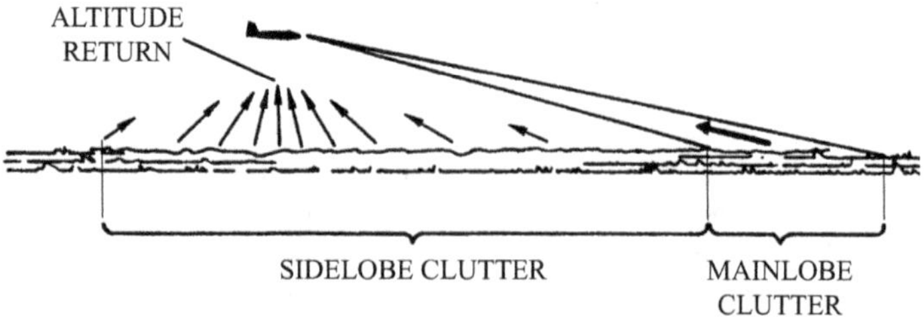


*Figure 2.33 Three types of ground return [1]*

Figure 2.34 shows the qualitative amplitude return as a function of range for an airborne or spaceborne moving target detection or SAR platform. By far the strongest return comes from directly beneath the radar platform, the altitude return. The altitude return can often be eliminated by range gating, that is, not receiving during the time of the altitude return (you must always know your altitude or you will crash!). Most of the sidelobe clutter will come from ranges between the altitude return and the antenna mainlobe intercept with the surface. Aircraft targets such as those at position, **C**, may be clear of mainlobe surface return. Most GMTs will be competing with the mainlobe surface return at positions **A** and **B**. Seeing those targets will depend on their relative return amplitude, Doppler difference or frame to frame position change for discrimination and detection.

Figure 2.35 shows the qualitative amplitude return as a function of Doppler for an airborne or spaceborne GMT or SAR platform. There will be some target returns that have an opening rate (going away from the radar platform by more than the velocity of the platform). There will be some target returns that have low opening or closing rates, which will be competing with sidelobe clutter. There will be some returns from targets, which have a zero rate that will be in the altitude return. There will be returns in which the target is traveling normal to the radar LOS and in mainlobe clutter as well as slow targets in the mainlobe but too slow to be out of mainlobe clutter. Finally, there will be targets whose Doppler frequency represents a closing rate higher than the platform velocity.

Even a GMTI radar focused on the ground will see commercial aircraft at long range because their RCS is so large and their velocities will fold about the PRF and appear to be large near moving targets. The actual amplitudes will be a function of radar observer altitudes, antenna sidelobes, antenna beamwidth, resolution cell size and range to target area. In addition, there will be some returns from quasi-stationary targets, which have high Dopplers such as windmills and ventilators.

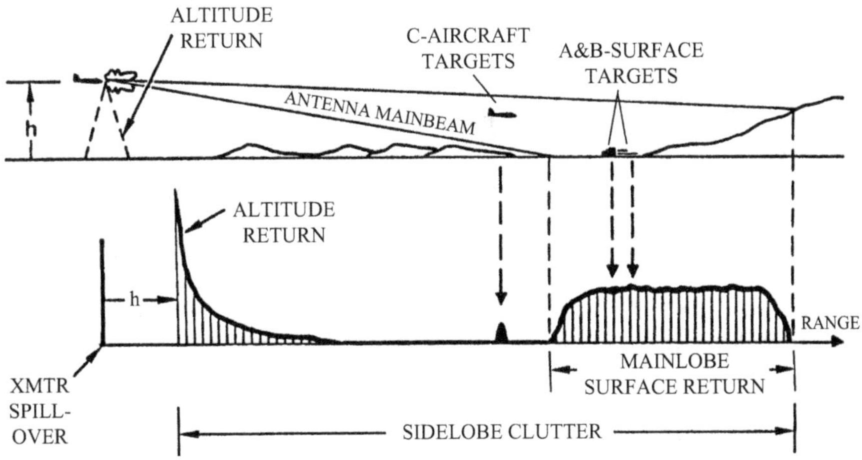


*Figure 2.34 Range return from clutter and targets [1]*

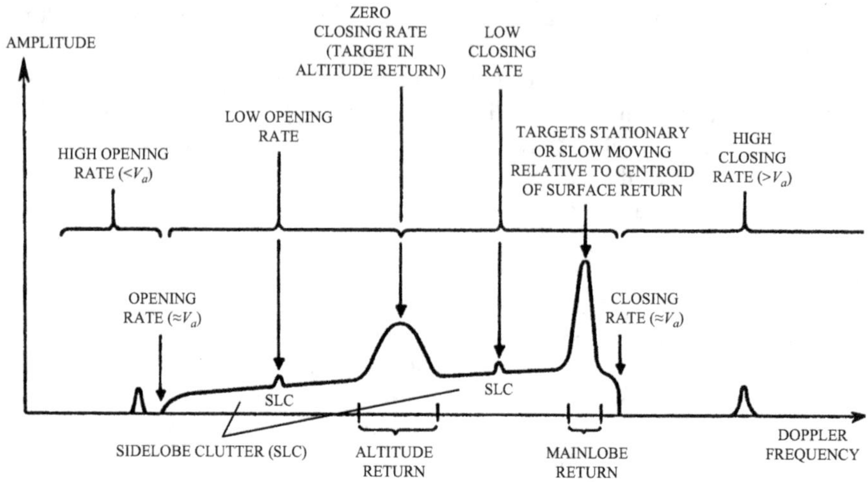


*Figure 2.35 Doppler return from sidelobes, mainlobe and targets [1]*

## 2.13 Ambiguities and folding

Since virtually all radars are sampled data systems (including FMCW), there is folding about each PRF line as shown in Figures 2.36 and 2.37. The effect of folding in the frequency domain is shown in Figure 2.36. The spectrum is replicated about each sideband. The first lower sideband is added to the central spectral return as well as the first upper sideband. This continues for the next upper and lower sidebands and so on until a composite Doppler spectrum is produced as shown in the bottom of Figure 2.36. Typically for low and medium PRF systems, there are so many foldings that the competing clutter and noise in the "clear" region is almost flat (like white noise, the noise your TV makes after the fat lady sings).

Similarly, the pulse repetition interval (PRI), which is the reciprocal of the PRF, causes folding in the range dimension. Figure 2.37 shows typical folding in the range dimension. In the example, there are three range zones or intervals including the mainbeam region. Zone 1 contains only the altitude clutter. Zone 2 contains an interval only containing intervening aircraft, not necessarily the object of the GMTI search or SAR map, but a very common occurrence. Zone 3 contains the mainbeam intercept with the surface and the GMTs of interest or terrain to be mapped. As each of these regions is folded into the composite, the targets or maps are all competing with some clutter both mainlobe and sidelobe. Some targets may appear farther away than their true location (target **C** for instance). The altitude return, even though it is in antenna sidelobes, may be so large that the receiver is turned off (blanked) for a short interval.

The clutter from all sidelobes in range and Doppler competes with both maps and movers. Poor sidelobes in SAR cause shadowed or very smooth regions, which should normally be dead black, to be filled with speckles or at worst ghosts of real

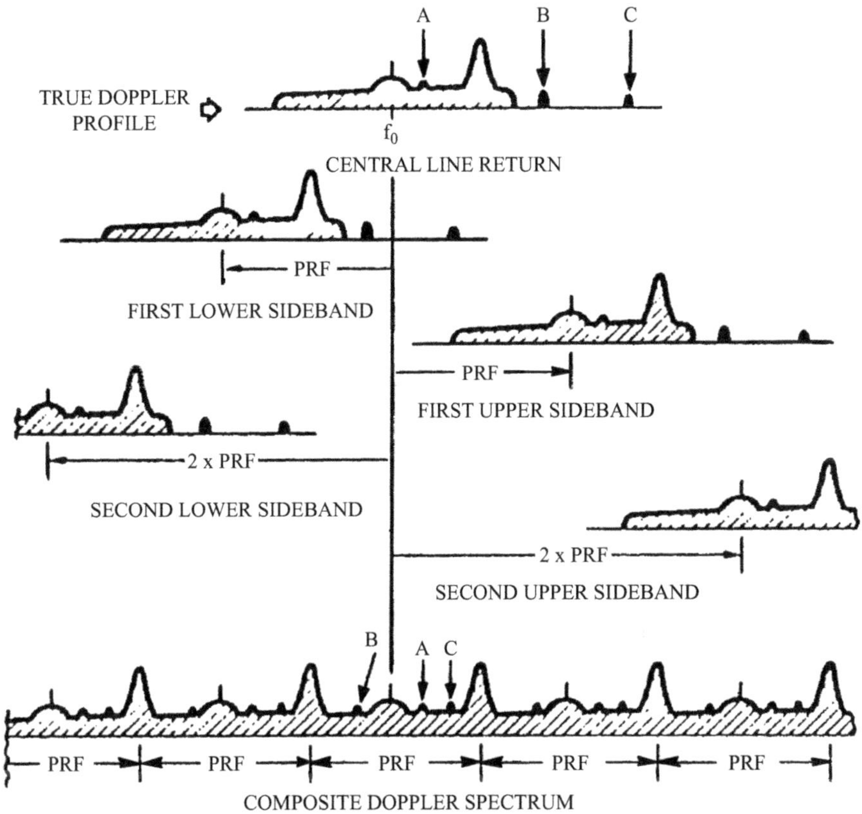


*Figure 2.36 Pulsed Doppler spectrum folding [1]*

features appearing in lakes or roadways. These ambiguous returns are usually eliminated by pulse-to-pulse coding over a coherent processing interval (CPI).

Poor sidelobes in GMTI cause large moving targets usually at nearer range to appear to be traveling at high speeds on roads, which aren't there. Large trucks such as over-the-road-18-wheelers can easily have cross sections of 1000 m$^2$ with the wheel tops traveling at 160 mph. Pulse-to-pulse coding doesn't work as well for these returns. These false targets can be eliminated only by tracking each return and applying heuristics, for example, you can't sustain greater than 20 mph in most surface vehicles if there are no roads. (If you doubt this, ride in a main battle tank in road-less country for 4 h at 30 mph.)

Surprisingly enough in a sparse moving target space (i.e. 1/100 of the cells contain a target), these ambiguities can be resolved. Furthermore, the blanked regions will not stay in the same place for different PRFs and so blanked regions are often filled in. Both the range and Doppler target ambiguities may have to be resolved to determine whether the threshold crossings are, in fact, targets of interest. This usually requires multiple looks at different PRFs, frequencies, angles or

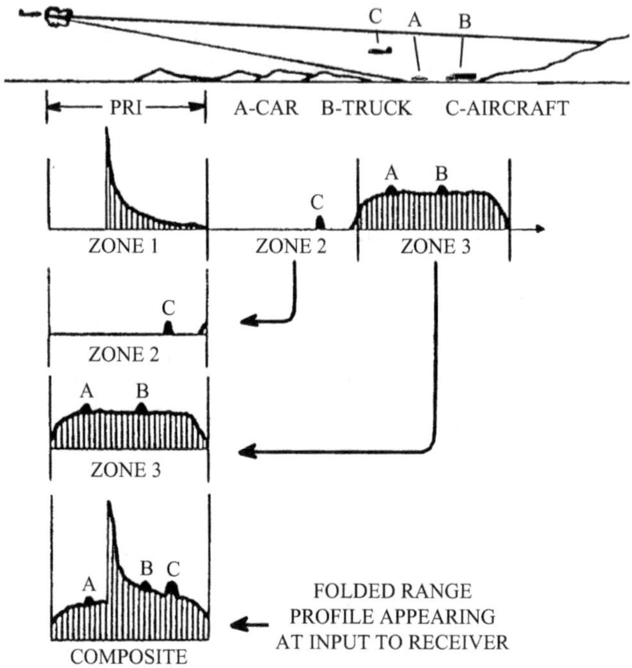


*Figure 2.37 Range folding [1]*

times (almost never single look!). These ambiguities can be resolved by using the Chinese Remainder Theorem. The theorem states that, if one knows the remainders of the division of an integer *n* by several integers, then one can determine uniquely the remainder of the division of *n* by the product of these integers, under the condition that the divisors are pairwise coprime, for example, multiple PRFs, which are coprime can uniquely resolve the ambiguities by unfolding each PRF and looking where a return is coincident in all unfolded PRFs.

In addition to the macroscopic features of moving targets, there are unique features of each class of moving target that complicate the signature of each moving target. Most moving targets have multiple scatterers, which are moving at apparently different rates. Radar waves bounce off road surfaces and allow scattering from partially covered moving parts such as wheels, tracks, cooling fans, drive shafts, vibrating structure and geometrical acceleration. The net result of these elements is to substantially broaden the return spectrum from a mover. Depending on the transmitter waveform, this may smear the target location in range and confuse the short-term estimate of speed.

Similarly, SAR returns exhibit features, which may smear (defocus) the image. Scatterers on tall structures will be in a different range-Doppler place than if they were on a flat earth and may not be correctly focused. Wind-driven movements of brush, trees and fences may cause defocusing. Moving objects inside ventilators and air conditioners will cause smearing. Smearing will also occur due to

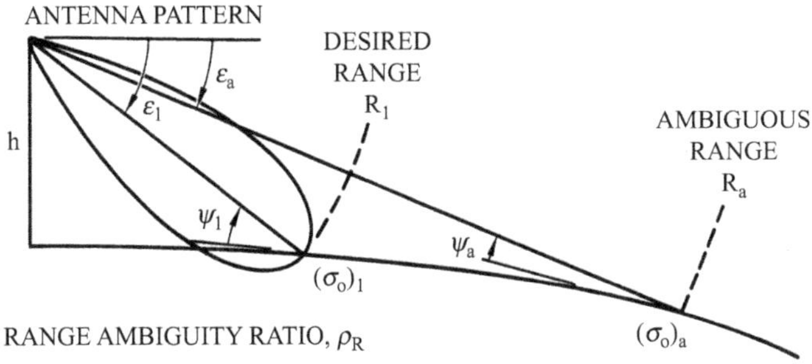


• RANGE AMBIGUITY RATIO, $\rho_R$

$$\rho_R \doteq \frac{\text{PROCESSOR OUTPUT POWER FROM ONE AMBIGUOUS RANGE}}{\text{PROCESSOR OUTPUT POWER FROM DESIRED RANGE}}$$

*Figure 2.38 Range ambiguity ratio [5]*

telecommunications transponders responding out of band. Multipath responses may also cause defocusing if the extra path length is long enough. Typically, sidelobe ambiguities in range or Doppler may not be fully focused.

Figure 2.38 shows an important feature associated especially with SAR mapping. Often in SAR the backscatter from the ground can be quite low in RCS. The back scatter is usually grazing angle dependent but sometimes the return from a range ambiguity can be nearly as large as the desired ground especially at steeper grazing angles. It is usual in the design process to calculate the worst-case range ambiguity ratio as suggested in Figure 2.38. Ambiguous power may come from diffuse terrain or from bright discretes (see Figure 3.24).

## 2.14 Range and Doppler resolution

The way moving targets or radar map pixels are resolved is by breaking the radar return into range-Doppler angle cells or bins. If the bins are just resolved, then each bin will have a dot of some brightness level and perhaps an associated phase. It is possible if the SNR is high enough to estimate the centroid and extent of the dot inside the bin but no further resolution is obtainable. This is also true of modern digital camera images. Once you are at the pixel size in the camera's CCD light sensor, no further resolution is obtainable no matter what you see on TV cop shows.

Just as in a modern digital camera, each radar pixel is arriving at a slightly different angle. A radar "camera" measures the time delay between the "flash" and the return photons (waves if you like). The time delay is, of course, the range to the reflecting object and back. If the radar antenna has multiple photon sensors like a digital camera (usually called phase centers) such as monopulse, then angle of arrival can be estimated. But if there is relative movement between the radar sensor and the illuminated patch, Doppler may also be used to estimate angular location. The slight difference in angle of arrival will result in a slight difference in measured Doppler. Then the range resolution and the Doppler resolution can be used to make a 2D image of the radar return from the surface as shown in Figure 2.39.

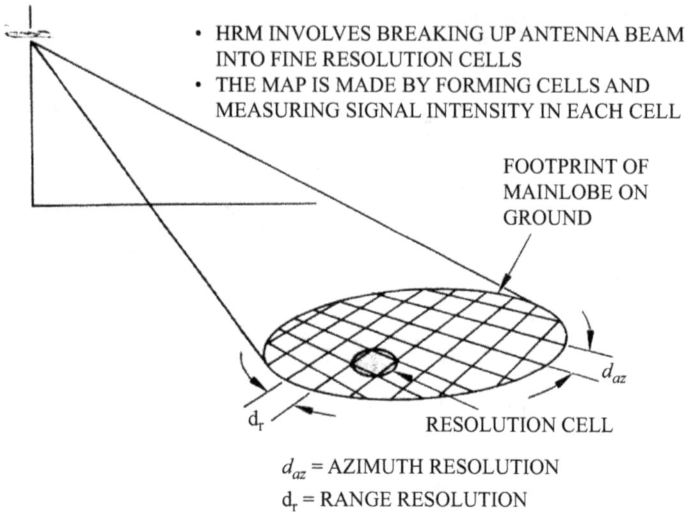


*Figure 2.39 Radar high-resolution mapping (HRM)*

The slight differences in angle and hence Doppler are what is exploited in both SAR and inverse SAR (ISAR) imaging. If there are enough phase centers in the antenna, these properties can be combined into what is called STAP. Adaptive processing cannot do anything that could not be done with *a priori* knowledge of target and clutter, that is, it still can't defy basic physics.

### *2.14.1 High-resolution mapping Example 2.5*

The original Venus mapping radar was a "gutless wonder." It depended on what is known as Doppler beam sharpening (DBS). The fact that different parts of the antenna beam see a slightly different Doppler (velocity) was already discussed in Section 2.9. That fact allows cross-range resolution by splitting the azimuth beam into multiple Doppler bins. That coupled with a simple pulse compression code will enable detection of multiple resolution cells in range and cross-range.

The radar planetary mapping notion is depicted in Figure 2.40 lower left. This allowed the beam to be broken up into $8 \times 8 = 64$ range-Doppler cells for each beam position. A simple eight-point weighted FFT was performed on each range bin inside the antenna beam. The surface echoing area in each bin was about $2.5 \times 2.5$ km or $6.25 \times 10^6$ m$^2$, so even with a flea-powered radar the surface features were easily discernable. Obviously there was some geometric distortion but the orbit and look angles relative to the surface were well known, so that could be rectified back on earth to create a composite map of Venus. One might observe that 2.5 km resolution isn't all that great, but for a planet's surface that has never been mapped it was spectacular. A radar map of Venus is shown in Figure 2.41.

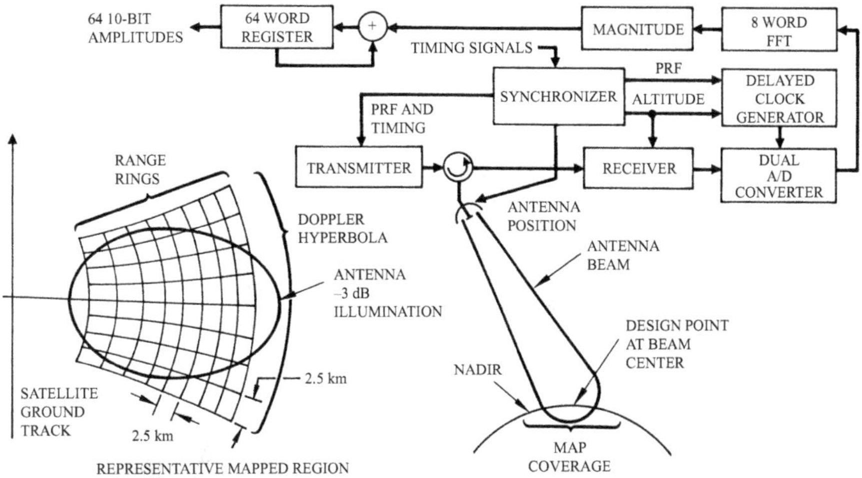


*Figure 2.40 Venus mapping radar [1]*

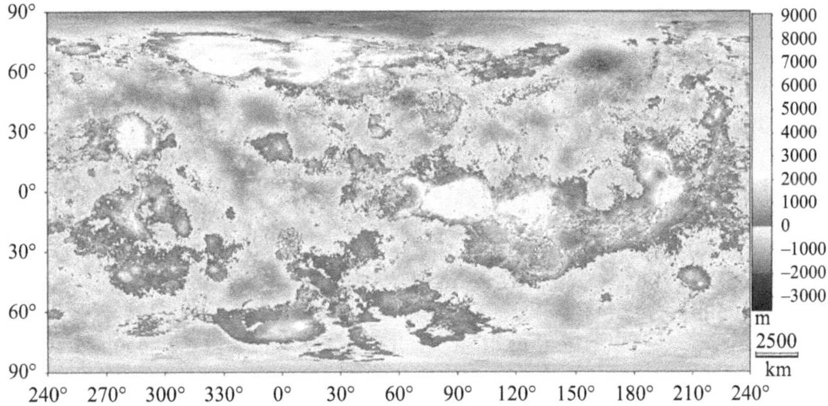


*Figure 2.41 Radar map of Venus ([6] Wikipedia™)*

## 2.15 Elements of geolocation

Most GMT detections are worthless without geolocation. Even with your eyes closed you can know that a patch of a few square kilometers will have moving objects anywhere on Earth. Both GMTI and SAR require the radar to "know" its true position about the earth as well as the corresponding position of the area illuminated with some accuracy. The altitude, latitude and longitude of the radar platform must be known with enough accuracy that the radar beam can be pointed to the correct region of interest or patch on the surface. The radar itself may improve the accuracy of its position and velocity as well as the accuracy of the

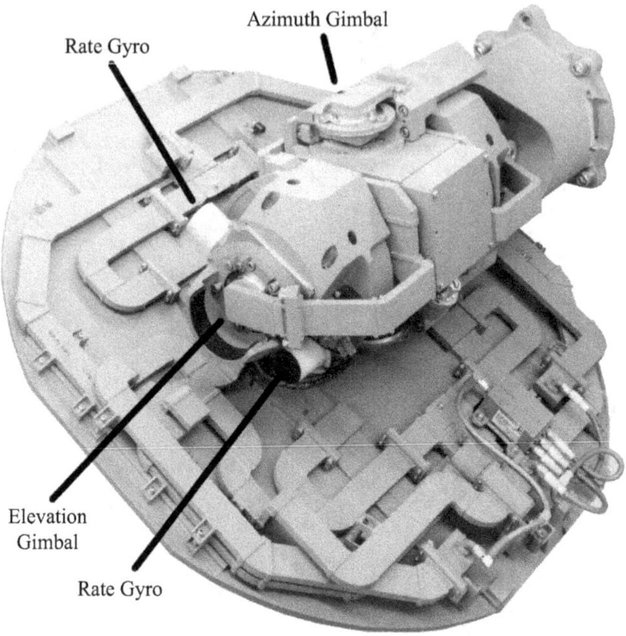


*Figure 2.42 Two axis gimbal and antenna [1]*

relative location of the region of interest. These improvements can come from Doppler aiding of the platform motion sensing and/or measurement of position from radar images with correction of predicted position compared to measurement of known landmarks.

Figure 2.42 shows the back of a radar antenna and a gimbal with two degrees of freedom. The antenna-gimbal is mounted to an aircraft bulkhead with the rear-most rotary joint axis vertical to allow azimuth scanning and the second axis scans in elevation. This gimbal also has motion sensing with two rate gyros and three accelerometers.

A set of target or position estimates such as slant range, $R$; range rate or Doppler, $\dot{R}$; azimuth, $\theta$ and elevation, $\varepsilon$, are measured in radar-referenced coordinates. The motion sensing is used to measure the antenna position in inertial space and compensate motion-induced phase errors. These measurements must be transformed from a radar-based frame to worldwide navigation coordinates, most commonly north-east-down (NED). NED coordinates allow the retrieval of stored map database information about the region of interest, not only terrain elevation versus latitude and longitude but also cultural features such as roads and bridges. The location of cultural features such as bridge abutments are often known to a few inches or less. This allows very accurate position estimates. Also, detailed knowledge of the terrain altitude in the antenna beam allows very accurate velocity measurements.

Once the area of interest is accurately illuminated by the radar, GMTs can be geolocated by their relative position in a map database. There are several common

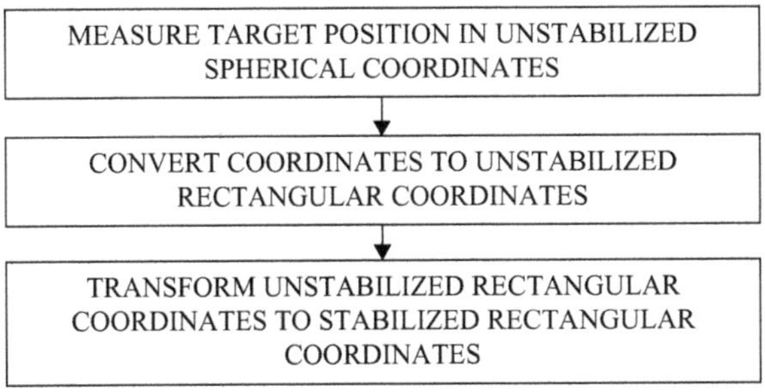


*Figure 2.43 Coordinate conversion and transformation [7]*

types of GMT geolocation; many of them are based on using one or more of the following: DTED, digital elevation model (DEM), digital line graphs (DLG) or other cartographic data. Modern Geographical Information System (GIS) databases usually have multiple layers that contain terrain elevation on a grid of ground stakes (DEM or DTED), cultural features such as buildings, roads and political boundaries, and natural features such as rivers and waterways (DLG).

In order to locate a radar target observation in an NED terrain database, the measurements must be transformed from radar-referenced estimates to Earth referenced estimates. This usually requires a transform from radar range and angles ($R$, $\theta$, $\varepsilon$) to Cartesian coordinates ($\boldsymbol{R}_{x,y,z}$) through the relative orientation of the radar platform, $\boldsymbol{O}_{\alpha,\beta,\gamma}$, into worldwide coordinates using direction cosines. The sequence is summarized in Figure 2.43.

Normally, the convention for Cartesian (rectangular) coordinates for a space-borne or airborne platform assumes the $x$-axis is approximately along the velocity vector in the plane of the platform horizontal with the $y$-axis normal to the $x$-axis in the same horizontal plane. The positive $z$-axis is normal to the $x$–$y$ plane and goes down. If the radar platform is an aircraft, typically the $x$-axis goes through the nose, the $y$-axis is parallel with the wing root and the $z$-axis goes out the bottom as shown in Figure 2.44. The azimuth angle, $\theta$, is measured from the $x$-axis and assumed positive to the right of the $x$-axis. Since the $z$-axis is assumed positive down, an elevation angle, $\varepsilon$, is positive for angles below the local horizontal. For example, the conversion of position and velocity to unstabilized rectangular coordinates is given in (2.22).

$$\begin{bmatrix} R_x \\ R_y \\ R_z \end{bmatrix} = \begin{bmatrix} \cos\varepsilon\cdot\cos\theta & \cos\theta\sin\varepsilon & -\sin\theta \\ \cos\varepsilon\cdot\sin\theta & -\sin\varepsilon\sin\theta & \cos\theta \\ \sin\varepsilon & \cos\varepsilon & 0 \end{bmatrix} \begin{bmatrix} R \\ 0 \\ 0 \end{bmatrix}$$

$$\text{and} \quad \begin{bmatrix} V_x \\ V_y \\ V_z \end{bmatrix} = \begin{bmatrix} \cos\varepsilon\cdot\cos\theta & \cos\theta\sin\varepsilon & -\sin\theta \\ \cos\varepsilon\cdot\sin\theta & -\sin\varepsilon\sin\theta & \cos\theta \\ \sin\varepsilon & \cos\varepsilon & 0 \end{bmatrix} \begin{bmatrix} \dot{R} \\ 0 \\ 0 \end{bmatrix} \tag{2.22}$$

The equations of (2.22) are sometimes expressed as direction cosines as shown in (2.23). Direction cosines are numbers, which often appear in navigation and antenna pointing equations. They are the cosine of the angle to a point in space from the origin of the coordinate frame or another reference point. For example, $\cos\varepsilon\,\cos\theta$ is the direction cosine of the $x$-axis in a radar antenna-based spherical coordinate frame.

$$R_x = R\cdot\cos\varepsilon\cdot\cos\theta = R\cdot\Lambda_x$$
$$R_y = R\cdot\cos\varepsilon\cdot\sin\theta = R\cdot\Lambda_y$$
$$R_z = R\cdot\sin\varepsilon = R\cdot\Lambda_z$$
$$V_x = \dot{R}\cdot\Lambda_x,\ V_y = \dot{R}\cdot\Lambda_y,\ V_z = \dot{R}\cdot\Lambda_z$$

where $\Lambda_x, \Lambda_y, \Lambda_z$ are the x, y and z direction cosines,

they are the projection of the line-of-sight direction of $R$

on to the x, y and z axes and $\dot{R}$ is the line-of-sight velocity (2.23)

The definition of direction cosines between two points in space in a rectangular coordinate system is given in (2.24).

$$\Lambda_x = \cos\alpha_x = \frac{x_2 - x_1}{r_{21}},\ \Lambda_y = \cos\alpha_y = \frac{y_2 - y_1}{r_{21}},\ \Lambda_z = \cos\alpha_z = \frac{z_2 - z_1}{r_{21}}$$

where x, y, z are the coordinate axes and

$$r_{21} = \sqrt{(x_2 - x_1)^2 + (y_2 - y_1)^2 + (z_2 - z_1)^2} \tag{2.24}$$

Similarly, the transform from unstabilized rectangular coordinates to NED coordinates requires a rotation in three axes. Sometimes, if the inertial reference is too far away from the antenna, a translation is also required. However, many times the distance between the motion sensors and the phase center of the antenna is small enough that the position offset can be ignored. Some radars and intercept systems have the motion sensors directly mounted on the antenna within a few inches of the phase centers. Often modern radars have full inertial sensors with worldwide navigation capability. Bright scatterers in the terrain database can be correlated with coarse radar DBS maps to improve the accuracy of patch center location using GMTI mainlobe clutter. This is often required over long radar path lengths due to atmospheric refraction [7].

The stabilized NED axes are platform heading, $\alpha_p$, in the local horizontal plane, platform pitch, $\beta_p$, in the plane containing the heading and the down axis, and platform roll, $\gamma_p$, about the heading axis. The heading and pitch define the platform velocity vector. The velocity vector seldom actually goes through the $x$-axis of aircraft or spacecraft. Figure 2.44 shows a Piogram that is a convenient graphical representation of the three rotations necessary to convert the unstabilized

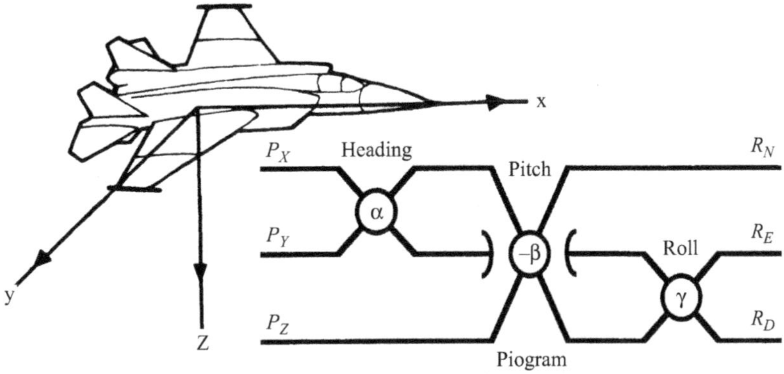


*Figure 2.44 Radar platform rectangular coordinates and NED Piogram [8]*

rectangular coordinates, $R_x$, $R_y$, $R_z$, of a radar measured point into stabilized coordinates for that same point in NED coordinates $R_N$, $R_E$, $R_D$.

The rotations represented by Figure 2.44 Piogram are shown in (2.25). Obviously, the velocity estimates use the same transformation. These multiple rotations can be represented by a single set of direction cosines, which multiply the range and range rate measurements directly into NED coordinates as shown in (2.25) for range.

$$\begin{bmatrix} R_N \\ R_E \\ R_D \end{bmatrix} = \begin{bmatrix} \cos\alpha_p \cos\beta_p & \sin\alpha_p \cos\beta_p & -\sin\beta_p \\ \cos\alpha_p \sin\beta_p \sin\gamma_p - \sin\alpha_p \cos\gamma_p & \sin\alpha_p \sin\beta_p \sin\gamma_p + \cos\alpha_p \cos\gamma_p & \cos\beta_p \sin\gamma_p \\ \cos\alpha_p \sin\beta_p \cos\gamma_p + \sin\alpha_p \sin\gamma_p & \sin\alpha_p \sin\beta_p \cos\gamma_p - \cos\alpha_p \sin\gamma_p & \cos\beta_p \cos\gamma_p \end{bmatrix} \begin{bmatrix} R_x \\ R_y \\ R_z \end{bmatrix}$$

where:

$\alpha_p$ is platform heading in the N-E plane,

$\beta_p$ is platform pitch in the plane which contains the D axis and heading,

$\gamma_p$ is platform roll about the heading. (2.25)

Equations (2.26) show the equivalent stabilized elevation angle, $\varepsilon_s$ and stabilized azimuth angle, $\theta_s$ as well as the stabilized range rate. These angles are *not* the radar referenced elevation and azimuth but rather the elevation and azimuth angles relative to latitude, longitude and altitude. The NED direction cosines are the projection of the radar LOS or "look" direction on to the NED axes.

$\Lambda_N = \cos\theta_s$ , $\Lambda_E = \sin\theta_s$ , $\Lambda_D = \sin\varepsilon_s$

$\varepsilon_s = \operatorname{asin}\Lambda_D$ and $\theta_s = \operatorname{atan}(\Lambda_E/\Lambda_N)$ and

NED range: $R_N = R\cdot\Lambda_N$ , $R_E = R\cdot\Lambda_E$ , $R_D = R\cdot\Lambda_D$

NED range rate: $\dot{R}_N = \dot{R}\cdot\Lambda_N$ , $\dot{R}_E = \dot{R}\cdot\Lambda_E$ , $\dot{R}_D = \dot{R}\cdot\Lambda_D$

where: $\dot{R}$ is $\dfrac{dR}{dt}$ and $\Lambda_N$ , $\Lambda_E$ , $\Lambda_D$ are the North, East,

down direction cosines respectively. (2.26)

Usually, the NED distance to the center of the patch is subtracted from each target NED distance so that every target can be tracked relative to the patch center. This improves target geolocation since each target can be placed in the terrain database. In addition, the reverse sequence must be calculated in many cases because an area of interest is often designated by off platform sources, sensors or preplanned missions. This normally starts by an in-radar or on-platform calculation of the range and desired LOS direction between the current position in latitude, longitude and altitude and the designated target area latitude, longitude and altitude.

The radar slant range, $R$, and ground range, $R_g$, are given for a spherical earth as shown in (2.27). The solid angle, $\beta$, between two latitude, longitude and altitude points is depicted in Figure 2.45. The ground range, $R_g$, between the two points is the great circle route between them, which is $R_g = \beta \times R_e$. As before, the earth radius is $R_e$, which is a local variable for the most accurate estimates.

Slant range:

$$R = \sqrt{h1^2 + h2^2 - 2 \cdot h1 \cdot h2 \cdot \cos(\beta) + 2 \cdot \left(R_e^2 + R_e \cdot (h1 + h2)\right) \cdot \left(1 - \cos\left(\frac{R_g}{R_e}\right)\right)}$$

The stabilized radar elevation angle below local horizontal:

$$\varepsilon_s = \mathrm{acos}\left(\frac{(R_e + h2)}{R} \cdot \sin\left(\frac{R_g}{R_e}\right)\right) \tag{2.27}$$

The angle, $\beta$, about the earth's center between two latitude, longitude and altitude points can be calculated as in (2.28). The stabilized azimuth, $\theta_s$, is given in

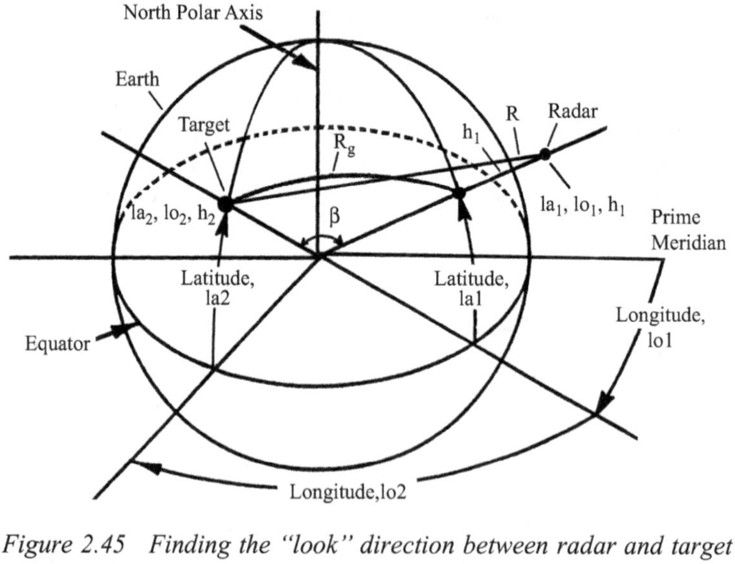


*Figure 2.45 Finding the "look" direction between radar and target*

(2.28) so everything is known to point the radar antenna beam, but it must still be transformed from NED coordinates into antenna $\varepsilon_r$, $\theta_r$ gimbal angles or electronic scan angles (not necessarily the same!). Also be aware that asin, acos and atan functions are ambiguous in angle and may require the addition of $\pi$ to the estimated angle to compensate for the ambiguity.

Earth center angle between points 1 and 2:

$$\beta = \text{acos}\left[\sin la_1 \cdot \sin la_2 + \cos la_1 \cdot \cos la_2 \cdot \cos\left(lo_2 - lo_1\right)\right]$$

where $la_1$ and $la_2$ are latitudes and $lo_1$ and $lo_2$ are longitudes for points 1 and 2 respectively, also $\beta = \dfrac{R_g}{R_e}$

The stabilized azimuth angle relative to North between the two points is:

$$\theta_s = \text{asin}\left[\cos la_2 \frac{\sin\left(lo_2 - lo_1\right)}{\sin\beta}\right] + \left(\pi \text{ as required}\right) \tag{2.28}$$

Using the slant range, $R$, and the desired "look" directions $\varepsilon_s$ and $\theta_s$ from (2.27) and (2.28), the transformation to the antenna base can be performed as in (2.29). This is just the transpose of the matrix given in (2.25). The next step is to calculate the look vector in antenna scan angles. The radar tentative scan angles, $\theta_r$ and $\varepsilon_r$, are shown in (2.30). Then the radar must determine whether this look vector is within the antenna scan limits. If it is, the antenna is commanded to that angle and a range window about the commanded range is established. Some number of pulses are transmitted to establish AGC, jamming (if any) and Doppler clutter centroid tracking (typically 16 pulses). Finally, the actual CPI is begun.

$$R_N = R \cdot \cos\theta_s \text{ , } R_E = R \cdot \sin\theta_s \text{ , } R_D = R \cdot \sin\varepsilon_s$$

$$\begin{bmatrix} R_x \\ R_y \\ R_z \end{bmatrix} = \begin{bmatrix} \cos\alpha_p \cos\beta_p & \cos\alpha_p \sin\beta_p \sin\gamma_p - \sin\alpha_p \cos\gamma_p & \cos\alpha_p \sin\beta_p \cos\gamma_p + \sin\alpha_p \sin\gamma_p \\ \sin\alpha_p \cos\beta_p & \sin\alpha_p \sin\beta_p \sin\gamma_p + \cos\alpha_p \cos\gamma_p & \sin\alpha_p \sin\beta_p \cos\gamma_p - \cos\alpha_p \sin\gamma_p \\ -\sin\beta_p & \cos\beta_p \sin\gamma_p & \cos\beta_p \cos\gamma_p \end{bmatrix} \begin{bmatrix} R_N \\ R_E \\ R_D \end{bmatrix}$$

where: $\alpha_p$ is platform heading in the N-E plane,
$\beta_p$ is platform pitch in the plane which contains the D axis and heading,
$\gamma_p$ is platform roll about the heading. (2.29)

The radar scan angles are

$$\theta_r = \text{atan}\left(\frac{R_y}{R_x}\right), \varepsilon_r = \text{asin}\left(\frac{R_z}{R}\right), R$$

as a check $R$ from (2.27) should equal

$$R = \sqrt{\left(R_x^2 + R_y^2 + R_z^2\right)} \tag{2.30}$$

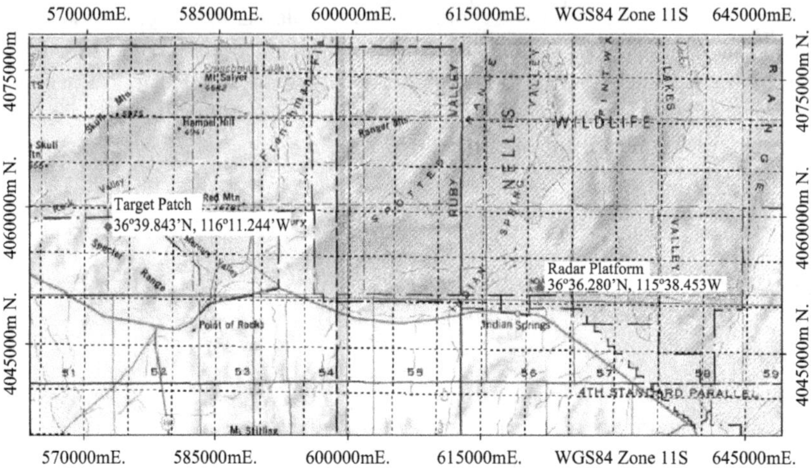


*Figure 2.46 Example 2.6 scenario (RadarLookAngle001)*

*Table 2.5 Example 2.6 initial values*

| **Parameter** | **Value** |
|---|---|
| Latitude 1 | 36°36.280′ |
| Longitude 1 | 115°38.453′ |
| Terrain altitude 1 | 7057 m |
| Latitude 2 | 36°39.843′ |
| Longitude 2 | 116°11.244′ |
| Terrain altitude 2 | 1230 m |
| Heading | $1.5\pi$ rad |
| Pitch | 0.035 rad |
| Roll | 0.050 rad. |

### *2.15.1 Radar antenna pointing Example 2.6*

Consider a surveillance radar aircraft platform orbiting over Creech AFB at the latitude and longitude shown in Figure 2.46. A designated target patch roughly 50 km away is the annotated by its latitude and longitude. The map underlay in Figure 2.46 is available in most GPS-based navigation systems and so the latitude and longitude also provides a terrain height. The radar platform is at an altitude 6096 m (20,000 ft) above ground level (AGL). The location and attitude of the radar carrying aircraft and the location of the target are summarized in Table 2.5. The first calculation is the slant range between the radar and the target patch using (2.27). The second calculation requires the conversion of latitudes and longitudes into radians.

Once the locations of target and platform are suitably converted, then $\beta$ can be calculated from (2.28). From that and (2.27), the stabilized look angles in NED coordinates can be calculated. This is shown in Mathcad program Radar Look Angle001.

*Table 2.6 Example 2.6 results*

| Parameter | Value | Parameter | Value |
|---|---|---|---|
| $\beta$, mr. | 7.7241 | $R_D$, km | 6.017 |
| $R$, km | 49.640 | $R_x$, km | −49.4185 |
| $R_g$, km | 49.265 | $R_y$, km | −6.8707 |
| $\theta_s$, rad | 4.57496 | $R_z$, km | 3.31345 |
| $\varepsilon_s$, rad | 0.12152 | $\theta_r$, rad | 0.13815 |
| $R_N$, km | 6.751 | $\varepsilon_r$, rad | 0.0668 |
| $R_E$, km | −49.179 | | |

In this example, the slant range is approximately 49640.4 m and the stabilized azimuth and elevation angles are $\theta_s = 4.57496$ and $\varepsilon_s = 0.12152$ radians. One caution with the slant range calculated using $\beta$ is that the equations in (2.27) depend on small differences in very large numbers. The vector $\boldsymbol{R_{NED}}$ can then be substituted into (2.29) to find the radar vector $\boldsymbol{R_{rad}}$. The x and y components are negative because the target patch is west of the radar platform. Finally the radar scan angles are calculated as shown in (2.30). This is well within the scan range of the radar. The results are summarized in Table 2.6.

## References


[1] Lynch, D., Aks, S., Kummer, W., Pearson, J.O., Shamash, E., "Introduction to Modern Radar", *Evolving Technology Institute Short Course Notes*, February 1988.

[2] Skolnik, M.I., *Introduction to Radar Systems*, McGraw Hill (1962), pp. 24, 28, 414.

[3] Lynch, D., *Introduction to RF Stealth*, SciTech Publishing (2004), pp. 98–114, 446–451, 82–84, 198–221, 467–470, 492–501, 504–531.

[4] Lynch, D. Kopp, C., Chapter 5, *Radar Handbook*, 3rd ed., M. Skolnik, Ed., McGraw Hill (2008), pp. 5.1–5.8.

[5] Tylka, J., Swiger, J.M., Pearson, J., Lynch, D., "Synthetic Aperture Radar", *ETS Short Course Notes*, 1977.

[6] Wikipedia, Venus Maps.

[7] Blackman, S., *Multiple-Target Tracking with Radar Applications*, Artech House (1986), pp. 49–81.

[8] Pio, R., "Algebra of Piograms or Orthogonal Transformations Made Easy", *Hughes Report M78-170*, October 1978, pp. 5-8–5-18.

# Chapter 3
# Nature of returns from the surface and tactical targets

## 3.1 Ground tactical target characteristics

In a military context, the objective is to automatically detect tactical vehicles and dismounts at beyond visual range such as those in Figure 3.1, while eliminating all the other uninteresting movers.

It is easy to find movers; what is hard is to find only the movers of tactical interest. There are numerous moving targets detectable in any significant FOR. They include people, insects, projectiles, missiles, animals, cars, trucks, trains, fences, power lines and suspended cables, leaves on trees and grass, ventilators, fans, clouds, rain, helicopters, UAVs, etc. What distinguishes many of these is that, although they have short-term movements and RCS that are similar to targets of interest, their ground coordinates do not change much. Usually a ground target must be observed for many seconds before its ground coordinates have changed enough for unequivocal discrimination. Even aircraft that have high ground speeds require seconds of observation for discrimination and trajectory determination. Trajectory discrimination over tens of seconds allows the sorting of projectiles from insects and birds, fixed wing aircraft from helicopters, wheeled vehicles from tracked vehicles, etc. These multiple observations if statistically independent are often called "Looks."

In addition, the cross section that a potential target presents to a sensor is very wavelength dependent. The signature that a tactical target presents to a long-wave IR sensor is quite different than the signature presented at microwaves. The typical manmade tactical target vehicle is made up of corners (called trihedrals and dihedrals), almost flat surfaces and edges. A good example is shown in Figure 3.2, an HMMWV with a TOW launcher. How many corners, almost flat surfaces (within a wavelength or two) and edges can you see?

Furthermore, the windows even when up are still relatively transparent to microwaves and inside the vehicle are many more similarly reflecting surfaces as shown in the plan view of an HMMWV in Figure 3.3. In addition, the ground or road surface acts as a mirror and there will be reflections from the underneath side of the vehicle including fender wells, which can be nearly as strong as the direct reflections.

*Figure 3.1 Typical tactical ground targets [1]*

*Figure 3.2 Typical tactical vehicle, HMMWV with a TOW launcher [1]*

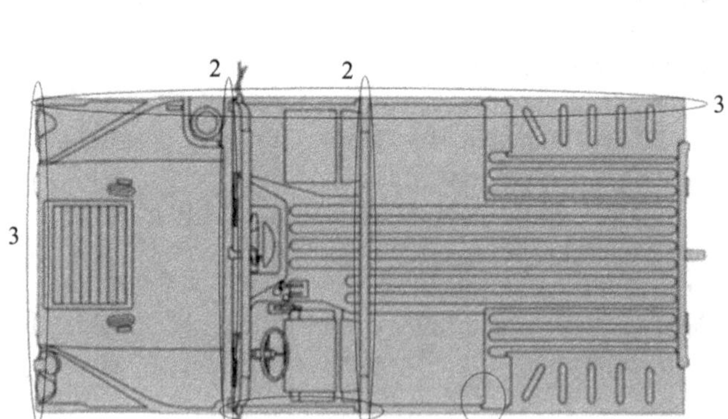


*Figure 3.3 HMMWV example speculars; adapted from [2]*

**Example 3.1** Most objects illuminated by radars are made up of more than one scattering center. These individual scatterers interact with one another to cause fading and speckle. Take a simple case of five scatterers of the same size, four separated by a fixed distance, **W**, and one in the center as shown in Figure 3.4. This might be typical of a modern vehicle with four scattering elements and a diffuse central scatterer (think a pickup truck bed plus cab). The example graph in the figure is for five scatterers, four separated by 2 m at X band ($\lambda = 0.03$ m). As the aspect changes, the RCS changes rapidly and goes through many deep nulls and some enhancements that are twice the combined RCS of $5\sigma_0$. This behavior is the norm not the exception and can only be minimized by averaging over multiple "looks" in angle or in wavelength. Also note that the apparent extent of the object is changing. In general, most objects are more complex with some scatterers changing their brightness with aspect. The result will be that the estimate of the target centroid will change with aspect and radar wavelength.

These features give rise to a few very bright reflectors (called speculars), which make up the RCS of the vehicle. Most car commuters have experienced speculars from the sun glint off a rear window of the car in front of them from time to time. These speculars make the vehicle RCS quite large for many angles. The three most prominent speculars are two types of corners called trihedrals and dihedrals, and large relatively flat surfaces. The Figure 3.3 plan view of an HMMWV shows an example trihedral (one circled) inside the vehicle but accessible through the windows. Trihedrals have very large cross sections even when only a few wavelengths across. There are three dihedrals (two) circled, one exterior to the vehicle and two inside. They are the next largest reflectors. There are two relatively flat surfaces (three) circled, which are the next largest. For any one viewing angle, only a few of

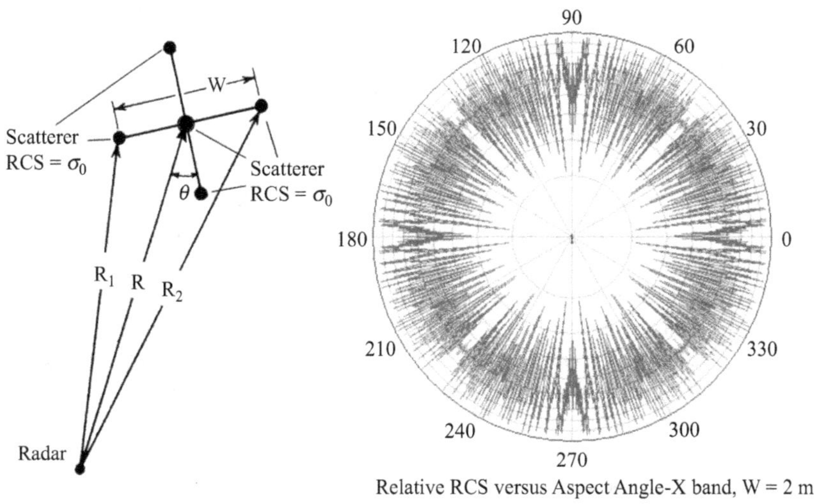


*Figure 3.4 Dumbbell scatterer fading Example 3.1 (5ScatterRCS1)*

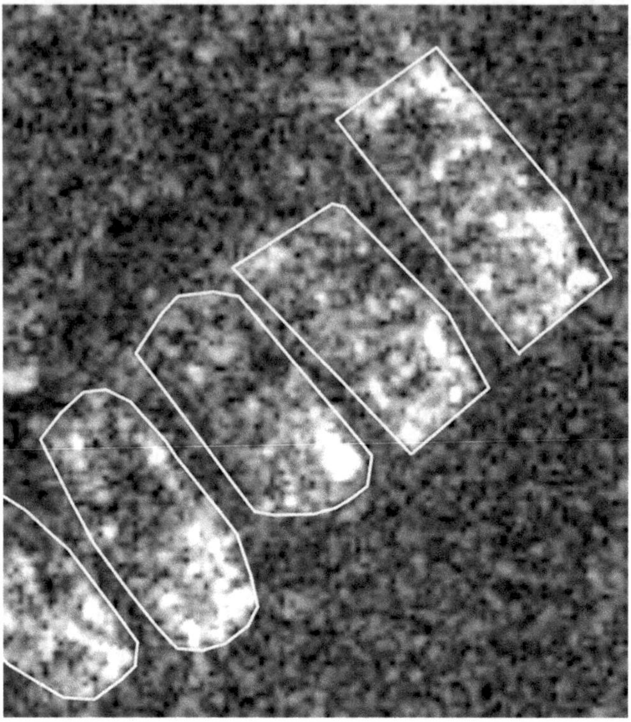

*Figure 3.5 Single-frequency SAR image of a row of cars, outlined in white (courtesy General Atomics) [3]*

these large reflectors will be significant. Figure 3.5 shows a row of parked cars and trucks illuminated from the right in the image (outlined in white). Even though individual radar pixels are quite small, the vehicles are just blobs due to speculars. Only multilook in frequency or angle will allow recognition.

The RCS for all of these surface features is proportional to the square of the illuminating frequency and the square of the area. For example, a trihedral only 4″ on a side has an RCS greater than ½ $m^2$ at X band. The net result is that on the average the RCS will be large but unfortunately a few large speculars can interact with one another causing what is known as destructive interference where the RCS can be quite small. Almost anyone who drives a car while listening to FM radio has experienced destructive interference when the radio fades in and out.

Since the orientation of a vehicle of interest can be anything relative to an airborne observer (as opposed to a police radar speed trap), there is a very reasonable chance that an observing sensor could be in a null or a peak. The average RCS of tactical vehicles is large but at any instant the RCS could be 100 times smaller. Hence multiple independent observations or Looks at different angles, times or frequencies are required to ensure detection and discrimination of interesting movers from everything else. When viewed for a full range of angles, the RCS of a tactical vehicle generally will appear as shown in Figure 3.6.

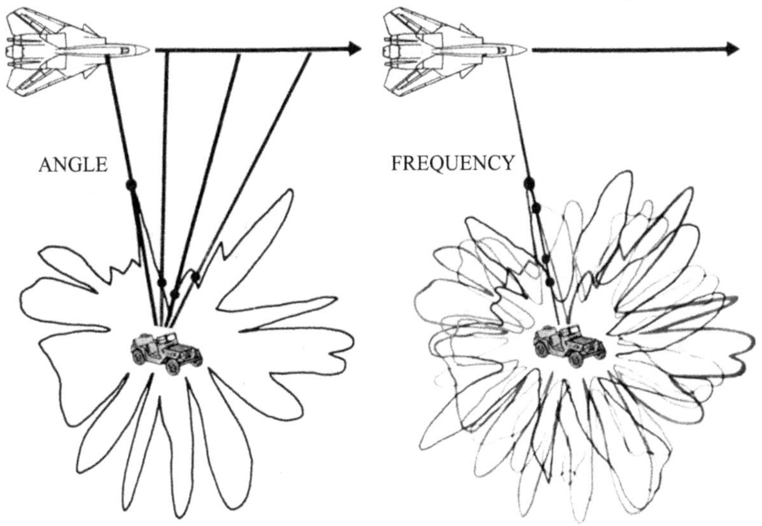


*Figure 3.6 Single radar looks on a tactical vehicle will vary dramatically [4]*

The radar characteristics of ground movers were studied in detail during the USAF Flamr program during the 1970s. There were many world firsts from this USAF program. It was a high-resolution Ku band electronically scanned digital radar with significant instrumentation and recording capability. It had very accurate motion compensation and worldwide navigation. It was housed in a B-47. The results of this program were classified until the end of 1987, at which time declassification made all the data available. One of the important sets of tests were measurements made of the Barstow-Nebo tactical target array at the Marine Logistics base and at the Twenty Nine Palms Marine base. A radar image of the target array is shown in Figure 3.7. In this image, the fenced-in array lies between Interstate 15 curving through the lower left and a major rail line filled with cars (the very bright streak in the upper part of the image). Another road, numerous buildings and power transmission towers can be seen in the image. It is outside Barstow, California. The array contains many types of tactical targets including jeeps, trucks, howitzers, APCs, tanks, cranes, etc. Each individual target RCS was taken from many different angles for all of the objects in the array. Generally these targets have large RCS but undergo deep nulls when the look angle changes by very small amounts.

During the tests at Twenty Nine Palms Marine Base, two targets were studied in detail to measure target RCS nulls through a series of small times and angles. This was primarily to measure target fading with time and angle in order to determine the number of observations necessary to insure detection in a tactical environment. These tests were conducted with and without frequency multilooks. In general, frequency change between independent looks must be greater than the velocity of light divided by the distance between major scatterers. For example, if principal scatterers are 10 ft apart then the frequency change between looks should be 100 MHz.

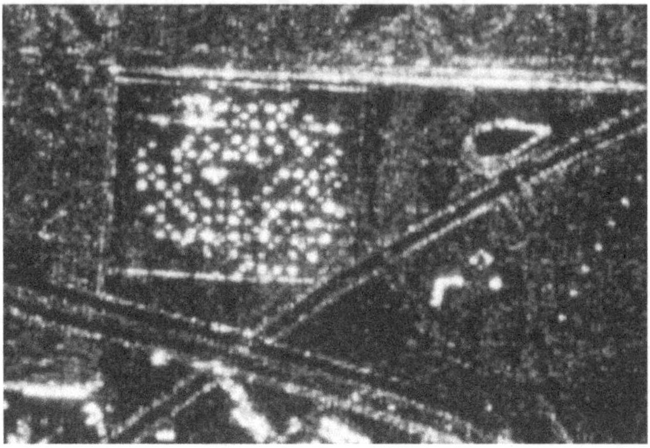

*Figure 3.7 Radar image Barstow-Nebo tactical target array [5]*

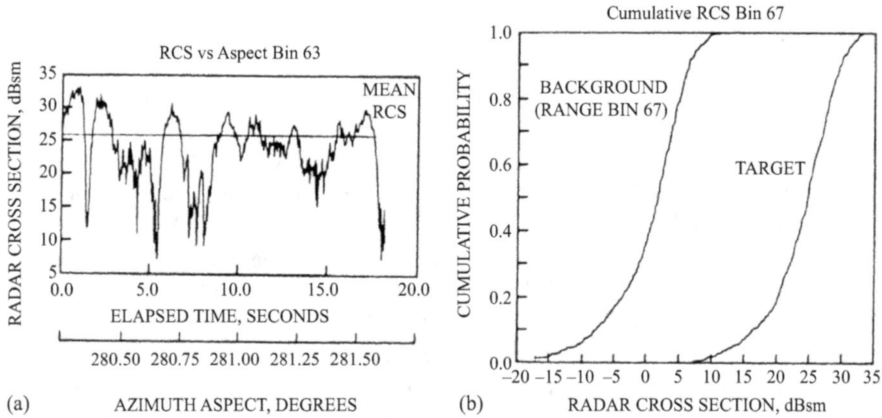


*Figure 3.8 RCS for 2½ ton truck – 20 ft resolution, single frequency [5]*

Figure 3.8(a) shows the RCS for time and angle for a 2½ ton military truck with 20 foot range and azimuth resolution with a single Ku band center frequency. Also shown in Figure 3.8(b) is the cumulative probability that a certain RCS will be observed for the full record. Note that the target goes through multiple deep nulls in a few tenths of a degree in aspect change. Depending on the velocity of the aircraft that was roughly 350 mph in this case, the nulls can last for several seconds. The nulls can last for tens of seconds if the aircraft is a helicopter or UAV.

Similarly, Figure 3.9 shows the RCS variation with angle and time for a self-propelled gun under comparable operating conditions. Again there is significant fading for very small angle changes and short time periods. These tests were done with several different resolutions and at many aspect angles to the tactical targets. The qualitative results were the same and the observed features are not unique.

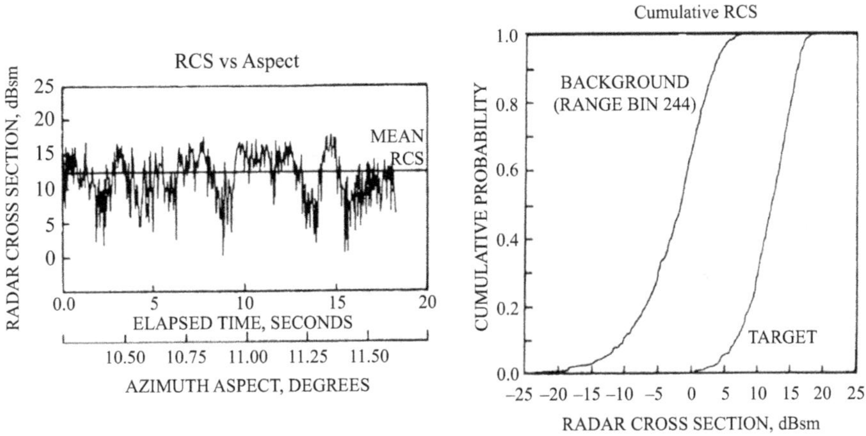


*Figure 3.9 RCS for self-propelled artillery – 20 ft resolution, single frequency [5]*

The conclusion from these tests is that either long observation times or frequency diversity is required to enable reliable tactical target detection. In summary for various military tanks and trucks, an experimental cumulative probability distribution versus RCS at Ku band is given in Figure 3.10. At X band the results are similar except shifted down by approximately 4 dBsm.

The targets were as seen in Figures 1.26, 1.27, 1.28, 3.11 and 7.39 positioned on dirt or grass. The Barstow-Nebo array is rearranged from time to time but always on dirt or grass. The distinct ground bounce that is observed in some surface target signatures is because they are on pavement where height is turned into layover. In rougher terrain, the bounce is still there but usually is much weaker. Shadows cast by targets are a much better method to estimate height and features. That is why range and azimuth sidelobes, that is, ISLR are so important to SAR. With higher resolution, it is usually easy to see vehicle tracks to the current target location if the target was moved recently.

General Motors Hughes developed automobile radars for automated cruise control station keeping in the late 1980s and early 1990s in the author's business unit. The field trials of those systems showed so much ground bounce from pavement that cars ahead of the closest cars could easily be detected as a confusor. Tests using AWACS in Europe showed multiple bounces from vehicle wheels, buildings and roadbed leading to many vehicle speed Doppler harmonics.

Figure 3.12 shows the mean and 1 sigma cross-sectional variation for a variety of tactical target vehicles using four multiple looks at different frequencies. Figure 3.13 shows the variation of cross section as a function of aspect angle. These results are for Ku band but similar results occur at most radar frequencies and at nearly the same radar cross section limited only by resolution cell size and scaling by the ratio of the wavelengths squared, i.e., Ku to X is 0.4 or −4 dB. These tests were all done under excellent weather conditions.

Humans, on the other hand, exhibit a relatively constant RCS of about 0.5–1 $m^2$ at most microwave frequencies. Humans often are modeled as a rough

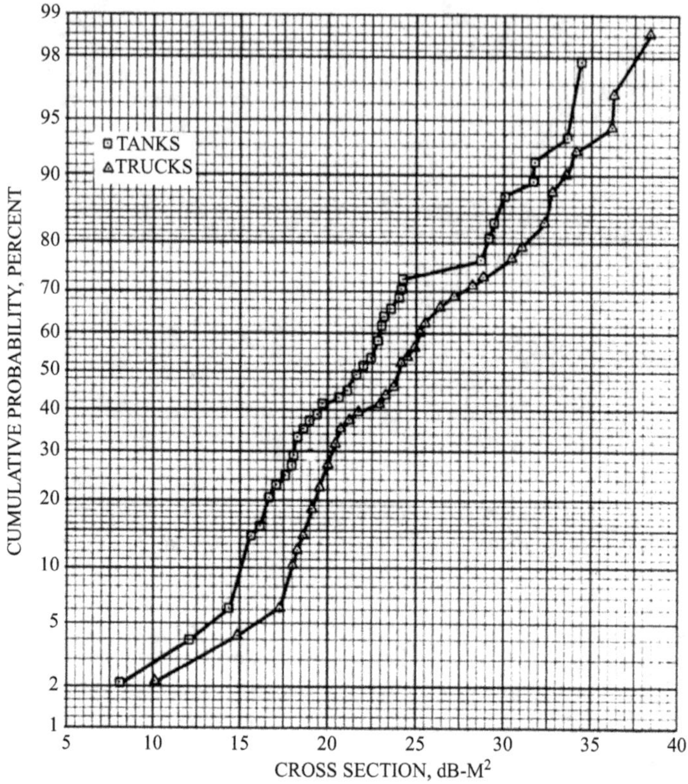


*Figure 3.10 Ku band tank and truck RCS distributions [5]*

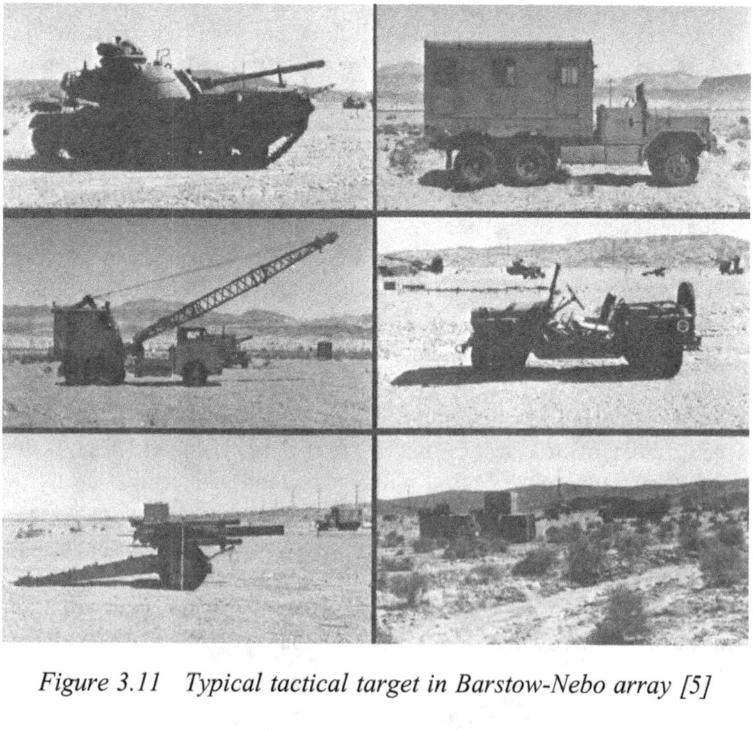

*Figure 3.11 Typical tactical target in Barstow-Nebo array [5]*

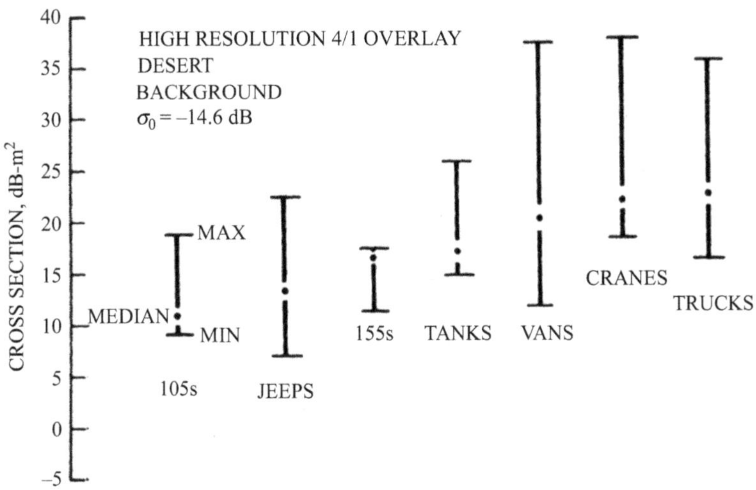


*Figure 3.12 Typical tactical target RCS [5]*

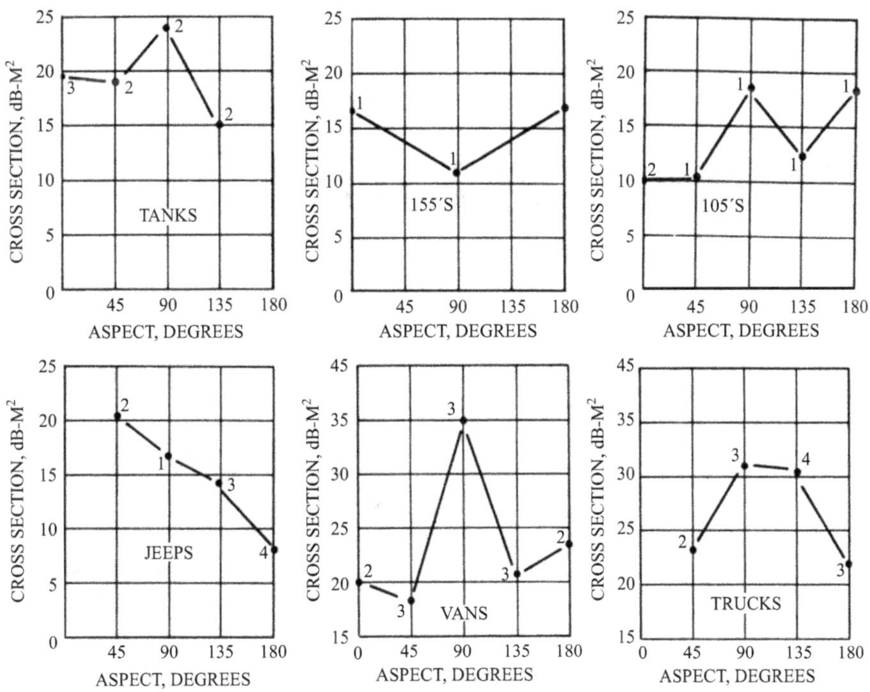


*Figure 3.13 Tactical target cross section with aspect [5]*

cylinder of water 0.5 m in diameter and 2 m high. Humans are readily detectable by high-resolution radars but competition from other objects with similar RCS and their irregular movement makes them difficult to detect except with coherent change detection or with space-time adaptive processing (more later in this chapter).

## 3.2 Environmental limitations

Everything discussed so far is highly idealized in the sense that the weather was perfect for the tests. The real environment is much more challenging. Every radar must cope with a wide variety of natural and manmade phenomena. Many of these are summarized in Figure 3.14. The atmosphere causes signal attenuation and refraction. Weather in the form of clouds, rain, snow, hail and fog not only cause attenuation but also scattering and ducting. The earth is not a smooth ball and intervening terrain often obscures visibility. This is especially true at low grazing angles. Sometimes there are places you just can't "see" due to weather and terrain. For very low-frequency radars, the diurnal cycling of the ionosphere can have a major impact on performance. The earth's curvature also limits visibility and surface scattering cross section.

Animals especially livestock, birds and insects can appear to be targets of interest for a short time. Multipath in which the radar signal travels two different paths to the reflector also may cause interference and target fading or false targets.

Another environmental effect is atmospheric refraction. We all have experienced this phenomenon when the setting sun apparently grows large as it nears the horizon. This magnification effect is the result of the air density gradient with altitude.

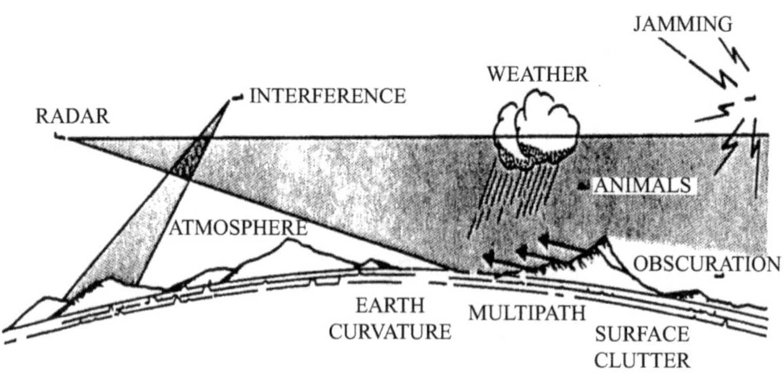


*Figure 3.14 Radar environmental limits [4]*

This causes all electromagnetic radiation to bend as it travels through the atmosphere. The graphs in Figure 3.15 show the effect for an airborne or near-space observer illuminating surface targets as a function of elevation angle and observer altitude. These curves are based on the exponential model of atmospheric refraction with altitude given in (3.1) and (3.2).

$$n(h) = 1 + 313 \cdot 10^{-6} \cdot \exp\left(-4.385 \cdot 10^{-5} \cdot h\right) \tag{3.1}$$

where $h$ is altitude above the Earth in feet.

$$\Delta\varepsilon_{el} \cong 313 \cdot 10^{-6} \cdot \cot\left(\varepsilon_{el}\right) \tag{3.2}$$

Above 100k feet range error doesn't change much and for elevation angles greater than 5° (3.2) is a good approximation for angle error. All of these estimates contain at least a 10% one sigma variability due to local temperature and humidity variations. For a spacecraft radar observer with a large antenna that is necessary to make high-resolution maps and detect GMTIs, these angle errors could be up to ½ beamwidth. Refraction error is important when a target is designated to an off-platform sensor or weapon.

There are propagation losses that are proportional to the altitude and range to the target and back. The atmospheric loss model consists of three elements, as shown in Figure 3.16. The clear air loss is exponential, with the exponent proportional to the ratio of the range, $R$, to the altitude, $h$, and loss constant $K_{CA}$. The losses due to cloud cover are path length dependent beginning at the upper altitude of the clouds, $h_{CL}$ (usually the freezing line) with loss constant $K_{CL}$. The losses due to rainfall are also path length dependent beginning at the upper altitude of rain, $h_r$

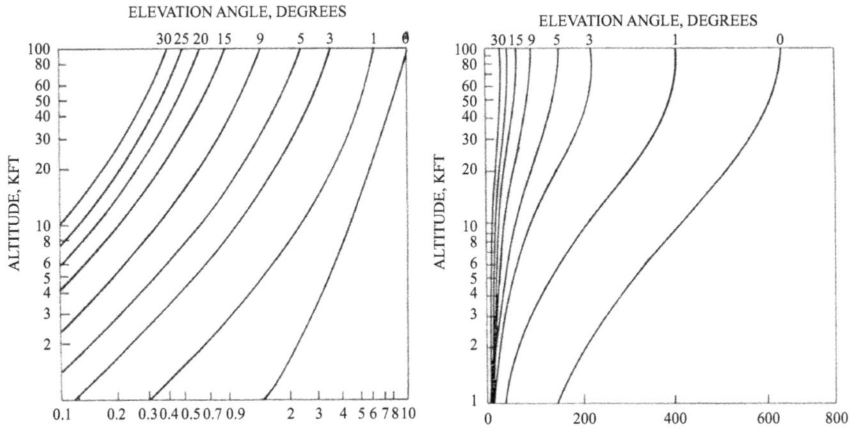


*Figure 3.15 Refraction-induced range and elevation error; adapted from [6,7]*

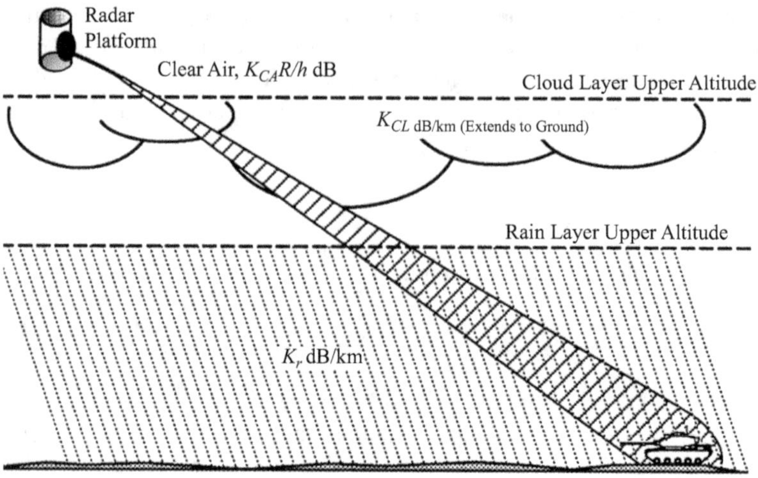


*Figure 3.16 Atmospheric loss model [8]*

with loss constant $K_r$ [6,8]. All of the Figure 3.16 elements in combination predict a one-way loss given in (3.3).

$$L = 10^{-R\cdot\left(\frac{K_{cl}\cdot h_{cl}+K_r\cdot h_r+K_a\cdot h_a}{h}\right)}$$

where: $R$ is range to target, $h$ is radar height,
$h_{cl}$ is maximum cloud height (6 km typ.),
$K_{cl}$ is cloud loss constant (0.02 typ.), $K_r$ is rain loss constant (0.2 typ.), $h_r$ is rain maximum height (3 km typ.),
$K_a$ is atmospheric loss constant (0.02 typ.),
$h_a$ is atmospheric height (20 km typ.). (3.3)

Large commercial aircraft, windmills and other large moving structures may cause interference or false targets. Almost every VHF, UHF and microwave band contains many emitters, which may interfere with surface surveillance. There also may be intentional jamming. Any surface surveillance design must cope with these as a matter of course.

## 3.3 Ground/surface return characteristics

### *3.3.1 Antenna footprints on the surface*

The earth's curvature causes the antenna footprint to be longer than the straight cone angle projection on a flat earth. Furthermore, the atmospheric density gradient with altitude causes refraction (bending) of the radar beam. As mentioned earlier, at very

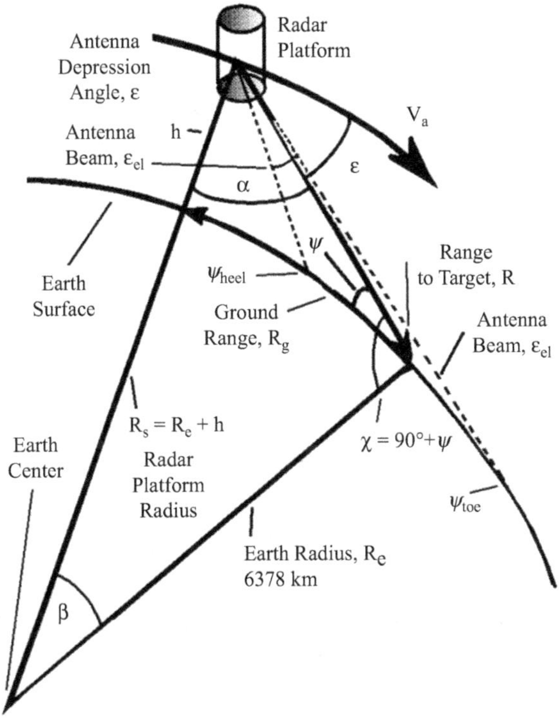


*Figure 3.17 Grazing angle geometry*

shallow grazing angles, the radar beam will "see" beyond the geometric horizon. Refraction causes the earth to appear to have a larger diameter, that is, it appears flatter. That's why the sun at sunrise and sunset appears so large relative to its size at zenith (local noon). Also, the atmosphere isn't uniform and the resulting refraction varies considerably over long path lengths. Equivalent apparent earth radius variations range between 1.25 and 1.45 at microwave frequencies. A typical value used for most analysis is that the refraction requires a 4/3 earth radius approximation.

There is some angle between the radar observer look direction and the local ground surface. This is commonly called the grazing angle. Consider the grazing angle geometry shown in Figure 3.17 for a radar platform at some altitude, $\boldsymbol{h}$, above the earth. The center of the radar beam is depressed from the local horizontal by $\boldsymbol{\varepsilon}$ and the beam intercepts the limb of the earth at some grazing angle $\psi$. The upper and lower edges of the 3 dB elevation beamwidth, $\boldsymbol{\varepsilon_{el}}$, intercept the surface at different ranges and grazing angles $\psi_{heel}$ and $\psi_{toe}$. The lower edge of the beam on the surface is often called the "heel" of the beam. Similarly, the upper edge is the "toe" of the beam. Usually, the region of interest is specified in ground coordinates. The approximate maximum width of the footprints in azimuth and elevation is given in (3.4). If the footprint is assumed elliptical, then the approximate footprint area is given in (3.4) bottom.

Also, the elevation footprint may contribute to the competing clutter in a range cell through range sidelobes. Even simple rectangular pulses when sent through the multiple filters in the radar front end will exhibit low-level-range sidelobes! Of course, all pulse compression waveforms have range sidelobes, which must be mitigated as well. For a given altitude, there is a maximum ground range, $R_{gmax}$, determined by the radar horizon as given in (3.4). Thus the radar footprint is limited by the horizon. The radar slant range is naturally a function altitude and ground range as given in (3.4). Given a desired ground range and altitude, a depression angle, $\varepsilon$, can be calculated as shown in (3.4) middle. From that and antenna beamwidth, the necessary calculations for elevation footprint can be obtained. There is a simple approximation for a flat earth footprint given in reference [8]. Sadly, on a curved earth this approximation only works for a few tens of kilometers. (Oh well, nothing is easy!). Even the equations given in (3.4) are an approximation since the earth is not actually a sphere but a geoid. The earth's diameter varies from approximately 6357 to 6378 km.

Some examples of slant range and elevation footprint versus ground range to the center of the FOR based on (3.4) are given in Figures 3.18–3.20. Figure 3.18 assumes a 0.3 m vertical aperture and a corresponding 130 milliradian elevation beamwidth at X band and a set of possible airborne operating altitudes. Note that at long ranges that as the grazing angle gets small the slant range asymptotically approaches ground range. In addition, once the toe of the beam goes over the horizon, the elevation footprint gets smaller. When desired ground range exceeds the horizon range, much of the power is being wasted if the objective is surface

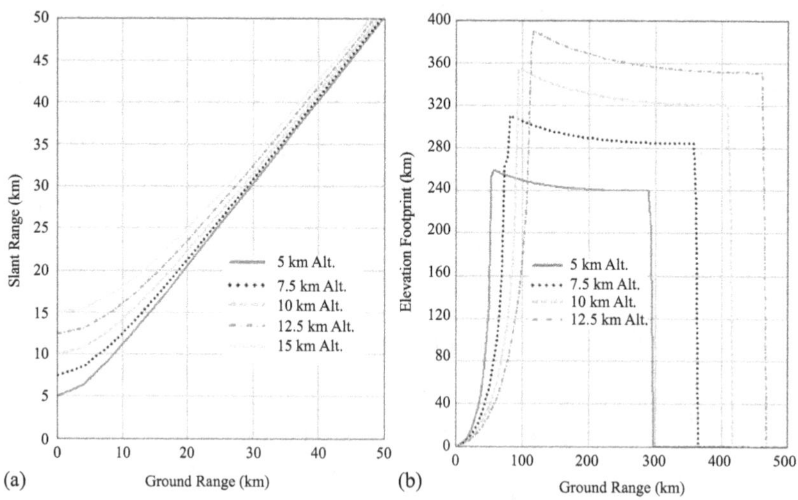


*Figure 3.18 Airborne radar altitudes: (a) slant range; (b) elevation footprint (SlantRange2a)*

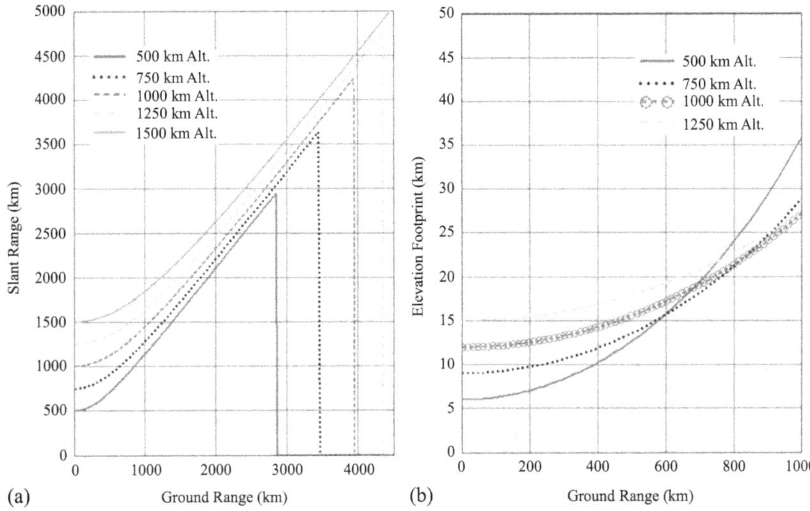


*Figure 3.19 Spaceborne radar altitudes: (a) slant range; (b) elevation footprint (SlantRange3a)*

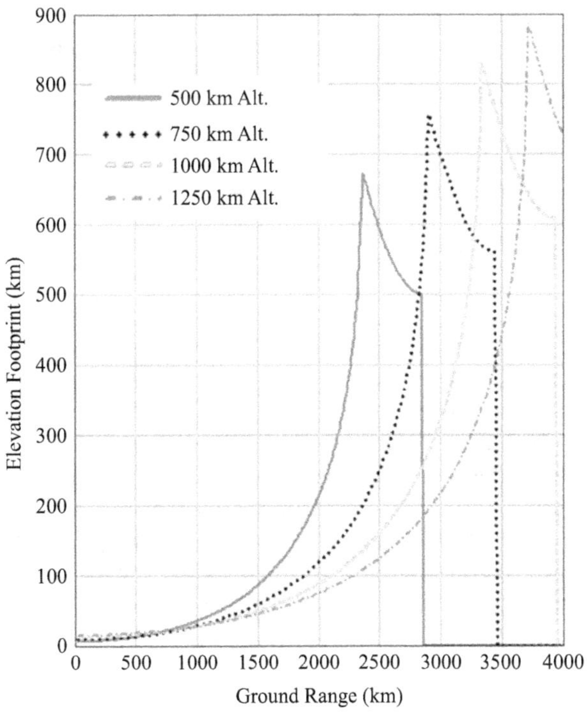


*Figure 3.20 Elevation footprint for spaceborne radar altitudes (SlantRange3a)*

target detection. Obviously, when the heel of the elevation beam reaches the horizon the surface footprint goes to zero.

Radar horizon: $R_{g\max} = R_e \cdot \operatorname{acos}\left(\frac{1}{1+h/R_e}\right)$

Slant range: $R = \sqrt{h^2 + 2\cdot\left(R_e^2 + R_e \cdot h\right)\cdot\left(1-\cos\left(\frac{R_g}{R_e}\right)\right)}$

Radar elevation angle below local horizontal:

$$\varepsilon = \operatorname{acos}\left(\frac{R_e}{R}\cdot\sin\left(\frac{R_g}{R_e}\right)\right) = \operatorname{acos}\left[\frac{R_e\cdot\sin\left(\frac{R_g}{R_e}\right)}{\left(h^2 + 2\cdot\left(R_e^2 + R_e\cdot h\right)\cdot\left(1-\cos\left(\frac{R_g}{R_e}\right)\right)\right)^{0.5}}\right]$$

Heel and toe grazing angles:

$\psi_{heel} = \operatorname{acos}\left[\left(1+\frac{h}{R_e}\right)\cdot\cos\left(\varepsilon+\frac{\varepsilon_{el}}{2}\right)\right]$ , check to insure argument $\leq 1$ and if $\psi_{toe} \to 0$

$\psi_{toe} = \operatorname{acos}\left[\left(1+\frac{h}{R_e}\right)\cdot\cos\left(\varepsilon-\frac{\varepsilon_{el}}{2}\right)\right]$ , check to insure argument $\leq 1$ and if $\psi_{toe} \to 0$

$\beta_{toe} = \varepsilon - \frac{\varepsilon_{el}}{2} - \psi_{toe}$ , $\beta_{heel} = \varepsilon + \frac{\varepsilon_{el}}{2} - \psi_{heel}$ , check to insure angles are positive

Heel and toe ground ranges:

$R_{g-toe} = \beta_{toe}\cdot R_e$ , $R_{g-heel} = \beta_{heel}\cdot R_e$ , check to insure $R_{g-heel} \leq R_{g-toe} \leq R_{g\max}$

Azimuth and elevation footprints and area: $FP_{az} \cong 2\cdot R\cdot\tan\left(\theta_{az}/2\right)$,

if $\theta_{az}$ is small, then $FP_{az} \cong R\cdot\theta_{az}$ , $FP_{el} = R_{g-toe} - R_{g-heel}$ ,

$$A_{FP} = \frac{\pi}{4}\cdot FP_{az}\cdot FP_{el} \tag{3.4}$$

The airborne altitudes are typical of platforms like Predator, Reaper, JSTARS, Astor, Global Hawk and TR-1. Similarly, Figures 3.19 and 3.20 assume a little over a 3 m vertical aperture and a corresponding 12 milliradian elevation beamwidth at X band for a set of possible operating altitudes. Note in Figure 3.19(a) that slant range stops growing if the ground range goes over the radar horizon (normally you can't see there). Also, once the heel of the elevation beam goes over the horizon the footprint disappears as in Figures 3.18(b) and 3.20. The spaceborne altitudes are typical of several proposed systems such as USAF SBR.

### *3.3.2 Ground/surface return area*

The RCS of the terrain (usually called clutter in this context) on which a ground mover travels competes with the RCS of the mover. The terrain must be eliminated in order to detect the mover. The natural question is, what are the properties of clutter? Also, how much clutter will compete with the mover? The largest contributor of clutter that competes with the mover is the mainbeam ground clutter RCS, which is made up of two factors. The first factor is the area of the ground, which is in the same range bin and antenna beam. This is basically a geometry problem and it is shown schematically in Figure 3.21. There are two cases: one in which the range bin and the beam azimuth determine the clutter area. In the first case, the competing clutter area is the product of the length of a range bin on the ground by the width of the antenna beam on the ground. In the second case, the competing clutter is determined by the azimuth and elevation beamwidth on the ground. Both are shown in (3.5).

The equivalent length of a range bin is related to its time length, $\tau$, by the velocity of light, c, and is $c\tau/2$ because radar is two way. The projection of the bin on the ground is the pulse length times the secant of the grazing angle. The width of the of the competing clutter assuming an elliptical approximation for the footprint is approximately the range, $R$, to the bin with the mover times $\sqrt{2}$ times the azimuth beamwidth, $\theta_{az}$. The ground projection may be significantly larger than the pulse length and is ultimately limited by the antenna elevation beam for very steep grazing angles as shown in the top of Figure 3.21.

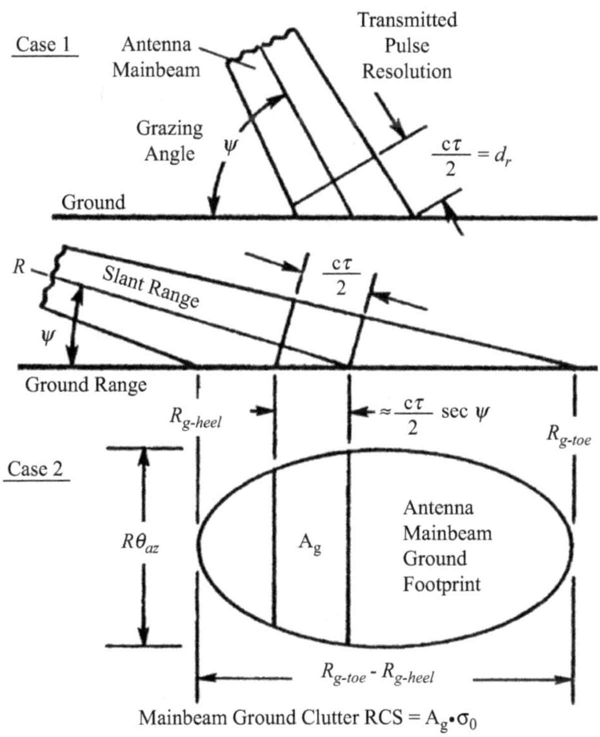


*Figure 3.21 Mainbeam ground return competing with mover [4]*

The second factor in the competing clutter is the nature of the backscatter of the ground terrain per unit area, $\sigma_0$. The area of the competing ground, $A_g$, is then multiplied by the backscatter coefficient, $\sigma_0$, to obtain the total clutter RCS.

$$\sigma_c \cong \frac{\pi \cdot R^2 \cdot \theta_{az} \cdot \varepsilon_{el} \cdot \sigma_0}{4 \cdot 2 \cdot \ln 2 \cdot \sin\psi} \text{ for } \tan\psi \geq \frac{\sqrt{2} \cdot R \cdot \varepsilon_{el}}{c \cdot \tau} \quad \text{case 1}$$

$$\sigma_c \cong \frac{c \cdot \tau \cdot R \cdot \theta_{az} \cdot \sigma_0}{2\sqrt{2} \cdot \cos\psi} \text{ for } \tan\psi < \frac{\sqrt{2} \cdot R \cdot \varepsilon_{el}}{c \cdot \tau} \quad \text{case 2} \tag{3.5.1}$$

where: $R$ is range, $\frac{c\tau}{2} = d_r$ range resolution,

$\sigma_0$ is backscatter coefficient, $\theta_{az}$ is azimuth beamwidth,

$\varepsilon_{el}$ is elevation beamwidth, $\psi$ is grazing angle,

$\sqrt{2}$ and $2 \cdot \ln 2$ are ellipitical pattern approximations,

$\tau$ is equivalent pulse width, c is velocity of light. (3.5.2)

### 3.3.3 Ground/surface backscatter

Table 3.1 gives land clutter reflectivity data for very shallow grazing angles between 0 and 1.5° for a variety of frequencies from UHF to Ka band.

Similarly, Table 3.2 shows land clutter for 5°–10° grazing angle. Parenthetically, ground clutter normalized RCS, $\sigma_0$, may be grazing angle dependent. Tabulated data for various types of terrain are shown for the median and the upper 1 SD point for both horizontal and vertical polarization, where available. These tables are the composite of multiple sources and the sources suggest that errors of 5 dB might be typical.

*Table 3.1 Land clutter reflectivity, 0°–1.5° grazing angle [8]*

| **Terrain type** | **Reflectivity, dB below 1 m²/m² @ band, GHz*** | | | | | | | | | | | | | |
|---|---|---|---|---|---|---|---|---|---|---|---|---|---|---|
| | ***UHF 0.5–1*** | | ***L 1–2*** | | ***S 2–4*** | | ***C 4–8*** | | ***X 8–12*** | | ***Ku 12–18*** | | ***Ka 31–36*** | |
| | $\sigma_0$ | $\sigma_{84}$ | $\sigma_0$ | $\sigma_{84}$ | $\sigma_0$ | $\sigma_{84}$ | $\sigma_0$ | $\sigma_{84}$ | $\sigma_0$ | $\sigma_{84}$ | $\sigma_0$ | $\sigma_{84}$ | $\sigma_0$ | $\sigma_{84}$ |
| Desert | 30 | | 45 | 35 | 46 | 31 | 40 | | 40 | 32 | | | | |
| Farmland | 28 | | 32V | 26 | 34 | 24 | 33 | 25 | 33 | 20 | 23 | 10 | 18V | 10V |
| Open Woods | | | 34H | | 33 | 26 | | | 30 | 22 | | | | |
| Wooded Hills | 24 | 18 | 28 | 20 | 28 | 15 | 27 | 14 | 26 | 13 | 20V | 17 | 13V | 8V |
| Residential | | | | | 35 | 26 | 35 | 26 | 30 | 24 | | | | |
| Cities | 22 | | 25 | 15 | 23 | | 21 | | 20 | >7 | 20V | 10 | | |

*Average of both polarizations except where noted.
$\sigma_0$ = median or mean backscatter coefficient in decibels below 1 $m^2/m^2$.
$\sigma_{84}$ = reflectivity that 84% of the cells are below [8].

The radar sensor designer is forced to minimize false alarms, and as a result, these clutter uncertainties lead to higher thresholds. Normalized reflectivity coefficients in meters per square meter are usually fairly low, especially for low grazing angles, but usually the number of square meters competing with a mover is large; hence, operating in a high clutter environment is a challenge. Operating at higher frequencies over areas that contain cities or heavily wooded terrain may degrade performance due to enhanced multipath and competing clutter.

Figure 3.22 shows a clutter summary as a function of depression angle at X band for a wide range of terrain types. As the grazing or depression angle gets more

*Table 3.2 Land clutter reflectivity, 5°–10° grazing angle [8]*

| ***Terrain type*** | ***Reflectivity, dB below 1 m²/m² @ band, GHz**** | | | | | | | | | | | | | |
|---|---|---|---|---|---|---|---|---|---|---|---|---|---|---|
| | ***UHF 0.5–1*** | | ***L 1–2*** | | ***S 2–4*** | | ***C 4–8*** | | ***X 8–12*** | | ***Ku 12–18*** | | ***Ka 31–36*** | |
| | $\sigma_0$ | $\sigma_{84}$ | $\sigma_0$ | $\sigma_{84}$ | $\sigma_0$ | $\sigma_{84}$ | $\sigma_0$ | $\sigma_{84}$ | $\sigma_0$ | $\sigma_{84}$ | $\sigma_0$ | $\sigma_{84}$ | $\sigma_0$ | $\sigma_{84}$ |
| Desert | 40 | 34 | 39 | | 36 | | 33 | | 30 | | 18V | 13V | 25 | |
| Farmland | 36 | 30 | 30 | 26 | 28 | | 26 | | 26 | | 22 | 11V | 18V | 10V |
| Open Woods | 23 | 20 | 23 | | 33 | 26 | | | 30 | 22 | 22 | 15V | 20 | |
| Wooded Hills | 22 | 18 | 26 | 20 | 24 | | 23 | | 23 | | 20 | 14V | 13V | 8V |
| Residential | 23 | 17 | | | 23 | | 35 | 26 | 30 | 24 | 18V | 11 | | |
| Cities | 9 | | 15 | | 18 | | 18 | | 16 | | 9V | 3V | −15◇ | |

*Average of both polarizations except where noted.
$\sigma_0$ = median or mean backscatter coefficient in decibels below 1 m²/m².
$\sigma_{84}$ = reflectivity that 84% of the cells are below.
◇This entry is above 1 m² [8].

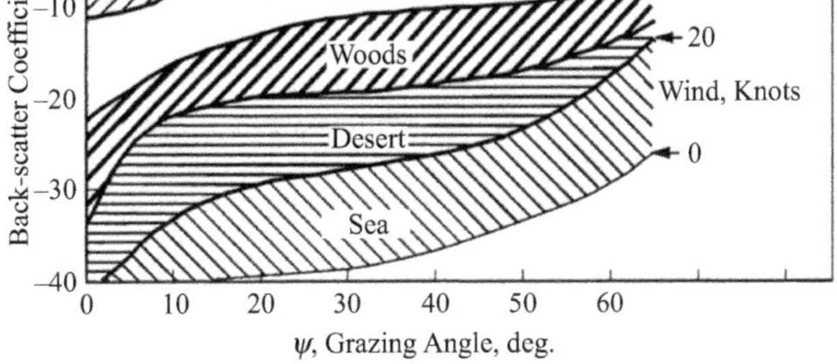


*Figure 3.22 Summary of X-band clutter reflection data; adapted from [8]*

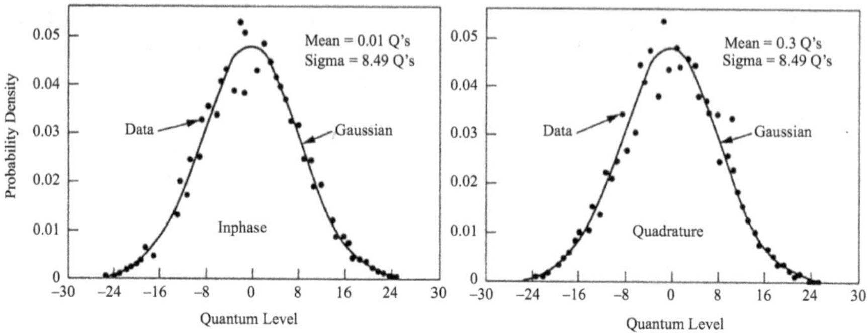


*Figure 3.23 Rural terrain raw I-Q data histograms [5]*

steep, clutter backscatter typically rises. There are some notable exceptions for this, such as commercial and industrial areas that may contain high-rise buildings and many vertical features with large RCS. Furthermore, commercial and industrial areas typically have dramatically higher cross sections than areas that are largely devoid of cultural features. In addition, backscatter from clutter is conservative: most of the energy that a radar places on the ground either intentionally or unintentionally is reflected, a fraction back in the direction of the radar observer, but much scattered in other directions. This multidirectional scatter can be a disadvantage in commercial, industrial and heavy vegetation areas because it confuses angle-of-arrival discrimination such as adaptive monopulse estimation of angles, of polarization, of interference, etc., using some form of STAP techniques [9]. Angle of arrival differences of 10° with only 5 dB of attenuation over path lengths of 50 miles are common occurrences. In areas with many cultural features, the scattering in other directions may be so strong as to punch through the sidelobes of the antenna and dramatically degrade dynamic range and signal discrimination characteristics. Over calm sea water or smooth desert sand, however, clutter is not a problem.

Another attribute of clutter that is an important property is the probability density function of RCS with respect to size. For example, Ku-band data taken over rural terrain at medium resolution (40 × 40 ft cell) can be approximated by a normal probability density function, as shown in Figure 3.23. The data shown is prior to pulse compression out of 6 bit A/D converters. A little over 20,000 sample points were used with a post-pulse compression dynamic range of 60 dB and a median reflectivity of −24.9 dBsm. The input is almost white noise but the output is most decidedly not as will be shown in the figures that follow.

Some representative statistics with and without smoothing for populated areas are given in Figures 3.24–3.27. This data is from the USAF FLAMR program, which was declassified in 1987 and 1988 [5]. One clutter feature important to persistent surveillance platforms is bright discretes. Figure 3.24 shows a summary collection of data for bright discretes in terms of density per square nautical mile versus equivalent RCS in square meters.

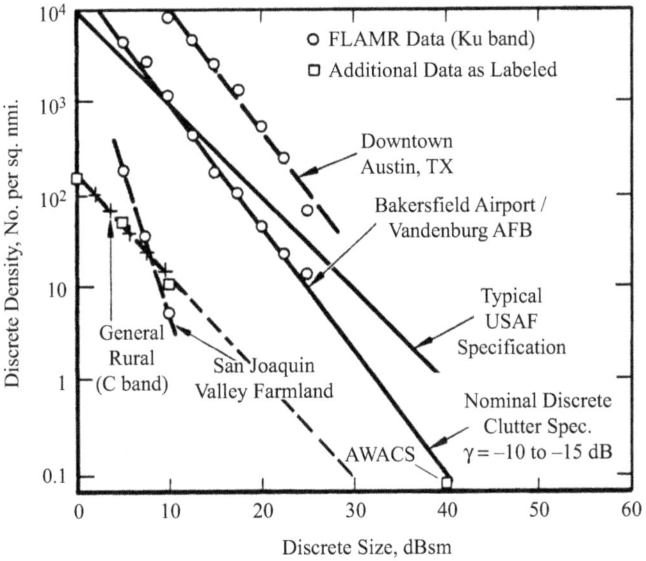


*Figure 3.24 Density of bright discretes [8]*

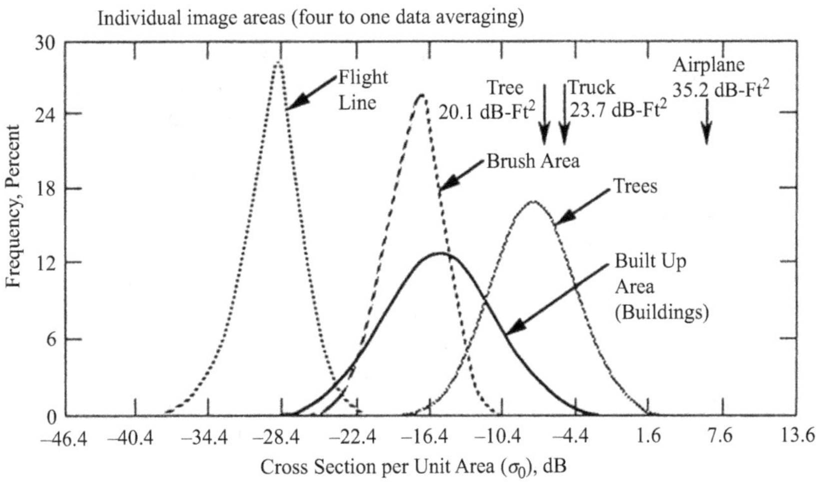


*Figure 3.25 Surface return histograms for Vandenberg AFB vicinity [5]*

There are several important features to the clutter in populated areas. The mean and the median clutter reflectivity can differ by as much as 10:1 (very much non-Gaussian!). Commercial areas have large statistical tails and hence a significant probability of very large cross section. This means that there is a small but finite

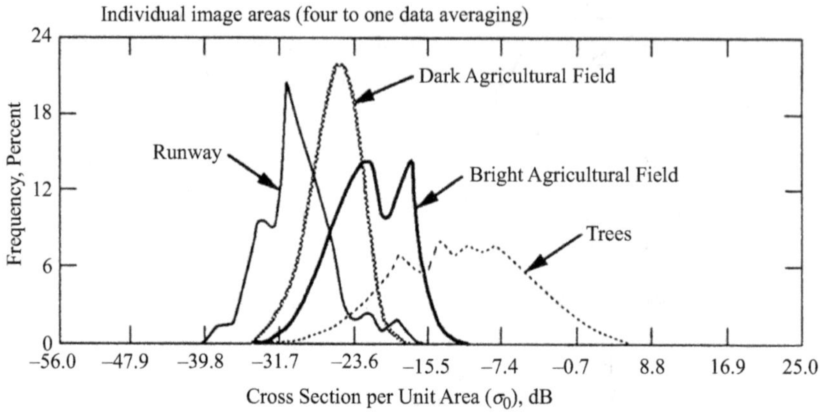


*Figure 3.26 Clutter histograms for Point Mugu, CA, vicinity [5]*

probability that there will be some very large RCS scatterers; over any reasonable observation space. Similarly, an RCS probability density function over heavily populated terrain (e.g., Los Angeles) has a greatly distributed density function, which has an even larger probability of a few extremely large scatters in any observation space. The probability of a $10^5$ $m^2$ scatterer can be as high as 0.001 in a city. There is a small but finite probability that a "bright discrete" scatterer will be in a range cell. A USAF specification used on many programs expects a bright discrete of $10^5$ $m^2$ in every 10 sq. nmi. and similarly a bright discrete of $10^6$ $m^2$ in every 100 sq. nmi., etc. The sidelobes of these scatterers can be in hundreds of cells, which can mask a small target for quite an area over densely populated areas.

The measurements of Ku-band reflectivity made with a SAR have high enough resolution so that individual terrain features can be separated, compared to ground truth and categorized as to type of clutter as shown in Figures 3.24–3.27.

These higher-resolution clutter histograms typically use thousands of data points to estimate the underlying statistics. Note that in Figure 3.27(a), when all data is collapsed, clutter statistics seem to be Rayleigh power distributed implying an underlying normal or Gaussian distributed amplitude variation. This is what one would expect based on the central limit theorem and very large sample sizes. Notwithstanding the large sample sizes and averaging, the four figures still show significant variability for terrain that does not seem that different in the SAR imagery. Therefore, when examining the fine structure of mixed terrain surface returns, the statistical distribution more closely resembles a log-normal distribution. The grazing angles in these cases were typically 5°–10°. Clutter has been measured by many researchers since the 1930s. There are many approximate equations that characterize the main aspects of clutter, which will be given in the following paragraphs but they are by no means the final word on the subject and the references throughout this document give many more examples.

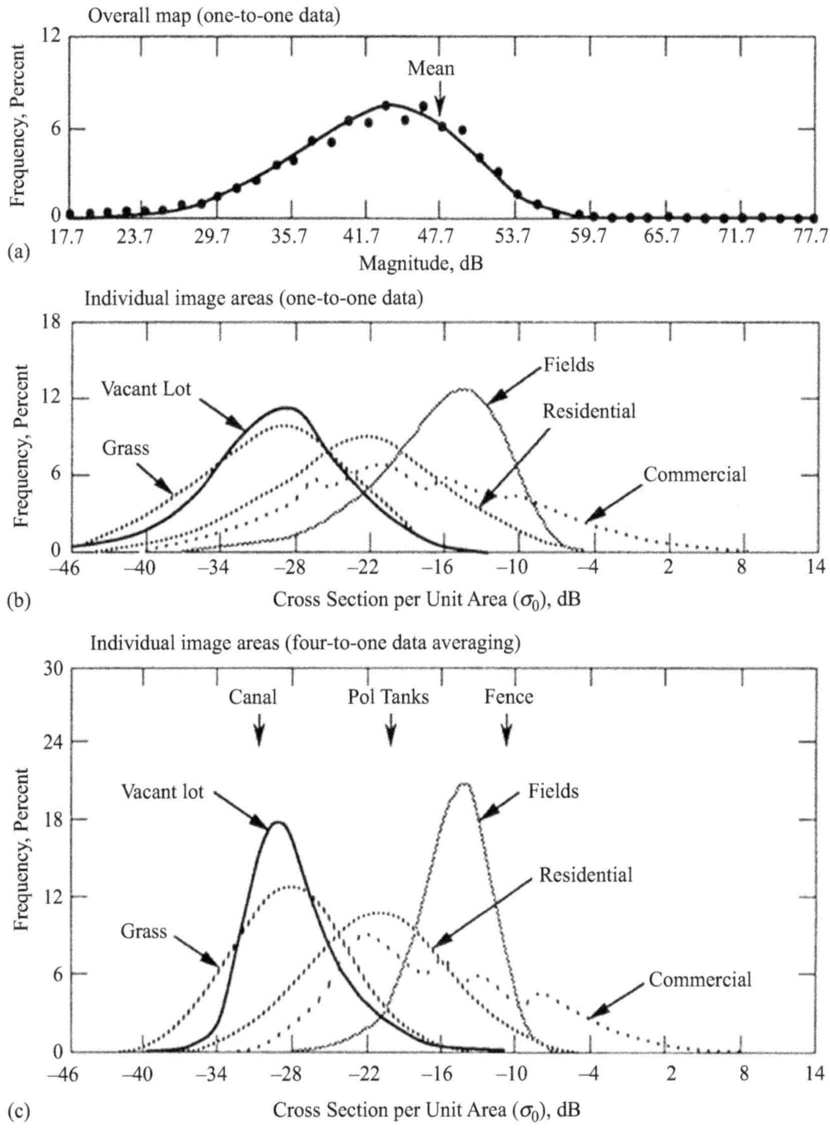


*Figure 3.27 Surface return histograms for Bakersfield, CA, airport vicinity [5]*

Because of the wide variation in terrain cross section and the nature of radar measurements, radar targets and clutter are typically measured in decibels as shown in Figures 3.8–3.10, 3.12–3.13, and 3.22–3.27. Since the random value of the surface terrain RCS can be approximated by a log-normal distribution, often it is both convenient and accurate to model clutter for GMTI as log-normal.

The log-normal probability density function for unit clutter cross section, $\sigma_0$, in linear units is given in (3.6).

$$P(\sigma_0) = \frac{1}{\sqrt{2 \cdot \pi \cdot s^2} \cdot \sigma_0} \exp\left( \frac{-\left( \ln\left( \frac{\sigma_0}{\sigma_{med}} \right) \right)^2}{2 \cdot s^2} \right) \quad \text{for } \sigma_0 \geq 0$$

$P(\sigma_0) = 0$ otherwise

where: $\sigma_{med}$ is the median cross section in linear units,

$\sigma_{mean}$ is the mean cross section in linear units,

which is always larger than the median and $s = \sqrt{2 \cdot \ln\left( \frac{\sigma_{mean}}{\sigma_{med}} \right)}$ (3.6)

This density function can be fitted to measured data by adjusting the mean and median cross-sectional coefficients. An example curve fit to the trees in Figure 3.25 is given in Figure 3.28. Note that $\sigma_0$ has been converted into dB to match the prior experimental data.

Land and sea clutter are proportional to illuminated area. They are strongly grazing angle dependent. Often grazing angle and depression angle are close

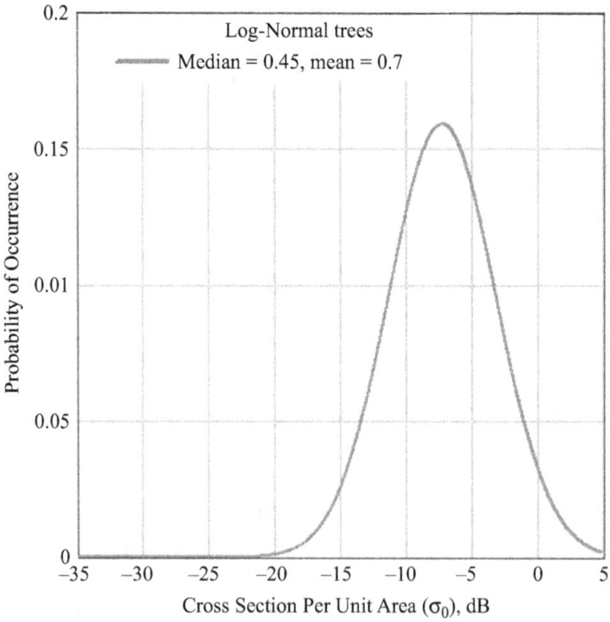


*Figure 3.28 Log-normal clutter model fit to trees of Figure 3.25 (LogNormal1)*

enough in value so that they can be used interchangeably but for extended clutter purposes grazing angle must take into account the earth's curvature. The grazing angle independent of atmospheric refraction is given in the first equation in (3.7). If the radar LOS is entirely in the atmosphere, then the 4/3 earth approximation for atmospheric refraction is often included in the grazing angle calculation. Under those circumstances the second equation in (3.7) is accurate to better than 0.1%.

$$\psi = \sin^{-1}\left(\frac{h_r}{R}\left(1+\frac{h_r}{2\cdot R_e}\right)-\frac{R}{2\cdot R_e}\right)$$

If line-of-sight entirely in atmosphere, then:

$$\psi \cong \sin^{-1}\left(\frac{h_r}{R}-\frac{R}{2\cdot 4/3\cdot R_e}\right) \tag{3.7}$$

where $h_r$ is the height of the radar, $R$ is the slant range to the target and $R_e$ is the average earth radius with a typical value of 6,370,880 m [6].

The grazing angle dependency of terrain cross section has been modeled by many different researchers. There are three simple models for the grazing angle dependence of reflectivity to be described in what follows. The emphasis in the equations is simple first-order approximations. The simplest is the constant $\gamma$ model [10]. A second more accurate model is the exponential model. It requires the selection of the minimum expected backscatter, $\sigma_{\text{min}}$. A third model called the Muhleman planetary model is probably the most accurate [11]. It is a "platelet" or facet model of the type used to predict first-order RCS. It assumes that the platelets are randomly oriented. Muhleman is the only one that matches the underlying physics of scattering [12]. It requires the selection of two parameters: a scale factor $K_0$ and a roughness factor $\alpha$. All three models are given in (3.8) in dB $m^2/m^2$ units. Then the unit area RCS of ground clutter is one of the following:

$$\text{Constant } \gamma \text{ Model: } \sigma_0 = 10\cdot\log(\gamma)+10\cdot\log(\sin\psi) \quad \text{(dB)}$$

$$\text{Exponential Model: } \sigma_0 = \sigma_{\min} + 8\cdot\psi - 4\cdot\log\left|\psi - \pi/2\right| \quad \text{(dB)}$$

$$\text{Muhleman Model: } \sigma_0 = 10\cdot\log\left(\frac{\sin\psi}{(\cos\psi+\alpha\cdot\sin\psi)^3}\right)+10\cdot\log(K_0) \quad \text{(dB)} \tag{3.8}$$

where $\gamma$ is a backscatter coefficient that is reasonably constant for a given frequency. When designing for GMTI clutter rejection, $\gamma$ is usually chosen high with a typical $\sigma_{84}$ value of 3. However, when designing for SAR mapping, $\gamma$ is usually chosen low with a value of 0.03–0.1 because interesting map features may be small.

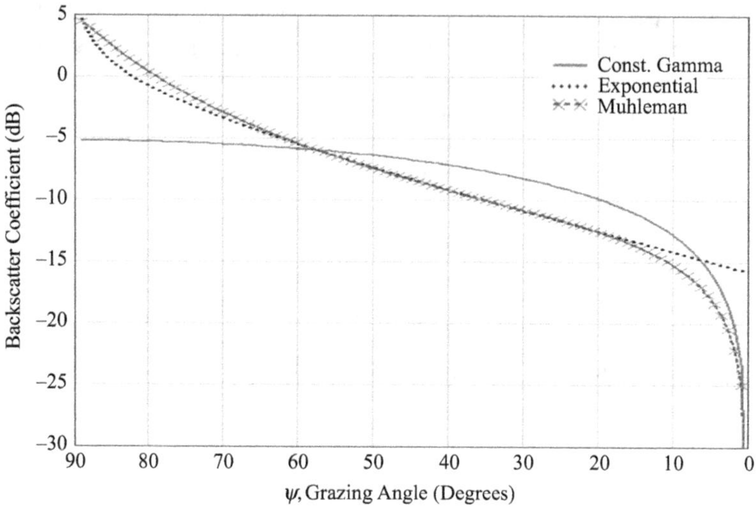


*Figure 3.29 Comparison of land clutter models (CluttervsGrazing1)*

When using the exponential model, the minimum backscatter, $\sigma_{min}$, in dB for a given frequency must be chosen, −15 dB is typical for forested, brushy terrain. However as can be seen from Tables 3.1 and 3.2 as well as Figures 3.22–3.27, $\sigma_{min}$ varies significantly with both terrain and frequency. When using the Muhleman model, the factor, $K_0$, can be thought of as the backscatter coefficient at mid grazing angles, that is, 45° and the roughness factor, $\alpha$, which might be between 0.2 and 0.4 dominates very steep grazing angles, that is, near 90°.

A comparison of the constant $\gamma$ model, exponential model and Muhleman model is given in Figure 3.29. In this graph, $\gamma = 0.3$, $K_0 = 0.2$, $\sigma_{min} = -15$ dB and $\alpha = 0.4$, which might be used for typical GMTI analysis. The constant $\gamma$ model provides a simple approximation for mid-range and low grazing angles. The exponential model provides a somewhat more complex estimate of grazing angle dependence, which is more accurate at medium and high angles. The exponential model fits recent data fairly accurately but provides no insight into the physics. The Muhleman model fits both space-based and land clutter data very well.

Sea surface return is strongly dependent on grazing angle, RMS wave height, ocean currents, wind direction and operating wavelength. Wave height is related to surface winds, sea bottom features and location on the earth. The simplest model is again a platelet or facet model based on geometrical optics and Rayleigh rough surface scattering. It is a function of roughness (wave height) relative to wavelength and the foreshortening of facets caused by grazing angle. Since the model is based on geometrical optics, it predicts nothing about polarization sensitive scattering. Although polarization sensitivity of sea clutter data is anomalous, historically overland radars have used vertical polarization and overwater radars have used horizontal polarization. Typically designers said,

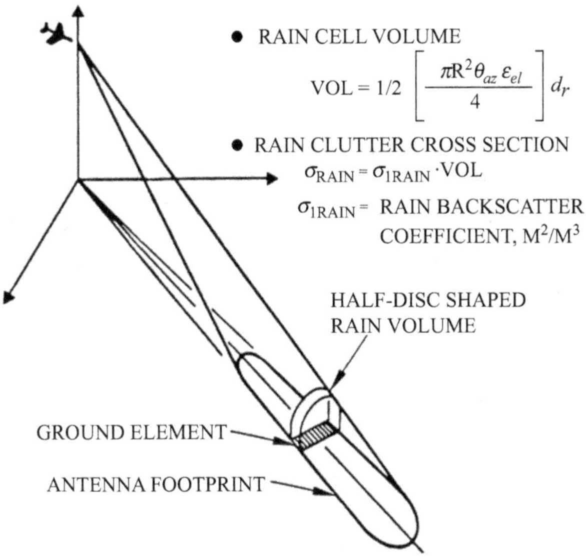


*Figure 3.30 Rain cell volume and RCS [4]*

"It might help who knows?" Equation (3.9) is one of many approximations to unit area sea clutter.

$$\sigma_{0SEA} = \exp\left[-8\cdot\pi^2\cdot\left(\frac{\sigma_h\cdot\psi}{\lambda}\right)^2\right] \tag{3.9}$$

where $\sigma_h$ is the RMS wave height and typically 0.107 m for sea state 1, 0.69 m for sea state 4, 1.07 m for sea state 5, and for the Perfect Storm, sea state 8, 6 m. The author has been in sea state 5 in a 27′ sailboat; all you can do is try not to be knocked down or capsized.

Similarly, rain clutter is proportional to illuminated volume. Equation (3.10) is one approximation to unit volume rain clutter.

$$\sigma_{1RAIN} \cong 7.03\cdot 10^{-12}\cdot f^4\cdot r^{1.6} \tag{3.10}$$

where $f$ is operating frequency in GHz and $r$ is rainfall rate in millimeters per hour.

Equation (3.10) must be coupled with the rain scattering volume to determine the total rain cross section competing with targets whether they are target movers or SAR maps. Figure 3.30 shows the rain volume that must be multiplied by (3.10). The final rain RCS is given in (3.11).

$$\sigma_{RAIN} = \sigma_{1rain}\cdot\frac{\pi\cdot R^2\cdot\theta_{az}\cdot\varepsilon_{el}\cdot d_r}{8}$$

Where: $d_r$ is the range resolution in meters, $R$ is slant range in meters,

$\theta_{az}$ is azimuth beamwidth in radians, $\varepsilon_{el}$ is elevation beamwidth in radians.

(3.11)

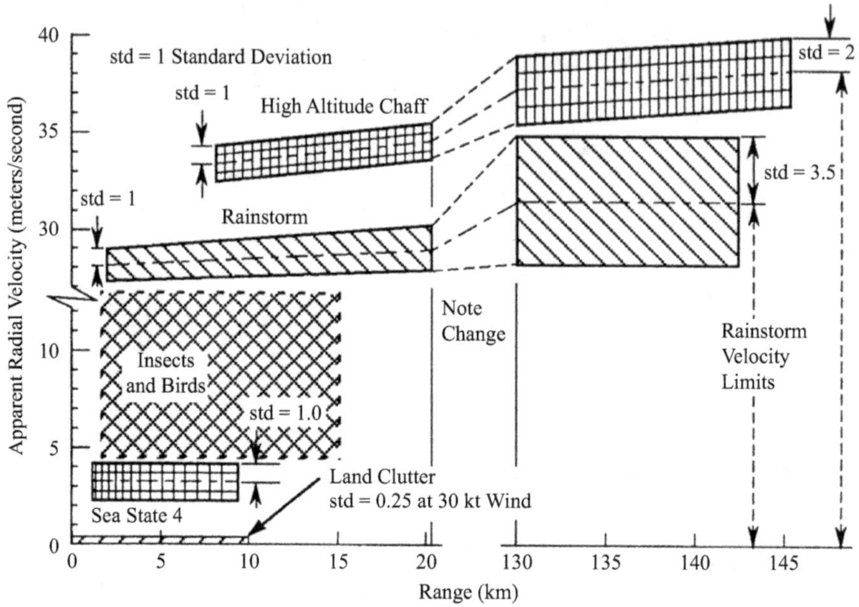


*Figure 3.31 Surface return velocity profile [8]; adapted from Nathanson [6]*

### *3.3.4 Ground/surface return Doppler characteristics*

Yet another aspect of clutter is its velocity distribution as a function of range. A clutter velocity profile for various types of backscatter is shown in Figure 3.31. This figure shows apparent radial velocity as a function of range. Notice that as range increases, the apparent radial velocity of chaff and rain gets larger and is determined by the velocity of either the storm cell or the chaff cloud. A similar condition exists at shorter range for sea and land clutter. There are many land clutter items such as rooftop ventilators and fans, which have apparent velocities similar to all land vehicles. The only way to sort these scatterers out is to track them long enough to determine that their location is unchanged along with the detailed Doppler signatures unique to an object class. In addition, there is a region of velocities occupied by insects and birds, which fills in additional parts of the range–velocity space at lower altitudes. Insects and birds can be especially problematic for radars attempting to detect and track dismounts.

As suggested by Figure 3.31, at low altitudes, there is significant natural moving clutter arising from insects and birds. In addition, there is substantial manmade moving but “stationary” clutter, most notably ventilators, fans and air conditioners on buildings. There are also synchronously moving targets from fixed locations, for example, cars at stoplights that seem to move at every other scan. Lastly there are surges of movers at quitting time and on freeways of dramatic proportions, for example, most segments of the San Diego freeway carry 15,000–20,000 cars per hour.

*Table 3.3 Moving animal clutter at low altitudes [8]*

| Type of bird or insect | Flock size/cell | Individual RCS (dBsm) | | | Typical net RCS/cell (dBsm) |
|---|---|---|---|---|---|
| | | UHF | S-band | X-band | |
| Sparrow | 10 | −56 | −28 | −38 | −18 to −46 |
| Pigeon | 100 | −30 | −21 | −28 | −1 to −10 |
| Duck | 10 | −12 | −30 | −21 | −2 to −20 |
| Grackle | 1000 | −43 | −26 | −28 | +4 to −13 |
| Hawkmoth | 10 | −54 | −30 | −18 | −8 to −44 |
| Honeybee | 1000 | −52 | −37 | −28 | +2 to −22 |
| Dragonfly | 10 | −52 | −44 | −30 | −20 to −42 |

Table 3.3 tabulates the RCS, flock size and aspect for three different radar bands for several common birds and insects. As can be seen from Table 3.3, even though the individual insect or bird cross sections are extremely low, their aggregate RCS can be significant depending on the quantity in a radar cell. Grackles, for instance, flock in large numbers at night to feed on insects and it is not uncommon for 1000 grackles to be in a single radar cell. A flock of grackles can have a 1 $m^2$ RCS in a single radar cell and have internal Doppler. These conditions have been observed regularly on radar programs such as the US Army/Marine AN/TPQ-36 and 37. The old saying "birds of a feather flock together" comes into play in fact, forcing higher false alarms, more processing or lower sensitivity. Furthermore, wing speed as well as velocity can be high enough, especially in certain kinds of hummingbirds and geese, that these birds can easily be confused for small aircraft or incoming projectiles. These sources of clutter can be rejected, of course, but a major increase in signal processing is required. The conclusions from Figure 3.31 and Table 3.3 are that there is significant moving clutter at lower altitudes, which results in a major increase in processing.

Land clutter has a velocity spread that has been extensively analyzed and modeled. The internal motion of brush and trees places a lower limit on both subclutter visibility and SAR imaging. The accepted model for velocity distribution is exponential as given in (3.12).

$$\sigma_{0v} = \frac{1}{\sqrt{2}\cdot\sigma_v}\cdot\exp\left(\frac{\sqrt{2}\cdot|v|}{\sigma_v}\right)$$

Where: $\sigma_v$ is the velocity deviation and

$v$ is the velocity variable. (3.12)

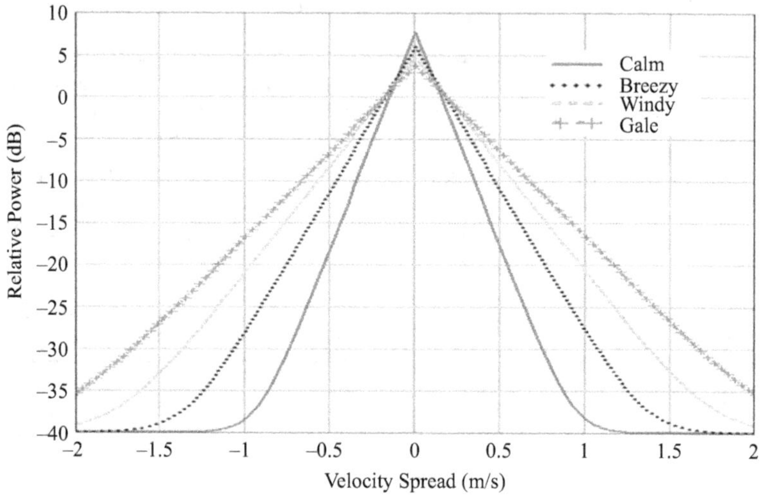


*Figure 3.32 Land clutter velocity spread (ClutterSpectralSpread)*

A plot of this model for standard deviation, $\sigma_V$, corresponding to wind conditions ranging from calm to gale force (0.12 to 0.3 m/s) is shown in Figure 3.32.

### *3.3.5 Rain Doppler spread*

Another issue especially for aircraft and spacecraft SAR imaging is rain, fog and dust. Usually the particle size for all of these obscurants is much less than a wavelength but they still cause Rayleigh backscattering and attenuation. The author has been in tests in which each of these obscurants reduced detection range to ½ of the expected in clear air at X, Ku band and above. Of course, obscuration under these conditions is much worse at IR, visible and UV wavelengths. In the Mideast at some times of the year, one can only see 100′ in a combination of haze and dust. One can't tell where the sun is even though it is daylight.

One rainfall Doppler model is given in (3.13) [6]. It consists of the root sum square of four terms: a term associated with wind shear in elevation and distance, a term associated with turbulence, a term associated with velocity spread across the azimuth beamwidth and a term associated with apparent distributions in falling droplet velocities with grazing angle. In most cases wind shear is the dominant term. The values for each of these terms have been approximated empirically.

$$\sigma_{VR}^2 = \sigma_{SHEAR}^2 + \sigma_{TURB}^2 + \sigma_{BEAM}^2 + \sigma_{FALL}^2 \ \ \mathrm{m^2/s^2}$$

Where: $\sigma_{SHEAR} \cong 0.84 \cdot R \cdot \varepsilon_{el}$, $\sigma_{TURB} \approx 1.0$; $\sigma_{BEAM} \cong 0.42 \cdot \Delta\overline{V} \cdot \theta_{az}$; $\sigma_{FALL} \cong \sin\psi$;
$R$ is slant range to scattering volume in kilometers, $\psi$ is grazing angle,
$\Delta\overline{V}$ is line of sight wind speed, $\theta_{az}$ is azimuth beamwidth [13]. (3.13)

**Example 3.2** For example, suppose the azimuth beamwidth is 50 mrad, platform velocity is 200 m/s, wavelength is 0.03 m, angle off the velocity vector is 45°, range is 120 km, elevation beamwidth is 105 mrad, grazing angle is 105 mrad and LOS wind velocity is 5 m/s. Then the Doppler spread for rain would be 832 Hz as shown in Figure 3.33.

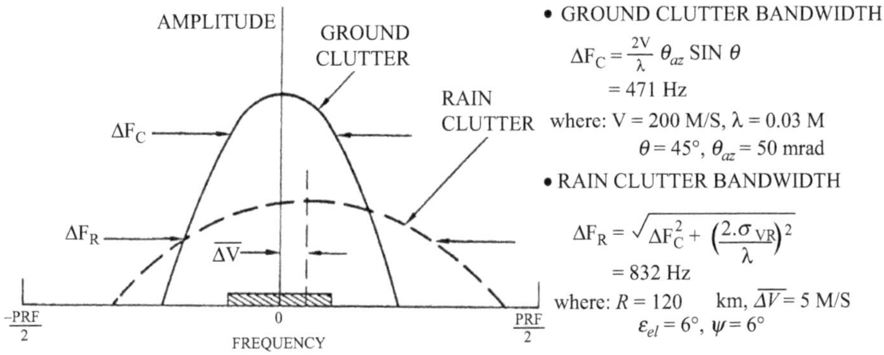


*Figure 3.33 Rain spectral situation*

Depending on the targets of interest, the rain will interfere with only a part of the spectrum. Nonetheless, rain makes a dramatic reduction in detection range. This is especially true in South East Asia.

## 3.4 Ground mover Doppler characteristics

A surface vehicle exhibits all the stationary RCS attributes shown in Section 3.1 when moving but also exhibits a range of complex Doppler features. The Doppler effect causes observed frequencies to be higher when a vehicle is approaching and lower when a vehicle is receding from the observer. This should allow most moving targets to be discriminated from the background, which can be motion compensated to appear stationary relative to vehicles and humans moving over the ground. As mentioned earlier, surface moving targets predominantly move in the local horizontal plane. Thus the projected velocity toward an observer will depend both on vehicle heading and the observer's grazing angle. For shallow grazing angles, the apparent velocity and the surface velocity are almost the same. A surface target can be moving in any direction relative to the observer. If one assumes that all directions are equally probable and that there are a range of velocities of interest such as 0.5–130 kmph, then (3.14) provide the probable velocity density and velocity distribution as a function of grazing angle.

$$P_2(v_r,\psi)=\frac{2\cdot\left(\operatorname{acosh}\left(\frac{v_{max}\cdot\cos(\psi)}{v_r}\right)-\operatorname{acosh}\left(\frac{v_{min}}{v_r}\right)\right)}{\pi\cdot\left(v_{max}\cdot\cos(\psi)-v_{min}\right)}\quad \text{for } 0<v_r<v_{min}$$

$$P_3(v_r,\psi)=\frac{2\cdot\operatorname{acosh}\left(\frac{v_{max}\cdot\cos(\psi)}{v_r}\right)}{\pi\cdot\left(v_{max}\cdot\cos(\psi)-v_{min}\right)}\quad \text{for } v_{min}<v_r<v_{max}\cdot\cos(\psi)$$

$$P_{cum}(\psi)=\int_0^{v_{min}}P_2(v_r,\psi)\,dv_r+\int_{v_{min}}^{v_{max}}P_3(v_r,\psi)\,dv_r \tag{3.14.1}$$

where: $v_{max}$ is the maximum expected GMT velocity,
$v_{min}$ is the minimum detectable GMT velocity,
and $\psi$ is the grazing angle. (3.14.2)

Figure 3.34 shows the cumulative probability of occurrence as a function of velocity for velocities 0.5 to 35 m/s. The 10%–90% region of velocities lies between a low of about 1.25 m/s and a high of 12.5–26 m/s depending on grazing angle as marked in the figure. The lower grazing angles would be characteristic of aircraft sensors and the higher grazing angles typical of spaceborne sensors.

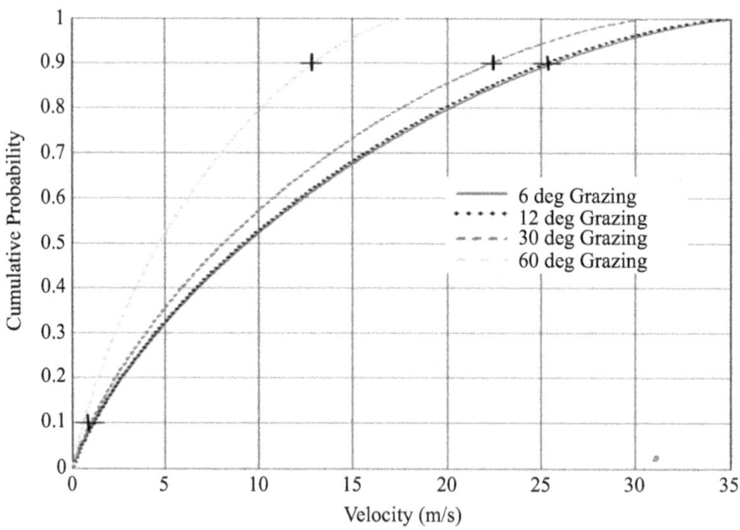


*Figure 3.34 GMT apparent velocity distribution with grazing angle (GMTIDopplers1)*

A significant fraction of GMTs could be competing with clutter in the near sidelobe or mainlobe region especially from space.

Unfortunately, since there are multiple paths for the reflections from the mover, objects on the mover appear to be traveling at different speeds and discrimination is not simple. There also are a range of Doppler frequencies, which are reflected from the ground over angles inside the antenna mainbeam that compete with the Doppler frequencies returned from a vehicle.

Whenever an observer views an object traveling in a straight line at any angle except straight at the observer, the motion can be resolved into a translation to-from the observer to the object's apparent center and a rotation about the object's apparent center. If the body is "rigid," each scatterer is moving along a circle that projects as an ellipse in the observer's direction. The apparent rotation gives rise to scatterers along the object appearing to move at different speeds and accelerations. The short-term apparent differences allow imaging by inverse SAR for such things as object recognition, space object imaging, ship imaging, etc. This has traditionally been called geometric acceleration in the air-to-air community. Parts of the object do appear to be accelerating. Because real apparent acceleration is occurring, often multiple subarray FFTs with inter-subarray autofocus are used to improve ISAR images [14]. The subarray time periods are typically 10–500 ms. It is possible to image synchronous satellites over spans of days using satellite station keeping movement to create ISAR images [15].

In addition, since most movers of interest have extent (length, width, height), there is geometrical acceleration. Geometrical acceleration is that property of rectilinear motion for extended objects, which causes some features of the object to appear to be traveling toward the observer at different speeds for short periods of observation again 10–500 ms.

It is the apparent rotation property of rectilinear motion that allows SARs to work. For example, a GMTI target may have an apparent velocity spread of approximately 0.16 m/s across a 20-ft-long vehicle traveling normal to the LOS at 15 km from an observer with 200 m/s relative velocity. There are many other internal motions in aircraft, surface vehicles, ships and people, which are not geometrical acceleration but subsystem velocities and accelerations, for example, jet engine compressor blades, surface vibration, swinging arms and legs. The remainder of the velocity spread comes from internal motion, for example, moving wheels, tracks, cooling fans, swinging arms, etc. This internal motion is velocity dependent, for example, maximum wheel and track speed is twice forward velocity, fan speed is related to engine speed, vibration is related to harmonics of engine speed. Slower vehicles will have much narrower velocity spread. Engine and fan speeds are typically 2–3 times ground speed.

Table 3.4 summarizes typical velocity spreads for categories of tactical target type. These velocity spreads can be converted into Doppler spreads using the model represented by (3.14). The Doppler spreads were taken at X band but are not qualitatively different at most microwave frequencies based on other tests. These spreads do not include geometrical acceleration, which must be calculated separately depending on vehicle extent and aspect.

*Table 3.4 Tactical target velocity width [16]*

| Target type | Velocity width (m/s) | | |
|---|---|---|---|
| | $\Delta V_t$ (−3 dB) | −10 dB | −20 dB |
| Moving man | 0.3 | 0.9 | 3 |
| Tank-tracked | 0.48 | 1.5 | 6 |
| Armored car-tracked | 0.48 | 1.5 | 5.7 |
| Passenger car | 0.6 | 1.8 | 6 |
| Heavy artillery tractor-wheels | 1.14 | 3.45 | 12–18 |
| Light artillery tractor-wheels | 0.9–3 | 3–9 | 9–16.5 |
| Truck-wheels | 0.48–1.32 | 1.5–4.05 | 6–13.5 |

These moving target spectra can be modeled using (3.15), where $\Delta V_t$ is taken from Table 3.4. Note that the spectrum at the −10 and −20 dB down points is quite wide and may interfere with other smaller targets, for example, large tractor trailers will interfere with nearby cars and humans.

$$G(f) \cong \frac{G_0}{1+\left|\frac{f-f_0}{\Delta f}\right|^2}, \quad f_0 = \frac{2\cdot V_{tlos}}{\lambda}, \quad \Delta f = \frac{\Delta V_t}{\lambda\cdot 2} \tag{3.15}$$

For example, shown in Figure 3.35 is the normalized spectrum for a moving tank at three different aspect angles with probable maximum speed of 20 km/h (not stated in reference). The Doppler center frequency shifts as expected and the half power width stays almost the same. However, the low-level spectrum broadens as more of the tracks are exposed.

Not surprisingly, spectra for a truck at nose-on and tail-on are quite similar in width and tails as shown in Figure 3.36 for a probable speed of 60 km/h (not stated in reference). The spectrum for a wheeled vehicle is much wider due to the other visible moving parts. For example, the drive shaft and cooling fan are running at different speeds than the vehicle's forward speed and are visible from road reflections. The top of the wheel is moving at twice the speed of the vehicle and the bottom is not moving at all. This is similar for tracks but usually the top of the track is protected from view by an armored skirt. In tests in Europe during Reforger exercises, AWACS found positive and negative double and triple Doppler harmonics of ground vehicle speeds. This implies conservative triple bounces of radar returns for moving targets in suburban/urban areas.

Vehicle acceleration versus time will determine the minimum Doppler filter bandwidth usable for detecting slow moving targets in clutter. The instant 3 dB Doppler spread is probably less than 10% of the dominant scatterers from a vehicle when in straight-line movement. Most of the large Doppler movement will come in turns associated with a change in direction. A "sporty" car can achieve about 1 g in lateral acceleration on a 100 ft radius skid pad. This calculates out to about 39 mph in a fast turn. A distant observer like a radar will see a maximum acceleration of 1 g occurring when the LOS velocity is almost zero and average acceleration over a 90°

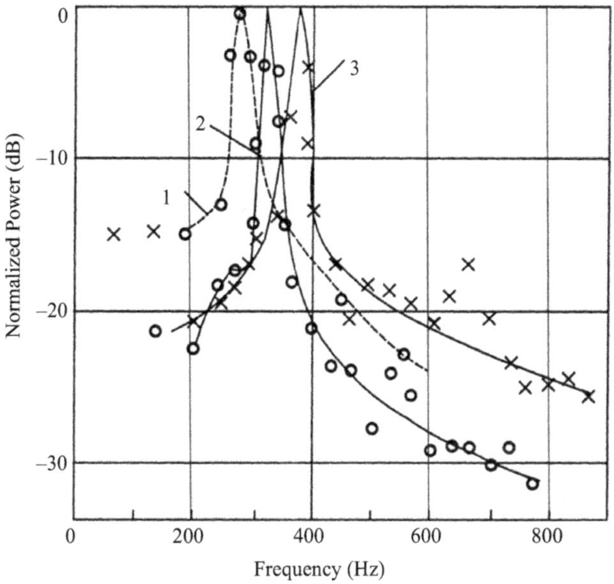


*Figure 3.35 X band spectrum from moving yank at 3 aspect angles [16]*

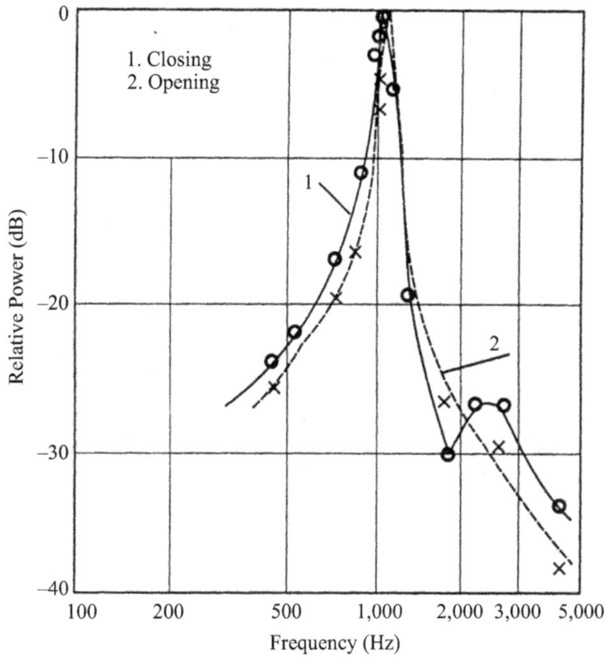


*Figure 3.36 X band spectrum from moving truck at two aspect angles [16]*

turn will be 0.5 g. Typical turn accelerations are more often 0.3 g or less based on tests on the FLAMR program using a vehicle with an inertial sensor. (In the United States you will be arrested for erratic driving if you make many turns over 0.3 g on city streets. Try it if you doubt this.) On a rough road or off road in open country, it is easy to create 0.5 g vertical acceleration for short periods. In fact, tank maximum speeds in open country are determined by suspension travel. Off-road racers use multiple shock absorbers to allow higher speeds. What does all of this mean? Filters narrower than 10% of expected velocity will not improve detection or tracking.

Moving humans also have a significant Doppler spread, primarily from moving arms and legs. Figure 3.37 shows the predicted X band spectrum for a moving man probably moving at 1.5 m/s (not stated in reference). This spectrum when converted to audio sounds like "whoosh-whoosh" as arms and legs move. Also note the very broad spectral tails for a moving man. Everything is moving on a human when she walks (if you doubt this, try your own experiment on yourself). In most walking and running, the load bearing/striding leg is moving slower than the body and the recovering leg is moving faster than the body. Similarly, the arms have alternate fast and slow motions to maintain balance and conserve energy. Human walking and running is amazingly energy efficient. Figure 3.37 shows the short-term velocity of a simulated moving man. The author observed spectra almost identical to the simulation during the U.S. Army MOTARDES program in the 1960s. Figure 3.38 shows Doppler measurements of a moving man.

Swimming humans also have a spectral spread for some of the same reasons. In addition, they usually have a wake, which contributes to their signature in Doppler and extent. Except for Olympiads swimming speeds are quite slow. An Olympic swimmer goes about 2.25 m/s for 100 m. Figure 3.39 shows Doppler measurements of a swimming man.

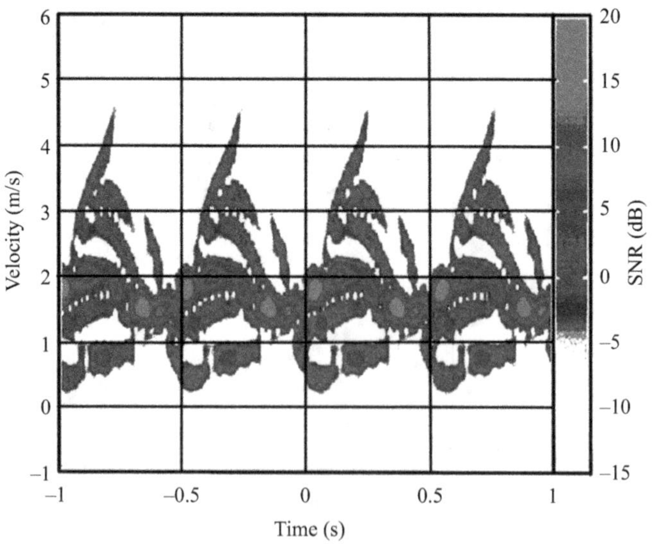


*Figure 3.37 Simulated moving man; adapted from [17]*

Table 3.5 summarizes some typical dimensions, radar cell sizes commonly used for detection and a range of RCSs for tactical targets useful for X and Ku band. After magnitude detection, range and Doppler bins are often combined before thresholding for the detection of larger extent targets.

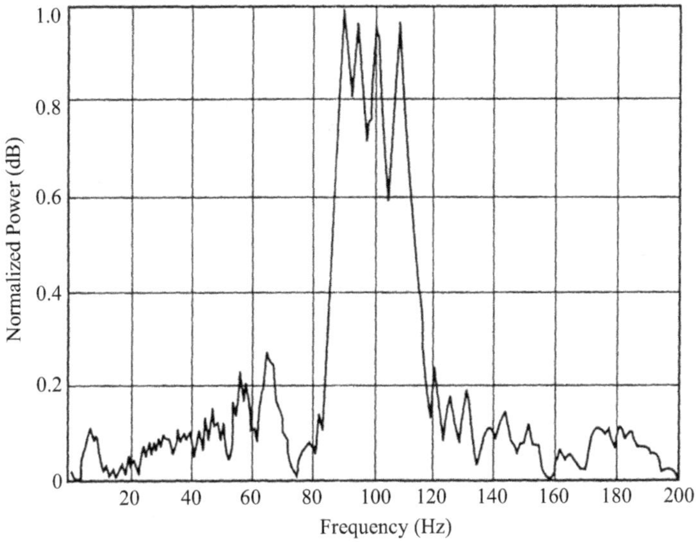


*Figure 3.38 X band average spectrum of a moving man; adapted from [16]*

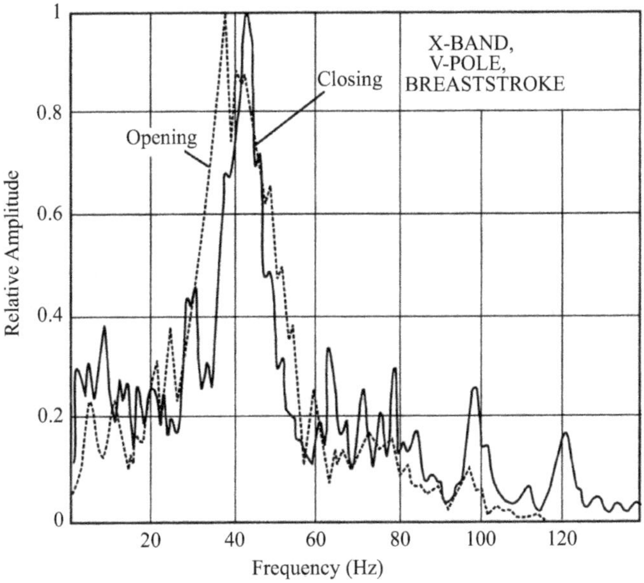


*Figure 3.39 X band spectrum of a swimming man; adapted from [16]*

*Table 3.5 Typical tactical target parameters*

| Type | Length (ft) | Width (ft) | Height (ft) | Internal motion (±ft/s) | Range bin (ft) | −1 Sigma RCS ($m^2$) | +/Sigma RCS ($m^2$) | Turn rate (deg/s) |
|---|---|---|---|---|---|---|---|---|
| Human | 2 | 2 | 6 | 0.5 | 3 | 0.3 | 1.25 | 30 |
| Passenger Car | 17 | 6 | 5 | 1 | 10 | 5 | 100 | 20–40 |
| HMMWV | 15 | 7.5 | 6 | 1.5 | 10 | 7 | 200 | 20–30 |
| Medium Truck | 26 | 8 | 10 | 1.5 | 20 | 60 | 3000 | 10–30 |
| Semi-Tractor-trailer | 55 | 8.5 | 13 | 1.9 | 20 | 200 | 6000 | 10–30 |
| Armored Personnel Carrier | 22 | 8.8 | 7.2 | 0.8 | 20 | 30 | 500 | 10–30 |
| Tank | 32 | 12 | 8 | 0.8 | 20 | 50 | 800 | 10–30 |
| Towed Howitzer | 20 | 8 | 10 | 3.2 | 20 | 10 | 80 | 5–15 |
| Self Propelled Gun | 30 | 10 | 10 | 0.8 | 20 | 50 | 800 | 10–30 |
| Destroyer | 500 | 65 | 100 | 10 | 3 | $10^3$ | $10^4$ | 1–6 |
| Freighter | 1000 | 100 | 190 | 7 | 50 | $10^4$ | $10^5$ | 0.1–0.4 |
| Fighter | 55 | 40 | 15 | 50 | 50 | 0.5 | 10 | 3–30 |
| Spacecraft | 100 | 30 | 50 | 0.5 | 1–10 | 10 | $10^4$ | 0.001–0.05 |

## References


[1] Army.mil, HMMWV Web page.

[2] AM General, Web page.

[3] General Atomics Aeronautical Systems, Reconnaissance Systems Group, Interim Report CDROMs and Web Page, 2008.

[4] Aks, S., Kummer, W., Lynch, D., Pearson, J.O., Shamash, E., "Introduction to Modern Radar", *Evolving Technology Institute Short Course Notes*, February 1988.

[5] Stipulkosky, T., "Statistical Radar Cross Section Measurements for Tactical Targets and Terrain Background, Phase I &II", *Hughes Aircraft Report* D6115, December 1976, declassified 9/9/1988.

[6] Nathanson, F., *Radar Design Principles*, 2nd ed., McGraw Hill, New York, NY, pp. 33–37, 64, 239–247, 256–257, 371.

[7] Barton, D., *Radar Systems Analysis*, Artech House, Norwood, MA, (1979), pp. 483–484

[8] Lynch, D., *Introduction to RF Stealth*, SciTech Publishing, Raleigh, NC (2004), pp. 112, 198–221.

[9] Guerci, J., *Space-Time Adaptive Processing for Radar*, 2nd ed., Artech House, Norwood, MA, (2014).

[10] Skolnik, M., ed., *Radar Handbook*, 1st ed., McGraw Hill (1970), pp. 25-2–25-10.

[11] Muhleman, D.O., "Radar Scattering from Venus and the Moon", *Astronomy Journal*, Vol. 69, February 1964, pp. 34–41.

[12] Katzin, M., "On the Mechanisms of Radar Sea Clutter", *Proc. IRE*, Vol. 45, January 1957, pp. 44–54.

[13] Skolnik, M.I., *Introduction to Radar Systems*, McGraw Hill, New York, NY, (1962), pp. 28, 414.

[14] Jain, A., Patel, I., "SAR/ISAR Imaging of a Non-Uniformly Rotating Target", *General Motors Hughes Electronics Report*, March 1991.

[15] Lynch, D., "Space Imaging Using the Allen Array", *DLS Report to Boeing*, Proprietary Restrictions, October 2003.

[16] Kulemin, G.P., *Millimeter-wave Radar Targets and Clutter*, Artech House, Norwood, MA, (2003), pp. 64–72, 121–122.

[17] Hersey, R., Melvin, W.L., Culpepper, E., "Dismount Modeling and Detection from Small Aperture Moving Radar Platforms", *IEEE Xplore Digital Library*, I-4244-1539-x/08.

# Chapter 4
# Sensor signal and data processing

## 4.1 Introduction

The suite of microwave and RF apertures in a fighter, surveillance aircraft or spacecraft might be as many as 20 apertures distributed throughout the vehicle performing radar, data link, navigation, missile warning, direction finding, jamming or other functions over a frequency range covering several decades [1]. There are apertures distributed over the platform that point forward and aft, right and left, as well as up and down. Some apertures will be shared for communications, radio navigation and identification (CNI) as well as identification, friend or foe (IFF) due to compatible frequencies and geometries. Data links such as JTIDS/Link 16 and Link 22 can share apertures with GPS and L-band satellite communications (L-SATCOM). There also may be dedicated data link apertures. EW apertures must be broadband by nature and can be shared with radar warning receivers (RWR), radar auxiliaries and some types of CNI.

The apertures are signal conditioned, controlled and interfaced through busses in the platform with remaining processing performed either in a common processor complex as shown in Figure 4.1 or in federated processors distributed throughout the platform or data linked to a surface-based processing and exploitation complex.

The functional block diagram and operation of a specific sensor mode is then overlaid on this hardware and software infrastructure. A specific mode is implemented in an applications program in the same sense that word processing is on a personal computer (PC). Carrying the analogy further, common experience with the unreliability of PC hardware and software requires that a system of the type depicted in Figure 4.1 must be redundant, error checking, trusted, fail-safe in the presence of faults and embody strict program execution security. This is a very challenging system engineering task. Exhaustive mathematical assurance and system testing is required, which is completely different from current commercial PC practice. There are multiple redundant processing arrays, which contain standardized modules connected in a nonblocking switched network. Internal and external busses connect the individual processing arrays to each other as well as the other suites, sensors, controls and displays.

Improper operation of many radar systems can be hazardous. As previously mentioned, the software must be exhaustively tested, error checked, mathematically trusted, fail-safe in the presence of faults and embody strict program execution

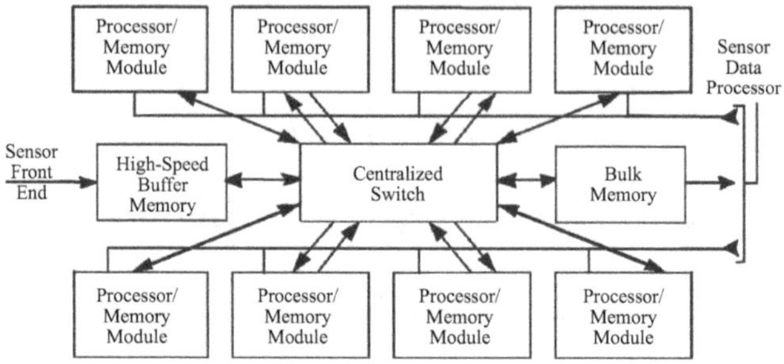


*Figure 4.1 Conceptual sensor signal processor hardware complex; adapted from [1]*

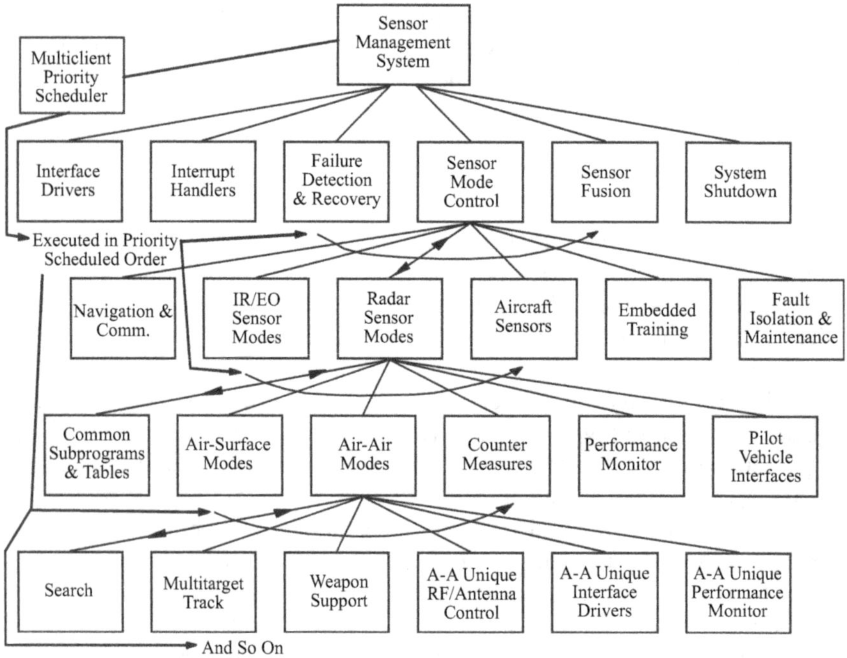


*Figure 4.2 Typical sensor structured software [2]*

security. One of the most important aspects is rigid adherence to a structured program architecture. An object-based hierarchical structure, where each level is subordinate to the level above and subprograms are called in strict sequence, is necessary. It also requires among other things that subprograms never call themselves (recursive code) or any other at their execution level. Subprograms (objects) are called, receive execution parameters from the level above (parent) and return results back to the calling level [3]. An example of such a software structure is shown in Figures 4.2 and 4.3. The software would be executed in the hardware of Figure 4.1.

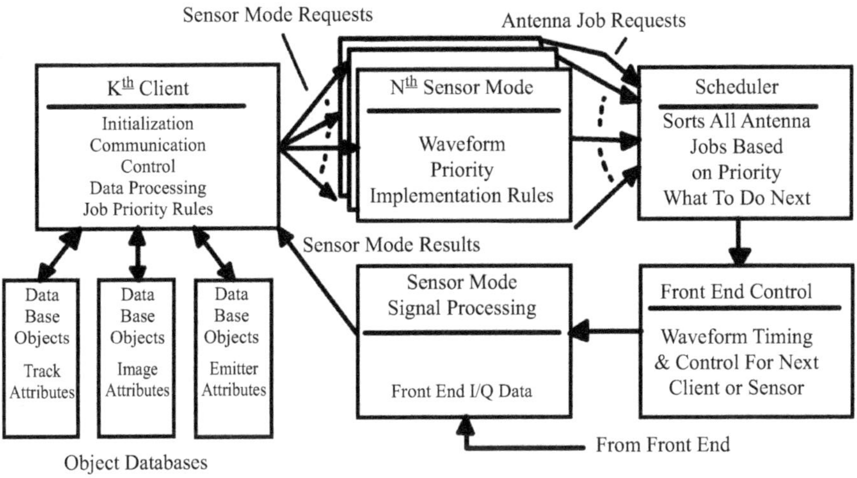


*Figure 4.3 Sensor task executive priority scheduling [2]*

Although this structure is complex and the software encompasses millions of lines of code, modern radar software integrity can be maintained with strict control of interfaces, formal configuration management processes and formal verification and validation software tools. In addition, most subprograms are driven by read-only tables as shown in Figure 4.3, so that the evolution of aircraft tactics, capabilities and hardware does not require rewrites of validated subprograms. Software versions (builds) are updated every year throughout the lifetime of the system, which may be decades. Each subprogram must have table driven error checking as well. Many lower levels are not shown in Figures 4.2 and 4.3; there may be several thousand subprograms in all.

Modern radars can support many activities (or modes) concurrently by interleaving their respective data collections. Surveillance, track updates and ground maps are examples of such activities. The software needed to support each activity is mapped to a specific Client module as shown in Figure 4.3. Each Client module is responsible for maintaining its own Object Database and for requesting use of the aperture. Requests are made by submitting Antenna Job Requests that specify both the waveform to be used (how to do it) and the priority and urgency of the request.

A scheduler executes during each data collection interval and decides what to do next, based on the priorities and urgencies of the Antenna Job Requests that have been received. This keeps the aperture busy and responsive to the latest activity requests. Following the selection of the Antenna Job by the Scheduler, the front-end (transmit and receive) hardware is configured and in-phase and quadrature (I/Q) data is collected and sent to the signal processors. There, the data is processed in a manner defined by the Sensor Mode and the signal processing results are returned to the Client that requested them. This typically results in database updates and/or new Antenna Job Requests from the Client. New activities can be added at any time using this modular approach.

## 4.2 Sensor signal processing architecture

The most important improvement in RF sensors since the 1950s is digital signal processing. It brought the advent of predicted performance closely approximating actual performance. The stability and accuracy of digital processing provided genuine breakthroughs in actual performance. Individual control of sensor subunits by local general purpose computer-based (Von Neumann machines) micro-control units (MCUs) facilitate real-time calibration, mode-by-mode and phase-by-phase parameter control as well as rapid built-in-test (BIT). These changes dramatically improved sensor performance. All of the apertures can be signal conditioned, controlled and interfaced through busses in the aircraft or spacecraft with remaining processing performed either in a common processor complex as shown in Figure 4.4, in ground-based processor stations or in federated processors distributed throughout the platform. The signal and data processor complex contains multiple processor and memory entities, which might be on a single chip or on separate chips depending on yield, complexity, speed, cache size, etc.

There are multiple redundant processing arrays, which contain standardized functions connected in a nonblocking switched network. Internal and external busses connect the individual processing arrays to each other as well as to other sensors, controls and displays. Usually there are both parallel electrical signal busses as well as serial fiber optic busses depending on speed and total length in the platform [2]. The number smashing following the A/D function is performed by programmable signal processing (PSP) chips embedded in a redundant processing array. Currently, there may be multiple PSPs and level 2 caches on a single chip with an integrated but expandable switch. The level 3 cache and bulk memory (BM) is usually on separate chips.

Each processor array may consist of PSP, general purpose processors (GPP), BM, input–output (I/O) and a master control unit (MCU). The PSPs perform signal processing on arrays of sensor data. The GPPs perform processing in which there are large numbers of conditional branches. The MCU issues programs to PSPs, GPPs and BM, as well as manage the overall execution and control. Typical processing speed is 8 GIPS (billions of instructions per second) per chip but might be 32 GIPS (billions of instructions per second) in the near future [4]. Clock frequencies are limited by on-chip signal propagation but are up to 4 GHz and could be 10 GHz in the near future. Sensor processing has arrived at the point where the conception of successful algorithms is more important than the computational horsepower necessary to carry them out.

One important class of standardized functions contains basic timing and programmable event generators (PEG), which create accurate timing for PRFs, analog-to-digital conversion (A/D) sampling, pulse and chip widths, blanking gates, beam repointing commands and other synchronized real-time interrupts. A second class contains RF and IF amplification and mixing. A third class contains low noise frequency synthesizers, which may include direct digital-frequency synthesis (DDS). A/D converters and control interface modules are the final class.

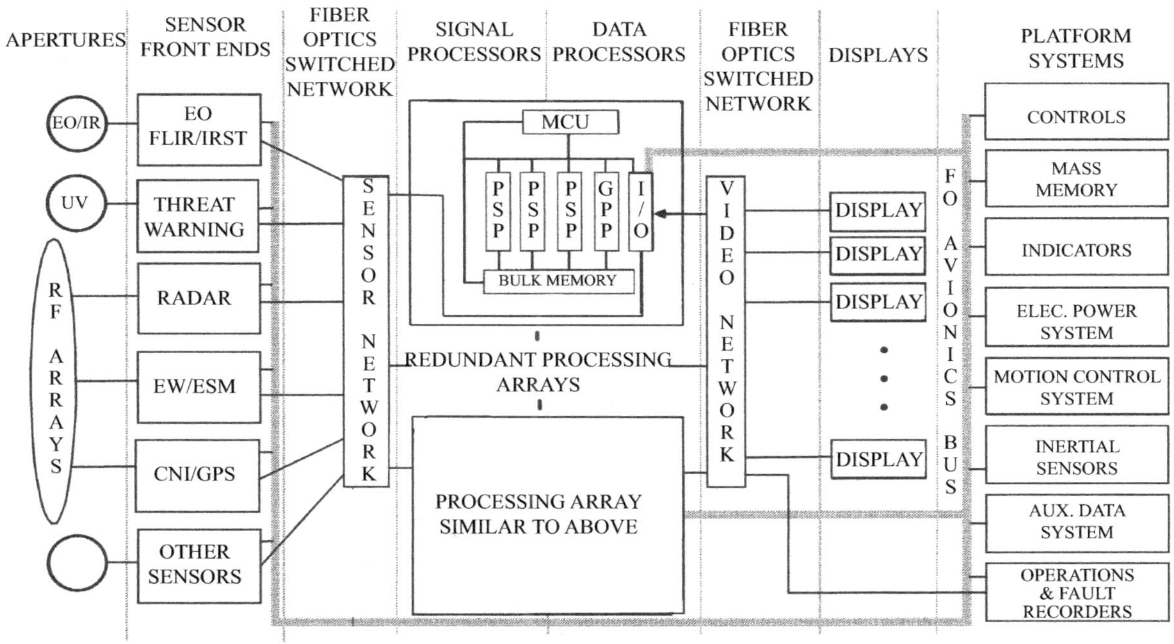


*Figure 4.4 Typical platform signal and data processor architecture; adapted from [2]*

Bussing protocols and speeds must have adequate reserves to insure fail-safe real-time operation.

Another new class of signal and data processing adaptable to sensor suites of the type shown in Figure 4.4 with general architecture like that of Figure 4.1 uses a graphics processor unit (GPU) coupled with a garden variety central processor unit (CPU). Although GPUs are not really optimized for signal processing, there are so many processors integrated onto a single chip that even with inefficient execution they can be very competitive. For example, the Nvidia Tesla K20X has 2688 separate processor cores running at 1.6 GHz clock rate. That unit managed by standard Intel/AMD CPU instruction set architecture can provide excellent performance [4].

## 4.3 Sensor data processing

Sensor data processing consists of two separate classes of functions. The first category contains traditional computer functions such as

1. Smoothes, predicts and edits sensor platform position data.
2. Performs antenna beam stabilization computations.
3. Performs sensor antenna scan program generation.
4. Performs sensor mode selection and control.
5. Provides standard bus interfaces and appropriate responses.
6. Performs online sensor diagnostics with software and hardware.
7. Creates soft and hard failure table and commands failure reconfiguration.
8. Controls the sensor based on computer inputs from multiple outside sources.

The second category contains special purpose equipment and functions imbedded in the sensor processor such as

1. Interfaces built-in-test-equipment (bite) and online diagnostics.
2. Provides special sensor bus interfaces (usually very high speed).
3. Provides computer with sensor system measurement inputs.
4. Converts platform position and attitude data from CNI/GPS/INS for sensor needs.
5. Provides and receives cueing data to/from other sensors.
6. Creates special display formats.
7. Monitors watchdog timer and reboot as required.

Figure 4.5 shows a functional block diagram of the receiver as viewed by the sensor computer. Each of these functions must be commanded by the sensor computer. Although the dedicated hardware performs the detailed settings, it is commanded parametrically for each separate sensor mode, platform velocity and spatial geometry. The antenna configuration, frequency, beam steering, phase center selection, STC, AGC offset, calibration tables and temperature compensation are usually provided by the computer for each mode for each coherent processing

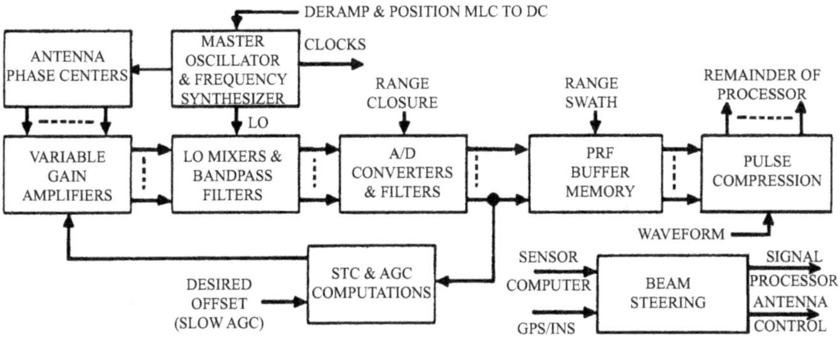


*Figure 4.5 Receiver as viewed from the sensor computer [5]*

interval (usually every few tens of milliseconds). Pulse compression may be partially done by the LO as well as partially prior to Doppler filtering and partially after filtering depending on waveform. The sensor computer would normally allocate the parts of compression depending on waveform and platform velocity.

Figure 4.6 shows a functional block diagram of the signal processor as viewed by the sensor computer. The sensor computer provides calibration tables, beam steering commands, motion sensing data and patch size to the signal processor. The signal processor usually provides all of the computations on the entire array of data in a patch. All of the compensations, resampling, rectification, amplitude weighting, Fourier transforms, adaptive filtering, correlations, magnitude formation and detections. Several different constant false alarm rate (CFAR) algorithms are used to provide reliable detections. There are typically millions of points to be processed. The signal processor provides mainlobe clutter parameters, correlated target hits and other housekeeping data to the sensor computer. When the patch is resolved to a few thousand hits, then the sensor computer takes over. Usually the target and clutter tracking Kalman filters are closed through the sensor computer. This is because all the motion compensation data is in the sensor computer and almost all of the target filters will require compensation for platform motion.

You may not realize it, but, if you are healthy, the images you see in the world are motion stabilized. When you turn your head or scan your eyes, the images that fall on your retina are different but your mind and body motion sensors compensate to provide a stabilized image. It is no different for a radar, EO or Elint sensor, all of the motion must be compensated. Some of the motion can be compensated on a predictive basis, such as a batter does looking at the pitcher's motion. Some of it must be done in real time based on signal measurements, such as a batter does when he senses the orientation of the stitches on a baseball in the first 30 ft from the mound. Just as batters often miss predicting all the motion, so also motion is usually only compensated for the centroid of the clutter return. Even that compensation dramatically improves the detection performance of significant movers whether in MTI or SAR modes.

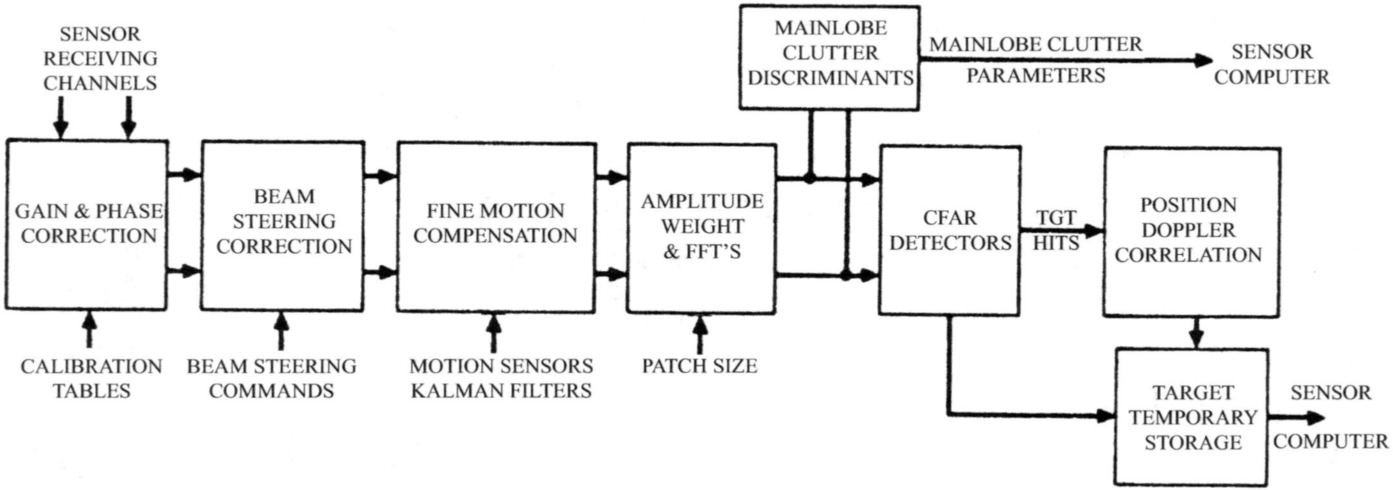


*Figure 4.6 Digital signal processor as viewed from the sensor computer [5]*

## 4.4 Basic digital signal processing

Modern radar processing is done digitally. Both the received signal and the matched filter will be sampled functions. Most digital signal processing can be characterized by a general finite difference equation given in (4.1).

$$E_{out}(n \cdot t_s) = \sum_{i=0}^{N} A_i \cdot E_{in}((n-i) \cdot t_s) - \sum_{i=1}^{N} B_i \cdot E_{out}((n-i) \cdot t_s) \tag{4.1}$$

where $t_s$ is the time between sampling instants, $E_{in}$ is the digitized input signal vector, $A_i$ and $B_i$ are complex coefficients which may be dynamic.

In its complete form (4.1) is a recursive or infinite impulse response (IIR) filter. When all the $B_i$s are zero, this is a finite impulse response (FIR) filter of which digital pulse compression and beamforming are subsets. Similarly when the $B_i$s are zero and the $A_i$s are equal to $\exp(-j \cdot 2 \cdot \pi \cdot m \cdot i/N)$, it represents a DFT. When all $B_i$s are zero and only the sign of $E_{out}$ is retained, then it represents an adaptive threshold. Thus the same form is used for beamforming, pulse compression, filtering and thresholding. Figure 4.7 is a canonical realization of (4.1).

Signal processors must be optimized to execute vector multiplication and addition recursively on large arrays of data. Since each time step is incremented for each new output, most data and coefficients are indirectly addressed not absolutely addressed. Signal processors must be optimized for indirect addressing and recursive looping. Often part of a signal processor has hardware specifically designed to perform loops with no execution time overhead for each iteration. Similarly, mass memory data flow rate often is a bottleneck and there will be hardware to initiate and manage data transfers in the background so that they are transparent to the filtering operations.

The matched filter convolution given in Figure 2.4 and (2.1) often is performed with sampled values of the desired impulse response, $h(t)$. Two common methods

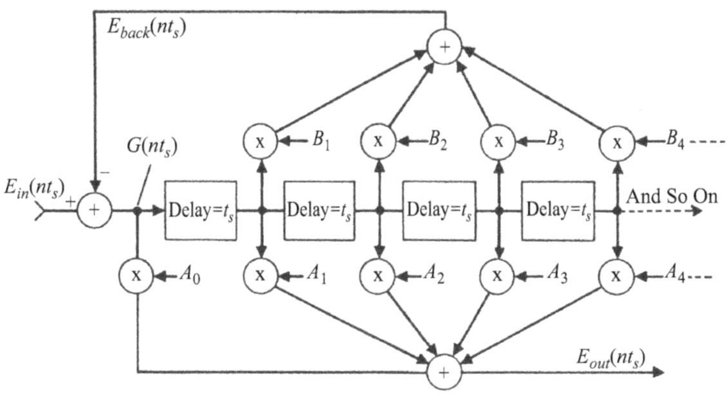


Figure 4.7 Canonical digital filter

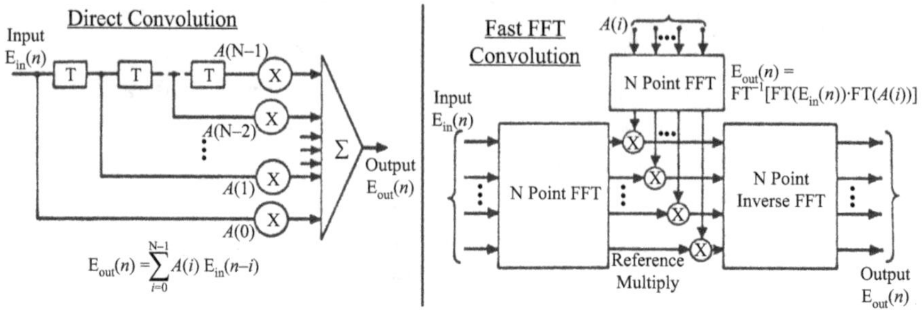


*Figure 4.8 Alternative digital convolution methods [5]*

of convolution are shown in Figure 4.8. If the filter function has only a few coefficients, then the convolution is usually direct as shown at the left of Figure 4.8. If on the other hand the number of filter coefficients or the size of the data array is large, then fast FFT convolution often is applied as shown in the right half of Figure 4.8. Both the input signal and the coefficients are Fourier transformed, multiplied and then inverse transformed to provide the output. This may not seem faster but it is often dramatically faster using modern digital programmable signal processor (D/PSP) chips.

### 4.4.1 Fast Fourier transforms

Most signal processing uses the fast Fourier transform (FFT) in one of its forms to create the bulk of the filters used in beamforming, pulse compression, Doppler filtering and image compression. In one sense the DFT is the matched filter for a sampled pulsed monotone of unknown phase. One form of DFT is given in (4.2), both input and output samples are represented by $A_\ell$ with $A_0$ representing the input data and $A_{\ell\max}$ representing the output data. The reason for this nomenclature will be obvious when recursive forms of the FFT are introduced in which the memory storing the data will be reused. It was observed that the redundancy in the exponential coefficients for a DFT could be exploited to greatly reduce the total number of additions and multiplications from order $N^2$ to the order of $N\log_2(N)$ or better first by Goertzel and two decades later by Cooley and Tukey. For large DFTs this difference is dramatic and hence the FFT.

$$A_{\ell=\ell\max}\left(\frac{n}{N\cdot t_s}\right)=\sum_{g=0}^{N-1}A_{\ell=0}\left(g\cdot t_s\right)\cdot\exp\left(-\mathrm{j}\cdot 2\cdot\pi\cdot g\cdot\frac{n}{N}\right)$$

Where: $N$= number of samples, $t_s$ = sampling time interval,

$g$ = sampling index, $A_{\ell=0}\left(g\cdot t_s\right)$ = input sampled data, $\mathrm{j}=\sqrt{-1}$, (4.2)

$A_{\ell=\ell\max}\left(\frac{n}{N\cdot t_s}\right)$ = output sampled frequency data.

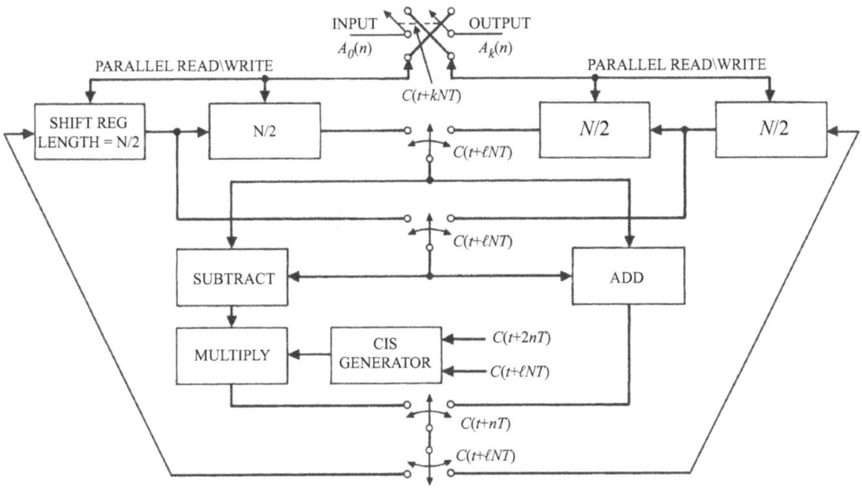


*Figure 4.9 Simple FFT processor [6]*

One of the simplest forms of FFT is shown in Figure 4.9. It is sometimes called a perfect shuffle or "slosh" FFT. There are forms that have ½ as many arithmetic operations but much more complex addressing and control. It is a designer's choice as to which form to use. This form adds and subtracts data values, which are always ½ the record length ($N/2$) apart. Only the difference term is multiplied by a complex phase shift by the cosine-i-sine (CIS) generator. The add and multiplier outputs are successively interleaved (perfect shuffle) and delivered to the register opposite to the source register (ping-ponged). After $k$ passes through the data, the resulting Fourier transform is read out from one side while the input data for the next transform is written into the opposite side. The clocking instant, $C(t)$, occurs at $C(0)$. This form is especially easy to use with large disk memories and $10^7$ point or larger transforms.

Several recursive forms of the FFT are given in (4.3). The perfect shuffle form of Figure 4.9 is shown at the top of the list of (4.3). Two other recursive FFT forms are also given in (4.3). The perfect shuffle form of the FFT was simultaneously discovered in several places including by the author in the late 1960s [6]. The recursive forms are especially efficient for use in most modern D/PSP (PSP or D/PSP). Usually, sample records of arbitrary length are filled with zeros to create a convenient easy to calculate FFT input record length. Most modern FFTs are dominated by the number of memory ports and memory port bandwidth but not the total quantity of memory words that hold data and coefficients.

All of the forms of FFT given in (4.3) are for records whose length is a power of two. There are also algorithms for base 3, 4, 5, 6, 7 and arbitrary factors [7]. There is roughly a 2:1 difference in the arithmetic hardware between the most

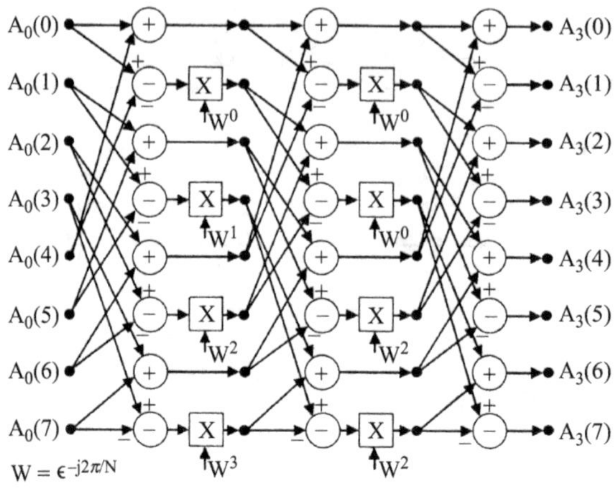


*Figure 4.10 Perfect shuffle/"slosh" eight-sample FFT flow diagram [6]*

arcane factored FFT and the base 2 algorithms. The most commonly used FFTs are base 2 and base 4.

Perfect Shuffle Bit Reversed Output Order:

$$A_\ell(2\cdot n) = A_{\ell\text{-}1}(n) + A_{\ell\text{-}1}(n+N/2)$$

$$A_\ell(2\cdot n+1) = \left[A_{\ell\text{-}1}(n) - A_{\ell\text{-}1}(n+N/2)\right]\cdot W^{2^{\ell-1}\cdot n}$$

Time Decimation Natural Output Order:

$$A_\ell(n+p) = A_{\ell\text{-}1}(n+p) + A_{\ell\text{-}1}\left(n+p+2^{k-\ell}\right)\cdot W^{m/2}$$

$$A_\ell\left(n+p+2^{\ell-1}\right) = A_{\ell\text{-}1}(n+p) - A_{\ell\text{-}1}\left(n+p+2^{k-\ell}\right)\cdot W^{m/2}$$

Frequency Decimation Bit Reversed Output Order:

$$A_\ell(n+m) = A_{\ell\text{-}1}(n+m) + A_{\ell\text{-}1}\left(n+m+2^{k-\ell}\right)$$

$$A_\ell\left(n+m+2^{k-\ell}\right) = \left[A_{\ell\text{-}1}(n+m) - A_{\ell\text{-}1}\left(n+m+2^{k-\ell}\right)\right]\cdot W^{2^{\ell-1}\cdot n} \tag{4.3.1}$$

Where: $k = \log_2(N)$; $W = \exp(-j2\pi/N)$; $\ell = 1,2,\ldots k$ (index of recursion);
$p = 0,1,2,\ldots 2^{\ell-1}-1$ (index of butterfly sequence); $m = p\cdot 2^{k-\ell+1}$;
$n = 0,1,2,\ldots 2^{k-\ell}-1$ (index of butterfly) (4.3.2)

Many engineers are more familiar with FFT flow diagrams than recursive equations and a flow diagram for the perfect shuffle 8 sample FFT algorithm is given in Figure 4.10. The flow diagram is identical for each recursion except for the

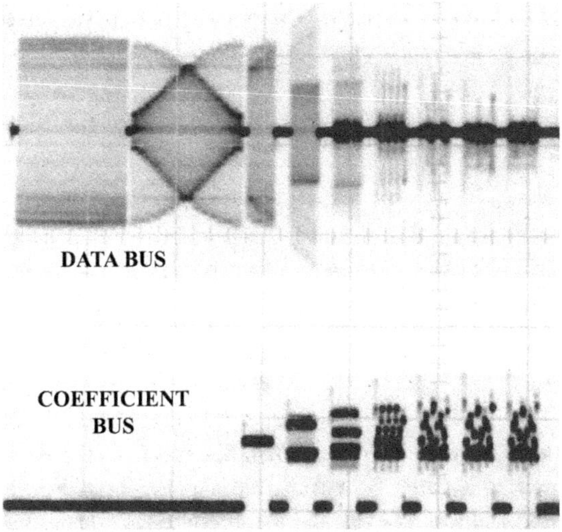


*Figure 4.11 Data and coefficient busses first eight FFT passes*

complex coefficients. There are many FFT textbooks that have other variations on FFT flow diagrams, input array properties, output array formats, etc. There are parallel forms of the FFT called polyphase filtering. The FFT is so efficient that almost anything that requires multiple angle beams, multiple filters or multiple range bins uses FFTs. There are relatives of the FFT such as the fast Hadamard transform and Hudson–Larson fast complementary code transform that are often used in signal processing. It is very likely that your cell phone uses FFT for compression and decompression of both sound and pictures.

Figure 4.11 shows oscilloscope images of the quadrature part of the vectors on the data bus and the coefficient bus in a PSP, which have been instrumented with D/A converters. The top trace shows first the unweighted input data. Second, the top trace shows amplitude weighting and data turning. Next on that trace is the first pass of the FFT and synchronized with it on the lower trace coefficient bus are the counterpart coefficients (quadrature only). The first seven of the FFT passes are shown with counterpart coefficients. As the FFT progresses, fewer filters are rung up at each stage on the top trace until finally only one filter has an output (not shown). At each FFT stage, successively smaller phase steps are used to form the final narrowband filters in this particular FFT embodiment.

## 4.5 Matched filtering and straddling losses

Since receive losses require more transmit power, it is essential to minimize losses. Most systems only allocate one sample per resolution cell. Since a received signal will be shifted by an amount in range and Doppler depending on its location, it will not be centered in a range bin or Doppler filter. When it is not centered, it will experience a loss in amplitude due to bin shape. This is called straddling loss. Filter bins or range samples must be more closely spaced or there may be high straddling losses.

---

**Example 4.1** For Example 4.1, assuming a perfect matched filter to a single chip width rectangular pulse, the peak straddling power loss as a function of sample spacing is given in (4.4).

$$L_s = {}^2\!/_{\pi} \cdot (t_s/t_c) \text{ for } t_s \le t_c \tag{4.4}$$

where $t_s$ is the time between sampling instants and $t_c$ is the equivalent time resolution for the time-bandwidth of the transmitted waveform. For the case $t_s = t_c$, the peak straddling loss is about −4 dB and the average loss is about −2 dB. Similarly, for a filter bank output using modified Taylor weight as shown in Figure 4.12, the frequency bin straddling power loss for filters ideally matched to the input spectrum is given in (4.5).

$$L_{s\max} \approx 0.5 - 0.5 \cdot \cos\left(\frac{\pi \cdot f_s}{2 \cdot B}\right) \tag{4.5}$$

where $B$ is the input signal matched bandwidth and $f_s$ is the filter spacing. If weighting is as shown in Figure 4.12 with $a = 0.54$ and if $B = f_s$ then the peak straddling loss is about −1.3 dB, but sadly the total average loss is still approximately −6 dB (no free lunch!). The filters can be placed closer together by making the record longer either by filling with zeros or dwelling longer (which may not be possible). Range bins often are spaced 10% closer than match to reduce straddling losses [5,7].

One other notion shown in Figure 4.12 is that filter sidelobe weighting can be implemented either before or after narrowband filter formation by a FFT or DFT. In order to reduce straddling losses almost to zero, beams, range or chip sampling bins and filters often must overlap 2:1 as suggested in Figure 4.13 for a typical SAR image formation process where very high-quality images are required. Shown in Figure 4.14 is a PSP data bus instrumented with a D/A converter, which shows the before and after of amplitude weighting on a frequency-modulated received pulse.

Figure 4.14 also shows the obvious disadvantage of weighting, that is, part of the signal power is attenuated and so there is some detection performance loss. The loss is summarized in Table 4.1. The weighting functions to reduce sidelobes mentioned in Chapter 2 work even better for digital filters since there aren't any manufacturing errors. There are several amplitude weighting functions (often

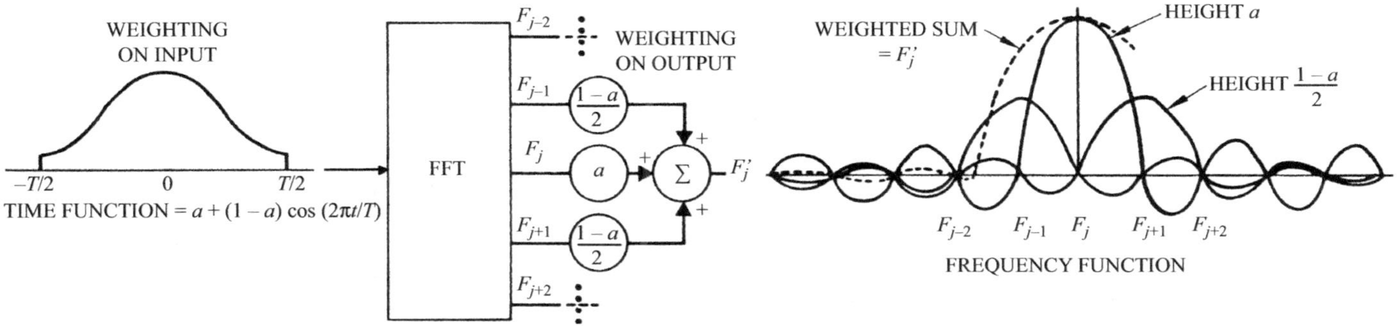


*Figure 4.12 Filter sidelobe weighting [6]*

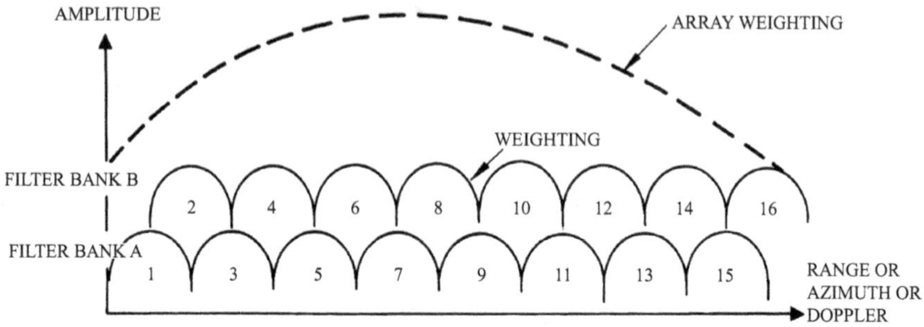


*Figure 4.13 Overlapped synthetic beams, in range, angle or Doppler bins to minimize straddling loss*

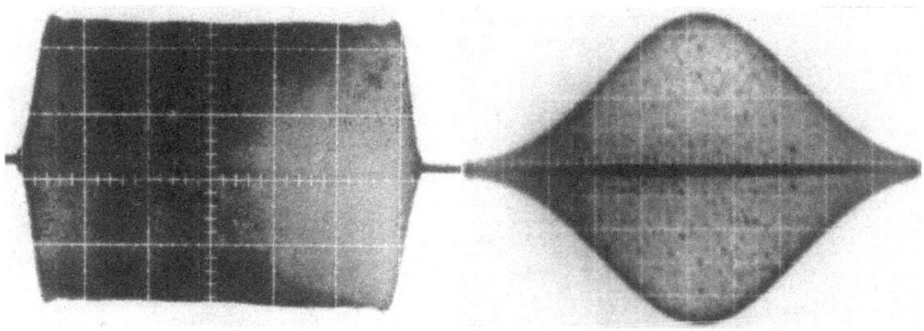

*Figure 4.14 Data bus before and after sidelobe weighting*

*Table 4.1 Filter weighting functions (repeated Table 2.2)*

| Name | BWEXP 3 dB | PSLR, dB | Sidelobe rolloff, dB/OCT. | Coherent gain | ISLR – 2.3 U 3 dB, DB |
|---|---|---|---|---|---|
| Rectangular | 1.0 | −13.3 | −6 | 1.0 | −10 |
| Parabolic | 1.3 | −21.3 | −12 | 0.83 | −20.7 |
| Bartlet | 1.44 | −26.5 | −12 | 0.5 | −23.2 |
| Hanning | 1.63 | −31.5 | −18 | 0.5 | −21.9 |
| Hamming | 1.48 | −42.6 | −6 | 0.54 | −22.4 |
| Cosine3 | 1.81 | −39.3 | −24 | 0.42 | −21.9 |
| Blackman | 1.86 | −58.1 | −18 | 0.42 | −23.7 |
| Parzen | 2.06 | −53.1 | −24 | 0.38 | −22.6 |
| Zero Sonine | 1.24 | −24.7 | −9 | 0.59 | −24.6 |
| Mod. Taylor | 1.22 | −28 | −6 | 0.7 | −22.2 |
| Dolph-Chebyshev | 1.49 | −50 | 0 | 0.53 | −22 |

called windows) that can produce 80–90 dB down sidelobes such as Blackman–Harris and Kaiser–Bessel. These can be found in reference [7].

**Example 4.2** Another way to reduce straddling losses is to form an FFT, which is larger than the number of samples by filling the remainder of the input to an FFT with zeros.

For Example 4.2, suppose the number of signal samples is 600 and one wants to form a 1024 point Fourier transform, then one would add 424 zeros. This has the effect of placing the filter outputs closer together reducing straddling losses, which is a very common technique. Figure 4.15 shows a comparison for two sinewaves of exactly the same amplitude separated enough to be resolved. The 600 filter version shows a straddle power loss of 0.3 dB for one of the signals, whereas the zero filled 1024 filter version shows no straddle loss. Recall that what is typically measured is amplitude proportional to voltage but decibels always are proportional to power, that is, voltage squared.

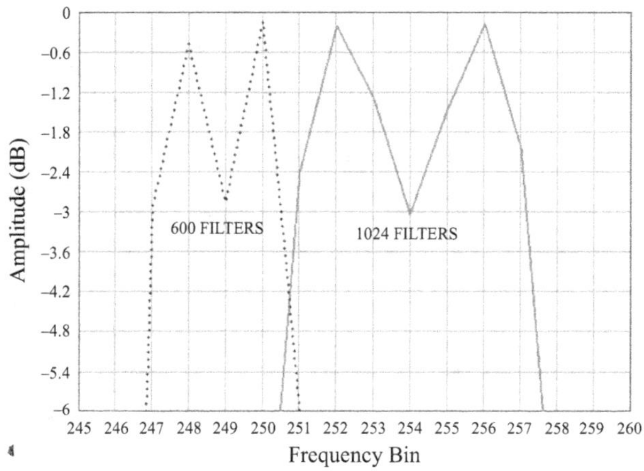


*Figure 4.15 Comparison of straddle loss with/without zero fill (ZeroFill001)*

## 4.6 Sliding windows

Another type of windowing function is the sliding window that averages multiple cells in one or two orthogonal directions to provide smoothing for imaging and thresholding. The presumption is that image discontinuities between CPIs or the occurrence of targets is a relatively low probability event. The idea is to create an ensemble average estimate of the background to provide coefficients for image equalization, clutter cancellation and target detection thresholding. Figure 4.16 shows the conceptual idea for thresholding.

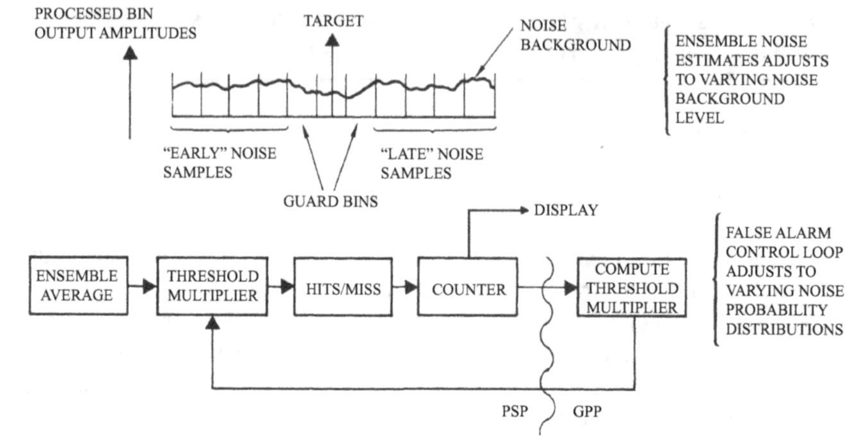


*Figure 4.16 Ensemble average around a potential target [5]*

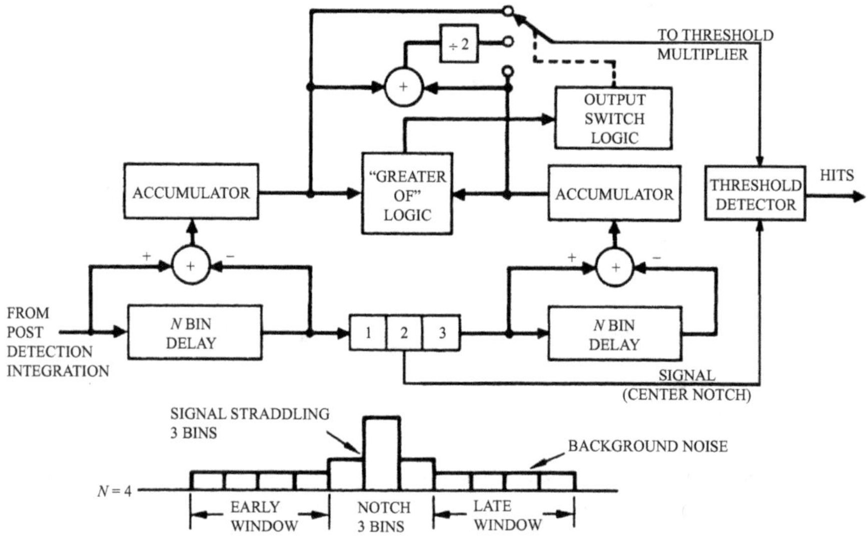


*Figure 4.17 1D sliding window used for detection in clutter [5]*

The ensemble average and, perhaps, its standard deviation (aka RMS value) is used as an estimate of the true statistics of the input range Doppler image. These statistics estimates are used to set a threshold to minimize false detections. This process is sometimes called a CFAR detection. The same general process can be used to adjust image intensity in a SAR map or to create STAP coefficients to cancel clutter or jamming.

Figure 4.17 shows a typical 1D sliding window for detection. One other attribute of sliding windows shown in Figure 4.17 is what is known as tilt logic.

The sum of the early and the late windows are compared. If one is significantly larger than the other, then the window must be encountering a dramatic rise in clutter or noise and just one side of the window is used to set the threshold. If both are reasonably the same, then the two are averaged to set the threshold. The early and late averages are shown in (4.6) for each bin. If the threshold is too low, it may cause false alarms so a minimum threshold is set based on the expected thermal noise and the overall RMS value of all the bins.

$$EA_{out}(n\cdot t_s)=\frac{1}{N}\cdot\sum_{j=0}^{N-1}E_{in}((n-j)\cdot t_s)+\frac{RMS}{TM}+Noise,$$

$$LA_{out}(n\cdot t_s)=\frac{1}{N}\cdot\sum_{j=0}^{N-1}E_{in}((n+N+3-j)\cdot t_s)=EA_{out}((n+N+3)\cdot t_s)$$

$$Th(n\cdot t_s)=TM\cdot 0.5\cdot\left[EA_{out}(n\cdot t_s)+EA_{out}((n+N+3)\cdot t_s)\right]$$

where: $n$ is the sample index, $j$ is the index of window samples relative to $n$,
$RMS$ is the rms of all bins in $E_{in}$ , $Noise$ is an estimate of random noise in a bin,
$N$ is the number of bins to be averaged on each side of the target bin,
$t_s$ is the sampling time interval and 3 is the cell under test +2 guard cells,
$TM$ is the threshold multiplier to control false alarms, typically 2.5 to 5. (4.6)

In general, sliding windows are 2D. A similar scheme may be used in the other dimension. The sliding window is calculated in 1D and then those calculations are subjected to a second sliding window in the orthogonal dimension. Figure 4.18 shows a conceptual 2D sliding window. Typically there are three range bins on each side of the test bin separated by the guard bins and a similar number in the Doppler dimension depending on the size of the FFT. If the FFT is big, then there might be as many as five Doppler bins averaged on each side of the test bin. If the FFT is small (256), then only two bins on each side might be averaged.

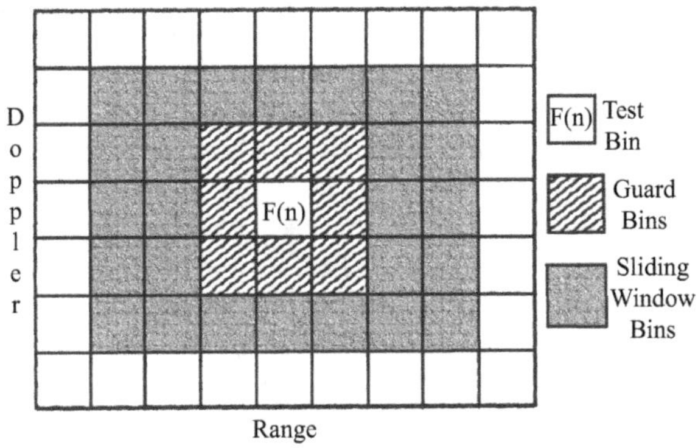


*Figure 4.18 2D sliding window concept*

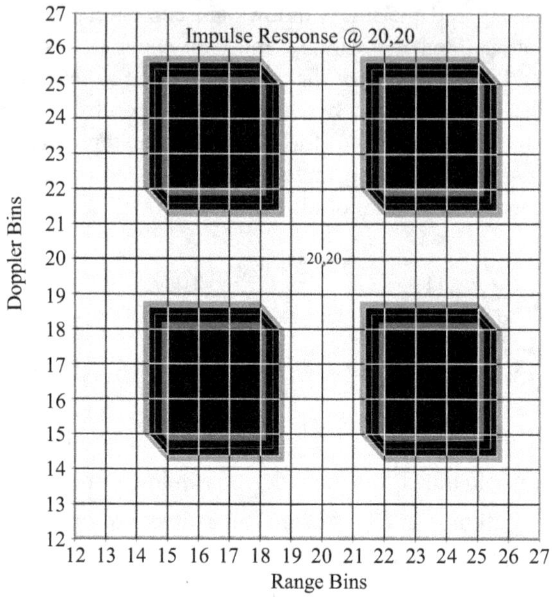


*Figure 4.19 2D impulse response sliding window 3-3-3 in range and Doppler (SlidingWindow001)*

Figure 4.19 shows the impulse response of a 2D sliding window centered at range Doppler test bin 20,20. There is no response at the test bin but there is a threshold raising response in the surrounding bins. This is a serious disadvantage if target density is high relative to resolution. If the surrounding area of a point target is reasonably quiet, then there will be a characteristic threshold pattern as shown in Figure 4.19.

An example 2D sliding window threshold for real clutter after STAP cancellation is shown in Figure 4.20. In the clutter dominated region close to the center of the mainbeam, the sliding window threshold is set by residual clutter. In the clutter sidelobe and clutter free frequency regions, one can easily see the four pulse threshold desensitization surrounding point targets in Figure 4.20. As long as targets are relatively sparse compared to range Doppler resolution, sliding windows are very effective in dramatically reducing false alarms.

For STAP the covariance matrix is averaged in each term for a number of range Doppler bins around the current bin before matrix inversion. The averaged inverted matrix is used to set the coefficients for the current bin cancellation. There is a fine tradeoff between the number of bins in range and Doppler to be used for the covariance matrix and the averaging versus the number of degrees of freedom available. Too many bins may actually provide poorer performance. Recently, knowledge-aided STAP techniques developed as part of the DARPA KASSPER project can improve performance in the presence of nonstationary clutter [8].

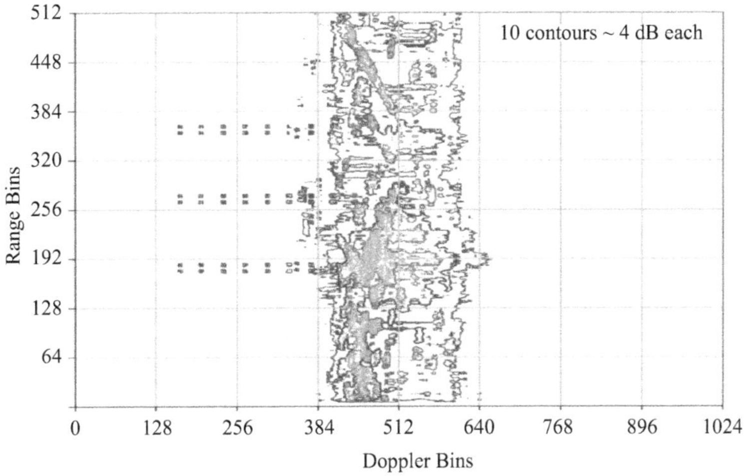


*Figure 4.20 Example sliding window threshold – Clutter @ PRF/2 (GMTIStLouis001)*

## 4.7 Analog to digital conversion [9,10]

The most common forms of A/D for radar and EW are flash converters and hybrid sigma delta converters. There are many other types of A/D including folding, successive approximation, pipelined multistage flash, ripple through pipeline and time interleaved. All have their attributes depending on the current device technology. There are methods using stretch waveforms and subbanding that relax the dynamic range or the sampling rate but, in general, the current sensor requirements lie between 1.3 bits-GHz and 160 bits-GHz.

The simplest form of A/D is a brute force flash converter. A typical block diagram for such a unit is shown in Figure 4.21. It consists of a parallel array of high-speed comparators whose reference is a distributed successively reduced voltage in a ladder network. The analog input is sampled and held for a major fraction of the clock interval. This gives the comparators time to settle and provide a unitary output to decode logic, which creates an unambiguous binary code output to subsequent processing. This architecture is quite good up to about 8 bits, which requires 255 comparators and fast decode logic. There are flash A/Ds up to at least 5 GHz sample rate. Often the sample and hold (S/H) is ping-ponged with two capacitors controlled by input and output switches. As previously mentioned, the S/H circuit may be quasi-resonant to minimize the equivalent aperture time. Eight bits is more than adequate for SAR in the absence of strong (saturating) RF interference because there are so many samples integrated providing output dynamic ranges of 60–70 dB.

Another type of A/D is the sigma-delta converter shown in Figure 4.22. It may use a multibit flash converter and multibit digital to analog converter

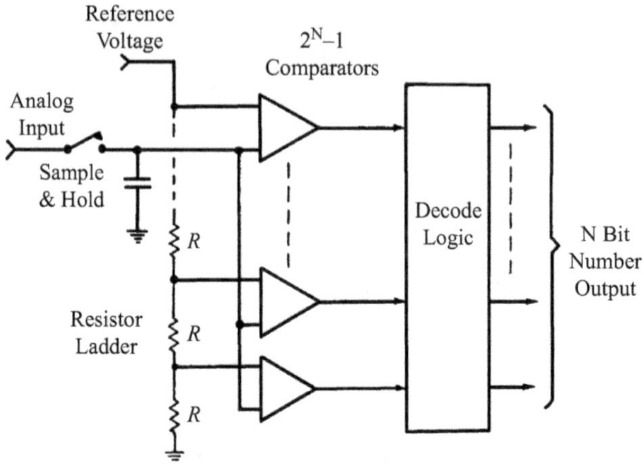


*Figure 4.21 Flash A/D converter*

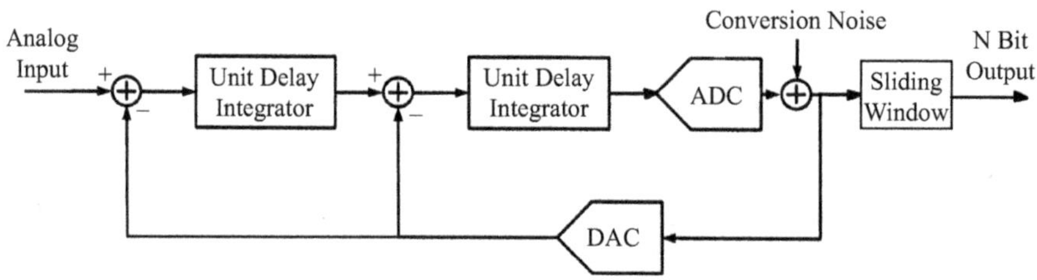


*Figure 4.22 Sigma-delta A/D converter*

(DAC) in its inner loop but may also use a single-bit converter and DAC. Obviously to provide 160 bits-GHz a sigma-delta converter must have a sampling clock much higher than the input bandwidth. To achieve a 13-bit output for detecting MTIs in clutter, it may have to run 32 times faster than the range bin rate using an inner loop A/D with 8 bits resolution. For 6 m resolution, that requires 800 MHz sampling rate. With modern device technology that is well within the state of the art.

Often there are 1–3 analog switched capacitor single-sample time integrators preceding the A/D. This has the effect of shaping the output noise spectrum so that a much smaller amount falls in the final desired band. Since the output may be at a much higher rate than the desired time resolution, there is often a digital sliding window lowpass filter following the ADC that is resampled at the desired output rate. In some cases, the output filter is an Hilbert filter, which has at least a 4:1 down-sample rate. Sigma-delta converters have been in use for digital communications differential PCM since the late 1960s. They have been used in radars and other sensor applications since the late 1980s.

## 4.8 Digital I/Q demodulation [11]

Mismatch between the I/Q components of a sensor vector measurement representing amplitude and phase gives rise to an image error term, part of which will be correlated and appear as ghost targets and the remainder will appear as background noise. The mismatch arises due to DC bias, amplitude errors and phase errors.

Beginning in in the late 1970s, Hilbert transform digital filters combined with single A/D converters made dramatic improvements in I/Q imbalance noise. A second element of these applications was the advent of fast sigma-delta ($\Sigma-\Delta$) A/D methods, which allow a small fast quantizer with digital integration and feedback to the A/D input to achieve large dynamic range as well as speed. The third element was the development of sample and hold circuits with picosecond aperture jitter. There are several ways that this general idea has been applied. These methods were incorporated into production radar and communications hardware in aircraft and spacecraft at General Motors Hughes Electronics in 1978 and beyond.

The idea is to choose the final IF center frequency to be the same as the desired receive bandwidth. (There are other choices of sample rate and bandwidth, which also work but not as well.) The IF signal is then digitized in a single A/D converter whose sample rate is four times the desired bandwidth. The A/D output is applied to a Hilbert transform filter followed by a 4 to 1 down sampler, which shifts the output to baseband. The basic idea is shown in Figures 4.23 and 4.24.

The spectral situation is shown in Figure 4.24. The center frequency of the output of the final IF is chosen to be at one quarter of the sampling frequency of the following A/D converter. The bandwidth of the final IF amplifier also is chosen to be

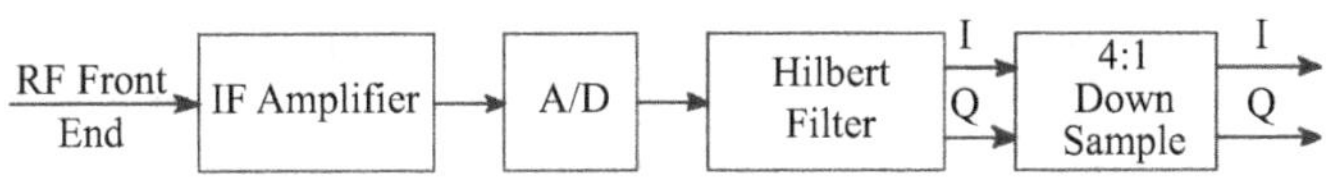


*Figure 4.23 A/D & Hilbert conversion to baseband*

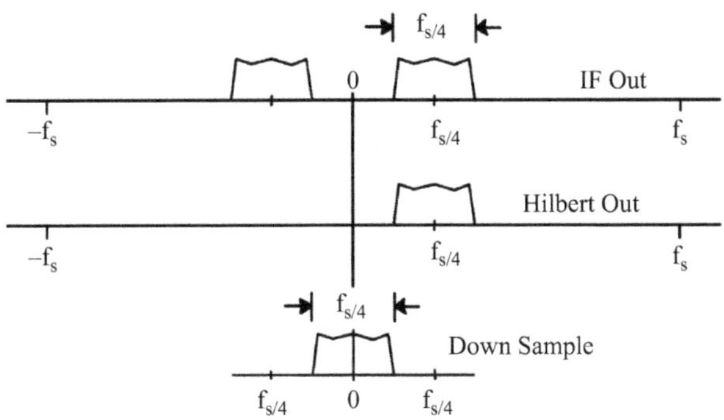


*Figure 4.24 Hilbert transform spectral situation*

approximately ¼ of the A/D sampling rate. As shown in Figure 4.24, the digital Hilbert transform filter selects the positive sideband of the IF output. The Hilbert filter output is down-sampled to ¼ of the A/D sample rate, placing the signal band at baseband as shown in Figure 4.24. Although it may not be obvious, this strategy significantly improves the SNR as well as the I/Q balance. The 4:1 integration doubles the SNR and reduces quantization noise by a factor of four (more in Chapter 5).

The typical Hilbert transform filter is what is known as a FIR half band filter shifted in center frequency to the sample rate divided by four. Such a filter configuration has many desirable properties including that half the filter coefficients are zero and the other half are anti-symmetrical about the center response.

---

**Example 4.3** For example, the impulse response of a typical 21-sample length half band Hilbert filter is shown in Figure 4.25. In addition, all of the filter coefficients (impulse response) are separable into even and odd samples, which are entirely imaginary (quadrature) or real (in-phase). Odd samples are real (in-phase) and even samples are imaginary (quadrature). This greatly simplifies the arithmetic as well as reduces the total number smashing. Note that the outermost values of the impulse response are zero.

The Hilbert transform filter impulse response of Figure 4.25 can be realized with the transversal filter shown in Figure 4.25. This implementation takes advantage of the symmetry of coefficients as well as the separability of I/Q coefficients of the transversal filter. Since a FIR filter has a linear phase, the in-phase component can have an exact phase match to the quadrature output with proper clock delay as shown in Figure 4.26.

Anticipating that the output of the Hilbert transform filter is to be resampled at ¼ of the input rate, only ½ of the calculations that occur on the even samples are required. However, those filter outputs are ½ of the input samples in the quadrature channel shown in Figure 4.26; therefore, only ¼ of the multiply and adds are required. There are even IIR filter versions of the Hilbert transform strategy that require no multiplies.

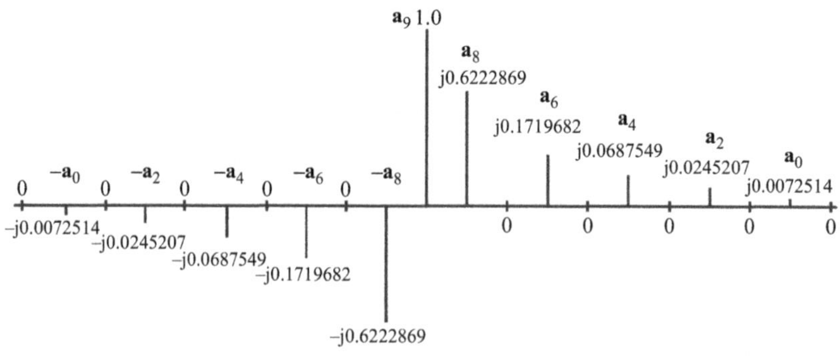


*Figure 4.25 A 21-sample Hilbert transform filter impulse response*

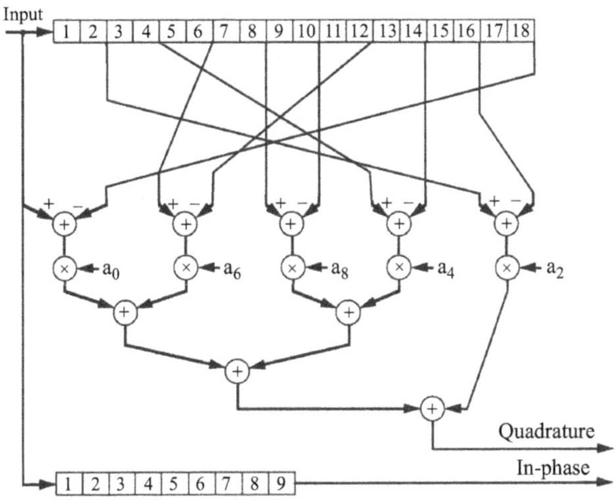


*Figure 4.26 A 21-sample Hilbert transform filter*

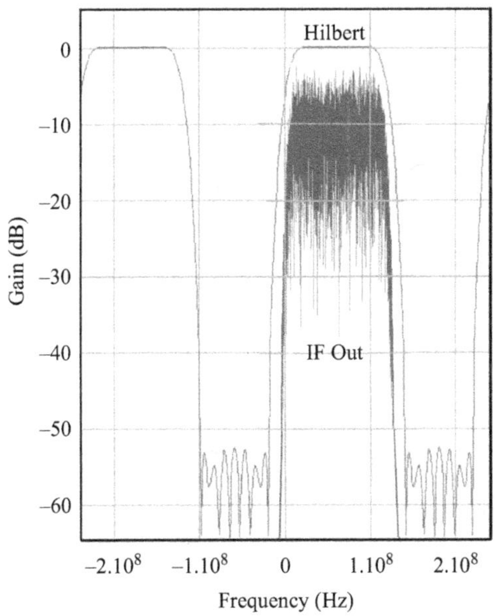


*Figure 4.27 IF output and Hilbert filter (HilbertTransform002, Noise004)*

The frequency response of the example impulse response shown in Figure 4.25 is shown in Figure 4.27 with a sampling rate of 240 MHz and an approximate IF bandwidth of 60 MHz. Note that it just selects one side of the two-sided spectrum at ¼ of the sampling rate. When that spectrum is selected by the Hilbert filter, it is

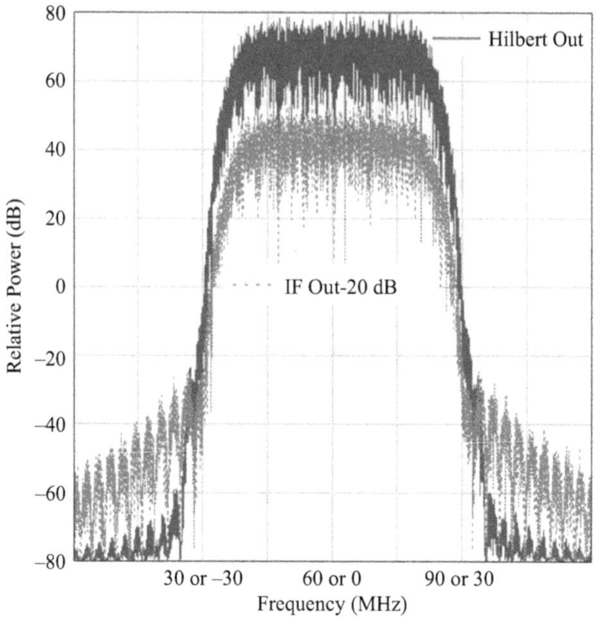


*Figure 4.28 Final baseband compared to IF output (Noise004)*

down-sampled to baseband providing the desired output. As a practical matter the actual IF bandwidth must be somewhat smaller than the center frequency so that the image is more completely rejected. Otherwise there will be "rabbit ears" in the transition band that will contribute to residual image noise. With careful design an additional reduction of 15 dB in IQ imbalance noise can be realized over analog techniques.

Figure 4.28 shows a random signal output from the final IF amplifier with a dynamic range of approximately 80 dB at a center frequency of 60 MHz with bandwidth a little less than 60 MHz, which was sampled at 240 MHz. A $-20$ dB offset is introduced into the plot to separate the two output responses. Also shown is the Hilbert transform output using the impulse response shown in Figure 4.25 and then down-sampled to zero center frequency with an approximate bandwidth of $\pm 30$ MHz. Note that the dynamic range has increased to about 130 dB in Figure 4.28.

## 4.9 Polyphase filtering

It often occurs at the high-speed input to a processor complex that a single channel of hardware cannot keep up with the data rate out of the A/Ds. One common solution to this challenge for Hilbert transformers and pulse compression is what is

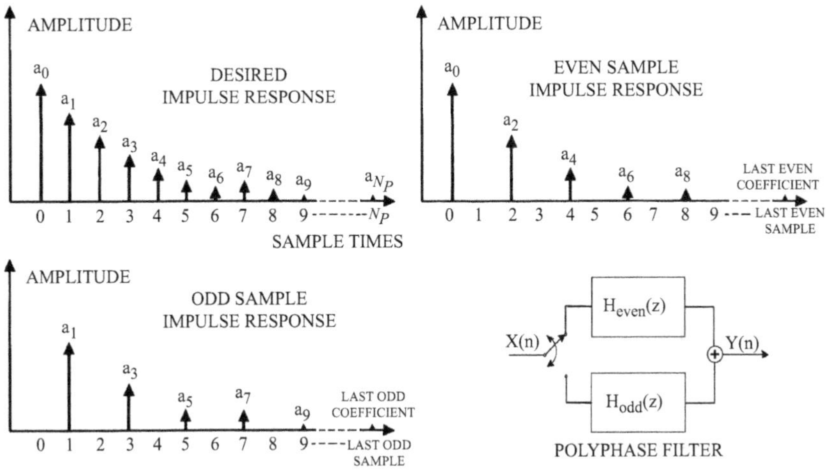


*Figure 4.29 Polyphase filter example*

known as a polyphase filter or filter bank. The idea is to distribute individual samples to multiple FIR filters or FFTs all running in parallel. Each is working on a subset of the data stream. It is not uncommon to have as many as 50 parallel channels commutated from the data stream. The number smashing rate in each channel is correspondingly reduced by the number of phases. There are many possible implementations of the polyphase filter idea. Often there is down-sampling after the filter bank, which reduces the computation rate still further. Figure 4.29 shows a two-phase polyphase filter. The overall desired impulse response is shown in the upper left in Figure 4.29. That impulse response is split into two responses one for even input samples and one for odd samples. They are shown in the upper right and lower left in Figure 4.29. Since there are ½ as many coefficients to multiply and add by only ½ as many input samples, the per processor work load is dramatically reduced.

The two parallel data streams are combined at the filter outputs creating a single data stream. Obviously when the number of filter phases is large keeping the timing straight is an important issue since some processes may not take the same amount of computation time. Polyphase FFTs are often used for very high-resolution linear FM pulse compression. Clearly this idea works for FFTs that are basically FIR filters; but it can also be made to work for IIR filters although it is more complex.

## 4.10 Pulse compression

### *4.10.1 Linear FM/chirp*

Half of the resolution achieved by a radar or Elint sensor is in the transmitted pulse bandwidth. Linear FM or "chirp" is one of the oldest pulse compression

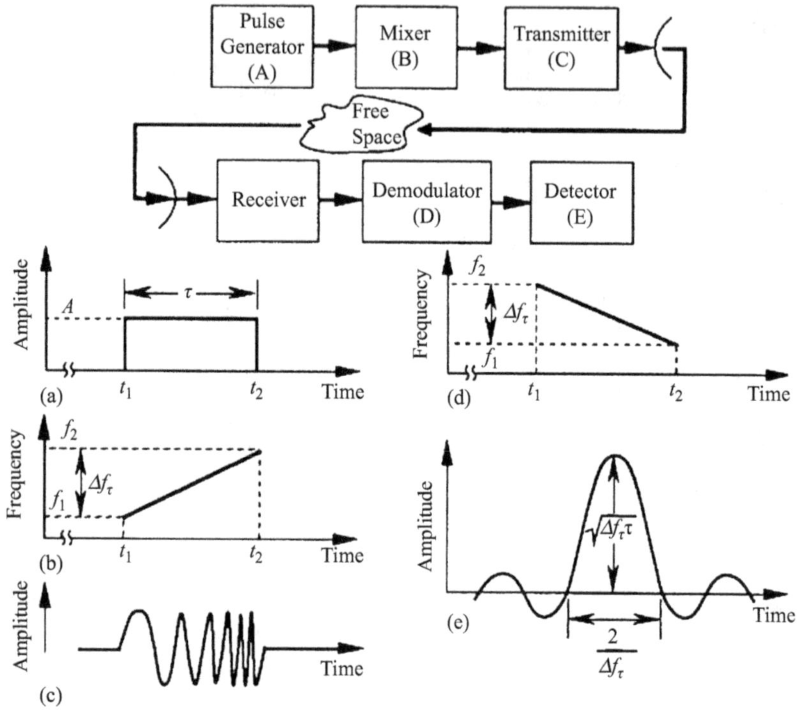


*Figure 4.30 Basic chirp process [6]*

or bandwidth expansion methods. The basic process is shown in Figure 4.30. The waveform output from each functional (lettered) block is shown in the bottom of the figure. A rectangular pulse of length, $\tau$, is generated by the pulse generator timing circuit. The pulse enables or selects a section of a steadily increasing or decreasing FM waveform. The linear FM or chirp bandwidth, $\Delta f_\tau$, from the mixer output is amplified by the transmitter, launched by the antenna and propagates to the target. The wavefront scattered from the target is received and amplified. This signal is then mixed in the demodulator with the opposite slope FM. This approximate matched filter results in an output pulse whose amplitude is proportional to the square root of the time bandwidth product $(\Delta f_\tau \tau)^{0.5}$. For large bandwidths and unweighted matched filtering, the null to null width of the output pulse is $2/\Delta f_\tau$. For high-resolution systems, this leads to very high bandwidth front-end signal processing. Since the processed range swath in such systems is small relative to the PRF, this high bandwidth was very inefficiently used and systems were invented to stretch the processing over the entire interpulse interval. Those approaches are called "stretch" processing.

To simultaneously achieve long-range detection and high-range resolution requires a very long pulse and large frequency excursion. The long, linear FM waveform in which the pulse length exceeds the swath width has some very nice processing characteristics, which can be exploited by stretch processing.

The chirp transmitted signal is of the form shown in (4.7).

$$s(t) = A \cdot \mathrm{PUL}(t,\tau) \cdot \cos\left(2 \cdot \pi \cdot \left(f_0 \cdot t + \frac{\Delta f_\tau}{2 \cdot \tau} \cdot t^2\right) + \varphi_0\right) \tag{4.7}$$

where $A$ is the transmitted amplitude, PUL(t,$\tau$) is a rectangular pulse of length $\tau$, $f_0$ is the center of the operating band, $\Delta f_\tau$ is the total frequency excursion and $\varphi_0$ is the starting phase. The output power from a filter matched to this waveform is given in (4.8).

$$\begin{aligned} P_{match}(t) = c_{match} \cdot SNR \cdot \exp\left(j \cdot 2 \cdot \pi\left(f_0 - f_D\right) \cdot (t - t_0)\right) \cdot \\ X\left((t - t_0), (f_0 - f_D)\right) \end{aligned} \tag{4.8}$$

where $c_{match}$ is a constant representing the transmitted amplitude, range equation and the receiver performance including the matched filter, $f_D$ is the received signal Doppler offset from the carrier $f_0$, $t_0$ is the time of the matched filter sampling instant and $\boldsymbol{X}$ is a chirp waveform output from its matched filter as shown in (4.9). Define $\Delta f_D = f_0 - f_D$ and $\Delta T = t - t_0$ then

$$\begin{aligned} X\left(\Delta T, \Delta f_D\right) = \tau \cdot \left(1 - \frac{|\Delta T|}{\tau}\right) \cdot \exp(-\mathrm{j} \cdot \pi \cdot \Delta f_D \cdot \Delta T) \cdot \\ \frac{\sin\left(\pi \cdot \left(\Delta f \cdot \Delta T + \Delta f_D \cdot \tau\right) \cdot \left(1 - \frac{|\Delta T|}{\tau}\right)\right)}{\pi \cdot \left(\Delta f \cdot \Delta T + \Delta f_D \cdot \tau\right) \cdot \left(1 - \frac{|\Delta T|}{\tau}\right)} \end{aligned} \tag{4.9}$$

---

**Example 4.4** For example, consider the case of a compression ratio of 20:1 and 100 foot resolution with $\Delta f$ of 5 MHz. Then the matched filter response as a function of time delay and Doppler offset using (4.9) is as shown in Figure 4.31. The figure is a top-view contour plot (i.e., like a topographic map) with 10 dB contour lines. The dark bands are very low sidelobes that are rapidly varying. The diagonal large central lobe shows one of the principal weaknesses of linear FM, which is the large ambiguity between range and Doppler. Another weakness shown in the plot is that any cut in Doppler has significant power for much of the range/time axis [12].

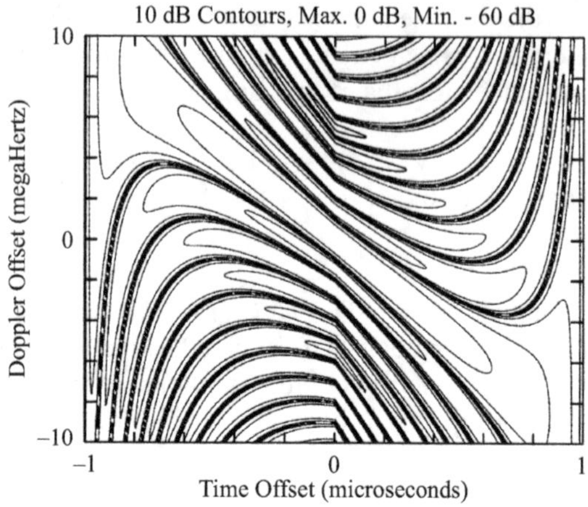


*Figure 4.31 20:1 linear FM ambiguity contour plot (LinFMAmb2) [6]*

## 4.10.2 Stretch processing

There are processing means available for linear FM that are so simple and advantageous that it is widely used. This simple processing method is known as stretch [13]. Some discussion of the fundamentals of stretch processing is appropriate to lay the groundwork for what follows. The basis for the derivation of the stretch processing parameters is shown in Figure 4.32, $T_{SW}$ is the swath width time and $T_E + 2T_{SW}$ is the receive window (the time during which returns from the entire swath come in; hence the time during which data is taken) [6,13]. The actual transmitted pulse length is $\tau = T_{SW} + T_E$.

Similarly, the actual frequency excursion is $\Delta f_\tau$. From simple geometrical arguments, the required IF bandwidth, $\Delta f_{IF}$, could be as small as shown in (4.10). Since the total frequency excursion can be resolved into range bins equal to the

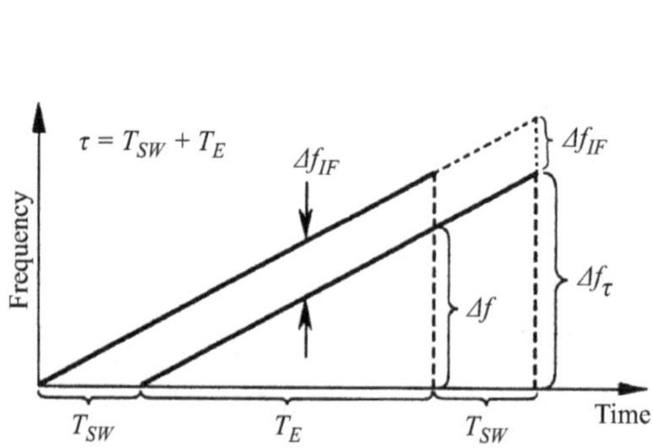


*Figure 4.32 Stretch processing basic parameter derivation [6]*

reciprocal of $\Delta f_\tau$, then $\Delta f_\tau = 1/T_{rb}$. Therefore,

$$\Delta f_{IF} = \frac{T_{SW}}{T_E + T_{SW}} \cdot \Delta f_\tau \quad \text{and} \quad \Delta f_{IF} = \frac{T_{SW}}{(T_E + T_{SW}) \cdot T_{rb}} = \frac{N_{rb}}{T_E + T_{SW}} \tag{4.10}$$

where $N_{rb}$ is the number of range cells across the swath and $T_{rb}$ is the range resolution or cell time.

A typical stretch mechanization is shown in Figure 4.33(a). The received signal is deramped with a center frequency matched to the center of the swath minus the IF frequency. The output of a narrowband IF matched to the frequency excursion over the swath width is heterodyned to baseband. The baseband is I/Q sampled at a sample rate of $N_s = 1/(2\Delta f_{IF})$ to meet the Nyquist criterion. This is typically followed by an FFT in one or two dimensions (Doppler and range). The total number of I and Q samples across $T_E + 2T_{SW}$ is then

$$N_s = (T_E + 2 \cdot T_{SW}) \cdot \Delta f_{IF} \tag{4.11}$$

which demonstrates one advantageous processing feature of stretch pulse compression processing. In particular, the number of samples required prior to range compression (which can be accomplished efficiently with an FFT) equals the number of samples after range compression, thus minimizing the precompression data rate. In addition, the pulse length is easily varied and preprocessing is minimal. The range compression can be accomplished by a simple FFT because after deramping, targets at a given range within the swath are characterized by an almost constant frequency within the $\Delta f_{IF}$ band.

Two functional approaches to mechanizing stretch are shown in Figure 4.33. The first approach (Figure 4.33(a)) is the one conventionally described. The second approach (Figure 4.33(b)) shows that the narrowband IF and A/D converter, whose

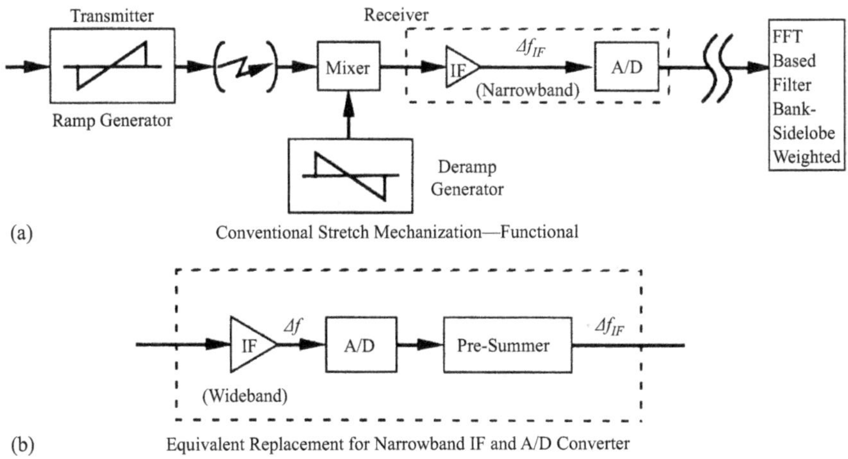


*Figure 4.33 Functional stretch mechanization [6]*

bandwidth and sample rate are $\Delta f_{IF}$, can be replaced by a wider band IF of bandwidth, $\Delta f$, followed by an A/D converter and a digital presummer. With the advent of higher-speed A/D converters, it often turns out that greater dynamic range can be achieved by digital filtering after conversion rather than analog filtering before conversion in the IF.

---

**Example 4.5** More details of the stretch technique are shown in the example of Figure 4.34. In the example given, the reference ramp from the deramp generator is a frequency of $f_0 + 1.539 + 0.2565$ GHz and has a total frequency excursion of 585.34 MHz. The received signal is at $f_0$ and three point targets (A, B, C) separated into the earliest, middle and latest range bins are represented by corresponding FM ramps. The time extent of the transmitted pulse in the example is 12 μs. The limiting resolution is the reciprocal of the time extent or 83.34 kHz. Since the transmitted pulse covers 500 MHz in 12 μs, the FM chirp slope is 41.67 kHz/ns. The dechirped signal at IF consists of continuous tones centered about the IF center frequency of 1.7955 (1.539+0.2565) GHz. The IF output is A/D converted using 513 MHz sample rate, which is slightly higher than the range resolution. If not for a frequency offset, the sample rate would fold the spectrum to DC. The digital presummer was chosen to be at 1/2 the sample rate (256.5 MHz) with an 85.5 MHz bandwidth. The output of the presummer is 4:1 subsampled to DC. That output is Fourier transformed into separate range bins. All of the frequencies, bandwidths and sample rates are "magic" in the sense that they bear an integral relationship to each other. The detailed simulation software is in the Chapter 4 appendix titled Stretch001. This processing is very simple because it allows a large transmitted bandwidth, a smaller receiver processing bandwidth, a small range window and a

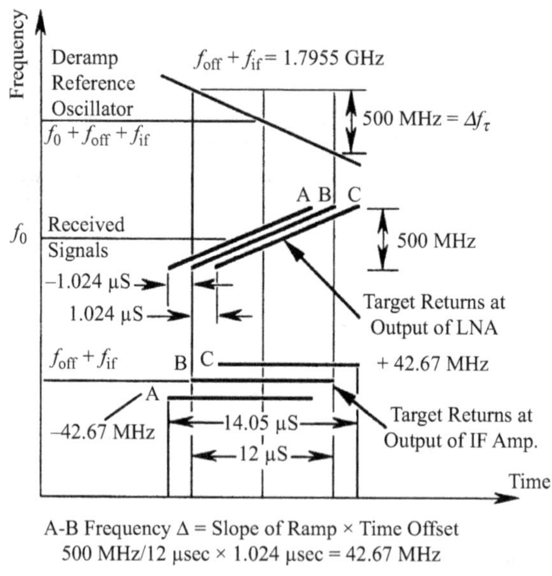


*Figure 4.34 Stretch technique converts time delay into frequency offset [14]*

wide range of usable PRFs (usually medium or low). One digital method that has compression as simple as linear FM, complementary coding, is described in a later section [13–15]. Often the processor must remove the time skew of the returns before further processing.

An important issue with any pulse compression waveform is the sidelobe performance since it limits the dynamic range in adjacent bins just as shown in Sections 2.6, 2.11, 4.5, 5.2 and 5.3 for different radar attributes but the same sidelobe limitation. There are several important pulse compression figures of merit including: the ratio of the peak compressed signal to the peak sidelobe (*PSLR*), the ratio of the peak signal to the RMS sidelobes (*RMSLR*) and the ratio of the peak signal to the integral of all the sidelobes (*ISLR*). If $X(t)$ is the output from the pulse compressor, then (4.12) defines the *ISLR*. The *RMSLR* is the *ISLR* divided by the PCR.

$$ISLR = \frac{\int_{\frac{-\tau}{2}}^{\frac{\tau}{2}} |X(t)|^2 \, dt - \int_{\frac{-2.3 \cdot T_{rb}}{2}}^{\frac{2.3 \cdot T_{rb}}{2}} |X(t)|^2 \, dt}{\int_{\frac{-T_{rb}}{2}}^{\frac{T_{rb}}{2}} |X(t)|^2 \, dt} \tag{4.12}$$

Equation (4.12) can often be integrated from 0 to the upper limit only if the sidelobes are symmetrical or Hermitian. With stretch processing time is converted to frequency (if one ignores the Doppler effects) and often the *ISLR* is calculated in the frequency domain. As a practical matter weighting must be applied to reduce sidelobes in the analysis filters so individual bandwidths are usually significantly larger than $1/T_{rb}$.

For a chirp waveform the swept frequency residue after deramp must lie in a single filter and the spurious FM noise covering the entire analysis band must be low enough that its integrated power doesn't limit the dynamic range of adjacent channels and vice versa. Therefore, linearity for the combination of the transmit waveform and its receiver demodulation must be on the order of $T_{rb}/\tau$ usually split evenly between transmit and receive. For the case of Figure 4.34 this requires 2 ns/ (2 × 12 μs)= 0.834 × $10^{-4}$ linearity each for transmit and receive. Furthermore, the integral of the FM noise and clutter in the analysis filter mainlobe and sidelobes must be as given in (4.13), otherwise distributed clutter, broadband noise jamming or FM noise will limit the desired Dynamic Range, $DR_d$, in every other bin since the FM spreads this signal over the entire band.

$$DynamicRange \leq \frac{P_{match}(t)\big|_{\max}}{\left(\int_{\frac{-\tau}{2}}^{\frac{\tau}{2}} \left(s_{FM}^2 + s_{clutter}^2\right) \cdot |h(t)|^2 \, dt\right)} \tag{4.13}$$

In addition, the pulse compression filter bank ratio of minimum signal to peak sidelobes for each filter must be greater than the desired Dynamic Range otherwise bright discretes (Figure 3.24) or repeater jammers will limit dynamic range. Thus,

$$DynamicRange \leq \frac{P_{match}(t)\big|_{\min}}{\left(\int_{\frac{-\tau}{2}}^{\frac{\tau}{2}} \left|BrtDis(\Delta T) * h(t)\right|^2 dt\right)} \tag{4.14}$$

For example, if the desired dynamic range is $10^6$, then the product of the filter *ISLR* and the noise variance must be $10^{-6}$. The *ISLR* of an unweighted FFT using the parameters of Figure 4.34 and 3 dB mainlobe width is −4.1 dB. So the FM noise must be approximately $2.5 \cdot 10^{-6}$ below the carrier if there is no interfering clutter. Assuming a weighted filter bank using a 4th power Parzen window and using the 10 dB mainlobe width yields an *ISLR* of −15.3 dB, which would provide some margin for interfering clutter as well as allow some relaxation of the FM noise requirement. Parzen window far sidelobes are low enough that a bright discrete 20 bins away will be attenuated by more than 60 dB. Over the bandwidth of the chirp in Figure 4.34, these requirements are very difficult to achieve [15]. One of the main weaknesses with chirp is that it is easy to detect and jam, even inadvertently. Another weakness is that linearity and long-term stability often degrade as hardware ages.

## 4.11 Discrete phase codes

Jamming resistance and aging degradation are significantly better for discrete phase codes than chirp. Another very important property of phase codes over chirp is the ability to create zero sidelobes for a few range ambiguities that may be inside the mainlobe footprint. Another important feature of phase codes in a crowded spectrum is the ability for two or more radars to use the same code family with almost no mutual interference as is done in CDMA cellular communications. No modern design should use chirp not withstanding its processing simplicity. Discrete phase codes include binary phase codes, such as Barker and compound Barker, random, pseudo-random and complementary codes; and polyphase codes, such as Frank codes and discrete chirp. Typically the *PCR* of a discrete phase code is the ratio of the chip time, $t_c$, to the pulse time, $\tau$, or alternatively the number of chips in the code [6,7,12–14,16–20]. The major problem with high-resolution long discrete phase codes is twofold. First is that the instant bandwidth is on the order of the reciprocal of the chip time. This requires real-time time delays across the AESA antenna, for example, a one-foot chip roughly requires real-time delays every foot. On the other hand, a reflector antenna easily accommodates a phase code with few or no real-time time delays. Second is that the number of range samples required before range compression is large; in fact, equal to the number of range cells in the swath plus the *PCR*, which could be 5:1 more than the final range swath.

**Example 4.6** For Example 4.6, for a swath width of 2048 range bins and a *PCR* of 2000:1, 4048 range samples are needed, which is more than the number required with stretch. This problem can be resolved for some codes such as the complementary codes by using a Hudson Larson decoder (a rough analog of the FFT).

If range pulse compression can be accomplished in a single ASIC, then this large number of samples is not a problem. Simple processing is possible for the more structured codes such as the Barker, compound Barker, Frank and complementary codes, but they do not lend themselves to incremental variation in pulse length. On the other hand, the random-like codes, which can be chopped off or extended to any desired pulse length, require complicated, inverse filter processing to improve their relatively poor *ISLR* resulting from matched filtering. (It is easy to show that the *ISLR* of a random discrete phase code approaches 0 dB as the code length increases.) Such processing could be done in a surface acoustic wave device for a few code lengths, but for large numbers of code lengths this approach is inappropriate. For example, a 10 to 1 pulse length variation would require 25 different code lengths in order to cover this range in approximately 10% increments. Digital compression is the normal method for phase codes.

---

**Example 4.7 Barker codes** The basic notions of discrete phase code can be illustrated with a 13 chip Barker code as shown in Figures 4.35 and 4.36. Figure 4.35 shows the waveform generation block diagram of a phase code. Figure 4.36 shows the waveforms at three points, A, B and C, in Figure 4.35. The phase code waveform shown at A in Figure 4.36 is applied to an RF phase shifter with a net phase difference between the two states of 180° at the source frequency.

The phase reversal CW signal at point C in Figures 4.35 and 4.36 is shifted upward in frequency in an RF amplifier and transmitted to the target. Usually the overall transmitted pulse envelope is determined separately and the modulation at B in Figures 4.35 and 4.36 is used to turn the RF amplifier on and off.

On reflection from the target and receipt by the radar, the signal is amplified, demodulated and presented to a correlation pulse compressor, as shown in Figure 4.37. The pulse compressor is often a "matched filter" and has the property that the output pulse width, $T_{rb}$, is the chip width, $t_c$, and the amplitude is equal to the number of chips (or bits) in the phase code. For many reasons such as sidelobe

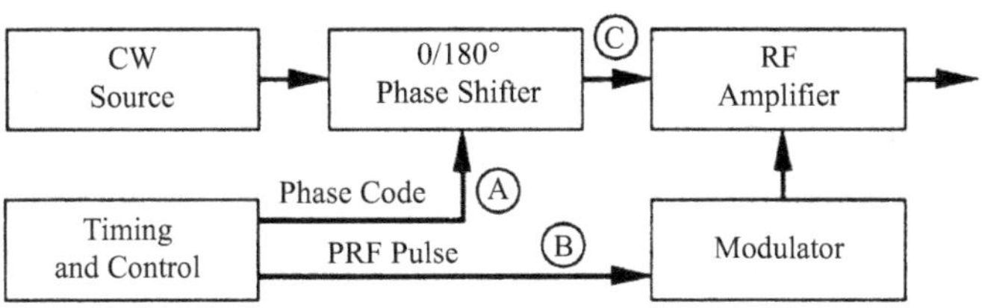


*Figure 4.35 Phase coded transmitter*

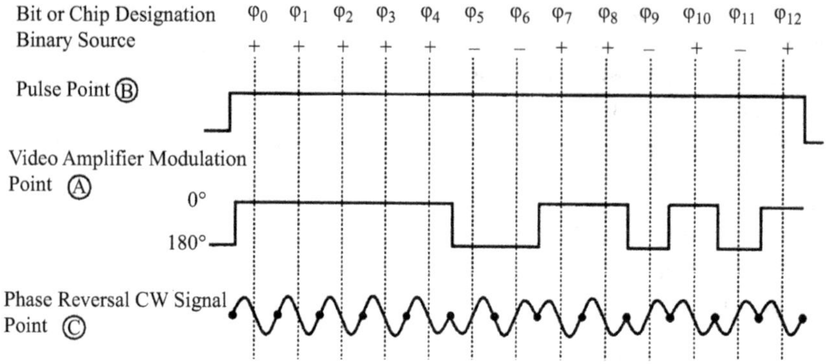


*Figure 4.36 Waveforms for a Barker code of 13 chips*

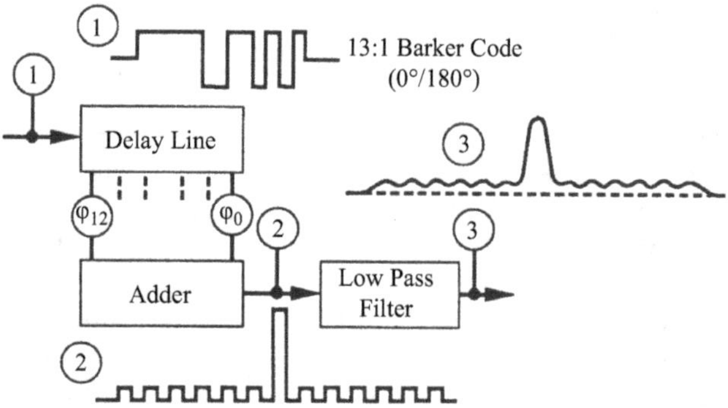


*Figure 4.37 Discrete phase Barker code pulse compression*

suppression and straddling loss, the compressed output only approaches the matched filter.

Returning, then, to the extended example of Figures 4.35 and 4.36, the output of a pulse compressor for a 13:1 Barker code is shown in Figure 4.37 point 2. Originally Barker code compressors were analog devices made up of delay lines and summing networks and as such completely inflexible. With the advent of high-speed A/D converters in the 1960s, Barker codes were some of the earliest examples of digital compression. The incoming waveform is shown idealized at point 1 in Figure 4.37. Each phase weight $\varphi_0$ through $\varphi_{12}$ is at a delay line tap one chip time apart. The outputs are summed and appear as shown at point 2. This output usually is quite noisy and so it was lowpass filtered with a bandwidth matched to the chip. The resulting output was as shown at point 3 in Figure 4.37.

The pulse compression process is the discrete autocorrelation function, of course, whose general form is shown in Figure 4.38. The Barker code cross-correlation calculation of Figure 4.37 follows the correlation model of Figure 4.38. The correlation is just the chip-by-chip product of the prototype waveform and the received waveform

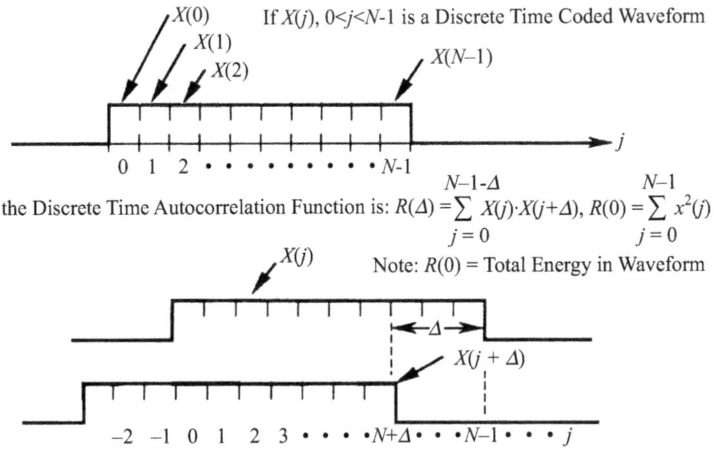


*Figure 4.38 Autocorrelation function for discrete time coded waveforms*

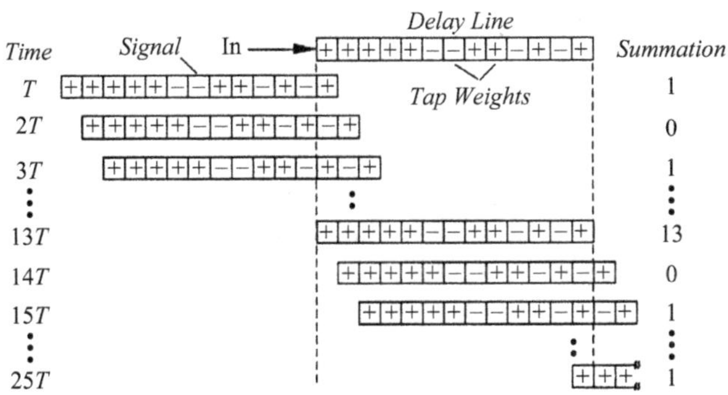


*Figure 4.39 13:1 Barker compression calculation*

summed over all chips with successive offsets of an integral number of chips. The zero offset case is just the sum of the square law detection of all the chips.

One example of a binary phase code pulse compressor correlation is shown in more detail in Figure 4.39. The figure shows a 13-chip Barker code decoder in which a shift register slides the received detected signal past a correlator, which multiplies the detected signal by phase weights of 1 or $-1$ corresponding to the transmitted phase waveform and sums the weighted results to produce a discrete output as shown by the output at location 2 in Figure 4.37. Since the returned waveform is not the transmitted waveform, the pulse compressor is performing a cross-correlation that digs out scattering from point targets in the range cell.

Note that the sidelobes for an ideal received waveform are identically 1 or 0 and extend $\pm$ 12 elements from the central response. Each successive time delay $T$ advances the waveform one chip and the product of the tap weights times the ideal signal when summed yields the successive summation outputs to the right in Figure 4.39.

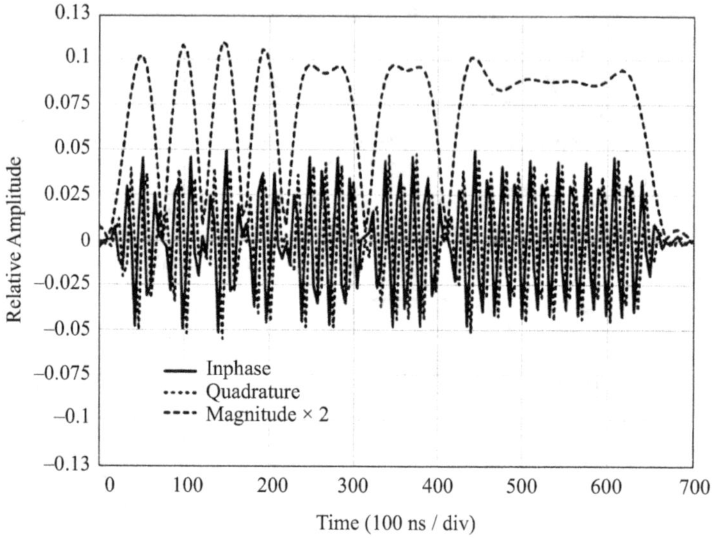


*Figure 4.40 Typical Barker code input to compressor (ChanSep1a) [6]*

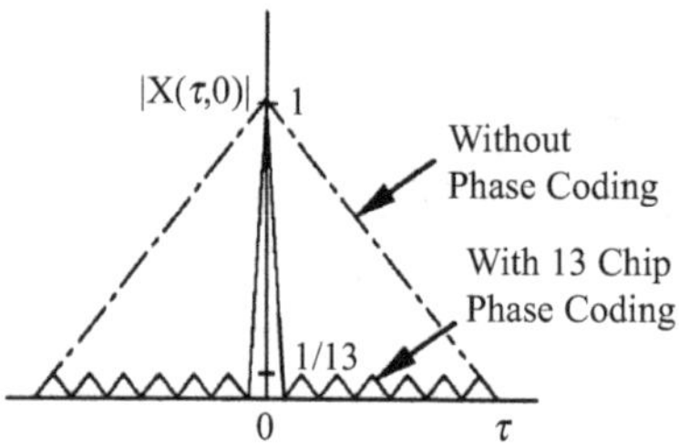


| *Length of Code N* | *Code Elements* | *Peak Sidelobe Ratio, dB (−20 log N)* |
|---|---|---|
| 2 | + −, or + + | −6.0 |
| 3 | + + − | −9.5 |
| 4 | + + − +, or + + + − | −12.0 |
| 5 | + + + − + | −14.0 |
| 7 | + + + − − + − | −16.9 |
| 11 | + + + − − + − − + − | −20.8 |
| 13 | + + + + − − + + − + − + | −22.3 |

*Figure 4.41 Barker code summary*

A more representative example of the sort of waveform at the receiver input to a 13 chip Barker code compressor is shown in Figure 4.40. Three curves are shown, the in-phase waveform, the quadrature waveform and the corresponding magnitude multiplied by 2 to separate it from its two components. The only waveform that is remotely recognizable is the magnitude wave. The chip time is 50 ns and the overall pulse length is 650 ns. Note that the front-end filters significantly degrade the rise and fall times as well as produce amplitude overshoot. These effects degrade total detection performance.

A summary of all the Barker codes and their peak to sidelobe ratios is given in Figure 4.41. The *ISLR* for Barker codes are roughly −10Log(*N*) with the 13:1 code obtaining a best value of −11.5 dB. No other codes do as well at any length. Barker codes are commonly used in many magnetic media synchronization, radar and communication applications.

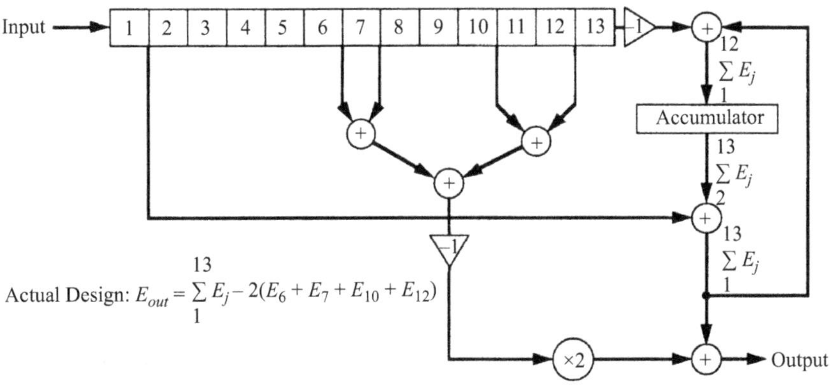


*Figure 4.42 Typical Barker pulse compression design [6]*

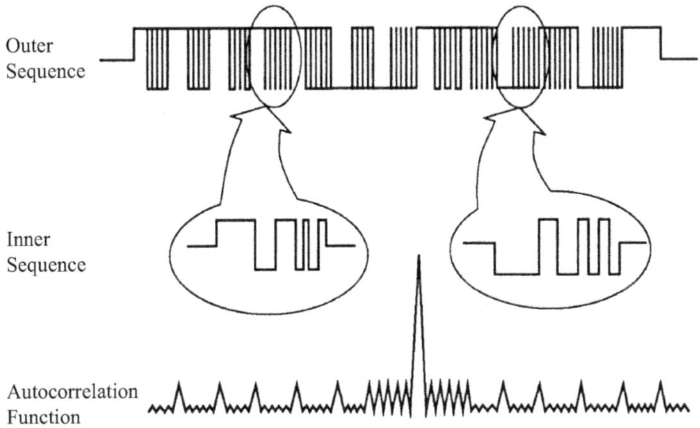


*Figure 4.43 Schematic of compound or concatenated Barker sequences [6]*

Numerous simplifications have been devised for Barker pulse compression, such as the one shown in Figure 4.42. This particular design was used in a single integrated circuit pulse compressor. The result is a very inexpensive and very fast pulse compressor that can be used alone or in cascade to create longer code compressors with chip times to the order of 1/3 nanosecond or less.

There are no known Barker codes beyond 13, but some of the advantages of Barker codes can be obtained by creating longer codes that are products of Barker codes. The inner Barker code is phase modulated by an outer Barker code such as that shown for 169:1 in Figure 4.43. Code compounding can produce quite an array of performance parameters as will be shown later in this chapter. In addition to reasonably good sidelobes, a compound Barker code is often much simpler to implement than the same length random code as shown in Figure 4.44. Sidelobe suppression filters for compound Barker codes are also very simple.

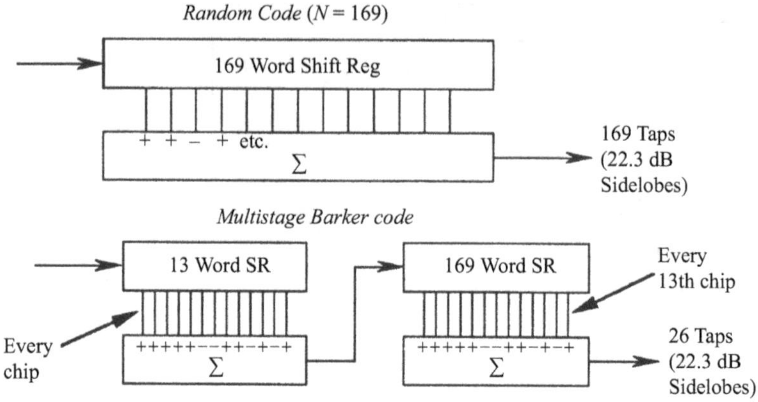


*Figure 4.44 169:1 multistage Barker code comparison [6]*

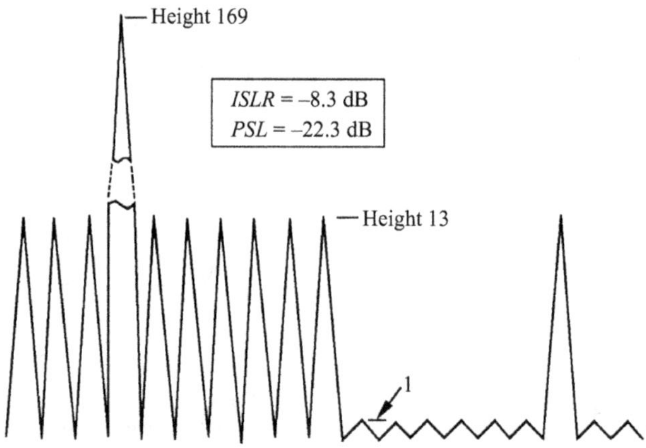


*Figure 4.45 Central portion of compound Barker autocorrelation function [6]*

The central portion of a compound or multistage Barker code is shown in Figure 4.45. Most of the sidelobes are of height 0 or 1 with 24 lobes of height 13. The very regular sidelobe pattern of compound Barker codes lends itself to simple sidelobe suppression filters, which will be explained later in this section.

It is interesting to compare the close-in power spectrum of a 13 chip Barker code and a compound Barker code of 169 chips. Figure 4.46 shows the normalized power spectrum of both codes. Obviously for a given peak power the compound code will have 11 dB more energy. The normalization is useful to see the relative similarity of the two in the frequency domain. Clearly, the Barker codes discussed here and polyphase Barker codes to be presented in later sections can be concatenated to create almost any length code and frequency occupancy.

Often the sidelobe performance of Barker codes is not adequate and sidelobe suppression is used to improve the peak and integrated sidelobe performance at the

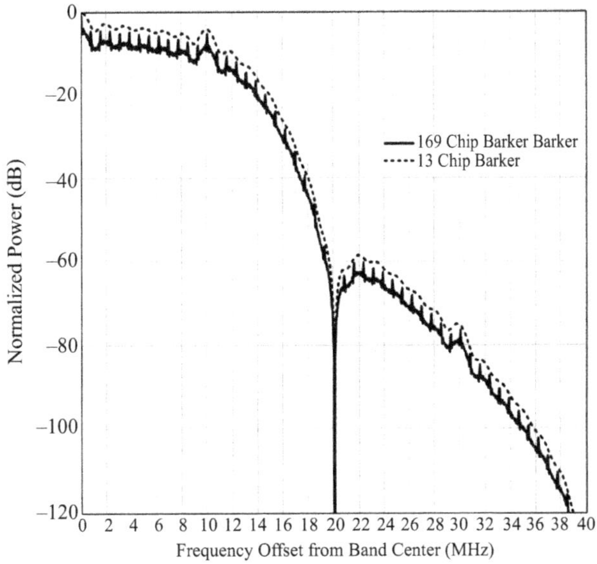


*Figure 4.46 Power spectrum 13:1 Barker and 169:1 compound Barker (ChanSep2a) [6]*

expense of some loss in *SNR*. The basic idea can be most easily understood by describing the pulse compressor output in the Z transform domain and then observing that there is an ideal desired part and an undesired part (the sidelobes), which should be completely cancelled. This approach can be used with almost any binary code.

---

**Example 4.8** For Example 4.8, the autocorrelation function of a 13-bit Barker code in the Z domain is as given in (4.15):

$$R(Z)=12\cdot\left(\underbrace{1}_{\text{Desired}}+\underbrace{C_B\cdot X(Z)}_{\text{Undesired}}\right)$$

$$\text{where: } X(Z)=\sum_{n=-6}^{6} Z^{2\cdot n} \text{ and } C_B=1/12 \tag{4.15}$$

The ideal sidelobe suppression filter transformation is then *H*(*Z*), which can be expressed as a power series in *X*(*Z*) as shown in (4.16).

$$H(Z)= 1\big/\left(1+C_B\cdot X(Z)\right)=1-C_B\cdot X(Z)+C_B^2\cdot X^2(Z)-\ldots \tag{4.16}$$

An approximate sidelobe suppression filter transformation contains the first two terms of the series in (4.16) as given in (4.17).

$$H_A(Z) = 1 - C_{app} \cdot X(Z) \tag{4.17}$$

where a near optimum $C_{app}$= 0.047= 1/32 + 1/64, was determined by computer simulation and a simple binary weight approximation. Then for convenience, $H_A$ can be written as shown in (4.18).

$$\begin{aligned} H_A(Z) &= Z^{-13} - C_{app} \cdot Z^{-1} \cdot \sum_{n=0}^{n=12} Z^{2 \cdot n} \\ H_A(Z) &= Z^{-13} - C_{app} \cdot Z^{-1} \cdot (1 - Z^{-26}) / (1 - Z^{-2}) \end{aligned} \tag{4.18}$$

The approximate sidelobe suppression filter, $H_A$, is used in cascade with the matched decoder pulse compressor as shown in Figure 4.47. Similar methods can be used with almost any discrete phase code. In some cases, it is as simple as adding adjacent samples from the digital pulse compressor. All of these schemes spread the mainlobe in trade for better sidelobes.

The filter, $H_A$, can be implemented either as a canonical finite impulse filter (FIR) filter or as a recursive filter as given in (4.18). Such sidelobe suppression filters can obviously be extended to compound Barker codes by applying the same type of filter to the outer code as well as the inner code, as suggested in Figure 4.47. Of course, the weighting terms are offset by 13 chips rather than 1 chip in the outer code. Figure 4.48 shows a FIR version of a Barker sidelobe suppressor.

The advantage of the FIR form of this sidelobe suppression filter is that it can be pipelined and thus handle smaller chip times. The IIR version uses fewer adders and may save cells in an ASIC or standard cell array. The zero Doppler sidelobes for a 13-bit Barker code with and without sidelobe suppression are shown in Figure 4.49. Note that the normal Barker sidelobes extend 12 bins on either side of the mainlobe, whereas with suppression the sidelobes extend twice as far but are down more than −10 dB. Although not shown, suppression performance is also improved in the presence of uncompensated Doppler.

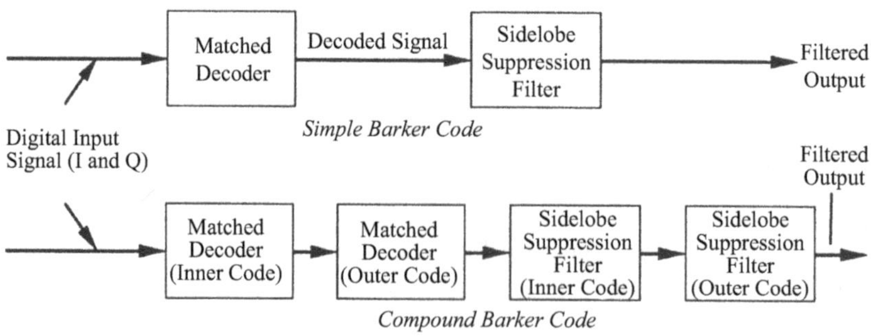


*Figure 4.47 Sidelobe suppression filters [6]*

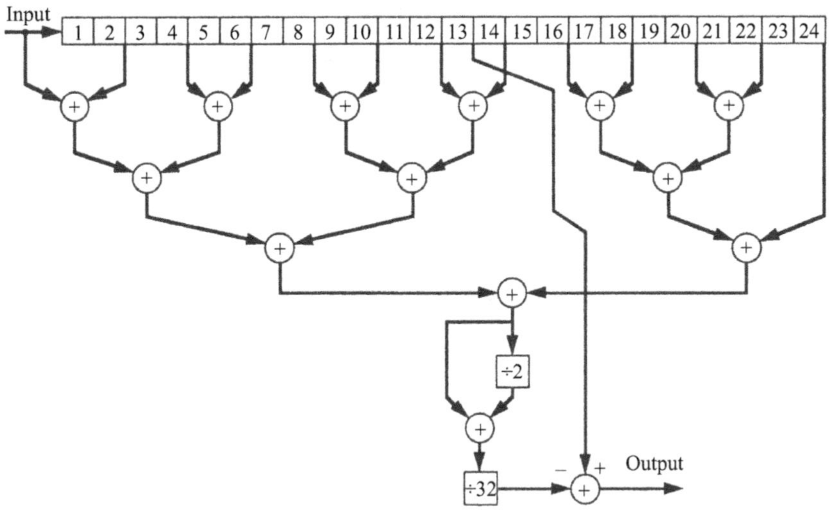


*Figure 4.48 Sidelobe suppression filter FIR form [6]*

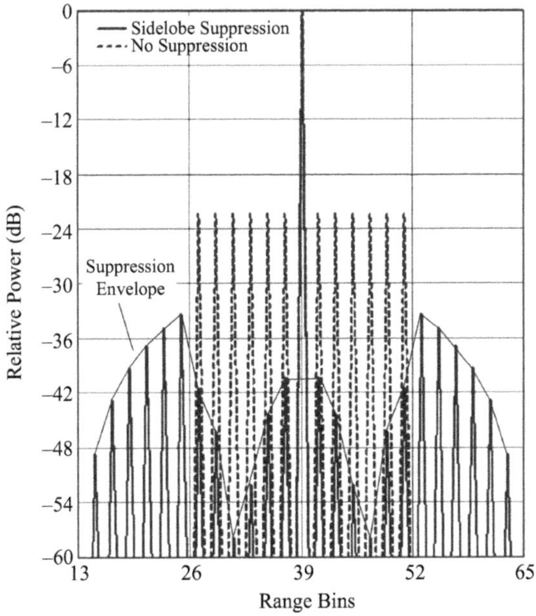


*Figure 4.49 Barker code sidelobes with and without suppression (BarkDopp2a) [6]*

*Table 4.2 Barker code sidelobe suppression summary [6]*

| Code length | Sidelobe suppression | ISLR (dB) | PSLR (dB) | SNR loss (dB) |
|---|---|---|---|---|
| 13 | Without | −11.5 | −22.3 | 0 |
| | With | −24.9 | −32.9 | −0.55 |
| 169 | Without | −8.3 | −22.3 | 0 |
| | With | −21.9 | −32.9 | −1.1 |

In summary, Table 4.2 shows the improvements to be gained in *ISLR* and peak to sidelobe ratio (*PSLR*) while sacrificing a small *SNR* reduction for both 13:1 and 169:1 codes. Obviously, similar improvements are obtainable with other Barker code lengths and code compounds. Appendix A.4 referenced by name in some of the figures contains Mathcad program and Adobe acrobat files with more details on Barker code as well as other pulse compression code sidelobe performance.

Binary and polyphase code compression and sidelobe suppression can be performed in a single dedicated chip. A digital pulse compressor or correlator chip can process at GHz rates in real time.

### 4.11.1 Effect of Doppler on phase codes

Another important consideration in any binary and polyphase code is its sensitivity to the Doppler shift. This can either be a result of relative motion between radar and target or between data link transmitter and data link receiver. Doppler sensitivity depends on the form of the code employed. Barker codes are often used as the synchronization code for simple data links. There are some Doppler invariant codes but they have not been widely used in radar or military data links. Chirp waveforms are sensitive to Doppler shift frequencies on the order of the bandwidth of the unmodulated transmitted pulse time reciprocal. Binary phase codes are sensitive to Doppler shifts on the order of the reciprocal of the uncompressed pulse length or the transmitted bandwidth divided by the *PCR*. Frank codes, which will be discussed in the following section, are less sensitive to Doppler shifts and are of the order of the reciprocal of the linear segment length or the transmitted bandwidth divided by the square root of the *PCR*. This is because Frank codes are approximations of chirp waveforms.

A way to visualize the result of Doppler shifting on a pulse compression waveform is shown in Figure 4.50. The figure shows a plot of the cross-correlation function between the transmitted waveform and its Doppler shifted return from a target at Doppler shift frequencies, $f_d \cdot T$, relative to the transmitted frequency of 1.2, 0 and 0.6. (These are three cuts through the ambiguity diagram at these offset Dopplers.) As is well known, the zero Doppler response is the waveform autocorrelation function, which is always greater than or equal to any other response. Doppler shifted responses typically have higher sidelobes and lower peak responses. If the objective is to reject Doppler shifted targets, this is ideal, of course; however, in most cases, the targets of interest are the Doppler shifted returns and

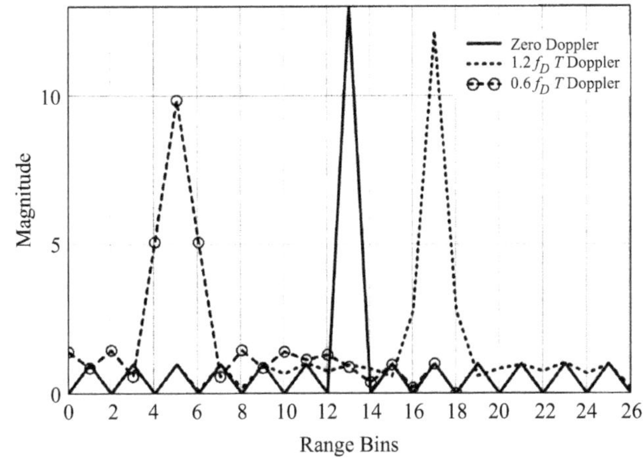


*Figure 4.50 Binary code response with Doppler shift (BarkDopp2a) [6]*

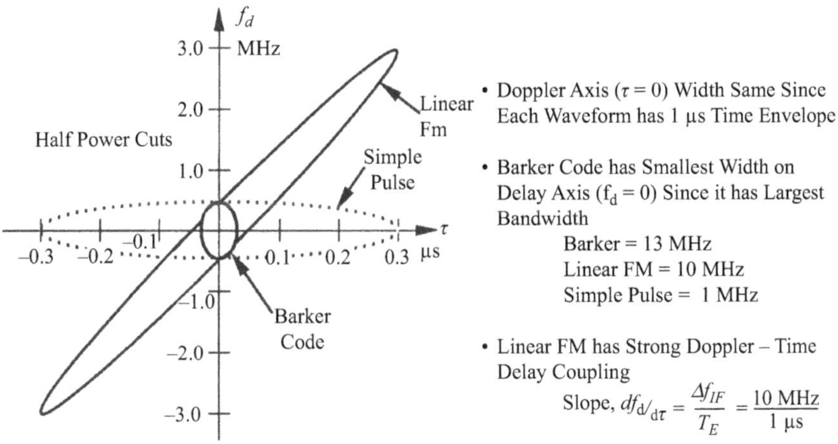


*Figure 4.51 Ambiguity function comparison [6]*

some compensation is required. These three cases represent *ISLR*s of −7 dB, −11.5 dB (theoretical best) and −1.2 dB, respectively. For short phase codes like the Barker code shown here with 50 ns chip widths, these Doppler offsets represent interplanetary velocities (~20,000 kmph) and just don't arise in most cases. For very long uncompressed pulse widths and large compression ratio codes, however, more normal Dopplers will cause sidelobe degradation if there is no Doppler compensation.

A widely published comparison of ambiguity functions for a simple pulse, a chirped pulse and a Barker code of equal time length and similar Doppler resolution is given in Figure 4.51. Each has a 1 µs pulse length in the example. One disadvantage of

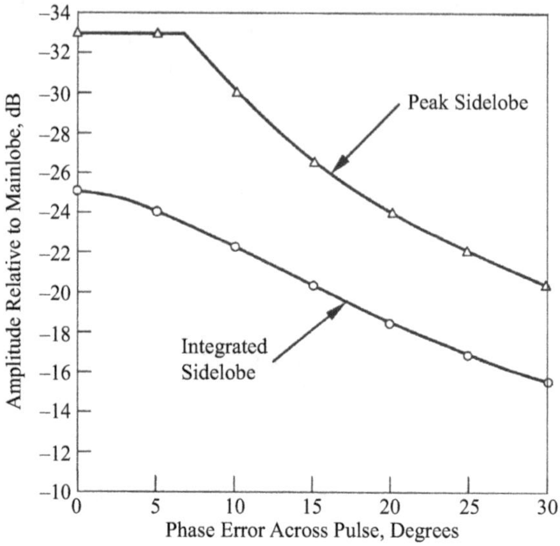


*Figure 4.52 Sidelobe decrease versus phase error 13:1 Barker – sidelobe suppression*

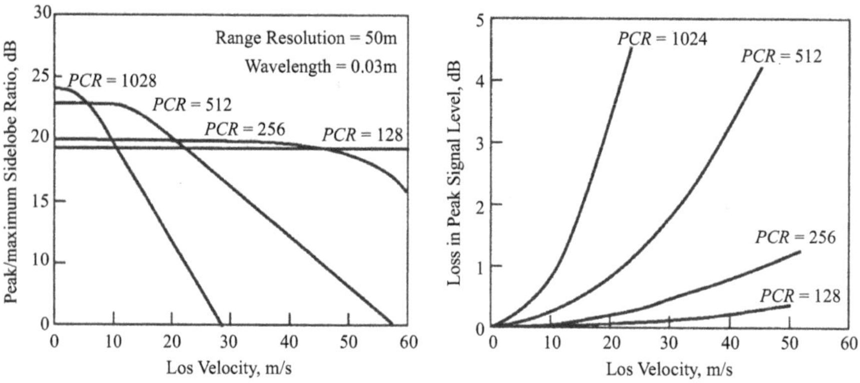


*Figure 4.53 Most binary codes have some Doppler sensitivity [6]*

a chirp waveform is that Doppler offset can be misinterpreted as a different target range because of its large half-power ambiguity area.

The necessity for Doppler compensation is graphically demonstrated in Figures 4.52 and 4.53. Figure 4.52 shows the degradation in peak and integrated sidelobes for a 13:1 Barker code with sidelobe suppression as a function of uncompensated phase error across the transmitted pulse. This behavior of the sidelobes in the presence of uncompensated phase warp across the pulse is similar for all binary phase codes. Figure 4.53 shows the effect for longer codes. Any successful binary code design must deal with Doppler compensation.

The simplest form of Doppler compensation is to apply binary phase pulse compression after Doppler filtering. Most modern radars apply Doppler filter bank processing to all target returns to improve SNR and false target rejection and so, by interchanging the order of pulse compression and Doppler filtering, Doppler compensation automatically occurs. If the code is long relative to the processed swath, compression is often done in two parts before and after Doppler filtering. Significant sidelobe degradation occurs when uncompensated Doppler phase error is between 30° and 60° across the total pulse. Similarly, data links require Doppler compensation. The most stringent case occurs during acquisition when all possible Dopplers must be searched. If fast acquisition is required, then all Doppler and time offsets must be match filtered simultaneously. Fortunately the number of bins to be processed is dramatically less than for a high-resolution radar. Usually the required computing power is already available.

## *4.11.2 Frank and digital chirp codes*

Another useful class of codes is digital approximations of chirp waveforms. Such schemes have been implemented in hardware for at least 35 years with varying degrees of success. Their main attraction is the simplicity of processing coupled with the stability of digital techniques. When chirp waveforms are used for very high (greater than 10,000:1) PCRs, performance is usually limited by linearity, stability and distortion in the transmitted and demodulating waveforms. This naturally leads to the use of polyphase digital techniques.

Figure 4.54 shows a quadratic phase curve for chirp and compares this to a 13:1 Barker code and a 16:1 Frank code. Note that Barker code is not a bad

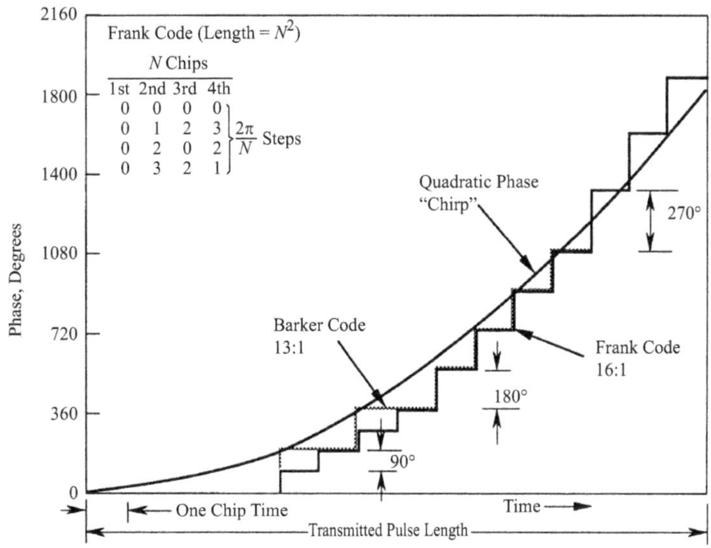


*Figure 4.54 Digital code phase histories approximate a chirp [6]*

approximation itself, but Frank codes are a better approximation for longer chirps. Frank codes of length $N^2$ have $2\pi/N$ phase states, which are visited first by 0 increments, then by 1, 2, 3, etc., as shown in Figure 4.54. Obviously, convenient length Frank codes have integer square roots. By phase modulating the transmitter carrier with digitally generated phase increments, large and very stable pulse compression waveforms can be generated and decoded [17]. Polyphase codes have recently been used in modern digital cell phone communication but most of the technology is many decades old.

Digital chirp is similar to Frank codes except that, at each bit or chip interval, the closest approximation to the desired continuous chirp phase history is selected, limited only by the quantization accuracy of the digital phase modulator, the accuracy of the chip clock and modulator phase accuracy (modulo $2\pi$, of course). Digital chirp systems with compression ratios of 50,000:1 and greater have been implemented in modern radars with phase quantization of 10–12 bits. These large compression ratios are often decoded by stretch techniques described earlier in this chapter.

Quite long Frank codes have been generated using a version of the algorithm in Figure 4.54 upper-left corner. Their sidelobes are somewhat worse than amplitude weighted linear FM but at large compression ratios the differences are small. Frank codes are complementary as will be shown in the next section. This property can also be exploited to reduce close-in sidelobes.

The "brute force" decode of a length 16 Frank code is shown in Figure 4.55. Figure 4.55 bottom also shows the tabular phase values of the transmit function of a Frank code of length 16. Note that the compression coefficients are the complex conjugate of the transmitted code. One of the beauties of four-phase modulation and decode is that multiplication by increments of 90° consists of digital complements and swapping in-phase for quadrature signal components and can be performed at very high speeds and with very little hardware. There are obviously simplifications

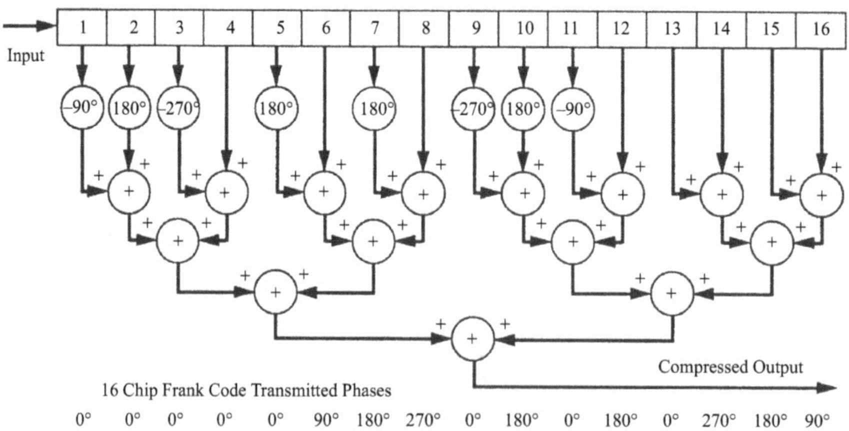


*Figure 4.55 Frank code pulse compressor length 16 [6]*

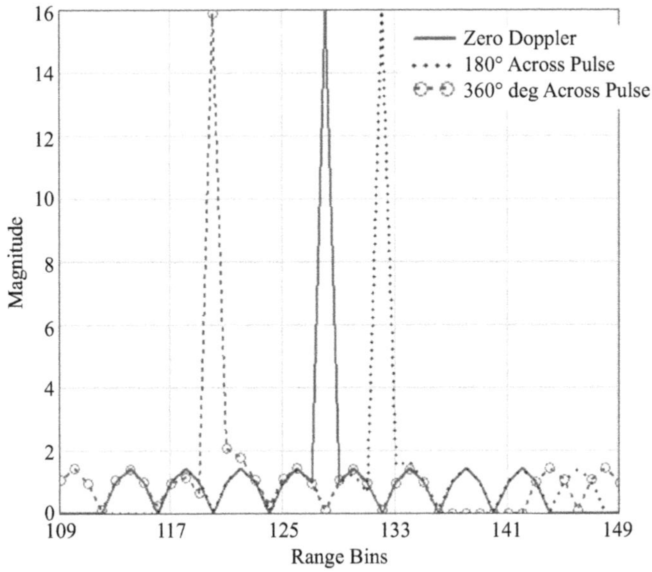


*Figure 4.56 Frank decoder output (Frank16)*

to reduce the total number of operations easily discernable in Figure 4.55. Longer Frank codes require progressively smaller phase increments. There are other codes that only require four phases that also have good properties such as P codes.

A plot of the cross-correlation function of Figure 4.55 with the points between the decoder outputs fitted with a straight line for visualization is shown in Figure 4.56. Figure 4.56 shows the response to Doppler represented as the total phase shift over the received pulse. The figure shows one of the attributes of Frank codes but also chirp waveforms, that is, range Doppler coupling. In the absence of Doppler compensation by a Doppler filter bank before compression or some other means, pulse compression places a target return in the wrong range bin. The good news is that within reason the amplitude is essentially the same independent of Doppler shift.

### *4.11.3 Complementary codes*

Complementary codes are a class of binary codes that have been used in radars because they are very easy to generate and decode. They have excellent properties, especially for air-to-ground waveforms. Complementary codes are defined as a pair of equal length codes having the property that the time sidelobes of the autocorrelation of one code are equal in amplitude and opposite in sign to that code's complement. Figure 4.57 shows an eight-chip code $A$, its complement $\bar{A}$, and their sum. Note the sum has sidelobes that are identically zero. By alternately transmitting $A$ and $\bar{A}$ and summing the range swaths bin by bin, *PSLR*s that are infinite could theoretically be achieved. In actuality, implementation errors limit

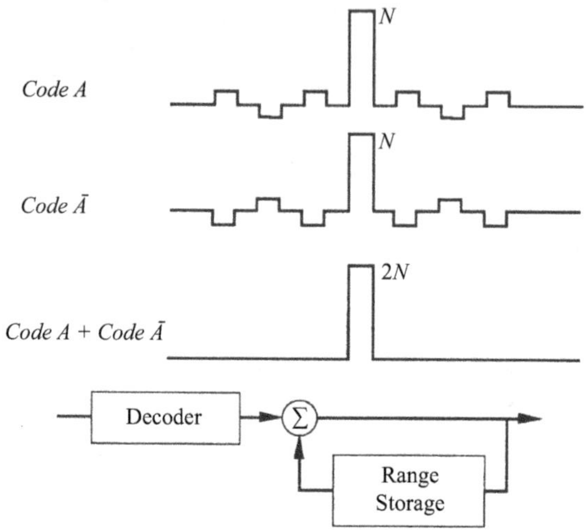


*Figure 4.57 Complementary codes [6]*

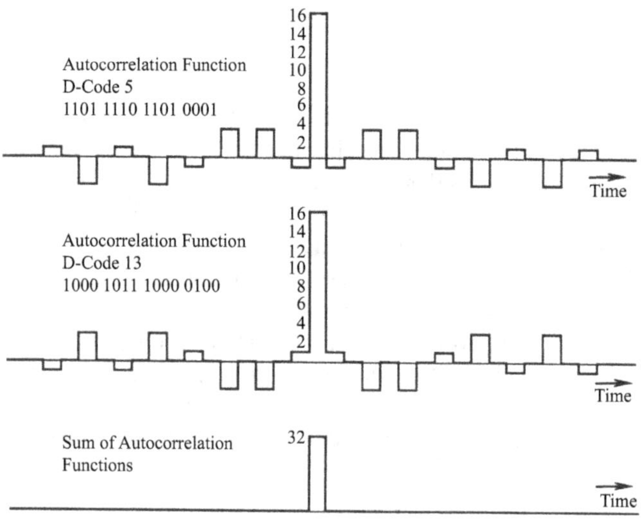


*Figure 4.58 Length 16 D-codes and corresponding idealized compressed outputs [6]*

performance, but the sidelobes of error terms generally are also complementary and partially cancel. The disadvantage is that alternating codes have an ambiguity at the PRF/2.

A number of complementary codes are known to exist. Golay [16] found codes of lengths $L = 2^n$ (e.g., 1, 2, 4, 16, 32...) as well as certain other lengths (10, 20, 26, 40 ...). Welti [20] found binary "D-codes" that are complementary pairs, have length $2^n$ and are formed from shorter complementary sequences. The rule is that a complementary code can be formed from a complementary sequence by dividing the sequence in half, designating the first half $A$ and the second half $B$ and then concatenating the sequence $AB$ to the sequence $A\,\bar{B}$; that is, $ABA\,\bar{B}$ and its complement is $AB\,\bar{A}B$. A D-code sequence is shown in Figure 4.58.

---

**Example 4.9** For Example 4.9, a 0.2 radian cumulative phase error and a 1/8 chip total timing error over the total transmitted pulse envelope in the received waveform yields an uncancelled residue of approximately −45 dB on peak sidelobes for a 1024:1 complementary code pair. Actual field performance in radars has demonstrated peak to far sidelobe ratios between −50 and −60 dB and peak to near sidelobe ratios between −30 and −40 dB. This field performance is unequaled by any other pulse compression scheme. Furthermore, the performance holds up for very long codes. Achieving the same performance from chirp systems requires very expensive linearizing schemes or non-real-time "massaging" of the data. More details are in the software in the appendices [20].

---

---

**Example 4.10** There are a large number of possible complementary codes. For Example 4.10, there are 16 D-code sequences of length 8 chips. Equations (4.19) and (4.20) show the 5th and 6th D-codes of length 8 and their complements. These 8 chip codes when concatenated produce 16 chip codes. D-codes 5 and 6 are obvious complements. Less obvious, until one thinks carefully about it, are D-codes 13 and 14, which are complements of 5 and 6, respectively. D-code 13 is the mirror image complement of 5 since the autocorrelation function of a real finite waveform is real and even; therefore, $R(-t) = R(t)$. The best *PSLR*s are obtained from D-codes 5, 7, 14 and 16.

$$\text{The fifth D-code sequence is:}\quad \overbrace{1101}^{\mathrm{A}}\ \overbrace{1110}^{B} \Rightarrow \overbrace{1101}^{A}\ \overbrace{0001}^{\bar{B}} \tag{4.19}$$

$$\text{The sixth D-code sequence is thus:}\quad \overbrace{1101}^{\mathrm{A}}\ \overbrace{1110}^{B} \Rightarrow \overbrace{0010}^{\bar{A}}\ \overbrace{1110}^{B} \tag{4.20}$$

---

---

**Example 4.11** Another type of complementary code is called a C-code. For Example 4.11, a length 16 C-code and its complement are shown in (4.21).

$$\overbrace{1011}^{\mathrm{A}}\ \overbrace{1110}^{B}\ \overbrace{1000}^{C}\ \overbrace{1101}^{D} \Rightarrow \overbrace{1011}^{\mathrm{A}}\ \overbrace{0001}^{\bar{B}}\ \overbrace{1000}^{C}\ \overbrace{0010}^{\bar{D}} \tag{4.21}$$

Note that D is the mirror image of A and C is the mirror image complement of B. C-codes have the property that every fourth bin after compression is zero, rather

*Table 4.3. Complementary code sidelobe levels [6]*

| Code type | Code length | Peak SLR (dB)* | ISLR (dB)* |
|---|---|---|---|
| D-codes | 8 | −8.5 | −4.3 |
| | 16 | −14.5 | −5.1 |
| | 32 | −11.0 | −4.6 |
| | 64 | −15.0 | −4.8 |
| | 128 | −15.7 | −4.7 |
| C-codes | 8 | −12.0 | −4.3 |
| | 16 | −14.5 | −5.1 |
| | 32 | −14.5 | −4.6 |
| | 64 | −16.1 | −4.8 |
| | 128 | −17.0 | −4.7 |

*Individual codes not pairs.

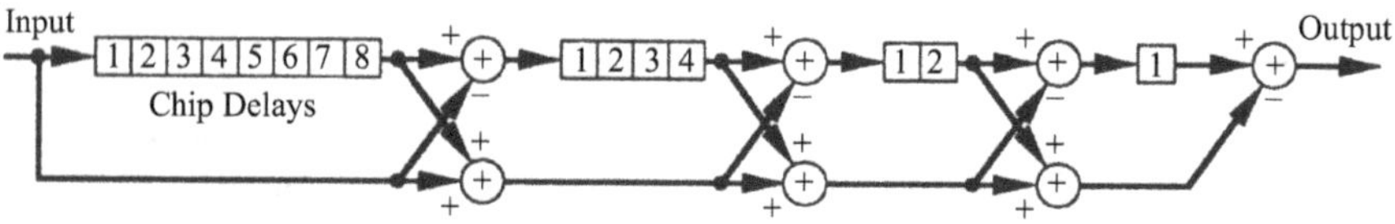


*Figure 4.59 Length 16 type II complementary code pulse compressor [6]*

than every other bin, and they produce lower *PSLR*s. Again there are a large number of possible codes. In addition as will be seen later in this section, there are a number of polyphase complementary codes.

Table 4.3 compares C and D codes of various lengths for each individual code; that is, not cancelled. Note that C-codes have slightly better *PSLR*s. Without code cancellation, none of these codes come close to Barker codes in *ISLR*. Although *PSLR* is important, *ISLR* is the real limiter of performance in almost all pulse compression applications.

What justifies the interest in complementary codes is the particularly simple decoding that can be used. Figure 4.59 shows one configuration of a pulse compressor for length 16.

Note that each unit delay can be replaced by a "seed" code, making the code an almost arbitrary length. The rule is extendable for any binary power or products of binary powers and smaller inner codes such as 10, 26, etc. Doubling the length of the code requires only the addition of one-half the code length of storage or delay and two additional adders. It is possible to employ very long codes with minor amounts of hardware. To give a practical advantage, the codes must be very long but since compression is simple, this is not hard. Code lengths over 20,000:1 often are required.

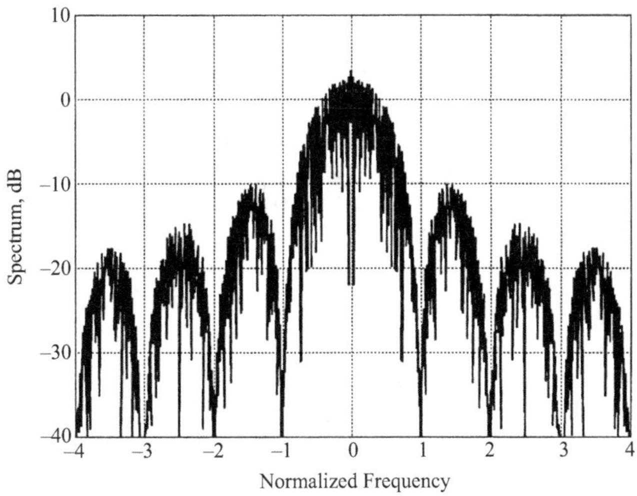


*Figure 4.60 Spectrum of alternating complementary code of length 64 [6]*

In addition, for codes that are much longer than the processed range swath, there is a further simplification called the Hudson–Larson pulse compressor that saves even more hardware or computation time. Another consideration for complementary codes is do they cover the band reasonably uniformly. Even though these waveforms seem completely predictable, they can be made very difficult to interfere with just by choosing long inner seed codes and selecting them at random at the time of the coherent array initiation. Further, by pseudo-randomizing the code length and chip width the spectrum can be filled very uniformly with very low interference with other emitters. The spectrum in Figure 4.60 shows an alternating complementary code of length 64 chips other than the double nulls near zero frequency offset even a short code looks like the transmitted chip width.

Well if complementary codes are so fabulous why aren't they used more often? There are two main weaknesses. First, the uncancelled sidelobes are shifted to ½ the PRF. Second, like all binary and polyphase codes they do have Doppler sensitivity, which must be compensated in most applications. Figure 4.61 shows the complementary code sidelobe shift attribute in the frequency and time domain. Figure 4.61 shows the actual sidelobes of a 128 chip length complementary code and the resulting sidelobes of a complementary pair when summed for a real radar return. In essence the sidelobes are shifted to the PRF/2. The simplest way around this weakness is to insure that PRF/2 Doppler lies outside the mainlobe of the antenna. The second way around this weakness is to transmit the code and its complement as simultaneous quadrature pairs. There are a small set of complementary codes in which the sidelobes of the A-code in the B-code compressor and vice versa are also complementary. Thus they can be transmitted simultaneously in quadrature and achieve the cancellation properties.

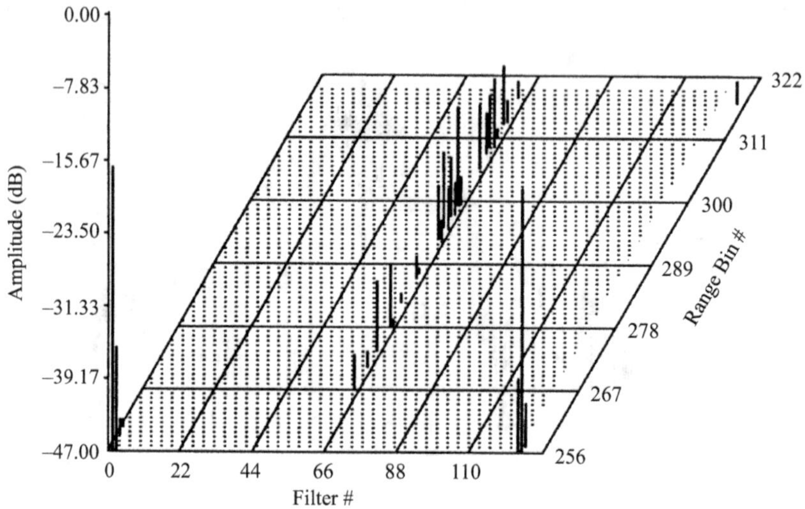


*Figure 4.61 Complementary pulse compression shifts range sidelobes PRF/2 [6]*

### 4.11.4 Type II complementary codes

There are many implementations of complementary codes, but the most popular today are the type II encoders and decoders. A type II code for a code of length $\ell$ is defined from a type I as

$$A = (a_1, a_2, \ldots a_{\ell/2}) \text{ and } B = (b_1, b_2, \ldots b_{\ell/2}) \tag{4.22}$$

That is, if $A\,B$ is the code and if $A$ and $B$ are complementary codes, then the bit-by-bit interleave are complements.

$$\begin{aligned} A_{II} &= (a_1, b_1, a_2, b_2 \ldots a_{\ell/2}, b_{\ell/2}) \text{ and} \\ B_{II} &= (a_1, \bar{b}_1, a_2, \bar{b}_2 \ldots a_{\ell/2}, \bar{b}_{\ell/2}) \end{aligned} \tag{4.23}$$

A set of recursive equation forms for generation and compression of type II complementary codes are shown in (4.24) and (4.25), respectively. One cannot help but notice the similarity to the fast Fourier and fast Walsh–Hadamard transforms and their recursive equations.

Type II complementary code generation

$$\begin{aligned} &A_i(p+m) = B_{i-1}(p) \text{ and } A_i(p) = A_{i-1}(p) \\ &B_i(p+m) = -B_{i-1}(p) \text{ and } B_i(p) = A_{i-1}(p) \\ &\text{Where for code length } 2N \text{ and recursion index } i\text{:} \\ &i = 0,1,2\ldots\log_2(N); m = 2^i; p = 0,1,2\ldots m-1 \end{aligned} \tag{4.24}$$

Type II complementary code compression

$$
\begin{aligned}
&A_i(p) = A_{i-1}(p) + B_{i-1}(p)\\
&B_i(p) = A_{i-1}(p+m) - B_{i-1}(p+m)\\
&\text{Where for code length } 2\cdot N \text{ and recursion index } i\text{:}\\
&i = \log_2(N), \log_2(N)-1, \log_2(N)-2, \ldots 1, 0;\\
&m = 2^i; p = 0,1,2,\ldots 2\cdot N + m - 2
\end{aligned}
\tag{4.25}
$$

---

**Example 4.12** An Example (4.12) transmitted phase code and its counterpart compressed output for a 32-chip code is shown in Figure 4.62. Codes as long as $2^{16}$ chips are easily generated and compressed with type II configurations. Usually the innermost 16 chips are compressed brute force to allow maximum flexibility of the innermost segment of the code. For example, the inner code might be a Frank code, a quadrature binary complementary code or a binary complementary code selected according to some rule. If one considers the simple 16-chip code flow diagram shown in Figure 4.63, it is easy to see how a very large compression ratio and a small range swath can be accomplished with minimal processing using type II complementary codes. This is very similar to stretch processing where long waveforms can lead to small processing load.

The observation that many intermediate results in the pulse compression did not have to be calculated when the swath was a fraction of the transmitted pulse led to the invention of the Hudson–Larson pulse compressor shown in Figure 4.64. Most of the complexity in the block diagram of Figure 4.64 is associated with the 4:1 multiplexing, which is only there to allow slower arithmetic to keep up with the small chip times. If the processing could keep up with the chips, only two of the

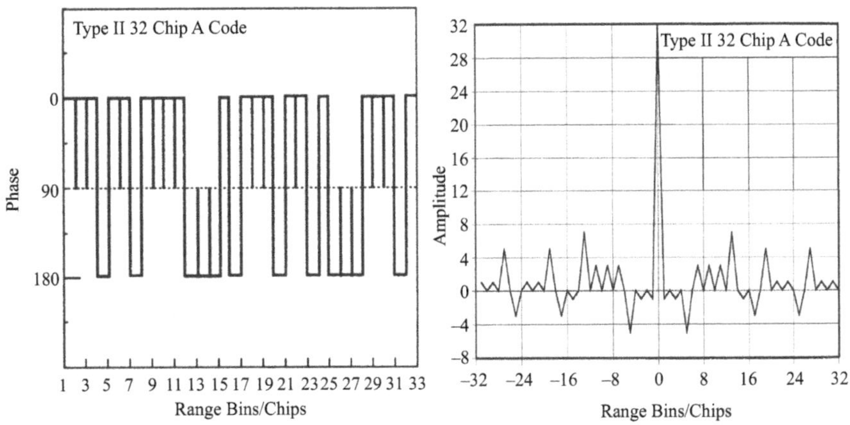


*Figure 4.62 Type II complementary code Example 4.12 [6]*

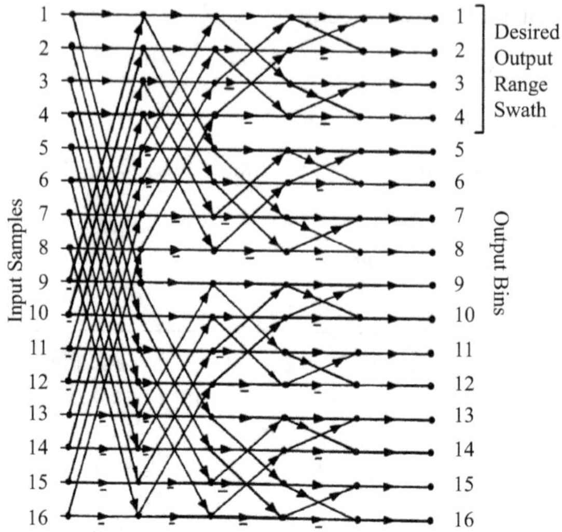


*Figure 4.63 Type II complementary code compression flow diagram [6]*

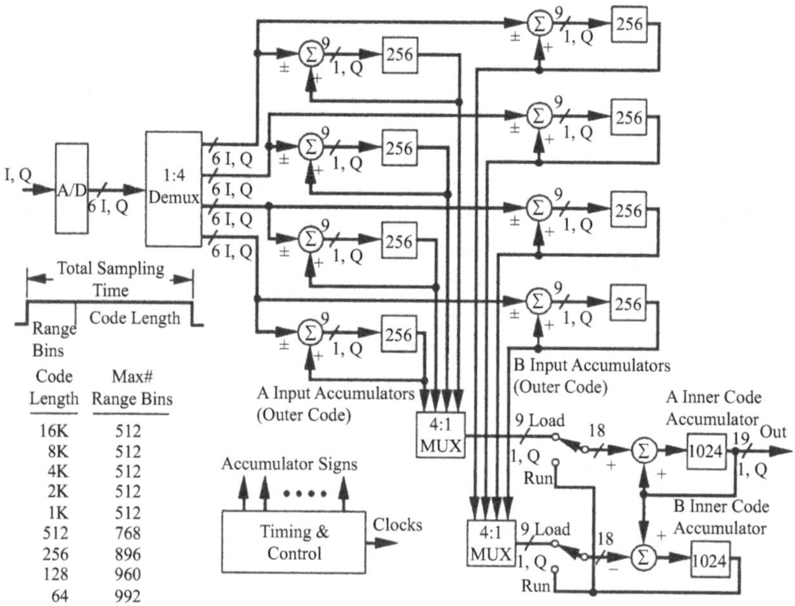


*Figure 4.64 Hudson–Larson complementary code pulse compressor [6]*

eight 256 bin accumulators would be required. Similarly, the inner code accumulators would not have to be "ping-ponged" if the hardware was faster. Although the drawing may seem complex, this is a very simple compression scheme allowing extremely long codes of up to $10^5$ chips.

### 4.11.5 Polyphase complementary codes

Polyphase complementary codes have several advantages. First, intercept exploitation is harder. Second, uncancelled sidelobes often are better, thus cancelled sidelobe residue in the presence of hardware or Doppler degradation is better. Third, cancelled sidelobes in the presence of jamming and front-end nonlinearities are often better. The easiest polyphase codes to implement for both transmission and compression are four phase (quadraphase) codes.

$$\text{If: } PC_2 = \begin{bmatrix} A_2 \\ B_2 \end{bmatrix} = \begin{bmatrix} 0 & 1 \\ 0 & 3 \end{bmatrix} \text{ and } \overline{PC}_2 = \begin{bmatrix} \overline{A}_2 \\ \overline{B}_2 \end{bmatrix} = \begin{bmatrix} 0 & 3 \\ 0 & 1 \end{bmatrix}$$

$$\text{Then: } PC_4 = PC_2 \text{ Join: } \overline{PC}_2 \cdot 90^\circ = \begin{bmatrix} A_2 & \overline{A}_2 + 1 \\ B_2 & \overline{B}_2 + 1 \end{bmatrix} = \begin{bmatrix} A_4 \\ B_4 \end{bmatrix}$$

$$\text{And } \overline{PC}_4 = \begin{bmatrix} \overline{A}_4 \\ \overline{B}_4 \end{bmatrix} = \begin{bmatrix} B_4^* \text{ (Bit Reverse } +2 \text{ Each Bit)} \\ A_4^* \text{ (Bit Reverse)} \end{bmatrix} \text{ Then:}$$

$$PC_4 = \begin{bmatrix} A_4 \\ B_4 \end{bmatrix} = \begin{bmatrix} 0 & 1 & 1 & 0 \\ 0 & 3 & 1 & 2 \end{bmatrix} \text{ and } \overline{PC}_4 = \begin{bmatrix} \overline{A}_4 \\ \overline{B}_4 \end{bmatrix} = \begin{bmatrix} 0 & 1 & 3 & 2 \\ 0 & 3 & 3 & 0 \end{bmatrix} \tag{4.26}$$

where each matrix entry of $A_N = a_0\ a_1\ a_2\ \ldots\ a_{N\text{-}1}$ and the matrix value equals $\exp(j \cdot 2 \cdot \pi \cdot a_i)$ with $A_N(i) = 0°, 90°, 180°, 270°$ for the entries in the matrices in (4.26), modulo $2\pi$ or $a_i$ is modulo 4. The rule above can be restated as

$$\overline{A}_N = -B_N^*(N-i)\ ;\ \overline{B}_N(i) = A_N^*(N-i) \quad \text{for } i = 0,\ 1, \ldots N\text{-}1 \tag{4.27}$$

Longer codes can be formed by concatenating shorter complementary pairs in which the second pair has a shift of 90° added to it to form a longer sequence pair as shown in (4.26). The complement of the longer pair is formed by interchanging A for B, reversing the chip order of $A$ and $B$, adding 180° to the new A complement and finding the complex conjugate of the new $B$ complement as shown in (4.26) and (4.27). Obviously for even codes this procedure can be continued indefinitely to create very long polyphase sequences.

Equations (4.28) show a few short polyphase code kernels from which codes of almost arbitrary length can be formed. Using the rule suggested above, these codes can be expanded in integral multiples of their length. They can also be used as the inner code in which the outer code is a binary complementary code.

$$\begin{aligned} PC_{13} &= \begin{bmatrix} A_{13} \\ B_{13} \end{bmatrix} = \begin{bmatrix} 0001200302031 \\ 0122212003203 \end{bmatrix} \\ \overline{PC}_{13} &= \begin{bmatrix} \overline{A}_{13} \\ \overline{B}_{13} \end{bmatrix} = \begin{bmatrix} 3203220100012 \\ 3102010023000 \end{bmatrix} \end{aligned} \tag{4.28.1}$$

$$PC_3 = \begin{bmatrix} A_3 \\ B_3 \end{bmatrix} = \begin{bmatrix} 010 \\ 002 \end{bmatrix} \qquad \overline{PC}_3 = \begin{bmatrix} \overline{A}_3 \\ \overline{B}_3 \end{bmatrix} = \begin{bmatrix} 022 \\ 030 \end{bmatrix}$$

$$PC_5 = \begin{bmatrix} A_5 \\ B_5 \end{bmatrix} = \begin{bmatrix} 01321 \\ 00013 \end{bmatrix} \qquad \overline{PC}_5 = \begin{bmatrix} \overline{A}_5 \\ \overline{B}_5 \end{bmatrix} = \begin{bmatrix} 31222 \\ 32130 \end{bmatrix}$$

$$\text{Also} = \overline{PC}_5 + [1] = \begin{bmatrix} 02333 \\ 03201 \end{bmatrix} \tag{4.28.2}$$

In addition, other lengths can be formed using combinations of the kernels as shown in (4.29). The equation shows the procedure for codes of $2{\cdot}N{\cdot}M$ in length. In all of these, the attraction is simple code generation and even simpler decoding than any other compression scheme. The simplicity is, of course, a weakness from a countermeasures point of view and that is why many codes must be used and many chip and pulse lengths, so guessing will not be an advantage to the interceptor. In (4.29), the example used is $M$=$N$=3 and so the resulting code is 18 long. The complement for length 18 is the one previously stated and repeated by example in (4.29).

$$PC_{2\cdot M\cdot N} = \begin{bmatrix} A_{2\cdot M\cdot N} \\ B_{2\cdot M\cdot N} \end{bmatrix} = \begin{bmatrix} (A_N \times A_M)(B_N \times \overline{A}_M) \\ (A_N \times B_M)(B_N \times \overline{B}_M) \end{bmatrix}$$

For example, for $M = N = 3$

$$PC_{18} = \begin{bmatrix} 010121010022022200 \\ 002113002030030212 \end{bmatrix} \tag{4.29}$$

where $A_N$= $a_0$ $a_1$ $a_2$ … $a_{N-1}$ and $B_M$= $b_0$ $b_1$ $b_2$ … $b_{M-1}$ and $A_N$ x $B_M$= $a_0\cdot(b_0\ b_1\ \ldots b_{M-1})\ a_1\cdot(b_0\ b_1\ \ldots b_{M-1})\ \ldots\ a_{N-1}\cdot(b_0\ b_1\ \ldots b_{M-1})$.

Lastly, it is also possible to form polyphase codes in which four members are required to obtain perfect cancellation. One example for length 16 is given in (4.30). These are obviously Frank codes and have all the good features of those codes plus the complementary sidelobe cancelling feature. Such codes are often used in conjunction with a binary complementary outer code to allow Hudson–Larson decode. The inner code is then compressed brute force. The disadvantage of a four-phase complementary code is that the ambiguities are at the PRF/4.

$$PC_{16} = \begin{bmatrix} 0000012302020321 \\ 0123020203210000 \\ 0202032100000123 \\ 0321000001230202 \end{bmatrix} \tag{4.30}$$

### *4.11.6 Polyphase P codes [17,19]*

P codes can be thought of as digital approximations to chirp. They exhibit some of the properties of complementary codes. A few PRF intervals can be made to have zero sidelobes to improve clutter rejection when there might be multiple ambiguities on the surface inside or near the mainbeam footprint. P4 codes are especially useful because they tolerate band limiting and very simple sidelobe suppression is available. The P4 phase modulation rule is given in (4.31).

$$A_N(i) = \exp\left( \mathrm{j} \cdot \pi \cdot \left( \frac{i^2}{N} - i \right) \right)$$
$$\text{where: } i = 0,1,2\ldots N-1, \quad \mathrm{j} = \sqrt{-1}$$
$$N \text{ is the code length} \tag{4.31}$$

Figure 4.65 shows the sidelobe pattern for a 256 chip P4 code with a 50 ns chip width with and without sidelobe suppression and a Doppler shift of 100 kHz. Note that Doppler doesn't shift the range location and doesn't do too much harm to the sidelobes. It is possible to make the sidelobes uniform with a simple two-range bin weighting. Two bin weighting does not improve the 3 dB width over 3 bin Hanning weighting and it has a somewhat poorer ISLR. The attribute of extremely low sidelobes over a limited range is often a very valuable attribute especially imaging a building complex.

Figure 4.66 shows another attribute of P4 codes besides their Doppler tolerance and that is very low sidelobes on long codes and low ISLRs (better than −30 dB typical for sidelobe suppressed codes).

With spacecraft radars there may often be many PRF intervals on the ground in the mainbeam. Some method must be used to cancel these PRF ambiguities. One method sometimes used is that each PRF interval has a unique starting phase. If that

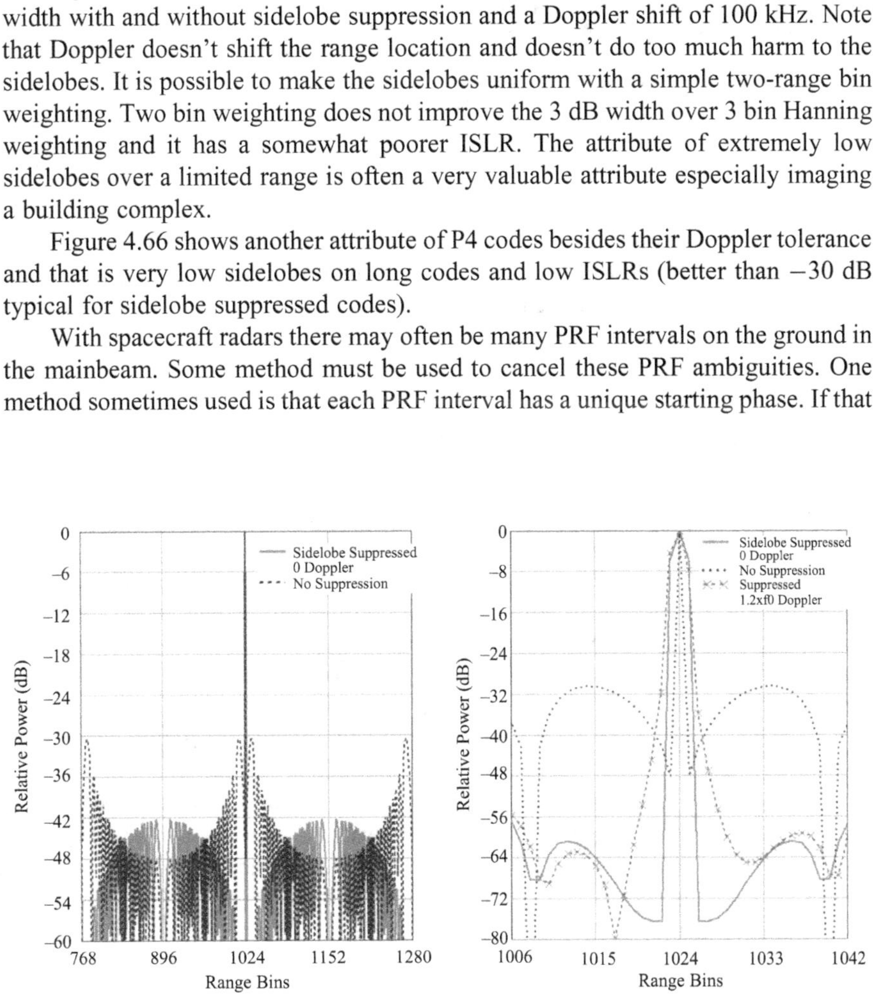


*Figure 4.65 P4 code 256 chips (ISATPcode4)*

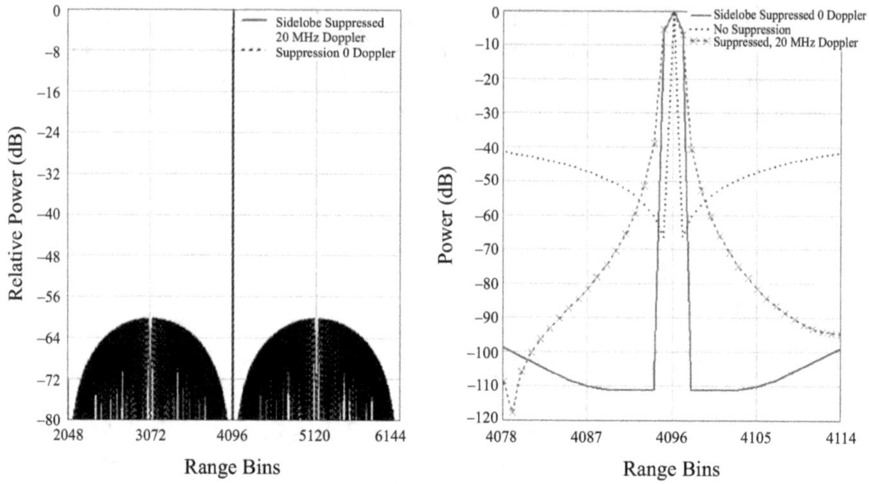


*Figure 4.66 P4 code 2048 chips (ISATPcode4d)*

starting phase is a polyphase code with very good cancelling or sidelobes, then only the region of interest will be detected. Clearly this causes additional Doppler ambiguities to show up but often the azimuth beamwidth is much narrower and the ambiguities are reduced by the antenna pattern.

## 4.12 Introduction to Kalman filters [22,23]

A Kalman filter is often applied for tracking in sampled data systems such as radars and navigation systems. It is an adaptive filter, which combines a set of current measurements with an accumulation of prior measurements in an optimum way (least square estimate) based on the relative uncertainty (SNIR) of the past accumulation to the present measurement. Part of its operation is to forecast the next measurement and compute the error between the measurement when it arrives and the forecast. It then corrects itself for the next forecast based on the noise statistics from the current measurement relative to the prior measurements.

Suppose there is an object moving in only one dimension X and only its position could be measured; then one might want to combine a new noisy measurement $X_m$ with a smoothed estimate $X_s$ of the anticipated true location of the object. This could improve the estimate of the location of the object through multiple measurements. One possible way to combine measurements with smoothed estimates would be to use the relative variances of the measurement and the smoothed estimate. The variance is a good representation of the relative SNRs of the two quantities to be combined in some optimal way. This might be as shown in (4.32). This is sometimes called an alpha tracker.

$$X_{Si} = X_{S(i-1)} + \alpha_i \cdot \left( X_{Mi} - X_{S(i-1)} \right)$$

Where $X_{Si}$ is the $i_{th}$ smoothed estimate of position,

and $X_{Mi}$ is the $i_{th}$ measured estimate of position,

with:

$$\alpha_i = \left( \frac{\sigma^2_{S(i-1)}}{\sigma^2_{S(i-1)} + \sigma^2_{Mi}} \right)$$

Where $\sigma^2_{S(i-1)}$ is the variance of the prior estimate $X_{S(i-1)}$ of position and $\sigma^2_{Mi}$ is the variance of the current measured estimate $X_{Mi}$ of position.

In order to start this iteration $\alpha_0 = 1$, $X_{S0} = X_{M0}$, $\sigma^2_{S0} = \sigma^2_{M0}$ so at the second measurement $i = 1$ the measurement and first measurement will be combined equally. (4.32)

Figure 4.67 shows the nature of the state estimate update or prediction equation. The new state estimate, $\hat{\mathbf{X}}_{\mathrm{NEW}}$ (target position, velocity and acceleration $(P, V, A)$), is a sum of the predicted state, $\hat{\mathbf{X}}_{\mathrm{PRED}}$, and some fraction ($K_k$) of the difference between the measured but noisy target parameters and the smoothed predicted measurement. The graph in Figure 4.66 suggests that the new state estimate at spot 1 in the figure may not coincide with the actual target trajectory but rather gets closer. The distance between spots 2 and 1 is the correction 4, which is based on relative signal to noise between the actual and predicted states. It is the "best" correction.

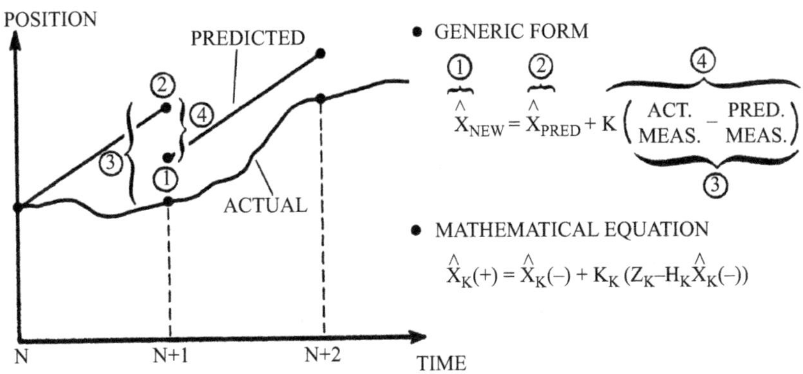


*Figure 4.67 Smoothed new state estimate*

Similarly, suppose that one could measure both position and velocity in one dimension then the smoothed forecasted position could take measured velocity into account. The equations of (4.32) can be extended to incorporate velocity as shown in (4.33). The smoothed estimate now includes both position and velocity.

$$X_{Si} = X_{S(i-1)} + \dot{X}_{S(i-1)} \cdot T + \alpha_i \cdot \left( X_{Mi} - X_{P(i-1)} \right)$$
$$\dot{X}_{Si} = \dot{X}_{S(i-1)} + \beta_i \cdot \left( X_{Mi} - X_{P(i-1)} \right)$$

Where $\dot{X}_{Si}$ is the $i_{th}$ smoothed estimate of velocity,

and $T$ is the time between measurements,

$$\text{with: } \alpha_i = \left( \frac{\sigma^2_{P(i-1)}}{\sigma^2_{P(i-1)} + \sigma^2_{Mi}} \right); \beta_i = \left( \frac{\sigma^2_{x\dot{x}P(i-1)}}{\sigma^2_{P(i-1)} + \sigma^2_{Mi}} \right)$$

Where $\sigma^2_{P(i-1)} = \sigma^2_{S(i-1)} + 2 \cdot T\sigma_{x\dot{x}S(i-1)} + T^2 \cdot \sigma^2_{\dot{x}S(i-1)}$ is the variance of the prior estimate $X_{S(i-1)}$ of position and

$\sigma^2_{x\dot{x}P(i-1)} = \sigma_{x\dot{x}S(i-1)} + T \cdot \sigma^2_{\dot{x}S(i-1)}$ is the variance of the prior estimate of velocity. (4.33)

---

**Example 4.13** For Example 4.13, let the initial position be zero, measurement error be a constant and the velocity be a constant; then Table 4.4 shows the evolution of the positional weighting coefficient, $\alpha_i$, and the velocity weighting coefficient $\beta_i$.

*Table 4.4 Example 4.13 Kalman filter α–β tracker*

| Iteration | Position coefficient | Velocity coefficient |
|---|---|---|
| 1 | 1.0 | 0.1 |
| 2 | 0.83 | 0.05 |
| 3 | 0.70 | 0.03 |
| 4 | 0.60 | 0.02 |
| 5 | 0.52 | 0.0143 |
| 6 | 0.46 | 0.0107 |
| 7 | 0.42 | 0.0083 |
| 8 | 0.38 | 0.0067 |
| 9 | 0.34 | 0.0054 |
| 10 | 0.32 | 0.0045 |
| 11 | 0.29 | 0.0038 |
| 12 | 0.27 | 0.0033 |
| 13 | 0.26 | 0.0028 |
| 14 | 0.24 | 0.0025 |
| 15 | 0.23 | 0.0022 |
| 20 | 0.18 | 0.0013 |
| 30 | 0.12 | 0.0006 |

A traditional $\alpha-\beta$ tracker might have fixed gains of 0.3 and 0.004. One thing should be obvious from the table and that is in the absence of constraints on unknown maneuvers the filter will ignore all new measurements before long. Therefore, a constraint is usually incorporated into the variance estimates that won't let them go below some possible target acceleration.

---

There is a somewhat standard body of nomenclature used for the elements of Kalman filters, which may seem arcane but the nomenclature, when mastered, helps one visualize a multidimensional complex process. The presumption is that target "state" is described by a system model in which the next state is related to the current state by a transition matrix and a perturbation that is partially random and partially driven by forces that are not known such as acceleration. Some of the nomenclature is summarized in Table 4.5.

$$\mathbf{X}_k = \mathbf{\Phi}_{k-1} \cdot \mathbf{X}_{k-1} + \mathbf{w}_{k-1}$$

Where $\mathbf{X}_k$ is the current state vector,

$\mathbf{X}_{k-1}$ is the last state vector,

$\mathbf{\Phi}_{k-1}$ is the state transition matrix

and $\mathbf{w}_{k-1}$ is a perturbation vector, which is assumed

to be a zero mean gaussian (not really ever true). (4.34)

The system being measured is assumed to have a current state, which is related to the future state through a transition matrix plus some noise as given in (4.34).

*Table 4.5 Kalman filter definitions [22]*

| Symbol | Definition |
|---|---|
| $\mathbf{X}_k$ | $N \times 1$ dimensional vector defining the system state at time $t_k$ |
| $\mathbf{\Phi}_k$ | $N \times N$ dimensional state transition matrix |
| $\mathbf{w}_k$ | $N \times 1$ dimensional vector defining the system uncertainty |
| $\mathbf{Z}_k$ | $M \times 1$ dimensional vector defining the measurements at time $t_k$ |
| $\mathbf{H}_k$ | $M \times N$ dimensional state-to-measurement transition matrix |
| $Q_k$ | $N \times N$ dimensional process noise covariance matrix |
| $R_k$ | $M \times M$ dimensional measurement noise covariance matrix |
| $\mathbf{v}_k$ | $M \times 1$ dimensional vector representing measurement noise |
| $\hat{\mathbf{X}}_k(-)$ | Kalman filter state estimate just before an update using measurements from time $t_{k-1}$ |
| $\hat{\mathbf{X}}_k(+)$ | Kalman filter state estimate just after an update using measurements from time $t_k$ |
| $P_k(-)$ | $N \times N$ dimensional error covariance matrix just before update using measurements from time $t_{k-1}$ |
| $P_k(+)$ | $N \times N$ dimensional error covariance matrix just after update using measurements from time $t_k$ |
| $\mathbf{K}_k$ | $N \times M$ dimensional Kalman filter gain matrix |

The first part of Kalman nomenclature is the estimated target state usually designated as $\hat{\mathbf{X}}_k(+)$, which contains the position ($\boldsymbol{P_{N,E,D}}$), velocity ($\boldsymbol{V_{N,E,D}}$), and acceleration ($\boldsymbol{A_{N,E,D}}$) of the target at sample time, k, after incorporating the measurement at time k. This is usually called the *a posteriori* (after the measurement) estimate of $\hat{\mathbf{X}}_k$. The *a priori* (before the measurement) estimate of the target state is designated as $\hat{\mathbf{X}}_k(-)$. $\hat{\mathbf{X}}_k(-)$ is the predicted state of the target vectors ($\boldsymbol{P}, \boldsymbol{V}, \boldsymbol{A}$) at the time of the measurement, k, but before incorporating the measurement. In general, the state of the target isn't known and often the target vectors can't be directly measured. Unfortunately, almost all real problems including radar, intercept, IR and inertial systems aren't linear as assumed in all the Kalman filter discussions above. Most sensor measurements are generally nonlinear functions of the state variables, that is, sines, cosines, exponentials or hard limited. Thus most Kalman filters are known as extended Kalman filters that are a piecewise fit to a linear model. The piecewise fit still works surprisingly well for a large range of applications. The radar measurement vector designated $\mathbf{Z}_k$ is related to raw state estimates through the measurement matrix, $\mathbf{H}_k$, which is usually a linearized approximation. There is always noise associated with every measurement and calculation that is assumed to be Gaussian noise with zero mean and covariance $R_k$. (This isn't really true but it makes analysis easier and in practical cases seems to make no difference.) The $R_k$ covariance is a function of the SNR and the measurement resolution, for example, antenna beamwidth and range resolution. Figure 4.68 shows the general block diagram of a Kalman filter.

There is a state transition matrix designated as $\mathbf{\Phi_k}$ that predicts the new state $\hat{\mathbf{X}}_k(-)$ from the last state estimate $\hat{\mathbf{X}}_{k-1}(+)$ at time k−1. Usually, there is a forcing function such as ownship motion for airborne or spaceborne radar platforms designated as $\mathbf{F}_k$, which often is rolled into the perturbation vector $\mathbf{w}_k$ with covariance $Q_k$. This vector describes the known radar phase center position, velocity, acceleration and attitude at time k as well as target motion. There also is a noise like term associated with the $\mathbf{w}_k$ perturbation. $Q_k$ also controls the minimum bandwidth of the Kalman filter since there might be unexpected maneuvers of the target. The gain of the actual Kalman filter is a vector or matrix whose coefficients

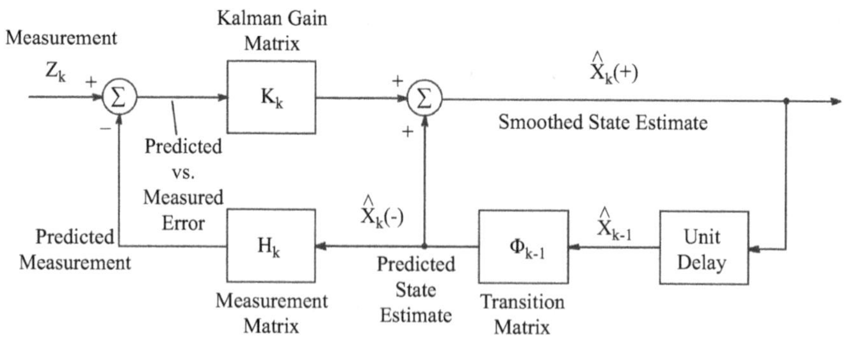


*Figure 4.68 General Kalman filter diagram*

are adapted by the relative noise between the current measurement and the smoothed state estimate of the past measurements designated $\mathbf{K}_k$.

State Estimate Extrapolation:

$$\hat{\mathbf{X}}_k(-) = \mathbf{\Phi}_{k-1} \cdot \hat{\mathbf{X}}_{k-1}(+)$$

Error Covariance Extrapolation:

$$\mathcal{P}_k(-) = \mathbf{\Phi}_{k-1} \cdot \mathcal{P}_{k-1}(+) \cdot \mathbf{\Phi}_{k-1}^{\mathrm{T}} + \mathcal{Q}_{k-1}$$

Kalman Filter Gain:

$$\mathbf{K}_k = \mathcal{P}_k(-) \cdot \mathbf{H}_k^{\mathrm{T}} \cdot \left[\mathbf{H}_k \cdot \mathcal{P}_k(-) \cdot \mathbf{H}_k^{\mathrm{T}} + \mathcal{R}_k\right]^{-1}$$

State Estimate Update:

$$\hat{\mathbf{X}}_k(+) = \hat{\mathbf{X}}_k(-) + \mathbf{K}_k \cdot \left[\mathbf{Z}_k - \mathbf{H}_k \cdot \hat{\mathbf{X}}_k(-)\right]$$

Error Covariance Update:

$$\mathcal{P}_k(+) = \left[\mathbf{I} - \mathbf{K}_k \cdot \mathbf{H}_k\right] \cdot \mathcal{P}_k(-) \tag{4.35}$$

There naturally is an error between the predicted state (e.g., position) and the latest measurement. Often, especially in NED tracking systems, errors (and noise) in 1D are correlated with others. An overall error covariance matrix is calculated before and after incorporating the latest measurement, which is designated as $\mathcal{P}_k$.

Equations (4.35) summarize the general Kalman filter equations. Very often the transition matrix, $\mathbf{\Phi}_k$, and the measurement matrix, $\mathbf{H}_k$, don't change with time. Each of the equations in (4.35) must be calculated with each new measurement and propagated to the next measurement. One will note that the equation to calculate the filter coefficients, $\mathbf{K}_k$, requires a matrix inversion. This in turn requires that the matrix in question is nonzero positive definite. There are strategies to insure that this matrix inversion will not blow up such as the Joseph method [22].

---

**Example 4.14** A simple example will help in understanding what is really going on. Example 4.14, suppose a target moves in two dimensions on the earth's surface inside the antenna mainlobe footprint from an initial point $R_0$ with constant velocity across the footprint and then changes direction and continues with constant velocity. Further assume that although the measurements are made in range and angle, the tracking will be done in x and y on the planet's surface. A number of matrices must be defined at time k= 0 for the Kalman filter to work including $\mathcal{Q}_k$, $\mathcal{P}_k$, $\mathcal{R}_k$, $\mathbf{H}_k$, $\mathbf{\Phi}_k$, $\mathbf{K}_k$. The equations (ex 4.14) summarize the initial matrix assumptions. Assume the following parameters summarized in Table 4.6.

The initial detection is at some range $R_0$ and angle $\eta_0$, which is transformed into Cartesian coordinates at $x_0$ and $y_0$. The subsequent motion in x and y is assumed linear in the example and constant except for one reversal 4 min into the tracking sequence. The angle and range measurement standard deviations are directly proportional to the antenna beamwidth, BW, the transmitted pulse width, PW, and inversely proportional to the square root of the SNR and some constant, 3,

*Table 4.6 Example 4.14 parameters*

| Parameter | Value |
|---|---|
| Range, $R_0$, km | 100 |
| Radar velocity, m/s | 0 |
| Angle, $\eta_0$, ° | 45 |
| Pulse width, PW, m | 6 |
| Beamwidth, BW, ° | 1 |
| Initial range rate error, m/s | 2 |
| Initial range error, m | 6 |
| $X$ target velocity, m/s | −9 |
| $Y$ target velocity, m/s | 9 |
| Signal to noise ratio, SNR | 30 |
| Sampling interval, s | 4 |
| Initial cross-range error, m | 223 |
| Minimum range error, m | 0.2 |
| Minimum angle error, ° | 0.03 |
| $x_0$, km | 70.711 |
| $y_0$, km | 70.711 |

associated with the discriminant method. There must also be some fixed term associated with systematic errors here assumed to be 1/30th of the resolution. The transformation into x- and y-coordinates results in standard deviations, which are a function of both the range and angle standard deviations. The x and y errors also are correlated as shown in the cross-term standard deviation. The measurement covariance matrix is made up of the individual variances. Also note that terms in the covariance can be negative. Usually the state matrix is initialized with the transformed first measurement. For the example here, it is assumed that only the x and y measurements are used in the Kalman tracking filter although a single estimate of x and y velocity will be used to initialize the velocity state.

Definition of measurement, transition matrix and initial Kalman gain matrix:

$$\mathbf{H} = \begin{bmatrix} 1000 \\ 0100 \end{bmatrix}, \ \mathbf{\Phi} = \begin{bmatrix} 1 & 0 & T & 0 \\ 0 & 1 & 0 & T \\ 0 & 0 & 1 & 0 \\ 0 & 0 & 0 & 1 \end{bmatrix}, \ \mathbf{K}_0 = \begin{bmatrix} 1 & 0 \\ 0 & 1 \\ 1 & 0 \\ 0 & 1 \end{bmatrix}, \ R = \begin{bmatrix} 17660 & 17660 \\ -17660 & 17660 \end{bmatrix}$$

Definition of process noise covariance and initialization of error covariance matrix:

$$Q = \begin{bmatrix} 15^2 & 15^2 & 1 & 1 \\ -15^2 & 15^2 & 1 & 1 \\ 1 & 1 & 2^2 & 2^2 \\ 1 & 1 & -2^2 & 2^2 \end{bmatrix}, \ P_0 = \begin{bmatrix} 500^2 & 500^2 & 1 & 1 \\ -500^2 & 500^2 & 1 & 1 \\ 1 & 1 & 50^2 & 50^2 \\ 1 & 1 & -50^2 & 50^2 \end{bmatrix} \quad \text{(ex 4.14)}$$

For this example, it is assumed that the measurement matrix, **H**, and the state transition matrix, **Φ**, are fixed as shown in equations (ex 4.14). The initial Kalman

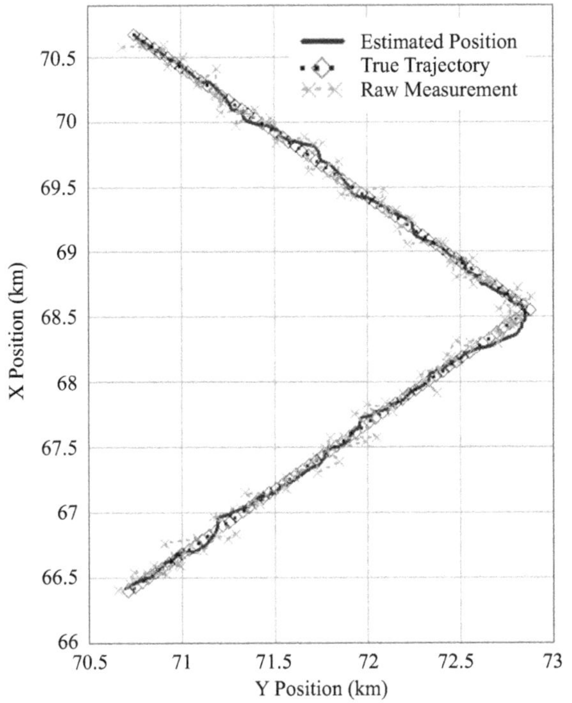


*Figure 4.69 Kalman filter tracker example (Kalman014)*

gain matrix, $\mathbf{K}_0$, is seeded with ones. The measurement covariance matrix values are computed from the given SNR, pulse width and beamwidth as given in (4.36). The knowledge of the process noise, $Q$, is assumed to be no better in x and y than 15 m and no better than 2 m/s in x and y velocity. The initial error covariance, $P$, is assumed to be 500 m in range and 50 m/s in velocity. $P$ will change as the measurements are made but can never get lower than $Q$. Also $R$ prevents the Kalman gains from getting too large. The off diagonal terms are seeded with 1 s for stability.

Figure 4.69 shows the raw measurement, true trajectory and the estimated position for the simulated target in this example. The vehicle speed is about 29 miles per hour. Both the raw data and the estimated position wander around the true trajectory but the filter has smoothed the estimated location dramatically. Obviously when the target makes a little less than a ½ g acceleration and reverses course at the 4 min mark both the position and velocity errors get large. The position and velocity tracking ultimately gets close after the maneuver as shown in Figure 4.70. When the SNR for each measurement is higher, the tracking accuracy is much better.

Some of the Kalman gain terms are shown in Figure 4.71. As can be seen after an initial transient, the gains are almost constant, which may cause loss of track without velocity aiding. Later in the book velocity aiding will be discussed. The x and y position error gain terms lie on top of one another in Figure 4.71. Similarly, the x and y velocity error gain terms lie on top of each other. The cross-correlation terms are very small and not shown.

Initial detection: $x_0 = R_0 \cdot \cos(\eta_0)$ , $y_0 = R_0 \cdot \sin(\eta_0)$

Subsequent motion: $x_k = k \cdot T \cdot \Delta vx + x_0$ , $y_k = k \cdot T \cdot \Delta vy + y_0$

SNR based standard deviations:

$$\sigma_{\eta_0} = \left[\left(\frac{BW^2}{3 \cdot SNR}\right) + \left(\frac{BW^2}{30^2}\right)\right]^{0.5}, \quad \sigma_{R_0} = \left[\left(\frac{PW^2}{3 \cdot SNR}\right) + \left(\frac{PW^2}{30^2}\right)\right]^{0.5}$$

Translation to x and y standard deviations and covariance:

$$\sigma_{x_0} = \left[\sigma_{R_0}{}^2 \cdot \cos(\eta_0)^2 + \sigma_{\eta_0}{}^2 \cdot R_0{}^2 \cdot \sin(\eta_0)^2\right]^{0.5}$$

$$\sigma_{y_0} = \left[\sigma_{R_0}{}^2 \cdot \sin(\eta_0)^2 + \sigma_{\eta_0}{}^2 \cdot R_0{}^2 \cdot \cos(\eta_0)^2\right]^{0.5}$$

$$\sigma_{xy_0} = \left[\frac{\sin(2 \cdot \eta_0) \cdot \left(\sigma_{R_0}{}^2 - \sigma_{\eta_0}{}^2 \cdot R_0{}^2\right)}{2}\right]^{0.5}, \quad \boldsymbol{R} = \begin{bmatrix} \sigma_{x_0}{}^2 & \sigma_{xy_0}{}^2 \\ -\sigma_{xy_0}{}^2 & \sigma_{y_0}{}^2 \end{bmatrix}$$

Initialization of state and measurement vectors:

$$\hat{X}_k = \begin{bmatrix} \hat{x}_k \\ \hat{y}_k \\ \hat{v}x_k \\ \hat{v}y_k \end{bmatrix}, \quad \hat{x}_0 = x_0, \; \hat{y}_0 = y_0, \; \hat{v}x_0 = \Delta vx, \; \hat{v}y_0 = \Delta vy, \; Z_k = \begin{bmatrix} x_k \\ y_k \end{bmatrix} \tag{4.36}$$

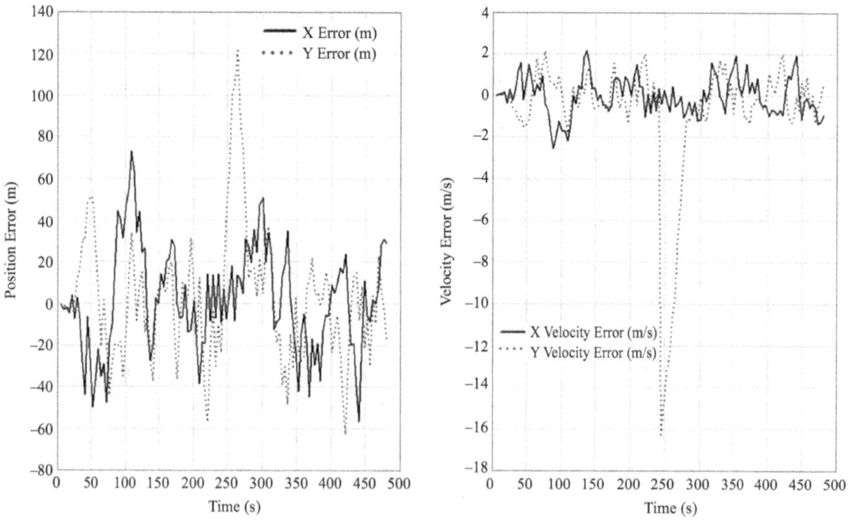


*Figure 4.70 Kalman filter position and velocity error versus time (Kalman014)*

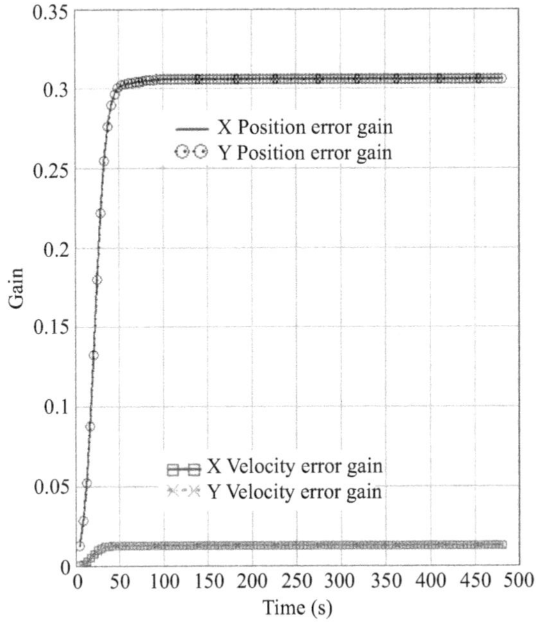


*Figure 4.71 Kalman filter gains versus time (Kalman014)*

```
f(Z, n): = | I ← identity (4)
           | K1 ← K
           | Z1 ← Z
           | Q1 ← Q
           | X1hat ← Xhat
           | P1 ← P
           | P3 ← 0
           | P4 ← 0
           | for  j ∈ 1, 2 . . . n
           |    | X2hat ← Φ · X1hat
           |    | P2 ← Φ · P3 · Φ^T + Q1
           |    | K1 ← P2 · H^T · [H·P2·H^T + R]^-1
           |    | Err_j ← [Z1_j – H · X2hat]
           |    | X3hat ← X2hat + K1 · [Err_j]
           |    | P3 ← (I – K1·H) · P2
           |    | X1hat ← X3hat
           |    | P4 ← P3
           |    | X4hat_j ← X1hat
           |    | K2_j ← K1
           | augment(X4hat,K2)
```

*Figure 4.72 Mathcad Kalman recursion subprogram segment (Kalman014)*

As can be seen in Figure 4.72, the per target filter computation is quite small but there may be thousands per beam dwell and most will not be interesting. The uninteresting movers can only be eliminated by discovering that they aren't actually changing locations, even though they exhibit short-term movement and Doppler.

---

## References

[1] Joint Advanced Strike Technology Program, "Avionics Architecture Definition 1.0", U.S. DoD public release, unlimited distribution and use, 1994, pp. 9, 11, 31, 32.
[2] Lynch, D., Kopp C., Chapter 5, *Radar Handbook*, 3rd ed., M. Skolnik, Ed., McGraw Hill, New York, NY (2008), pp. 5.1–5.16.
[3] Hughes Aircraft, "Pave Mover TAWDS Design Requirements", November 1979, unclassified specification.
[4] Lindholm, E., Nickolls J., Oberman S., Montrym J., "NVIDIA Tesla: A Unified Graphics and Computing Architecture", *IEEE Computer Society Micro Magazine*, March 2008, pp. 39–54.
[5] Aks, S., Lynch D.D., Pearson J.O., Kennedy T., "Advanced Modern Radar", *Evolving Technology Institute*, Short Course Notes, November 1994.
[6] Lynch, D., *Introduction to RF Stealth*, SciTech Publishing, 2004, pp. 446–451, 82–84, 198–221, 467–470, 492–501, 504–531.
[7] Eliot, D., ed., *Handbook of Digital Signal Processing*, Academic Press, San Diego, CA (1987), pp. 173–286, 253–287, 364–464, 527–589, 590–593, 594–631.
[8] Guerci, J., *Cognitive Radar*, Artech House, Norwood, MA (2010), pp. 23–26, 135–153.
[9] Walden, R., "Analog to Digital Converter Survey", *GM Hughes Report*, 5 March 1992.
[10] Kester, W., *Analog-Digital Conversion*, Analog Devices, 2004, Chapter 2.
[11] Harris, F.J., *Multirate Signal Processing for Communication Systems*, Prentice Hall, Englewood Cliffs, NJ (2004), pp. 210–214, 396–409, 446–458.
[12] Richards, M., *Fundamentals of Radar Signal Processing*, McGraw Hill, New York, NY (2005), pp. 197–198.
[13] Cohen, M., Nathanson, F., *Radar Design Principles*, 2nd ed., McGraw Hill, New York, NY (1990), pp. 33–37, 64, 371, 533–582, 587–634.
[14] Perdy, R.J., "Signal Processing Linear Frequency Modulated Signals" in *Radar Technology*, E. Brookner, ed., Artech House, Norwood, MA (1977), pp. 155–162.
[15] Peebles, P., *Radar Principles*, John Wiley & Sons, New York, NY (1998), pp. 288–318, 557–614.

[16] Golay, M.J.E., "Complementary Series", *IRE Transactions on Information Theory*, Vol. IT-7, April 1961, pp. 82–87.

[17] Frank, R., "Polyphase Complementary Codes", *IEEE Transactions on Information Theory*, Vol. 26, No. 6, November 1980, pp. 641–647.

[18] Sinsky, A., Weil, T., Brookner, E., ed., *Radar Technology*, Artech House, Norwood, MA (1977), pp. 132–137, 379–383.

[19] Lewis, B., Kretschmer F., Shelton W., *Aspects of Radar Signal Processing*, Artech House, (1986), pp. 7–63.

[20] Grettenberg, T., Jeffers, R., "Complementary Codes for Hughes", *Applied Systems Laboratories Report*, unclassified report, April 1979.

[21] Fahey, M.D., *et al.*, *Interception of Bi-Phase Coded Radar Signals*, U.S. Army Missile Command Technical Report RE-77-e, originally classified confidential, declassified 12/31/1990.

[22] Grewal, M., Andrews A., *Kalman Filtering Theory and Practice*, Prentice Hall, Englewood Cliffs, NJ (1993), pp. 108–115.

[23] Lynch, D., Bogdanovic P., Pearson J., Mellama J., Graham G., "Radar Data Processing", *Evolving Technology Institute*, 1986.

# Chapter 5

# Noise in radar and intercept systems

## 5.1 Introduction

Noise of all kinds limits ultimate detection performance for all electromagnetic sensor systems whether cosmic, UV, visible, IR, microwave or RF. Thermal noise is often thought of as the ultimate limitation on detection and tracking but often it is self-noise in a sensor system, which limits performance. Low-frequency systems with seemingly very low noise figures (NF) are often limited by sky noise from the residual of the "Big Bang," from auroras and from what used to be called "Monkey Talk." Great strides have been made in signal processing in the last 50 years. This is the result of dramatic improvements in signal processing hardware as well as a better theoretical understanding. In spite of this, many systems provide disappointing performance.

The two main reasons for shortfalls are sensor signal processor self-noise and the physical world's continuing refusal to be normal (Gaussian). Although we have undreamed of device performance and number smashing today, we are inevitably forced by economics to limit hardware complexity that gives rise to significant self-noise. Noise and interference in the real world are "bursty," "lumpy," discrete, not stationary, anomalous and even malevolent! This chapter is an introduction to analog and digital signal processing self-noise analysis and heuristic means to control ill-behaved internal and external noise.

Coherent radar and intercept systems also depend on accurate Doppler information. The Doppler information is used in Doppler navigation, passive emitter location, target tracking, surface moving target detection (SMTD), DBS and SAR. One of the principal limitations in both radar and intercept systems is the spectral purity in the master oscillator (MO) used as the time and frequency reference in these systems. Example analytical software is included in Chapter 5 appendix.

Modern intercept and radar systems often use active array antennas, which have a transmit and receive (T/R) or receive only channels at each radiating element. Figure 5.1 shows the block diagram of a typical front end. Those channels are driven by a phase-locked exciter containing a MO chosen for the best short-term (<10 s) stability. In turn, the MO combined with synthesized offset frequencies produces a transmit frequency oscillator (TO), which is used to drive the transmit side of the T/R channels through a signal distribution manifold. On receive, each

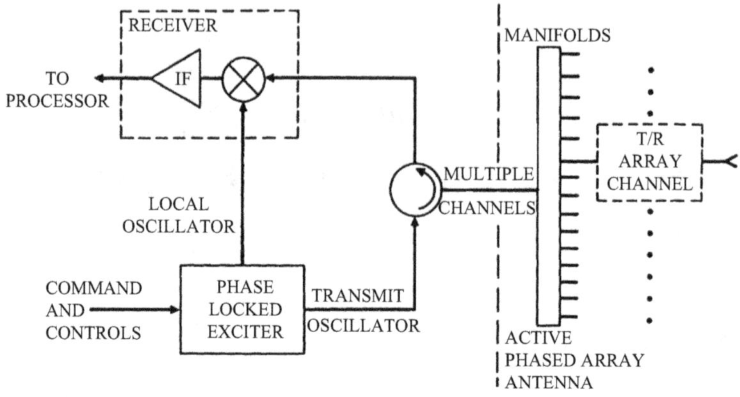


*Figure 5.1 Active array RF front end (repeated Figure 2.16)*

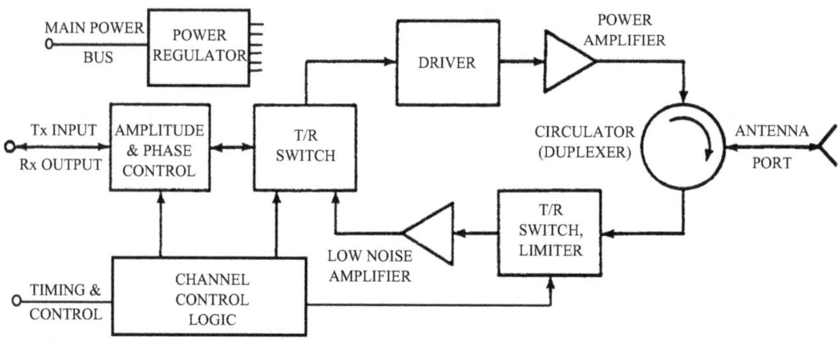


*Figure 5.2 T/R channel block diagram (repeated Figure 2.17)*

channel may drive multiple manifolds, which are configured to provide desirable angle-space properties. Those multiple manifold channels are converted to an intermediate frequency with the local oscillator (LO) reference allowing more gain control and frequency selectivity in the receiver block.

Subsequently, the channels are digitized in an A/D converter and further processed. The digital processor introduces other sources of noise that could limit performance. In each case, the design objective is to specify the sources of noise other than thermal noise to be substantially lower ($<-10$ dB) than thermal. Obviously this requirement is bandwidth dependent. In order to determine what self-noise is acceptable, one must define the detection bandwidth and the acceptable noise-induced false alarm rate. Then working backward from this requirement, a designer can allocate an acceptable noise contribution from each potential noise source. The first piece of the puzzle is the sensor front end.

A typical T/R channel is shown in Figure 5.2. There is power regulation, amplitude and phase control, a T/R switch, channel control logic, a driver and

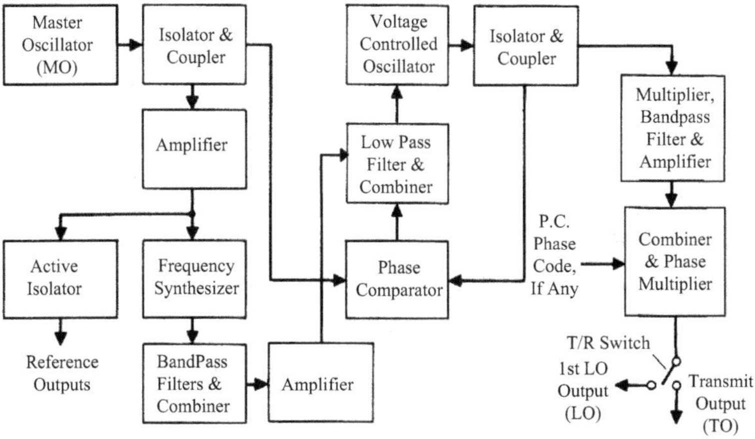


*Figure 5.3 Typical exciter block diagram*

power amplifier, a circulator or duplexer switch, a LNA and protection for the LNA in T/R switch and limiter. One output of the circulator drives an antenna radiating element. The channel control logic is embodied in a single logic chip, which is quite complex. Part of the control logic function is adapting commands from the overall control computer in the sensor to the nonlinearities in each channel. The nonlinearities may occur with commanded frequency, phase and amplitude, as well as array temperature. These calculations must usually be updated as often as each pulse, coherent dwell interval or second. All of these functions give rise to some self-noise in addition to the thermal noise, which arises from the environment that the signal propagates through. There is nothing to be done about the natural noise in the environment. The external manmade noise and sensor system self-noise can be mitigated with careful design and adaptive techniques. However, a better strategy is to provide premeasured corrections to every known inaccuracy so that the adaptive degrees of freedom are used to compensate for external degradation rather than hardware and software deficiencies.

The phase-locked exciter shown in Figure 5.1 gives rise to frequency modulation (FM) and AM noise on both targets and clutter. Often this is a greater limit than thermal noise. More details of the exciter subsystem are shown in Figure 5.3. There is a MO, which provides reference signals to the rest of the exciter as well as references to the rest of the sensor system. One portion of the MO output drives a synthesizer. The frequency synthesizer must switch output waveform between transmit and receive. The synthesizer output drives one or more voltage-controlled oscillator (VCO) loops. The VCO must be fast enough to track to the new frequency by the time the receive signal arrives. This is usually done by having two or more VCO loop bandwidths to allow fast acquisition and then low noise narrower frequency tracking. Both the frequency synthesizer and the multiplier chain (if any) will raise the phase (and frequency) noise power by the square of the multiplication ratio. Usually during receive, there is no signal anywhere in the exciter, which is at the transmitter (TO) or transmitter derivative frequencies in order to prevent dynamic range limiting cross-talk in the receiver. Pulse compression phase codes are usually applied close to the last circuit in the exciter. Linear FM (if any) is usually applied earlier in the exciter chain but again errors are magnified by the frequency multiplication ratio.

The following sections suggest ways to estimate and mitigate self-noise. It is by no means exhaustive. The noise is both analog and digital. Analog limits in frequency, jitter, drift, linearity, dynamic range, EMI, reproducibility, temperature control, etc., are often the hardest to counter. Digital limits in word length, record length, coefficients, mathematical stability, sampling rate, limit cycle oscillation, sidelobes, math execution rate, anomalous inputs, hard and soft failures, single event upset (SED), compression, etc., must also be controlled for best performance.

Table 5.1 tabulates the types of noise encountered in sensor systems. The noise comes in four flavors: amplitude modulation (AM), FM, line type (spurs/birdies) at a single or a few frequencies, and broadband random noise. A birdie is typically self-interference in a single or small band of frequencies caused by intermodulation and/or harmonic distortion. It occupies almost all range bins at the same frequency. If it appears in at least 1/3 of the range bins at the same frequency, then the threshold must be raised in that Doppler bin above a normal threshold multiplier. Since there may be clutter in the mainlobe frequencies, birdie killing is usually applied outside mainlobe clutter. Birdie killing usually consists of replacing the birdie frequency bins output with the average of the bins just outside the birdie region. Although birdies are usually only at a single frequency for a single transmitter channel, they usually spill over to a few adjacent bins through filter

*Table 5.1 Sources of noise*

| **Source** | **AM** | **FM** | **Line type** | **Random** |
|---|---|---|---|---|
| *Exciter* | | | | |
| Master oscillator | | Yes | | Yes |
| Vibration | | Yes | Yes | Yes |
| Power supply ripple | Yes | Yes | Yes | |
| Jitter | Yes | | Yes | Yes |
| *T/R channel* | | | | |
| Thermal | | | | Yes |
| Power supply ripple | | | Yes | |
| *Receiver* | | | | |
| Thermal | | | | Yes |
| Intermodulation | | | Yes | |
| Birdies/spurs | | | Yes | |
| Power supply ripple | Yes | Yes | Yes | |
| *Analog processing* | Yes | Yes | Yes | Yes |
| *A/D converter* | Yes | Yes | | Yes |
| *Digital processor* | Yes | Yes | | Yes |

sidelobes. Depending on source they may move for each transmit channel and operating mode. Much design work is required to minimize them but sadly they still appear.

The system viewpoint is to deal with the frequency spectrum at the output of the receiver assuming that spectral folding associated with a pulsed waveform has already occurred. Usually total noise both AM and FM at the receiver output is the most important because all the gain in front of the remaining analog and digital signal processing will have suppressed the noise contribution of the latter when referred to the front end. One must make a distinction between the line type and broadband noise because line type may only mask a single range, Doppler or azimuth bin.

## 5.2 Noise and gain in signal selective networks [1]

When a composite spectrum containing both continuous and discrete signals and noise is filtered, the relationships between peak signal, RMS signal and RMS noise are substantially altered. One well-established criterion of processor performance may be interpreted in terms of Figure 5.4 (similar to Figure 2.7). Here a signal spectrum is depicted with a strong point target at one frequency (or angle or range) and a "hole or shadow" in the spectrum at another frequency (or angle or range), representing a signal of very low *SNR*. The dynamic range criterion is the measure of how much weaker than the average background a weak target can be without being obscured by noise and/or sidelobes of surrounding signals, and of how much stronger than the surrounding average background a target can be before its sidelobes obscure, the background signals.

In a situation where the system performance is not limited by saturation, quantization or receiver noise, the weakest area signal that the processor could handle would be determined by the frequency (or range or angle) *ISLR* as mentioned in Sections 2.6 and 2.9. This is a measure of the total sidelobe power of the array or filter and represents the interference to a point signal from all adjacent signals. A typical *ISLR* design goal for high-performance signal processors is $<-40$ dB (1/10,000). A processor achieving this goal could recognize signals whose power is $-40$ dB below the surrounding average background power, assuming no receiver, quantization or saturation noise.

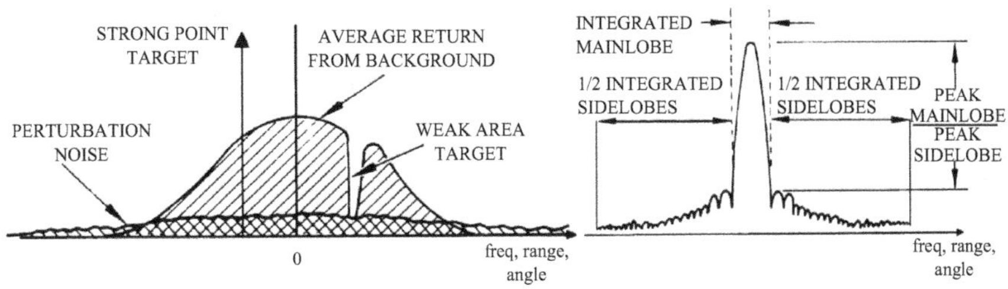


*Figure 5.4 Sidelobe and dynamic range considerations [2]*

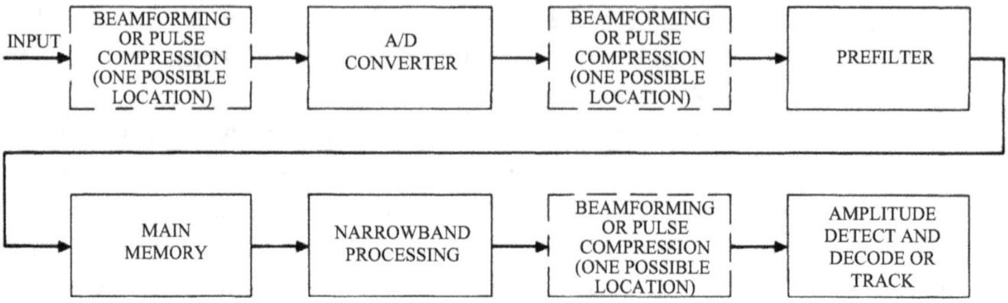


*Figure 5.5 Signal processing order considerations [1]*

A processor whose design *ISLR* is −40 dB has this fundamental performance limit for weak signals. In the analysis of the effects of saturation and quantization, and in the determination of memory size, scale factors, roundoff procedures, etc., it should be remembered that the total level of noise is established by the processor ISLR (i.e., noise, *ISLR* and other error sources are combined root-sum-square (RSS) to form overall performance predictions). Figure 5.5 also illustrates the situation in which the noise spectral density in the region of the weak point target determines the noise interference to that target. Because spectral density is significant and not simply total noise power, the distribution throughout the sampling spectrum of the noise added by saturation, quantization, FFT's, multiplication, magnitude calculation, etc., is important and often equals noise at the RF front end.

Signals much greater than the surrounding background level will have range, angle and frequency sidelobes that compete with the average background level. A target stronger than the average background by the peak sidelobe ratio will produce interference with targets at the location of that peak sidelobe. If the average or RMS sidelobe level of the pulse compressor, angle beam or filter sidelobe pattern is taken, rather than the peak sidelobe level, then targets stronger than the background level by the RMS sidelobe ratio, on the average, will compete with all targets in the area. Thus, once the processor design is established, for a certain RMS sidelobe level, a fundamental performance limit is also established. This limit determines the maximum target strength for which signal handling capacity must be provided. In other words, targets so strong as to generate sidelobe interference at all other angle/frequency/range locations may be allowed to saturate the processor as well since the detection performance is degraded in either case [1,3].

## 5.3 Effects of processing order on dynamic range [1]

Pulse compression, Doppler filtering and beamforming can be done in several locations in the overall signal processing sequence. Since these processes affect dynamic range, they require gain changes, more memory, and perhaps saturation, truncation or roundoff errors. For an 11-bit A/D converter, the value of $K_{AD}$ (see Section 5.9) that minimizes the total perturbation power is about 4.76, and for this value, the ratio of total perturbation power to RMS signal is seen to be about $1.95 \times 10^{-6}$ or $-57.1$ dB. The largest signal that would not saturate the A/D converter is 4.76 times the RMS signal level or about $+13.6$ dB. The largest signal is assumed to be concentrated at a single spectral point; the RMS signal has the spectral distribution impressed by the real sensor pattern, and the perturbation noise is partly white and partly colored. The total dynamic range at the A/D for this example is approximately 70.4 dB. Pulse compression or beamforming will change this dynamic range substantially. Figure 5.5 is a generalized block diagram of a signal processor that will serve as a model for the following discussion [2].

### *5.3.1 Pulse compression and beamforming*

Pulse compression or beamforming can be performed at three points in the signal processing sequence: before A/D conversion, immediately following A/D conversion and following narrow band processing. Angle or range sidelobes in a beamforming or pulse compression system are a reality whether done by analog or digital methods. Angle, range and frequency processing are to a first-order orthogonal operations.

It can be shown that when the A/D converter input signal level is set to minimize the ratio of total perturbation power to average signal power, the ratio does not improve when the pulse compression and/or beamforming circuitry is moved from ahead of the A/D converter to behind the A/D converter in the signal processing sequence. Clearly, an improvement occurs in peak signal-to-RMS noise (and, therefore, peak signal-to-RMS signal) when such a rearrangement is made. A similar argument applies when the pulse compression and/or beamforming circuitry is relocated to a position following the narrowband processor. It usually turns out that the total number of cells to be processed by subsequent operations is reduced by each processing block. So it often happens that there is a processing order, which yields the lowest total operations and or required dynamic range (word length).

The pulse compression or beamforming circuitry, being relatively wideband compared to the ultimate processed bandwidth, does not alter the relative spectral distribution of perturbation noise and average clutter return. The peak compressed pulse or angle response is larger than the peak amplitude of the uncompressed pulse or angle response by the compression ratio or angle gain; a pulse compression or beamform circuit operating on 11-bit input data words could produce output requiring $11 + \log_2 N$ bits to represent them, where $N$ is the compression ratio or array gain. Hence, it is apparent that, for a specified processor peak signal-to-RMS noise capability, a larger main memory is required for prestorage pulse compression or beamforming than for the same process performed after narrowband processing. Other factors may make prestorage pulse compression or beamforming more desirable on an overall basis. On the other hand, pulse compression may require Doppler compensation before compression processing but pulse compression after narrowband filtering requires none since the compensation occurs naturally in the Doppler filtering.

### *5.3.2 Scale factor changes, saturation and truncation*

When the analog dynamic range or digital word size grows as a result of pulse compression, Doppler filtering and/or beamforming (or other signal processing), extra bits or higher spur-free dynamic range (SFDR) must be provided in the circuitry to handle the larger range. Alternatively, the saturation level or word size must be reduced by discarding MSBs or LSBs or some combination of the two. Each of these processes adds noise that may create harmonics or "almost white noise." Another possibility is the use of floating point or AGC to handle a large dynamic range with a fixed word size or linear region. Discarding LSB is equivalent to dividing the signal by $2^N$ (where $N$ is the number of LSBs discarded) and then dropping the fractional part below the least significant bit, that is, to integer denominate fractional values. The process of truncating or rounding the rescaled data to integer fraction values is almost equivalent to quantizing in the A/D converter. The slight difference arises at the A/D converter where the data before quantization may take on any value between quantum steps, whereas in the truncation or roundoff process, the fractional part, which is dropped, has only $2^N$ discrete values.

Discarding MSBs or saturating is equivalent (assuming the use of proper overflow or saturation logic) to allowing the signal to saturate at peak values and the analysis applied to the A/D converter is usable. It can be shown that the output SNR of a digital pulse compression or beamform circuit is the SNR input signal times the square root of the compression or gain ratio; hence, the output SNR can be computed when the RMS input SNR is known.

## 5.4 Noise in prefiltering [1]

Prefiltering is the process of combining a set of successive samples to form a single data record before narrow band processing. Prefiltering has several effects. Since it is a filtering and amplitude normalizing process, it is equivalent to forming a noise whitening filter. The prefilter is often read out at a fraction of the sample rate or PRF; therefore, sampling ambiguities are possible (sometimes called multirate filtering). Since the prefilter passband width is some fraction of the total sampling interval, the effect of the prefilter on the signal is different from its effect on perturbation noise. The input signal is concentrated in a region of the sampling interval determined by the real signal space. The perturbation noise is the collective effect of receiver noise, saturation noise, quantization noise and truncation noise. Receiver, quantization and truncation noise may be considered white, that is, the total noise power has a relatively even spectral distribution throughout the sampling interval. Saturation noise is colored with a bandwidth somewhat wider than the input bandwidth. The saturation noise spectrum for an input signal with a Gaussian-shaped input spectrum is shown in Figure 5.25. When those noise spectra are applied to a digital filter, the output noise can be estimated using the filter noise bandwidth. The frequency response transfer function of a sampled data filter is given in (5.1).

$$H(f) = \sum_{n=0}^{N_p} A_n \cdot \exp\left(\frac{-\mathrm{j} \cdot 2 \cdot \pi \cdot n \cdot f}{f_s}\right) \tag{5.1}$$

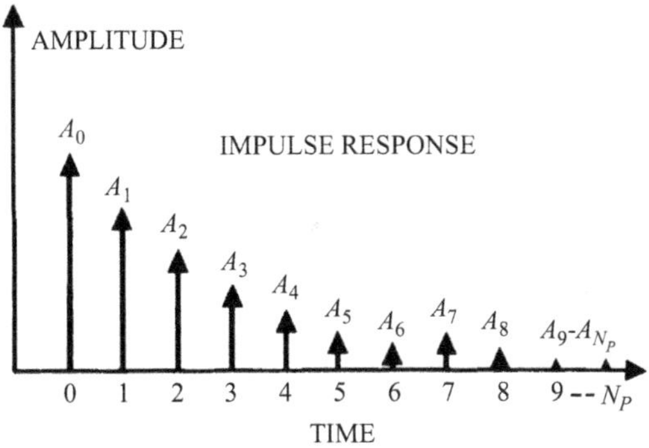


*Figure 5.6 Digital filter impulse response [1]*

where $A_n$ are the impulse response coefficients, $N_p$ is the number of significant terms in the impulse response and $f_s$ is the sampling rate. The meaning of (5.1) can be better understood by considering Figure 5.6. When an impulse is applied to a digital filter, it sweeps out the time-sampled coefficients of the impulse response of that filter. If the filter is feed forward only, the impulse response coefficients can usually be written by inspection. Every stable filter has a finite number of significant (greater than the LSB) impulse response coefficients. There are two effects for a set of samples as shown in Figure 5.6. First, there is a frequency response ambiguity at the sampling rate. Second, since the response is practically limited in the time domain, it has a sinc($x$) form convolved on it in the frequency domain that gives rise to frequency ripples.

$$B_N = \frac{\sum_{n=0}^{N_p} A_n^2}{\left( \sum_{n=0}^{N_p} |A_n| \right)^2} \cdot f_s \tag{5.2}$$

When a filter operates on a white noise input, the output may be calculated by multiplying the input noise spectral density by the noise bandwidth of the filter times the square of the filter's center frequency gain. The noise bandwidth of a digital filter of the type used in prefiltering is given by (5.2).

The filter center frequency maximum gain is given in (5.3).

$$Gain_{f_0} = \sum_{n=0}^{N_p} |A_n| \tag{5.3}$$

When the white noise part of the perturbation power is $\sigma^2$, the spectral density is $\sigma^2/f_s$. If spectral density, noise bandwidth, and square of center frequency gain are multiplied, then the total output power due to white noise input is given in (5.4).

The output noise power is just input noise variance times the sum of the squares of the impulse response coefficients as shown in (5.4).

$$P_{N\text{out}} = \frac{\sigma^2}{f_s} \cdot \frac{\sum_{n=0}^{N_p} A_n^2}{\left(\sum_{n=0}^{N_p} |A_n|\right)^2} \cdot f_s \cdot \left(\sum_{n=0}^{N_p} |A_n|\right)^2$$

$$\text{and: } P_{N\text{out}} = \sigma^2 \cdot \sum_{n=0}^{N_p} A_n^2 \tag{5.4}$$

Colored noise can be handled similarly for most cases of interest. The saturation noise occupies a spectral region slightly wider than the input signal bandwidth and may be regarded to a first approximation as having uniform spectral density over the prefilter bandwidth. Filter output noise due to saturation noise input may then be calculated by a formula similar to (5.4) by using an adjusted value of $\sigma^2$ to account for the greater spectral density of the input noise in the filter passband. If saturation noise power is defined as $\sigma_{sat}{}^2$ and the noise is assumed to be concentrated entirely in the bandwidth of the prefilter, $B_s$, then the value $\sigma_{sat}{}^2 \cdot (f_s/B_s)$ should be used in (5.4). The signal can be handled identically when calculating average signal power at the prefilter output; the signal power at the input is treated as if it had uniform spectral density across the prefilter bandwidth and zero density outside.

Sometimes, the above approach to colored noise is unsatisfactory. Under these circumstances, the procedure is to obtain the Fourier transform of the impulse response, find the magnitude squared of this frequency response and multiply it by the energy density spectrum of the input. The resultant is the output energy density spectrum for subsequent input to the next filter in the system. The Fourier transform of the energy density spectrum is the autocorrelation of the signal whose value at $\tau = 0$ is the variance in each time slot. The DFT frequency response, $H(f_i)$, of a time sampled filter is given in (5.5).

$$H(f_i) = \sum_{n=0}^{N_p} A_n \cdot \exp\left(\frac{-\text{j} \cdot 2 \cdot \pi \cdot n \cdot i}{N_p}\right) \tag{5.5}$$

And the continuous output spectrum is given in (5.6).

$$S_{out}(f) = S_{in}(f) \cdot |H(f)|^2 \tag{5.6}$$

where $S_{in}(f)$ is the input energy density spectrum at frequency $f$, $S_{out}(f)$ is the output energy density spectrum. So that the normalized sampled total output power is given in (5.7).

$$\sigma_{out}^2 = \frac{1}{N_p} \cdot \sum_{i=0}^{N_p} S_{in}(f_i) \cdot |H(f_i)|^2 \tag{5.7}$$

### 5.4.1 Saturation and truncation introduced by prefiltering

Because amplitude weighting is done digitally in the prefiltering, the digital word sizes grow according to the number of bits used to represent a coefficient. If the extra bits so generated are truncated immediately, the error introduced in the overall noise total must be included. An expression similar to (5.4) with $A_n = 1$ describes the output noise of the prefilter contributed by this source.

Because the prefilter has gain, the dynamic range or digital words at its output require more bits to represent them than did those at its input. The prefilter output usually is stored in the main memory before narrowband processing; hence, the word size of the memory determines whether or not truncation is necessary. If the prefilter output word has more bits than the main memory word, the saturation and truncation noise considerations that prevailed at the pulse compression or beamforming output apply equally to the prefilter output.

### 5.4.2 Coefficient noise and stability of prefilters

Coefficient approximations caused by limited word length or analog component tolerances can lead to significant degradation of performance and in some cases limit cycle oscillation. The most serious problem associated with analog or digital filters is stability. Since a digital filter can exhibit the same behavior as active analog filters, digital filters can produce ringing and oscillation. Obviously undesired oscillation destroys performance. The same rules used for analog circuit stability must be applied to digital filters. In addition, there are forms of oscillation that are unique to time and amplitude quantized systems. Limit cycle oscillation is the result of the feedback in IIR filters and amplitude quantization. It can result from both saturation and quantization. Even low-level oscillation will desensitize a system. A necessary but not sufficient condition for stability is that (5.8) is less than infinity as $N_p$ goes to infinity.

$$\sigma^2_{coeff} \approx \frac{2 \cdot N_p - 1}{3} \cdot 2^{-2 \cdot M} \tag{5.8}$$

where $N_p$ is the number of significant terms in the impulse response and $M$ is the number of bits in the coefficient words.

One common technique to minimize the effect of coefficient errors in filters is to group coefficients pairwise so that complex conjugate coefficients lie on the same radius circle about the origin [1,4,5]. Alternatively, design coefficients can be restricted to exact binary weights and an exhaustive search for the best set about the desired pole locations can be used. The noise power at the output of a filter due to coefficient quantization can be estimated as shown in (5.8) [1,4].

Normally, coefficient errors are designed to be a small fraction (<0.1) of the total expected noise at the filter output. For example, suppose the data is quantized to 16 bits and $N_p$ is 50 then if coefficient noise is to be 0.1 of quantizing noise, coefficients must be quantized to 19 bits ($16 + \log_2(0.1 \times 50) = 18.322$) and no roundoff can occur until after the final sum at the filter output.

## 5.5 Noise and signal response in narrowband filtering [1]

Often narrowband filtering is realized with a discrete (often fast) Fourier transform. Its elemental frequency response make it somewhat different from prefiltering, but the same analysis used earlier still applies. Narrowband processing is performed on the prefilter outputs as stored in the main memory. A sum (complex weighted) of the stored elements is formed to represent a single narrowband filter element. The complex weight is regarded as a phase rotation followed by an (real) amplitude weight. The phase rotation, performed simultaneously with the amplitude weight, is in effect a combination of real and imaginary channel data with no net amplitude change; therefore, no change in signal-to-noise relationship is produced and no noise is added. The narrowband processor, therefore, may be viewed in the same light as the prefilter as far as its effects on signals and noise are concerned. The noise bandwidth is calculated and used to compute the power at the output of the narrowband filter processor, after proper account is taken of the differences in noise densities at the input (i.e. that the truncation noise introduced at the previous stage is spread uniformly over the effective sampling bandwidth, whereas the signal and other noise components are more heavily concentrated in that bandwidth about the stage bandwidth in which the narrowband filter is centered).

Narrowband filtering obtained through the use of discrete or FFTs has certain distinctive properties that affect the spectral density relationships. Since the output from each discrete Fourier transform filter is an unweighted but phase rotated sum of input samples, it is not surprising that the frequency response has a sins($x$) form (see antenna equations (2.10) in Chapter 2). The frequency response of a fixed interval discrete Fourier transform, $G_{DFT}(f)$, to a complex sinusoidal input can be found by assuming an input, $g(t)$, as follows in (5.9).

$$g(t) = \exp(\mathrm{j}\cdot 2\cdot \pi \cdot f \cdot t) \tag{5.9}$$

where $t$ is time and $f$ is frequency. The counterpart sampled signal is given in (5.10).

$$g(n\cdot \Delta t) = \exp(\mathrm{j}\cdot 2\cdot \pi \cdot f \cdot n\cdot \Delta t) \tag{5.10}$$

where $\Delta t$ is the sample period, $n$ is the index of each sample and the maximum number of samples is $N$. Now, substituting into the DFT equation yields the frequency response of each of the $M$ DFT outputs as a function of frequency as shown in (5.11).

$$G_{DFT}(f,m) = \sum_{n=0}^{N-1} g(n\cdot \Delta t)\cdot \exp(-\mathrm{j}\cdot 2\cdot \pi \cdot m\cdot n/N) \tag{5.11}$$

And solving (5.11) yields

$$G_{DFT}(f,m) = \sum_{n=0}^{N-1} \exp(-\mathrm{j}\cdot 2\cdot \pi \cdot n\cdot (m/N - f\cdot \Delta t)) \tag{5.12}$$

Each DFT output is a narrowband filter, which is a matched filter for a single frequency signal of time length $N\Delta t$ at each frequency $m/(N\Delta t)$. But the form of

(5.12) was already solved for the linear array in (2.10) so the counterpart closed form for the series of equation (5.12) is given in (5.13).

$$G_{DFT}(f,m) = N \cdot \exp\left(\mathrm{j} \cdot \pi \cdot (N-1) \cdot (m/N - f \cdot \Delta t)\right) \cdot \left( \frac{\sin\left(N \cdot \pi \cdot (m/N - f \cdot \Delta t)\right)}{N \cdot \sin\left(\pi \cdot (m/N - f \cdot \Delta t)\right)} \right) \tag{5.13}$$

Let $m = 0$ in (5.13). then it is obvious that each filter is of the sins($x$) form as previously stated. Also, all of the weighting functions described earlier for antennas can be used to improve filter sidelobes in the DFT. There is one important weighting function difference and that is realizability is greatly expanded – limited only by word length (e.g., −80 dB sidelobes *are* achievable, which is not possible in a real antenna).

## 5.6 Transmitted noise sources

Noise on the transmit oscillator (TO) and LO will be associated with the signal itself and does not get better with more power from the transmitter. This attribute demands low noise sources. Amplitude noise usually comes from power supply ripple. Phase noise arises from phase locked loops, swept filters, multipliers, intrinsic phase stability, microphonics, poor temperature control and delay line references. The multiplicative noise on TO and LO arises from the intrinsic capabilities of the low noise MO but may also be caused by power supply ripple and time jitter. Thermal noise on the transmit signal also may be caused by amplitude and phase control, drivers and power amplifiers. Some of this noise will be correlated and potentially cancelable; some will be uncorrelated and performance limiting.

$$E_o(t) = P(t) \cdot \left(1 + n_a(t)\right) \cdot \cos\left(2 \cdot \pi \cdot f_0 \cdot t + \varphi(t) + n_\varphi(t)\right)$$

where: $P(t)$ is pulse train waveform, $\varphi(t)$ is the phase modulation of the transmit sinewave, $f_0$ is transmitted frequency, $n_a(t)$ is the total amplitude noise and $n_\varphi(t)$ is the total phase noise. (5.14)

The typical transmit waveform is given in (5.14). It consists of the intentional phase modulation and AM as given in Chapter 2 as well as the unwanted amplitude and phase noise modulation. This noise arises from multiple sources in the exciter and transmit chain.

The gross features of an actual transmitted spectrum are shown in Figure 5.7. The pulse train and transmit frequency give rise to a power spectrum approximating a sine x/x envelope caused by the pulse width and sampling ambiguities. An example of actual transmitter FM noise is shown for both narrowband and broadband noise in Figure 5.8. Around each spectral line in the spectrum is a fine spectrum of the type shown in Figure 5.8. It is the result of the various contributors

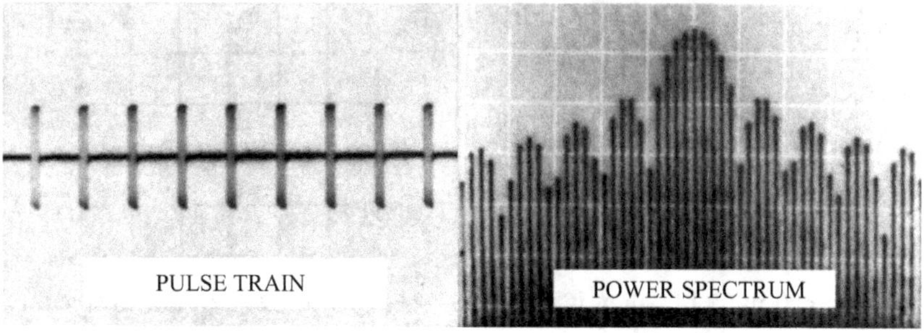


*Figure 5.7 Transmit waveform (repeated Figure 2.25)*

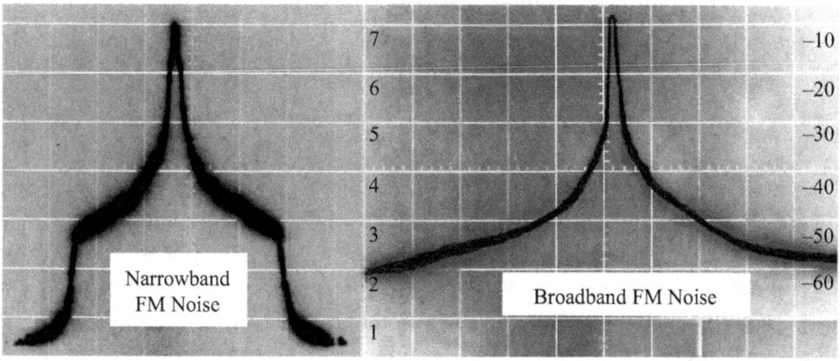


*Figure 5.8 Actual transmitter FM noise spectra*

to phase and amplitude noise. The spectrum of the basic MO is shown schematically in Figure 5.9 top. The noise in the center frequency comes from the bandwidth of the oscillator circuit itself plus the so-called 1/f noise plus the far out conventional random noise. Because the pulse sampling causes folding about the ambiguities, the noise floor is raised by the sum of all the folds out to the point where the bandwidth is limited by front-end analog filters. This is shown in Figure 5.9 bottom. If the amplitude and phase noise modulation is small relative to the intentional waveform, then amplitude and phase noise can be treated similarly.

If the FM is small relative to the carrier, then the first terms of the power series expansion of the carrier in (5.14) assuming no intentional phase modulation or FM is given in (5.15).

Assume: $n_\varphi(t)$ is of the form $n_\varphi(t) = \dfrac{\Delta f}{f_m}\cos(2\cdot\pi\cdot f_m\cdot t)$

$$\text{then: } \cos(2\cdot\pi\cdot f_0\cdot t + n_\varphi(t)) \approx \cos(2\cdot\pi\cdot f_0\cdot t) - \frac{\Delta f}{f_m\cdot 2}\sin(2\cdot\pi\cdot(f_0 - f_m)\cdot t)\ldots$$

$$-\frac{\Delta f}{f_m\cdot 2}\sin(2\cdot\pi\cdot(f_0 + f_m)\cdot t) \tag{5.15}$$

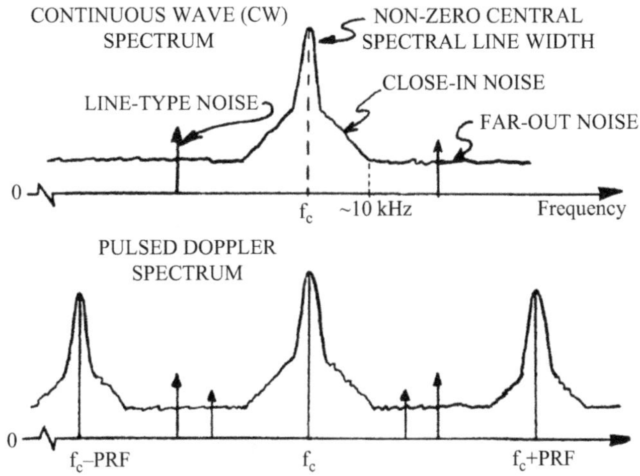


*Figure 5.9 Noise spectrum components*

Similarly, for small signal AM then the first terms of the power series expansion carrier of the (5.14) are given in (5.16). Each of these modulations gives rise to upper and lower sidebands of similar amplitudes relative to the center frequency carrier. However, because there is amplitude limiting in the MO and subsequent limiting amplifiers in the transmit chain, the AM other than the pulse train itself is usually much smaller than the FM.

Assume: $n_a(t)$ is of the form $n_a(t) = a \cdot \cos(2 \cdot \pi \cdot f_a \cdot t)$

$$\text{then: } (1 + n_a(t)) \cdot \cos(2 \cdot \pi \cdot f_0 \cdot t + \theta) \approx \cos(2 \cdot \pi \cdot f_0 \cdot t) + \frac{a}{2}\cos(2 \cdot \pi \cdot (f_0 - f_a) \cdot t - \theta) \ldots$$

$$+ \frac{a}{2}\cos(2 \cdot \pi \cdot (f_0 + f_a) \cdot t + \theta) \tag{5.16}$$

The natural question to ask is, "What is the required noise level to allow successful detection?" The simple but evasive answer is that it depends on the bandwidth at detection and where in frequency relative to the PRF ambiguities. Referring to Figure 5.9 the spectral line width, the close-in noise, the far-out noise, the line type (birdies) noise must all be specified to allow detection. Clearly, if the PRF is low then the far-out noise will fold many more times and it will dominate. If detections are to be at low Dopplers, then close-in noise will dominate. If there are too many birdies, then they will dominate.

Referring back to Figure 5.7 each spectral line at the right of the figure will have the transmitter noise folded around it. This raises the noise floor dramatically as well as folding all the birdies back into the region between the PRF frequency ambiguities. Figure 5.10 shows the transmit noise sidebands reflected around each PRF line. They all add up in each PRF interval.

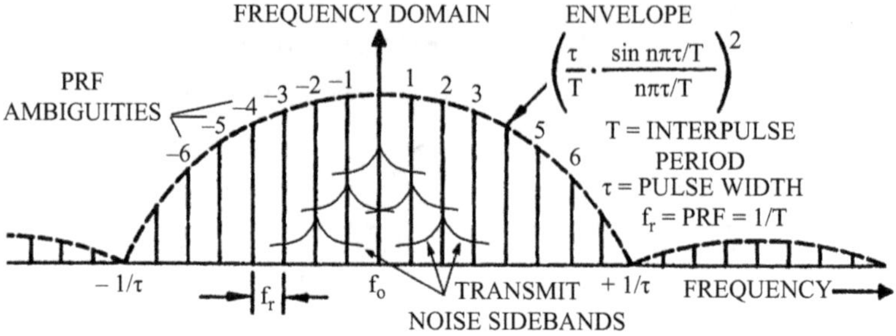


*Figure 5.10 Transmit noise reflected around each PRF ambiguity (similar to Figure 2.26)*

For example, since there will be 2000 PRF ambiguities arising from a 1 kHz PRF and a 1 μs transmitted pulse in its main spectral lobe, the noise floor will be raised by roughly 1000 (actually ~904 null to null). The far-out transmitter noise modulates the clutter and spreads the clutter spectrum across the entire band between the PRF ambiguities.

The folding factor can be approximated by (5.17).

$$FF = 2\cdot\left[\sum_n\left[\left|\frac{\sin(n\cdot\pi\cdot c_0)}{n\cdot\pi\cdot c_0}\right|\right]^2\right]$$

Where $n$ is the number of PRF ambiguities,

and $c_0$ is the ratio between the pulse width and the PRI, $\tau/T$. (5.17)

The ideal objective is for the transmitted noise to be substantially less than the thermal noise in the receiver. This is usually accomplished by selecting the MO with the greatest short-term stability regardless of frequency and then multiplying and dividing its output to create the desired references, LO and TO for a system.

Referring back to Figures 5.1 and 5.3, typically the MO phase locks a VCO to generate either the final transmit frequency or a signal, which must be multiplied further in frequency to generate the final frequency output. If the VCO output must be multiplied, then the output waveform is as shown in (5.18).

$$E_o(t) = P(t)\cdot(1+n_a(t))\cdot\cos\left(2\cdot\pi\cdot M\cdot\left(f_{VCO}\cdot t+\varphi(t)+n_\varphi(t)\right)\right)$$

where: $P(t)$ is pulse train waveform, $\varphi(t)$ is the intentional phase modulation of the VCO output, $f_{vco}$ is the VCO output frequency, $n_a(t)$ is the total amplitude noise, $n_\varphi(t)$ is the VCO phase noise, and $M$ is the frequency multiplication factor. (5.18)

Unfortunately this multiplies the FM noise voltage by a factor of $M$ and the FM noise power by a factor of $M^2$. It would be desirable for the highest practical VCO

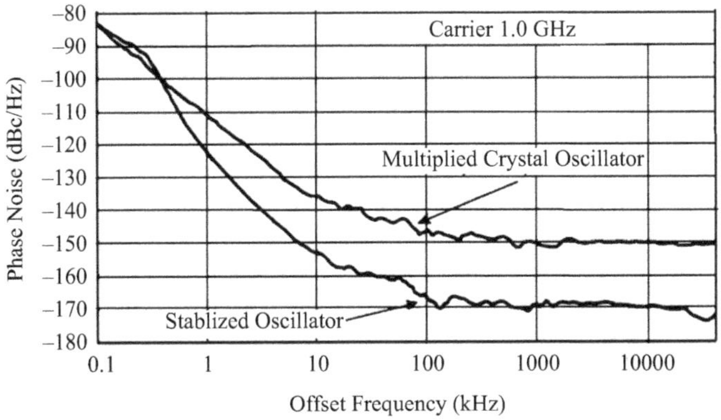


*Figure 5.11 Phase noise comparison – two methods; adapted from [6]*

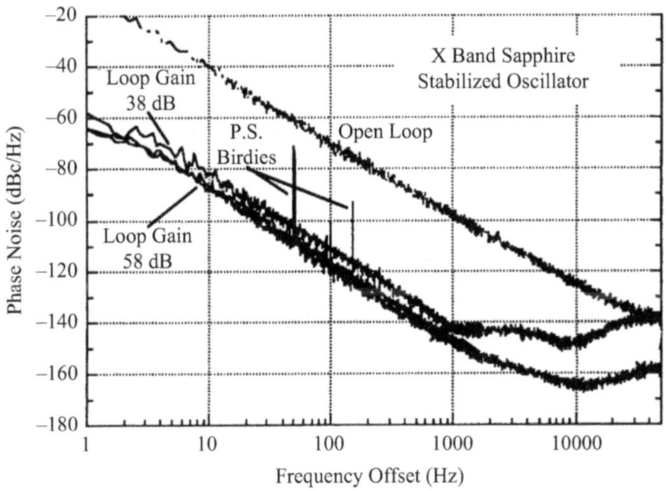


*Figure 5.12 X band stabilized VCO phase noise; adapted from [7]*

frequency to be used. Figure 5.11 shows some comparison examples between a stabilized VCO at 1 GHz and a 10:1 multiplied crystal oscillator. As expected the far-out phase noise is 20 dB higher for the multiplied crystal output. At X band both of these phase noise profiles would be 20 dB higher if multiplied up from 1 GHz.

Figure 5.12 shows the phase noise profile of a typical X band stabilized MO with various loop gains. Once outside the closed loop bandwidth, the phase noise follows the intrinsic noise performance of the oscillator and its following amplification, in this case about −140 dBc/Hz out to the transmit pulse bandwidth. The near-in noise is rolling off at approximately 20 dB per decade. Note also in Figure 5.12 power supply ripple induced "birdies" at 60 and 120 Hz. These birdies are at frequencies where slow-moving target Dopplers are common, for example, a radial velocity of

6.5 km/h corresponds to 120 Hz at X band. The birdies cause false alarms and also will mask any real targets at that Doppler. If one assumes that adaptive filtering in the processing will eliminate these birdies, then degrees of freedom that might be used to cancel clutter or interference will be diverted to this noise.

The general form of the folded transmit spectrum will be approximately as shown in (5.19).

$$NO = B_{\text{det}} \cdot \left[ \left( \frac{Near}{f} \right)^2 + FF \cdot Far^2 + Birdies^2 + \left( \frac{Near}{PRF - f} \right)^2 \right]$$

Where: $B_{\text{det}}$ = detection bandwidth, $f$ = frequency relative to the carrier,

$Near$ = near in noise voltage at 1 Hz, $Far$ = far noise voltage floor,

$FF$ = folding factor, $Birdies$ = discrete frequency voltages from all sources. (5.19)

For example, suppose the MO of Figure 5.12 was used for a GMTI system with a 3 kHz PRF, which is $T = 333 \times 10^{-6}$, a 1/8 s time on target ($T_{OT}$), an 88% integration efficiency ($eff_n$) and a 25 MHz transmit bandwidth ($\tau = 40 \times 10^{-9}$ s). Those parameters provide an ultimate detection noise bandwidth of approximately 9 Hz and a folding factor of 5229 or about a 37.2 dB increase in the noise floor. The corresponding overall transmitted spectrum including birdies in that detection band will look like Figure 5.13 around each PRF line.

The FM noise floor is about −95 dB below the clutter peak. Typical slow-moving target specifications require detection of targets moving at 4 km/h radial to the observer. At X band a 4 km/h radial velocity has a Doppler of about 37 Hz. Reading from Figure 5.13, the corresponding FM noise on a single clutter patch will be approximately −84 dB down from the clutter peak at antenna beam center. Unfortunately clutter has its own Doppler spread and not all of it can be compensated. Adjacent clutter patches in the assumed 8 Hz filter spacing have their phase noise down by only −70 dB so the sum of transmitter noise bleeding from the adjacent filters on each side of a specific low-speed MTI inside mainlobe clutter will be approximately −67 dB down. The birdies will also create clutter sidebands that are not easily cancelled. Those Doppler bins will have higher thresholds to prevent false alarms and hence poorer sensitivity. As will be shown in Section 5.7, the higher power TO noise is at lower frequencies and even at relatively long distances it is partially correlated with the LO assuming they come from the same MO. For example, the transmit noise at an FM frequency of 10 Hz will be partially correlated with returns that are 1/10 s away, that is, ~15000 km! The correlated fraction will cancel on receive. This cancellation will cause the 1/f noise to be flattened across most of the band (see Figure 5.16).

A common phase modulation scheme to create a chirp waveform for pulse compression is direct digital synthesis at a convenient intermediate frequency. To the extent that there are nonlinearities during the chirp waveform (inevitable), those nonlinearities will create two or more sidebands of the type given in (5.15), that is, each will have a power of $(Mn_\phi(t)/2)^2$. Unintentional nonlinearities in chirp will

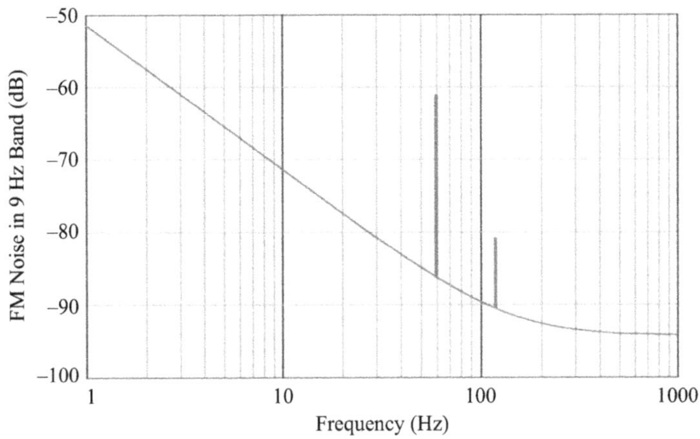


*Figure 5.13 Typical transmitted FM noise including folding (NoiseSidebandFold)*

prevent achievement of the desired pulse compression ratio as well as poorer detection sensitivity.

### *5.6.1 AESA unique noise contributions*

An AESA gives rise to some new forms of antenna sidelobe noise. Normally there is angular filtering between the transmitter and free space. When the last circuit element in the transmit chain is just behind the radiator, the thermal and other random noise in the T/R channel power amplifier gives rise to degraded sidelobes. Although some fraction of this noise will be correlated channel to channel, the majority will be uncorrelated and will not go away on receive. An example of the effect of uncorrelated random phase and amplitude noise on transmit sidelobes is shown in Figure 5.14. It shows an antenna pattern of a $50 \times 50$ element active electronic scanned array in an inter-cardinal plane with a random error of $\pm 1\%$ in amplitude and $\pm 0.025$ radians in phase arising from all sources of T/R noise. Obviously the receive pattern is randomly different than the transmit pattern because temperature and ripple will be different. Usually, the NF of the final T/R amplifier chain is poor because the amplifiers are hot and the noise bandwidth is large. Nonetheless, this effect is minimal compared to FM noise from the MO.

## 5.7 Receiver noise sources

### *5.7.1 Thermal noise*

Any sensor receiver whether electro-optical, Elint, Comint or radar is limited by the external noise from the environment around us as well as the internal noise in the system. Thermal noise arises from the temperature dependent radiation from any

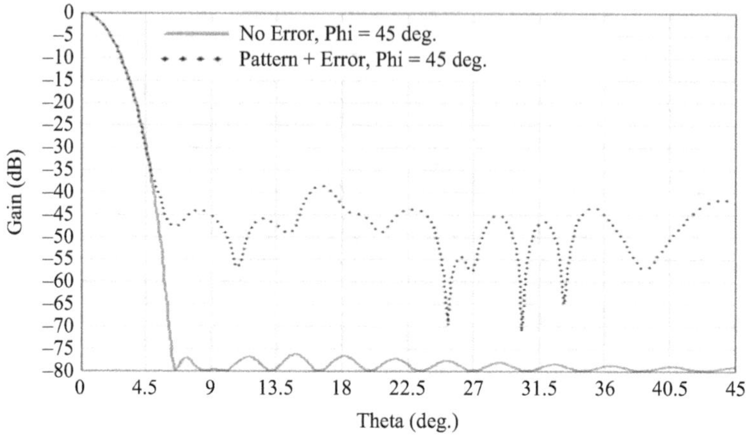


*Figure 5.14 AESA transmit noise sidelobes (AESANoiseSidelobes001)*

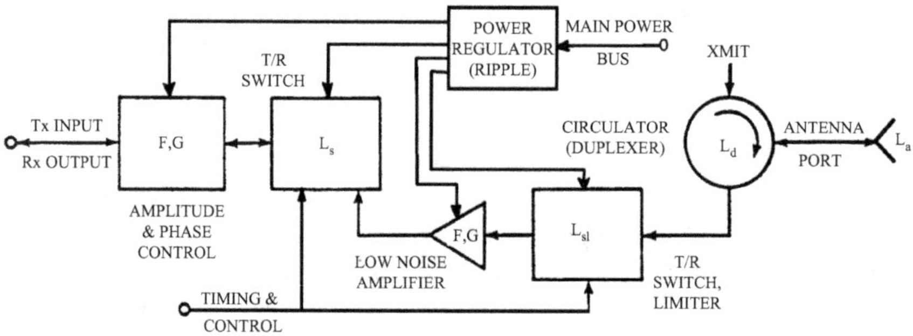


*Figure 5.15 T/R channel with emphasis on thermal noise*

material substance including air. Conventional additive thermal noise in a receiver (NF) will be the ultimate limit on system sensitivity. The NF is defined as the ratio of the input SNR to the output SNR as shown in the first line of (5.20). Recall the radar equation (2.21) had a term, $k_B T_e B$, which is the product of Boltzmann's constant, the equivalent temperature and the noise bandwidth. It represents thermal noise at the input to a sensor system.

Figure 5.15 shows a T/R channel with focus on the sources of noise. Each source of loss in the receive chain increases NF directly. Each element of the receive chain, which has gain, suppresses the NF of all the following components. The losses in the antenna, duplexer and T/R switch-limiter set a lower bound on the noise floor. Although these sources of noise must be kept as low as possible, that must be balanced against the potential burnout of the front-end LNA. Also shown in the last line of (5.20) is a tabulation of the NF associated with the cascade of circuits in the T/R receive channel of Figure 5.15. This begins with the losses in the antenna itself. It is followed by losses in the circulator and/or T/R switch limiter before getting to the first gain stage the LNA. The LNA is usually followed by another switch and the phase and amplitude control circuits. Often but not always the system NF is established by this point. Unfortunately, to achieve best performance both dynamic range and self-noise are still serious issues that must be dealt with in very careful design.

Thermal noise is overcome by transmitting more power with a larger antenna aperture or in a passive system by getting closer to the emitter with a larger antenna aperture. Receivers also are characterized by their equivalent noise temperature as shown in Chapter 2, Figure 2.28. Since thermal noise is temperature dependent, some front ends have been cooled with liquid nitrogen to improve NF but that is not feasible with active arrays. The short path length between the radiating element and the LNA more than compensates for this limitation. As long as there is significant gain in the LNA, the overall NF will be dominated by the losses before the LNA and the LNA itself. A typical overall system NF might be 1.5–4. The rest of the thermal noise will come from the atmosphere and surrounding warm terrain. Equation (5.20) losses are greater than 1 because they appear in the denominator of the radar equations (2.21).

$$F = \frac{SNR_{in}}{SNR_{out}} = \frac{S_{in}/k \cdot T \cdot B}{S_{in} \cdot G/N_{out}} = \frac{N_{out}}{k \cdot T \cdot B \cdot G} = \frac{N_{in} + \Delta N}{k \cdot T \cdot B \cdot G} = 1 + \frac{\Delta N}{k \cdot T \cdot B \cdot G}$$

For multiple circuits in cascade, then define front end losses as:

$$L_l = L_a \cdot L_d \cdot L_{sl}$$

$$F_o = L_l \cdot F_{LNA} + \frac{L_l \cdot (L_s - 1)}{G_{LNA}} + \frac{L_l \cdot L_s \cdot (F_{a\varphi} - 1)}{G_{LNA}} + \frac{L_l \cdot L_s \cdot (F_{RX} - 1)}{G_{LNA} \cdot G_{a\varphi}} \tag{5.20}$$

Power supply ripple in the T/R channel will induce amplitude and phase modulation on the received signal. During the transmit main bang, the power supply ripple is large, which results in some phase pulling. During receive cycle, the energy storage in the T/R channel is replenished, causing ripple at the PRF and its harmonics. Since the range to the region of interest is almost always changing, this ripple will cause varying phase and amplitude noise. The ripple is partially filtered with bypass capacitors but it is usually not quite enough.

---

**Example 5.1:** For example, suppose the T/R channel had the parameters shown in Table 5.2 and a block diagram like Figure 5.15. Then the overall NF would be as given in equations (ex 5.1).

What one can see in the example is that the thermal noise is increased 2.84 times by the T/R channel network shown in Figure 5.15. The other important item to note is that loss directly raises the system NF. The lower the front end loss the

*Table 5.2 Thermal noise example 5.1*

| Parameter | Value | Parameter | Value |
|---|---|---|---|
| Antenna loss, $L_a$ | 1.1 | T/R switch loss, $L_s$ | 1.1 |
| Duplexer loss, $L_d$ | 1.1 | Amplitude and phase NF, $F_{\alpha\phi}$ | 2 |
| Switch limiter loss, $L_{sl}$ | 1.4 | Amplitude and phase gain, $G_{\alpha\phi}$ | 20 |
| LNA noise figure, $F_{LNA}$ | 1.67 | Receiver NF, $F_{RX}$ | 4 |
| LNA gain, $G_{LNA}$ | 160 | $kTB$ | $10^{-12}$ |

better. That is one of the major reasons for the popularity of AESAs; they reduce front end loss dramatically.

$$L_l = 1.1 \cdot 1.1 \cdot 1.4 = 1.694$$

$$F_o = 1.694 \cdot 1.67 + \frac{1.694 \cdot (1.1-1)}{160} + \frac{1.694 \cdot 1.1 \cdot (2-1)}{160} + \frac{1.694 \cdot 1.1 \cdot (4-1)}{160 \cdot 20} = 2.84 \qquad \text{(ex 5.1)}$$

### 5.7.2 Dynamic range and self-noise

Following the T/R channels on the array and the beamforming manifolds are the receiver–exciter channels mentioned in Chapter 2, Figure 2.21. This is usually where dynamic range and self-noise are very important. The elements of dynamic range and self-noise interact to limit performance. The key issues are signal saturation, 1 dB compression power, single tone linearity, second-order intermodulation (2nd IM) spurs, 2nd IM external interference, third-order intermodulation (3rd IM) due to cross-modulation from internal sources, 3rd IM spurs, 3rd IM due to external interference, pulse on pulse (POP) external interference, internal cross-talk and finally thermal noise level. All of these elements set the largest and smallest signal that can be detected.

As was shown in Section 5.6 and the example of Table 5.3 which follows, the transmitter will have folded noise sidebands. Nonetheless, there will be some correlation between the transmitted noise sidebands and the modulation of the received signal by LO FM noise, which will partially mitigate the transmitted noise. Figure 5.16 shows the effect based on an assumed range of 150 km for the return. Comparing the Figure 5.13 spectrum with the spectrum of Figure 5.16, one can see that the close-in spectrum has been dramatically reduced. Unfortunately, the spectrum of the transmit noise rises as the assumed range of the target gets further away and can rise as much as 7 dB. A careful study of Figures 2.32 and 5.16 suggests that a MO could have 23 dB worse performance than Figure 5.13 and just cause clutter sidebands to be at thermal noise.

The design objective for a receiver is to provide the largest dynamic range consistent with minimum additional distortion self-noise. Ideally self-noise should

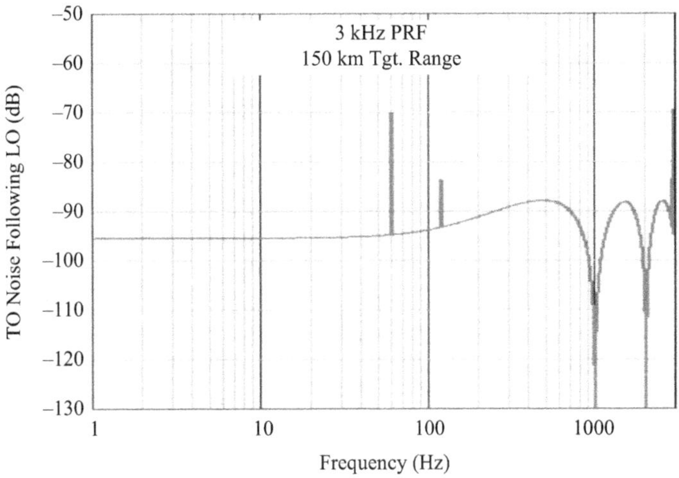


*Figure 5.16 Partial cancellation of transmitted FM noise by LO FM noise (NoiseSidebandFold)*

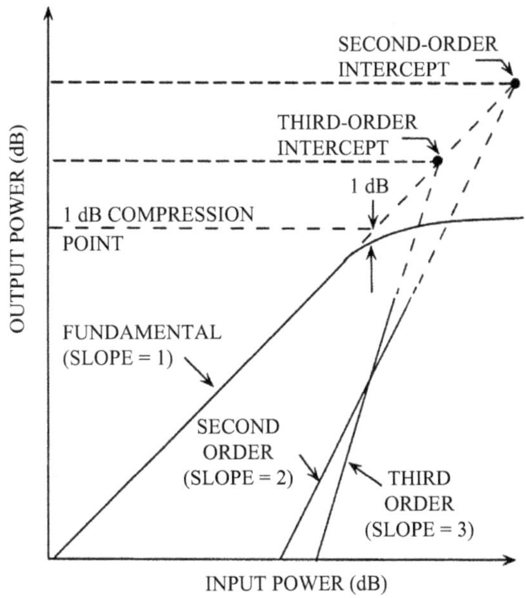


*Figure 5.17 Intermodulation distortion (IMD) definitions; adapted from [8]*

be kept below thermal noise. There always will be some spectral spreading of received signal by the nonlinearities in the receiver chain. The typical definitions of distortion, that is, nonlinearity induced self-noise are given in Figure 5.17.

The circuit nonlinearities can be approximated by the first few terms in a power series as shown in (5.21) first line. One common way of understanding the effect of these nonlinearities is to assume two input frequencies as shown in the second equation in (5.21). The last lines in (5.21) show the output amplitudes and frequencies of the first few intermodulation products. Obviously, real input signals are hardly ever just two frequencies but one can imagine how much distortion will be produced.

For example, assume that two frequencies of similar amplitude at 50 and 51 MHz are applied to the nonlinearities of the receiver chain. Figure 5.18 shows the qualitative output spurs from those two signals. The figure identifies those outputs, which are associated with the voltage squared power series intermodulation term as well as those associated with the voltage cubed term. In a broadband receiver front end, many or all of these spurious outputs may be in-band and lead to false target detections or reduced detection sensitivity. The power series coefficients $a_2$ and $a_3$ in (5.21) may need to be less than $10^{-4}$ for good performance. Also note in Figure 5.18 that there will be products both above and below the interacting signals, which may limit low Doppler detection sensitivity.

First three terms of nonlinearity power series:

$$V_{out} = a_1 \cdot V_{in} + a_2 \cdot V_{in}^2 + a_3 \cdot V_{in}^3$$

Assume the input consists of two frequencies:

$$V_{in} = A_i \cdot \cos(2 \cdot \pi \cdot f_i \cdot t) + A_j \cdot \cos(2 \cdot \pi \cdot f_j \cdot t)$$

Then the first few output distortion terms will be:

$$DC \Rightarrow \frac{a_2 \cdot \left(A_i^2 + A_j^2\right)}{2}, \qquad 3 \cdot f_i \Rightarrow \frac{a_3 \cdot A_i^3}{4}$$

$$f_i \Rightarrow a_1 \cdot A_i + 3 \cdot a_3 \cdot \left[\frac{A_i^3}{4} + \frac{A_i \cdot A_j^2}{2}\right], \qquad f_i \pm f_j \Rightarrow a_2 \cdot A_i \cdot A_j$$

$$2 \cdot f_i \Rightarrow \frac{a_2 \cdot A_i^2}{2}, \qquad 2 \cdot f_i \pm f_j \Rightarrow \frac{3 \cdot a_3 \cdot A_i^3 \cdot A_j}{4} \tag{5.21}$$

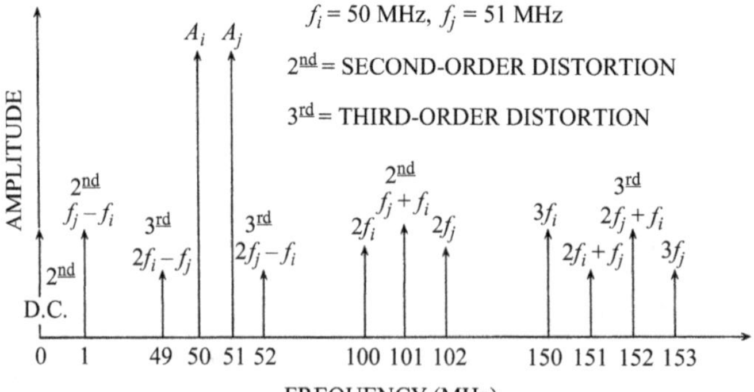


*Figure 5.18 IMD example; adapted from [8]*

Many analog circuits are characterized by their SFDR as defined by Figure 5.19. This is usually found experimentally by introducing one or more large amplitude signals and measuring the largest output spurs. Full scale is usually taken at the 1 dB compression point. One only need measure this out to half the sampling rate or half the PRF since the spectrum is ambiguous past that point. It often occurs that the spur may be associated high-order harmonics (20–30 times the base frequencies).

### 5.7.3 *Intermediate frequency analog processing*

Figure 5.20 shows a single receiver channel. As an incoming signal is processed and analyzed, the required dynamic range can grow to 120 dB ($10^{12}$ to 1). In order to adapt to this very large range, CAGC and fine automatic gain control (FAGC) as well as STC (gain adjustment with range) may be applied to the incoming signal. Normally the power at the output of the T/R channel is 30 dB above the input as shown in Figure 5.21 with a peak perhaps of −20 dBw. The 1 dB compression point might be at −20 dBw or lower. However, depending on antenna pointing direction, the total power could easily be higher, which would put the first stages of the receiver into saturation so the first CAGC might require 30 dB of attenuation before the first LO.

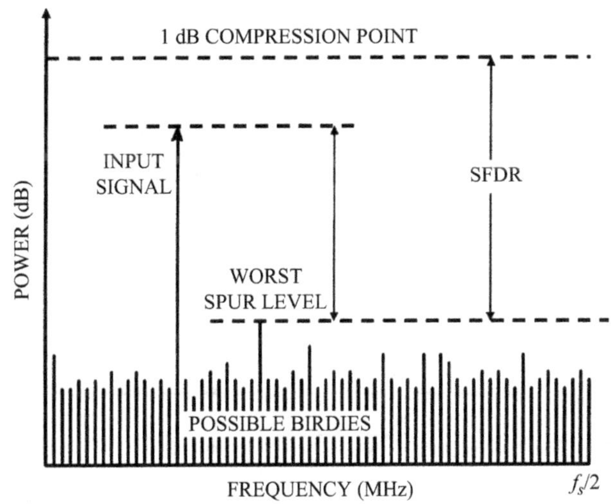


*Figure 5.19 Spur-free dynamic range; adapted from [8]*

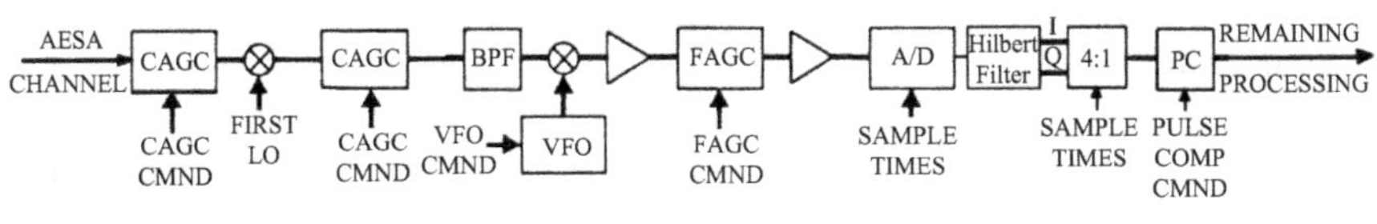


*Figure 5.20 Receiver channel with emphasis on noise*

If the first mixer has some gain, then there might be a second 30 dB CAGC prior to bandpass filtering matched to the pulse compression chip width or chirp swath width.

The last step in analog processing is translation of the RF signal to baseband. The baseband is represented as in-phase and quadrature components of a complex vector for each sample. This translation may create imbalance terms that can cause false targets and noise at the image frequency. Often digital Hilbert transform techniques are used to select a single sideband of the analog waveform to provide better baseband performance. Hilbert transform methods were described in Section 4.8. The noise advantage will be described in Sections 5.8 and 5.9.

---

**Example 5.2** For example, using the parameters of Table 5.3 including parameters for a typical barrage noise jammer, transmitter FM noise could be as shown in Figure 5.21.

*Table 5.3 Receiver channel noise example 5.2*

| Parameter | Value | Parameter | Value |
|---|---|---|---|
| $P_T$, *transmit power* | 10 kW | $\lambda$, *wavelength* | 0.03 m |
| $D_T$, *transmit duty* | 0.2 | *Antenna area*, $A_{eR}$ | 1.8 m$^2$ |
| $T_o$, *standard temperature* | 290 K | $k_B$, *Boltzmann's constant* | $1.38 \times 10^{-23}$ |
| *L, losses* | 2.5 | *Noise factor, NF* | 1.6 |
| *Antenna face normal*, $\theta_0$ | 0° | $\sigma_t$, *target RCS* | 10 m$^2$ |
| $G_T$, *target gain* | 25130 | $T_{OT}$, *time on target* | 0.125 s |
| *PRF* | 3 kHz | $R_{max}$, *maximum range* | 150 km |
| $eff_n$, *integration efficiency* | 0.88 | $B_T$, *transmit bandwidth* | $25 \times 10^6$ Hz |
| $c_{min}$ | 0.05 | $T_{OT}$ *x PRF* | 375 |
| *PCR, pulse compression* | 1670 | *n, number of samples* | 626,250 |
| $\sigma_o$, *clutter RCS* | 0.001 | *Antenna beam, radians* | $0.13 \times 0.0065$ |
| $\eta$, *velocity vector look angle* | 90° | *ISLR* | $6 \times 10^{-3}$ |
| $d_r$, *range resolution* | 6 m | *Elevation footprint*, $FP_{el}$ | 150 km |
| *T/R channel bandwidth*, Hz | $250 \times 10^6$ | *Jammer bandwidth, Hz* | $500 \times 10^6$ |
| *Jammer power*, $P_J$ | 1 kW | *Jammer gain*, $G_{TJ}$ | 300 |
| *T/R channel gain*, $G_{T/R}$ | 1000 | *A/D converter bits* | 9 |

Using the parameters of Table 5.3, the return power with range for target, clutter, typical jammer and noise is given in Figure 5.21 (left) referred to the input of the T/R channel and at the output from the beamforming manifold including the antenna pattern (right) but before bandlimiting, pulse compression and Doppler filtering. The dynamic range presented to the T/R channel is very small even in the presence of jamming. The jammer is assumed to be in the antenna mainlobe. Thermal noise at the bottom and main bang leakage at the top may set input dynamic range. Note that the clutter sidebands from the transmitter FM noise are below thermal noise everywhere. The PRF chosen in the example is 3000 Hz (no others are better for many reasons), which has an ambiguity every 50 km. Thus as the clutter goes below thermal noise, a PRF ambiguity shows up and higher clutter repeats in the second ambiguity, etc. Fortunately, the 10 km wide main bang

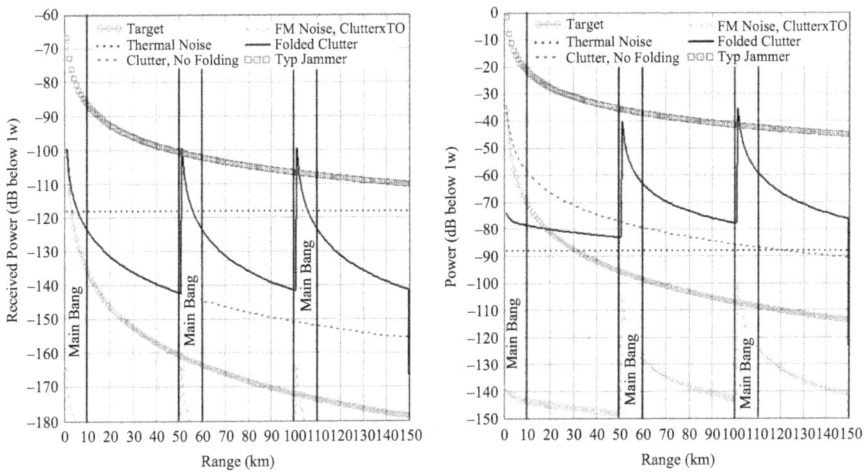


*Figure 5.21 Power at T/R channel input (left) and AESA manifold output (right) (GMTIPower006)*

blanking in this example masks most of the clutter peaks so that the dynamic range presented to the T/R channel LNA is less than 40 dB in the presence of jamming. (It would require a jammer at very close range to get to the probable 1 dB compression point at each T/R channel.) However, at the output of the beamforming manifold, the jammer power could be close to the 1 dB compression point. In the presence of high-power jamming, as much as 18 dB of attenuation might be applied in the first CAGC before the first LO mixer.

The receive channel noise and dynamic range can be estimated by using versions of the radar equation specific to signal, clutter, noise and jamming as given in (5.22), which are similar to (2.21). The first line of (5.22) is the radar equation for the assumed target. The second line is the estimate of noise power through the A/D assuming a 4:1 oversampling Hilbert conversion. The third line is the estimate of clutter power, which decreases as the third power of range. The fourth line is the jammer equation, which decreases as the square of range. Each of these must be solved for each range bin to estimate dynamic range.

$$\text{Target: } P_{tr} = \frac{P_T \cdot D_T \cdot G_T \cdot A_{eR} \cdot \sigma_t \cdot eff_n \cdot N \cdot G_{T/R}}{(4\cdot\pi)^2 \cdot R^4 \cdot L},$$

$$\text{Noise: } P_N = k_B \cdot T_0 \cdot B \cdot NF \cdot N \cdot ISLR \cdot G_{T/R} + \frac{10^{-2} \cdot 2^{-2\cdot(Bits-1)}}{eff_n \cdot N},$$

$$\text{Mainlobe Clutter: } P_{mlc} = \frac{P_T \cdot D_T^2 \cdot G_T \cdot A_{eR} \cdot d_{az} \cdot d_r \cdot \sigma_0 \cdot eff_n \cdot N \cdot G_{T/R}}{(4\cdot\pi)^2 \cdot R^4 \cdot L},$$

$$\text{Jammer: } P_J = \frac{P_J \cdot G_{TJ} \cdot A_{eR} \cdot B_{Jeff} \cdot G_{T/R}}{(4\cdot\pi)^2 \cdot R^2 \cdot L} \tag{5.22}$$

In addition, to properly represent the contributions of the antenna pattern, PRF ambiguities, clutter and jamming to the power levels at various locations in the receiver, the following equations (5.23) are also required. These equations are similar to (2.20). The actual calculation for Figure 5.21 is given in the Section 5 appendices GMTIPower006, etc.

$$amb = truncate\left(\frac{2 \cdot R_{max} \cdot PRF}{c}\right),\ Fold = \frac{R_{max}}{amb},\ \Delta R = \frac{D_T \cdot c}{2 \cdot PRF}$$
$$FP_{az} = R \cdot \theta_{az} = \frac{R \cdot 1.3 \cdot \lambda \cdot (1 + c_{min})}{L \cdot (\cos(\theta_0) + c_{min})},\ d_{az} = \frac{FP_{az} \cdot B_{filt}}{B_{amlc}},\ P_{smlc} = P_{mlc} * \mathrm{PUL}\left(\frac{R}{\Delta R}\right)$$
$$B_{amlc} = \frac{2 \cdot V \cdot \theta_{az} \cdot \sin(\eta)}{\lambda},\ B_{filt} = \frac{1}{T_{OT} \cdot eff_n},\ B_T = \frac{PRF \cdot PCR}{D_T}$$
$$W_{el} = \left(0.5 + 0.5 \cdot \cos\left(\frac{-\pi \cdot R}{R_{max}}\right)\right),\ B_{Jeff} = \frac{B_T}{B_J \cdot N \cdot ISLR \cdot eff_n} \tag{5.23}$$

The first line of (5.23) shows the calculation of the number of ambiguities out to the chosen maximum range, next the distance represented by each ambiguity and the third equation on line 1 estimates the length of the range covered by the transmitted pulse envelope where clutter will compete with the target prior to pulse compression. Before pulse compression, the return from each range bin is spread over many ultimate resolution cells. This causes averaging of the return over many range bins, which reduces the dynamic range presented to the AESA and the receiver up to the pulse compressor. The second line of (5.23) shows an estimate of azimuth footprint, the fraction of the azimuth footprint, which will fall in a single Doppler filter and the averaging of range bins prior to pulse compression for mainlobe clutter. Mainlobe clutter is "smoothed" across the length of the transmitted pulse of length $\Delta R$ by the convolution of a rectangular pulse (PUL($R/\Delta R$)) of width $\Delta R$ by the bin-by-bin return clutter power from each range cell. The third line of (5.23) calculates the clutter Doppler spread across the antenna beam for a given look angle ($\eta$), the final target detection bandwidth and the approximate bandwidth of the transmitted pulse compression waveform. The last line of (5.23) calculates the amplitude of the elevation footprint on the surface as a function of range and finally the fraction of the jammer power, which will occupy the ultimate target detection bandwidth. Both in the noise power equation of (5.22) and the jammer power of (5.23) use the ISLR of the pulse compression waveform rather than the noise bandwidth of the equivalent filter because that is more representative of the competing random noise in the presence of jamming and clutter.

The FAGC of Figure 5.20 is often used on a smoothed bin-by-bin basis to provide STC. In the example here, no FAGC-based STC is used ahead of the A/D as shown in Figure 5.22 (left). Once pulse compression is performed, one can see a significant increase in dynamic range from about 40 dB to about 55 dB depending on the location of the jammer as shown in Figure 5.22 (right). Note also that a return from a target has a positive SNR for the first instance in the receiver chain.

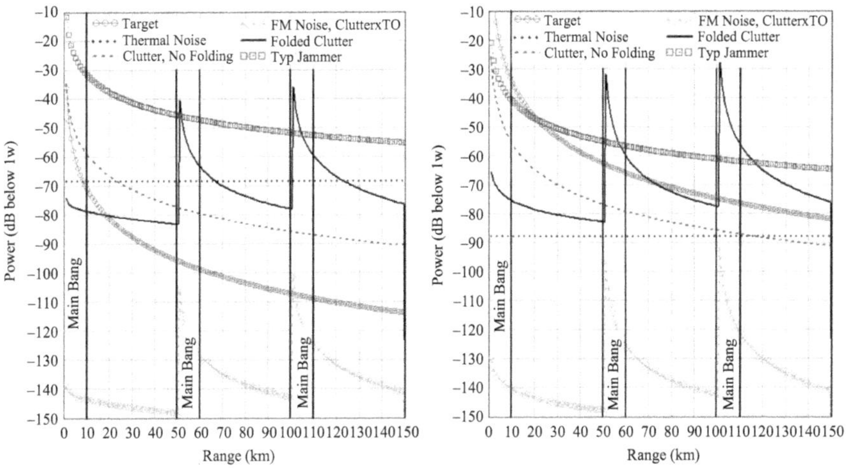


*Figure 5.22 Receiver channel noise example at A/D output and after pulse compression (GMTIPower006)*

The challenge in the design of a receiver chain is "birdie" interference and self-noise cancellation by chasing each spur to its source and dramatically reducing it. This includes power supply ripple reduction as well as suppression of oscillation in high gain circuits. The suppression usually requires multiple layer shielding, filtering at all inputs and outputs, careful grounding to prevent ground loops, conductor routing/rerouting to minimize cross talk in cables and connectors. Both signals and reference frequencies as well as reference voltages require isolation. Such isolation often requires boxes within boxes within boxes to achieve adequate noise reduction. For example, conventional coax cable only provides about 20 dB isolation. Double-shielded cable or solid coax may reduce leakage to 60 dB. Recall that Figure 5.16 shows birdies that are only 70 dB down relative to the carrier but at the front end of the AESA the target is 140–180 dB down. Thus 70–110 dB of reduction/isolation will be required to prevent false targets. STC often is based on averaging some number (16) of pulses prior to the onset of the CPI.

## 5.8 I/Q imbalance

Mismatch between the in-phase and quadrature components of a sensor vector measurement representing amplitude and phase gives rise to an image error term part of which will be correlated and appear as ghost targets and the remainder will appear as background noise. The mismatch appears due to DC bias, amplitude errors and phase errors. Figure 5.23 shows the approximate relative image noise due to imbalance in the I and Q channels.

Phase and amplitude imbalance noise power can be estimated as shown in (5.24).

$$P_{N\,\mathrm{IQ}} \approx \frac{\sigma_a^2}{4} + \frac{\sigma_\phi^2}{4} \tag{5.24}$$

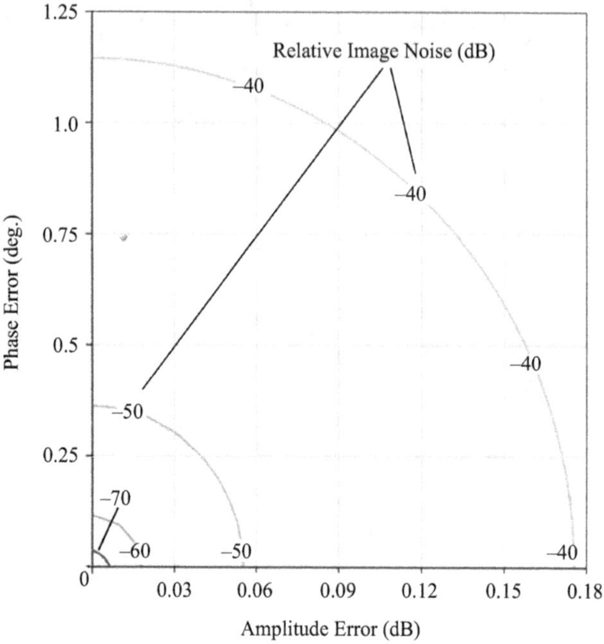


*Figure 5.23 Image noise versus phase and amplitude imbalance (IQ Imbalance001)*

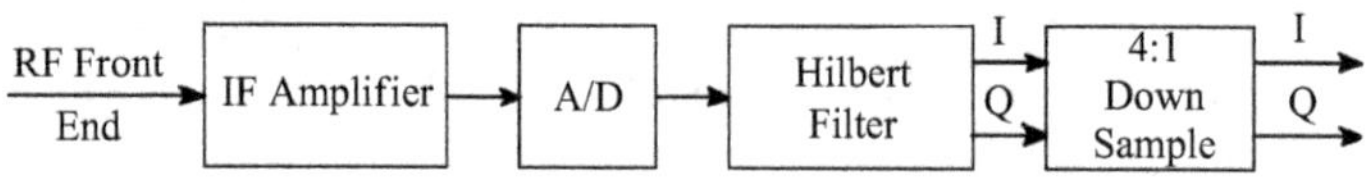


*Figure 5.24 Hilbert transform concept (repeated Figure 4.22)*

About the best that can be done with analog I/Q demodulation is −40 dB imbalance. If detections must be done in a high clutter environment, such an imbalance will greatly limit performance. The imbalance can be corrected by measuring the errors every few seconds using a pilot signal and then compensating each received signal. Alternatively, digital I/Q demodulation can be achieved with Hilbert filter techniques shown in Figure 5.24.

For the extended example of Table 5.3 using the 21-point Hilbert filter of Figure 4.25 with 16-bit coefficient length and (5.8), the I/Q error will be approximately −84.8 dB down from full scale. This error will be less than the recommended 0.1 of all other noise.

## 5.9 A/D noise

The first element in a digital processor is the A/D converter. Several types of A/Ds and their limitations were described in Chapter 4. Some typical performance of

*Table 5.4 Typical A/D converters*

| Resolution (bits) | Conversion rate | Architecture |
|---|---|---|
| 3 | 20 GHz | Dual Flash |
| 8 | 12.5 GHz | Quad Flash |
| 12 | 800 MHz | Hybrid D-S |
| 16 | 10 MHz | Delta Sigma |
| 20 | 100 kHz | Delta Sigma |

A/Ds is given in Table 5.4. Sampling and quantizing are used to convert analog signals to numerical form. An A/D converter with, say 11 bits capacity, produces numerical output data that may have integer values ranging from $-1024$ to $+1023$. Voltage inputs corresponding to noninteger output values are assigned the nearest integer value in the quantization process, with a corresponding error being introduced. Voltage values corresponding to numerical outputs outside the range $-1024$ to $+1023$ are assigned the values $-1024$ or $+1023$, depending upon their polarity, with a corresponding saturation error being introduced. The greater the gain ahead of the A/D converter, the greater the likelihood there will be saturation. Saturation error, therefore, may be minimized by using low gains.

Quantization errors, on the other hand, are determined by the quantum level or step size, and are relatively independent of signal level. To maximize signal-to-quantization error ratio, it is desirable that gain be high. These conflicting requirements are best met by a compromise that maximizes the signal-to-total distortion power ratio. This compromise, which is outlined later, dictates that the gain be set so that the RMS signal level is approximately 13–14 dB ($\sim$1/5) below the saturation level of the A/D converter.

In a common signal processor configuration, the functional block following the A/D converter is the pulse compression and/or beamforming circuitry. In other arrangements, pulse compression and/or beamforming are performed after filter processing to simplify Doppler compensation. When beamforming and/or pulse compression are performed directly following A/D conversion, the arithmetic procedure consists of adding a number, $N$, of range and/or sensor samples (sometimes called chips) together, perhaps with complex weights, to form a single compressed pulse or angle output per bin.

Modern sensors operate at relatively high bandwidths and demand both fast A/Ds as well as ones with high dynamic range. Various strategies have evolved to meet these needs. There are A/Ds operating at 20 GHz sampling rates and 3-bit resolution to 100 kHz sampling at 20-bit resolution. Each has a different architecture. Many have multistage pipeline conversion architectures.

There are several ways of characterizing A/D performance metrics. The most common metrics are effective number of bits (*ENOB*), *SFDR* and signal-to-noise plus distortion ratio (*SNDR*). The *SFDR* is a measure of the harmonic distortion relative to a sinewave signal input. The *SNDR* takes into account all of the distortions relative to full scale – quantization noise, spurious distortion, saturation noise and any other nonlinearities. The *ENOB* is related to the *SNDR* by (5.25). Basically the *ENOB* compares quantization noise to all noise to provide an estimate of the A/D performance against the ideal. Most A/Ds are 1.5 bits or worse than ideal.

$$ENOB = \frac{SNDR - 1.76}{6.02} \tag{5.25}$$

### 5.9.1 Saturation, quantization and optimum signal level [1]

When analog data are passed through an A/D converter and sent to a digital filter, the proper measure of the effect of the saturation and quantization noise introduced is the proportion of the noise power spectrum within the filter passband. As long as there are signals significantly larger than a quantization step size, quantization noise may be considered, to a very good approximation, white or uniformly distributed in the spectral region between sampling spectral lines. Thus, the proportion of quantization noise power that falls within a filter bandwidth is given by the ratio of that bandwidth to the sampling frequency.

On the other hand, saturation noise tends to be concentrated in the spectral region of the input signal. The spectral relationships of signals (assumed to have a Gaussian-shaped lowpass spectrum) and saturation noise for four different ratios of sample rate-to-signal bandwidth ($f_s/B_s$) and for different degrees of saturation are shown in Figure 5.25. The parameter, $K_{AD}$, is the ratio of saturation level to RMS signal. Note that the case $K_{AD} = 0$ corresponds to hard limiting or 1-bit quantization; in this case quantization and saturation are one, and the curves represent the total perturbation noise spectrum. If the input spectrum has a lobe structure (bump shape) spectrum and the sampling ambiguity spacing is comparable to the lobe bandwidth (Figure 5.25(a)), the saturation noise spectrum is approximately white. In a nonminimum sampling saturation (Figure 5.25(b)–(d)), the saturation noise density is not constant and the variations must be considered. It is reasonable to treat saturation noise, in this case, as having a bandwidth approximately equal to the input bandwidth.

The idealized transfer function of an A/D converter is shown in Figure 5.26, where the saturation level has been set at $K_{AD}$ times the input RMS voltage, $\sigma_c$. The resulting error signal $e(x)$ is shown to the right in the figure as well as the assumed input probability density function $P(x)$. The output noise power resulting from some input signal probability density function can be calculated by integrating the errors for the complete range of possible inputs. Since the error function is very nonlinear, the integration needs to be performed numerically or by the use of simple approximations.

$$P_N = \int_{-\infty}^{\infty} e^2(x) \cdot P(x)\, dx = Q_N + S_N$$

$$P_N \cong 2 \cdot \frac{Q^2}{12} \cdot \int_0^{K_{AD}\cdot\sigma_c} P(x)\, dx + 2 \cdot \int_{K_{AD}\cdot\sigma_c}^{\infty} (x - K_{AD}\cdot\sigma_c)^2 \cdot P(x)\, dx$$

$$P(x) = \frac{1}{\sigma\cdot\sqrt{2\cdot\pi}} \exp\left(\frac{-x^2}{2\cdot\sigma^2}\right) \quad \text{and} \quad Q = \frac{K_{AD}\cdot\sigma_c}{2^{M-1} - 1} \tag{5.26}$$

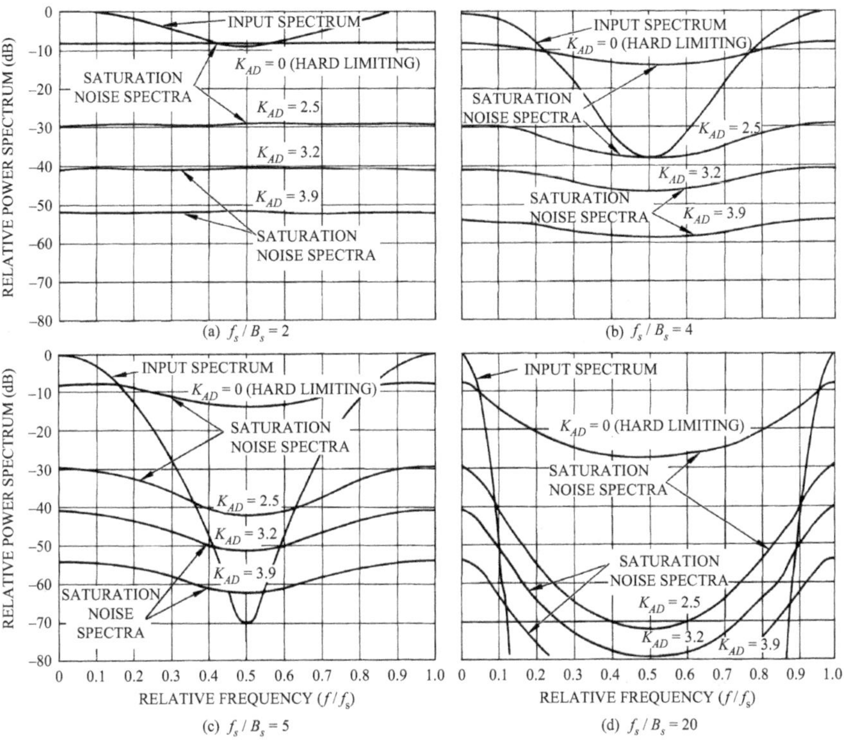


*Figure 5.25 Typical saturation noise spectra [1]*

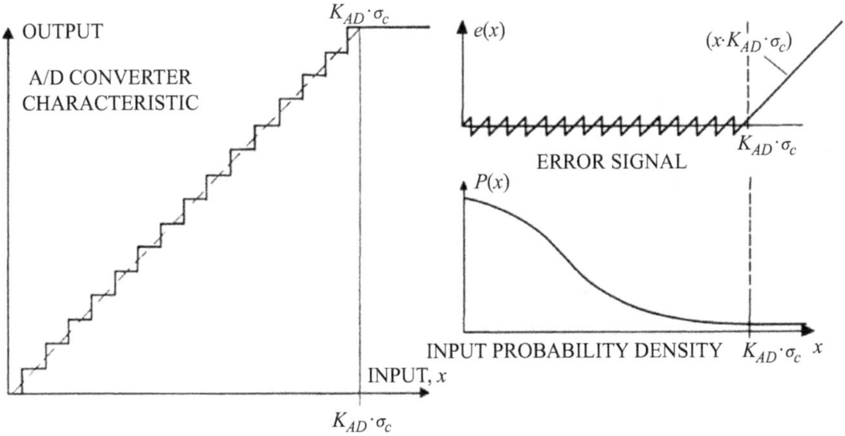


*Figure 5.26 A/D converter transfer characteristics [1]*

The total noise power $P_N$ due to the error term, $e(x)$, can be computed from (5.26). This noise power can be broken into two parts: one due to quantization ($Q_N$) and one due to saturation ($S_N$). These terms also are given in the second line of (5.26). Consider the first integral on the second line of (5.26). If $x_n$ represents the value of $x$ at each zero crossing of $e(x)$ in Figure 5.26 and assuming that over any one interval $-Q/2 < x < Q/2$, that $P(x) \cong P(x_n) =$ a constant, then it can be shown that the first part of the second line of (5.26) is approximately correct. The second integral of the second line of (5.26), the saturation noise term, steadily increases with increasing input signal and can be computed as shown. If for example the expression for a Gaussian (normal) probability density function with zero mean as given in the third line of (5.26) is used, then the total noise as a function of the number of A/D bits and the setting level of the RMS input relative to full scale can be calculated. $M$ is the number of bits (including sign) in the A/D converter, $K_{AD}$ is the ratio between full scale and the RMS signal level at the A/D input and $\sigma_c$ is the RMS signal level at the A/D input.

The optimum values of $K_{AD}$, as well as the corresponding values of $Q_N/\sigma_c^2$, $S_N/\sigma_c^2$ and $P_N/\sigma_c^2$ for a Gaussian probability density function for various values of $M$, are given in Table 5.5. The curves from which these tabulated values were derived are shown in Figure 5.27. The curves show the sensitivity of the total noise power to a change in $K_{AD}$. It is also seen that at optimum $K_{AD}$, $Q_N$ is always several times larger than $S_N$.

It is interesting to compare the total noise power $P_N$ with the value $Q^2/12$, which is often assumed as an approximation. If one writes $P_N = \zeta\, Q^2/12$, then the parameter, $\zeta$, is found to vary between 1.38 and 1.09 for $M = 3$ to 13. Also, it is found that over this range of $M$ values, the normalized total noise can be approximated within about 3 dB by (5.27).

Equation (5.27) is the saturation and quantization noise to signal ratio ($1/SNR_{qs}$).

$$P_N \Big/ \sigma_c^2 \approx 7 - 6 \cdot M \quad \text{(dB)} \tag{5.27}$$

*Table 5.5 Noise at $K_{ADopt}$ versus A/D Bits [1,3]*

| $M$ | $K_{ADOPT}$ | $Q_N/\sigma_C^2$ | $S_N/\sigma_C^2$ | $P_N/\sigma_C^2$ | $P_N$ (dB) |
|---|---|---|---|---|---|
| 3 | 1.9 | $3.15 \times 10^{-2}$ | $1.5 \times 10^{-2}$ | $4.65 \times 10^{-2}$ | −13.3 |
| 4 | 2.5 | $1.05 \times 10^{-2}$ | $0.24 \times 10^{-2}$ | $1.29 \times 10^{-2}$ | −18.9 |
| 5 | 2.9 | $3.1 \times 10^{-3}$ | $0.6 \times 10^{-3}$ | $3.7 \times 10^{-3}$ | −24.3 |
| 6 | 3.26 | $0.92 \times 10^{-3}$ | $0.13 \times 10^{-3}$ | $1.05 \times 10^{-3}$ | −29.8 |
| 7 | 3.6 | $2.7 \times 10^{-4}$ | $0.4 \times 10^{-4}$ | $3.1 \times 10^{-4}$ | −35.1 |
| 8 | 3.9 | $7.9 \times 10^{-5}$ | $1.0 \times 10^{-5}$ | $8.9 \times 10^{-5}$ | −40.5 |
| 9 | 4.24 | $2.3 \times 10^{-5}$ | $0.2 \times 10^{-5}$ | $2.5 \times 10^{-5}$ | −46 |
| 10 | 4.5 | $6.5 \times 10^{-6}$ | $0.5 \times 10^{-6}$ | $7.0 \times 10^{-6}$ | −51.6 |
| 11 | 4.76 | $1.8 \times 10^{-6}$ | $1.42 \times 10^{-7}$ | $1.95 \times 10^{-6}$ | −57.1 |
| 12 | 5.02 | $5.01 \times 10^{-7}$ | $3.46 \times 10^{-8}$ | $5.36 \times 10^{-7}$ | −62.7 |
| 13 | 5.24 | $1.36 \times 10^{-7}$ | $1 \times 10^{-8}$ | $1.46 \times 10^{-7}$ | −68.4 |

Figure 5.27 shows total A/D noise for three different probability density functions common in radar: uniform, arc sine and Gaussian as a function of RMS signal to saturation level for various numbers of A/D bits. From these curves, the values of $K_{AD}$ corresponding to minimum total noise power can be determined for different values of $M$. For all values of $M$ shown, a $\pm 10\%$ change in $K_{AD}$ will result in a very small change in total noise power. Hence, the A/D converter input gain need not be closely controlled.

## 5.9.2 *Aperture jitter*

In one way or another, every A/D has a sample and hold (S/H) function to allow the converter the longest period of time available consistent with sampling rate to make the conversion to digits. The behavior of a S/H is shown in Figure 5.28. During the finite S/H switching time $t_a$ and depending on the input signal slope, there will be an error in the held signal relative to the "true" value. This is called aperture jitter because the apparent held value will correspond to a slightly different time instant on the input waveform depending on slope.

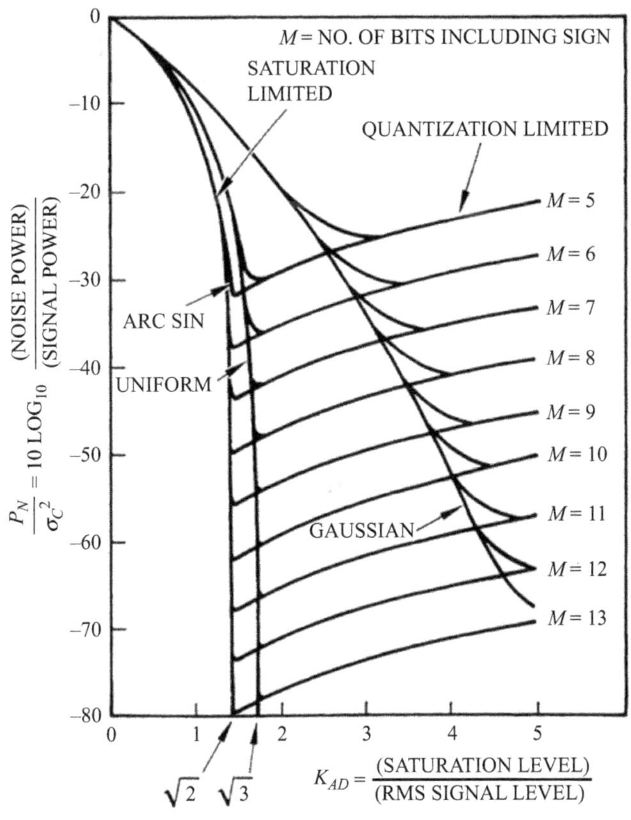


*Figure 5.27 A/D conversion noise versus operating level; after Harris [3]*

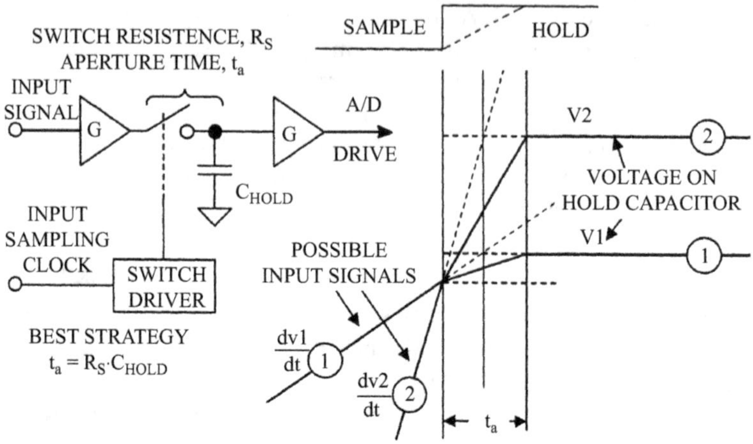


*Figure 5.28 Sample and hold aperture jitter; adapted from [9]*

It has been found that if the time constant of the signal drive amplifier, switch resistance and hold capacitor is equal to the switching time $t_a$, the error is least. This allows S/H aperture jitter to be fractions of a picosecond. The SNR for aperture jitter error is given in (5.28).

$$SNR_{aj} = 20 \cdot \log\left(\frac{1}{\pi \cdot B_s \cdot t_a}\right)$$

where: $B_s/2$ is the maximum frequency at the S/H input and $t_a$ is rms aperture jitter time. (5.28)

## 5.9.3 Other sources of noise in A/Ds

There are several other sources of noise in A/Ds including nonlinearities, which cause harmonic distortion that are no different than in analog circuits. There are some fundamental limits such as thermal noise in the resistance in the drive amplifier and S/H switch as well as gain variations caused by amplifier cutoff frequency. The SNR for the thermal noise associated with the driver and switch is given in (5.29).

$$SNR_{temp} \approx -10 \cdot \log(3 \cdot k_B \cdot T_o \cdot f_{s\max} \cdot r)$$

where: $k_B$ is Boltzmann's constant, $T_o$ is standard temperature, $r$ is amplifier output impedance matched to S/H switching impedance and $f_{s\max}$ is allowable maximum sample rate. (5.29)

The sample and hold drive amplifier also has a limited gain and bandwidth. The ultimate possible bandwidth is limited by the $f_T$ of the underlying circuit technology. The maximum current $f_T$s are about 300 GHz. The maximum current

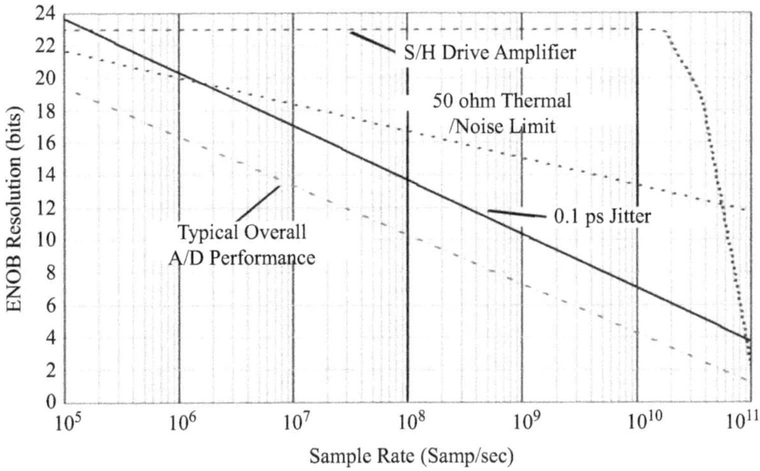


*Figure 5.29 Typical A/D noise contributors (ADNoise002)*

stable gain may be as high as $10^7$ V per volt. A 20-bit A/D needs that much open loop gain in order that the drive amplifier does not limit overall accuracy. Long before the 3 dB rolloff of the closed loop amplifier the passband begins to droop with frequency. In order to stay within the accuracy of the following A/D, the passband may be orders of magnitude wider than the A/D sampling rate. Many amplifiers also exhibit very long overload recovery times. In addition, all of these problems also apply to the circuits in the A/D itself. Figure 5.29 shows the A/D noise contributors and typical performance.

The SNR for the drive amplifier and amplifiers inside the A/D is given in (5.30). The maximum normalized sampling frequency relative to the amplifier rolloff is also given in (5.30).

$$SNR_{amp} = 20 \cdot \log\left(\frac{\text{Open Loop Gain}}{\text{Frequency Rolloff}}\right)$$

$$f_{s\max} = \sqrt{\frac{1}{\left(1 - 1/2^{M+1}\right)^{Npoles}} - 1}$$

$$SNR_{amp} \approx 20 \cdot \log\left(G_{open} \Big/ \sqrt{\left(1 + \left(\frac{f}{f_T \cdot f_{s\max}}\right)^{Npoles}\right)}\right) \tag{5.30}$$

## 5.9.4 *Advantages of A/D oversampling*

The largest single contributor to A/D noise is usually quantizing noise. There is a significant reduction in quantizing noise by oversampling a band limited signal. The oversampling spreads out the ambiguities and the quantizing noise that allows

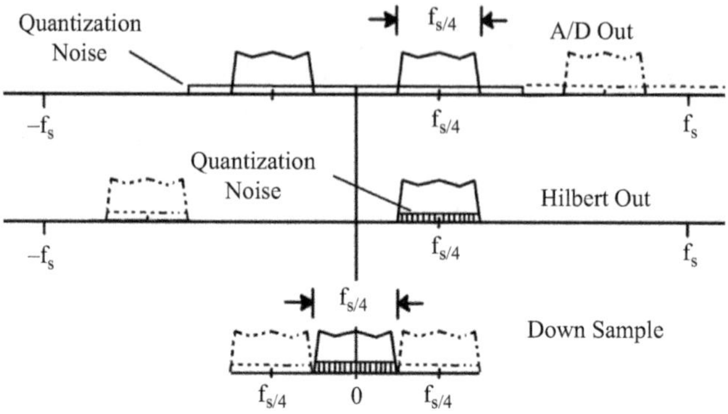


*Figure 5.30 Oversampling A/D and Hilbert filter advantage*

filters to better reject out-of-band signals as well as reducing the in-band fraction of quantizing noise. The ideal signal-to-quantizing-noise ratio (*SQNR*) is often approximated by (5.31).

$$SQNR \approx 6.02 \cdot M + 1.76 \text{ dB}$$

If there is over sampling of the input bandwidth

$$SQNR_{os} \approx 6.02 \cdot M + 1.76 + 10 \cdot \log\left(\frac{f_s}{B_s}\right) \tag{5.31}$$

The available improvement can be seen in Figure 5.30 using a Hilbert transform filter. If the sampling rate is four times the signal bandwidth, then there is an SQNR improvement of 4:1 or 6 dB, which is the equivalent of 1 additional bit. Since an Hilbert filter sums four samples of the coherent signal, the dynamic range could increase by as much as 12 dB but at minimum by 6 dB. Often Σ–Δ A/D converters shape the input signal and noise such that there is more than $N^{1/2}$ improvement with oversampling. The equivalent improvement of 2 or more bits is not uncommon.

Returning to the extended example of Table 5.3, the final SNR for the receive chain from T/R channel input to Doppler filter output is shown in Figure 5.31. The target signal at 150 km is 35 dB above random noise, 8 dB above a mainlobe jammer and a few dB above clutter. This scenario would still require some angular clutter cancellation (sometimes called STAP) to allow low false alarm GMTI detections.

## 5.10 Noise in digital signal processing systems [1]

Saturation, roundoff and filter impulse response influence digital filtering performance. In the design of digital signal processing equipment, it is common to base first-order performance predictions on linearized models of the actual digital

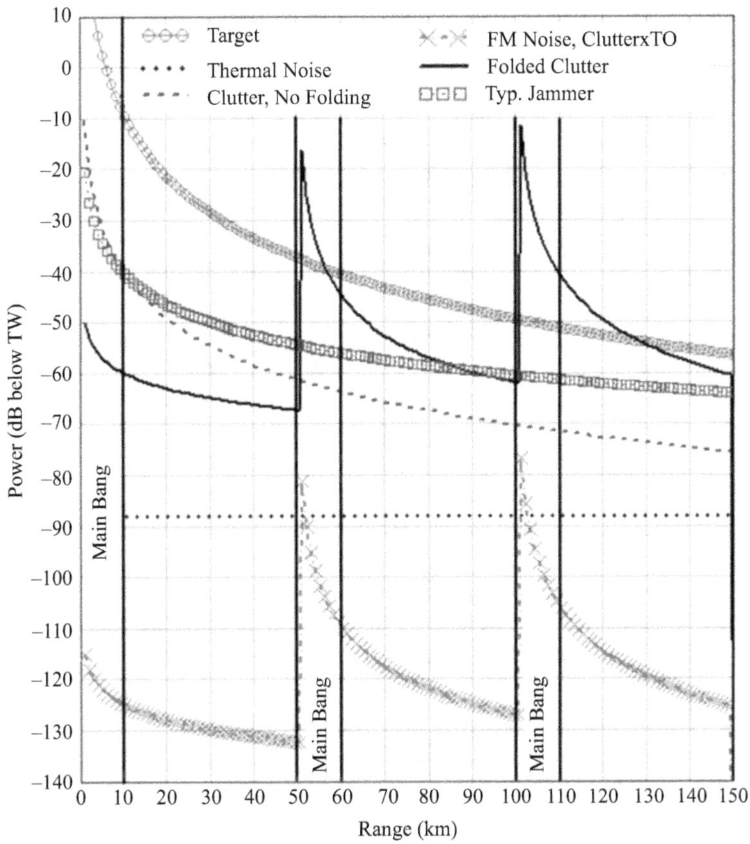


*Figure 5.31. Receiver channel noise example after Doppler filtering (GMTIPower006)*

circuitry. These models provide fairly accurate information on such characteristics as filter bandwidth, passband ripple, sidelobe response, etc. When more accurate information is desired, however, the processes of signal quantization and saturation must receive more attention, because at each point where these effects occur, they are equivalent to the addition of noise into a linearized model [1,3,4].

A very strong point source or reflector may distribute its energy through multiple range, frequency and/or angle samples. The largest possible signal that would not saturate the example (11 bit) A/D converter could have amplitude $-1024$ or $+1023$ in each of $N$ adjacent sensor samples. Summing over $N$ sensor samples with uniform weighting would produce an output with amplitude $-1024 \times N$ or $+1023 \times N$. Having increased the possible amplitude range of the data by filtering, beamforming or pulse compression, provision must be made for $\log_2(N)$ additional bits of data handling capacity to accommodate it, or introduce errors from saturation (discarding the most significant bits (MSB)) or truncation (discarding the least significant bits (LSB)). Saturation and truncation errors introduced in the digital processor are equivalent in effect to saturation and quantization errors introduced at the A/D converter.

Before digital data are entered into the main memory, they are usually prefiltered into subbands with expected common signal statistics. The prefilter combines many individual samples into a single sequence of data for each subband associated with range, frequency or angle. As with pulse compression, filtering or beamforming, weighting and summing several digital words to form one filter output necessitates providing additional bits of data handling capability to avoid errors of saturation and truncation. Since the word length is an important cost-determining factor in subsequent processing, it is generally cost effective to accept some saturation and truncation errors to reduce word size.

Each time saturation and/or truncation occurs, noise is introduced. The optimal processor from a minimum noise standpoint would provide whatever additional bits of data handling capacity signal buildup necessitated. Since large fast buffer memories and large-word fast arithmetic capacities are expensive, it is desirable to reduce the sample rate to reduce the arithmetic rate. Furthermore, it is obvious that the sidelobe performance of the linearized error-free processor model establishes a design limit beyond which nothing is gained by reducing the noise contributions of saturation, quantization and truncation, and at which point the processing configuration can be optimized with respect to other factors.

### *5.10.1 Roundoff, saturation and truncation*

When the digital word size grows as a result of pulse compression (and/or beamforming or other signal processing), extra bits must be provided in the circuitry to handle the larger words or the result must have the word size reduced by discarding MSB, LSB or some combination of the two and/or the use of floating point. Discarding LSBs is equivalent to dividing the signal by $2^N$ (where $N$ is the number of LSBs discarded) and then dropping the factional part (where the data are regarded as having only integer values before being divided), as shown in Figure 5.32.

The process of truncating or rounding the rescaled data to integer values is almost equivalent to quantizing in the A/D converter. The slight difference lies in the fact that at the A/D converter, the data before quantization may take on any value between quantum steps, whereas in the truncation process, the fractional part that is dropped has only $2^N$ discrete values. The noise contributed by truncating or rounding various numbers of LSBs is tabulated in Table 5.6. The noise is of two types: an AC component and a DC bias term. Often the DC bias term is the most problematical because subsequent filter integration just makes it bigger, thus diminishing dynamic range.

### *5.10.2 Multiplier roundoff noise*

Errors generated in multiplier roundoff for equally probable multiplicands and multipliers are not exactly equal to the normal roundoff noise just discussed. The

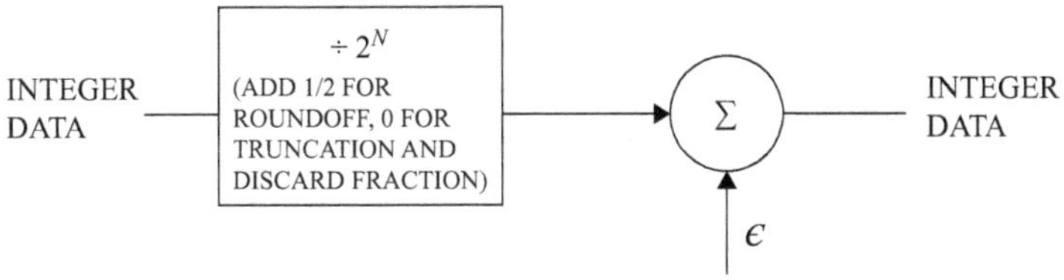


MEAN ERROR $\quad \epsilon = -\frac{1}{2} + \frac{1}{2^{N+1}}$ FOR TRUNCATION

$\bar{\epsilon} = \frac{1}{2^{N+1}}$ FOR ROUNDOFF

ERROR VARIANCE $\quad \sigma_\epsilon^2 = \frac{1}{12}\left[1 - \frac{1}{2^{2N}}\right]$ FOR BOTH

MEAN SQUARE ERROR $\quad \overline{\epsilon^2} = \frac{1}{12}\,\frac{(2^N - 1)(2^{N+1} - 1)}{2^{2N-1}}$ FOR TRUNCATION

$\overline{\epsilon^2} = \frac{1}{12}\left[1 + \frac{1}{2^{2N}}\right]$ FOR ROUNDOFF

*Figure 5.32 Truncation and roundoff noise model [1]*

*Table 5.6 Truncation and roundoff noise [1]*

| **No. of bits discarded** | **Truncation** | | | **Roundoff** | | |
|---|---|---|---|---|---|---|
| | **Mean error** | **Mean square error** | **Error variance** | **Mean error** | **Mean square error** | **Error variance** |
| 1 | −0.25 | 0.125 | 0.063 | 0.25 | 0.125 | 0.063 |
| 2 | −0.375 | 0.219 | 0.078 | 0.125 | 0.094 | 0.078 |
| 3 | −0.438 | 0.273 | 0.082 | 0.063 | 0.086 | 0.082 |
| 4 | −0.469 | 0.302 | 0.083 | 0.031 | 0.084 | 0.083 |
| 5 | −0.484 | 0.318 | 0.0832 | 0.016 | 0.0835 | 0.0832 |
| 6 | −0.492 | 0.326 | 0.0833 | 0.0078 | 0.0834 | 0.0833 |

usual assumption made in determining errors due to truncation and roundoff is that all bits of the number under consideration (e.g., the product of a multiplication) have equal probabilities of being 1 and 0 and are uncorrelated with each other (i.e., that all possible values are equally likely). Therefore, it follows that all possible values of the $N$ LSB are also equally probable and that the errors resulting from truncating or rounding these $N$ bits are uniformly distributed. The mean error, mean-square error and error variance resulting from these assumptions for several values of $N$ are listed in Table 5.7.

*Table 5.7 Multiplication noise summary [1]*

| No. of bits discarded | Truncation | | | Roundoff | | |
|---|---|---|---|---|---|---|
| | Mean error | Mean square error | Error variance | Mean error | Mean square error | Error variance |
| 1 | −0.125 | 0.0625 | 0.04687 | 0.125 | 0.0625 | 0.04687 |
| 2 | −0.25 | 0.14062 | 0.07812 | 0.125 | 0.07812 | 0.0625 |
| 3 | −0.34375 | 0.20703 | 0.08887 | 0.09375 | 0.08203 | 0.07324 |
| 4 | −0.40625 | 0.25488 | 0.08984 | 0.0625 | 0.08301 | 0.07910 |
| 5 | −0.44531 | 0.28638 | 0.08807 | 0.03906 | 0.08325 | 0.08173 |
| 6 | −0.46875 | 0.30597 | 0.08624 | 0.02344 | 0.08331 | 0.08276 |

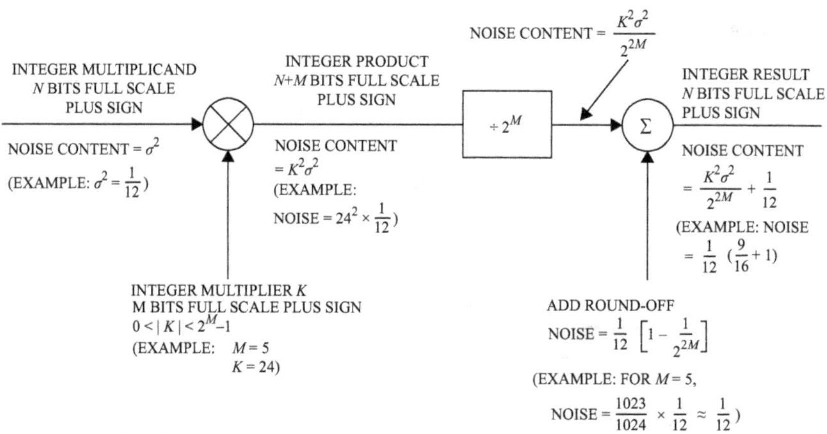


*Figure 5.33 Multiplication noise model [1]*

The binary point is considered to be to the left of the $N$ LSB, and the error is defined as

$$Error = Truncated\ or\ Rounded\ Product\ -\ True\ Product \tag{5.32}$$

The model described earlier is summarized in Figure 5.33, including an example roundoff of $M=5$ bits and a coefficient of $K=24$.

However, if the multiplicand and multiplier (rather than the product) are assumed to be uniformly distributed and uncorrelated, somewhat different results are obtained. In this case, the values of the $N$ LSB (and the errors) are not equally probable. The results of a computer program written to investigate this case are listed in Table 5.7. The significant point to be observed in Table 5.7 is that the mean error does not approach zero or 0.5 as rapidly with increasing numbers of discarded bits. Thus, slightly more DC bias may occur in the filter or FFT output than classically predicted, depending on the signal statistics.

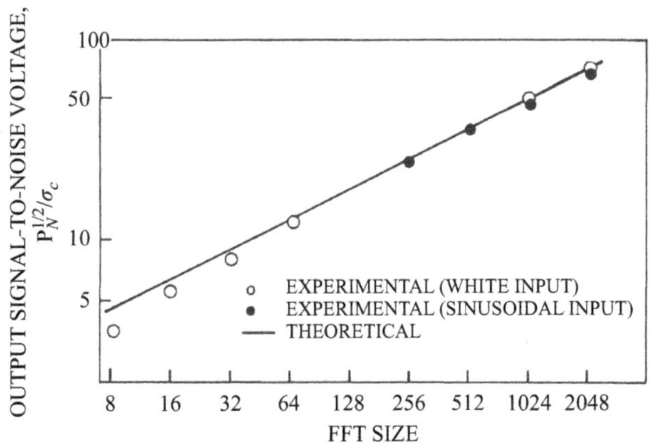


*Figure 5.34 FFT noise – scaled by ½ after each pass [1]*

## 5.11 Saturation and truncation during FFT processing

The truncation and saturation noise introduced during FFT processing may be viewed in identical fashion to those introduced in the proceeding sections. FFT noise is summarized in this section. It has been shown both theoretically and experimentally with white noise input or sinusoidal input that fixed point rounded coefficient errors can be calculated as shown in (5.33) [1,3,10].

$$P_N \Big/ \sigma_c^2 = \frac{2}{3} \cdot 2^{-2 \cdot M} \cdot \log_2(N) \tag{5.33}$$

where $M$ is the number of coefficient bits including sign, $N$ is the record length or FFT size, $\sigma_c^2$ is the variance of the input signal and $P_N$ is the noise power generated in the FFT operation. Similarly, for the same signal input types, assuming scaling after every transform pass by 1/2 for a fixed point rounded system, the noise can be calculated as given in (5.34).

$$P_N = \frac{5}{3} \cdot N \cdot 2^{-2 \cdot M} \tag{5.34}$$

where the same variable definitions as in (5.33) apply to (5.34). Obviously both of these equations are approximate but their correspondence with actual FFT noise outputs in real signal processors is very good. The agreement between theory and the experimental results is shown in Figure 5.34.

Once narrowband processing is completed, it may be desirable to reduce the digital word size if further processing, such as beamforming, pulse compression, post detection integration (PDI), tracking, is to be done. The same approaches described for pulse compression, prefiltering, narrowband filtering and even tracking apply to these cases.

## 5.12 Self-noise performance of binary and polyphase codes

All pulse compression waveforms are subject to real-world degradations from hardware and software limitations. There are at least three significant internal sources of sidelobe cancellation degradation for complementary codes. First, transmit phase modulation may not be orthogonal so that one phase state is less than 90° from another. A phase modulation error of 10° total will not be completely cancelled on receive, thus raising sidelobes as summarized in Figure 5.35.

Similarly, transmitter power supply droop during high-power operation can cause phase pulling of the transmitter during a pulse. This pulling causes phase warp across the code, which is similar to Doppler, which, if not compensated, increases sidelobes and reduces mainlobe after compression. The transmitter has thousands of electrical degrees of delay and so a small droop can easily cause 60° phase warp across the uncompressed pulse (Table 5.8). Lastly, receiver AGC and STC can cause AM across the received uncompressed pulses, which also gives rise to increases in uncancelled sidelobes. Such modulation, if not compensated, can easily be 5% across the pulse. The effect of all three of these sources for various code lengths is summarized in Figure 5.35.

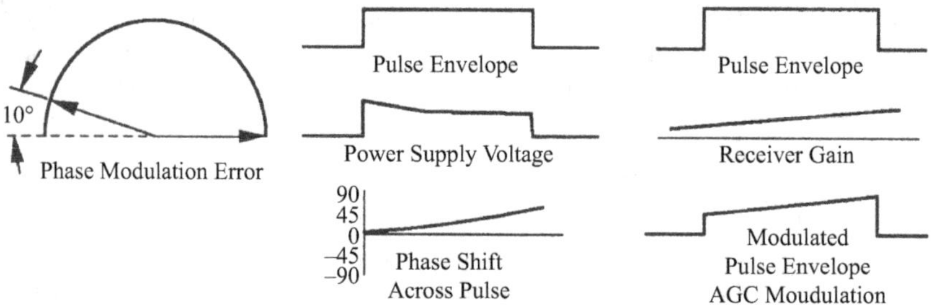


*Figure 5.35 Effects that reduce the sidelobe cancellation property of complementary codes [1]*

*Table 5.8 Complementary code sidelobe degradation [1]*

| Code length | 10° phase modulation error, dB | 60° phase shift across pulse, dB | 5% AGC amplitude modulation, dB |
|---|---|---|---|
| 16 | −26.2 | −19.5 | −44 |
| 20 | −28 | −12.5 | −43.7 |
| 32 | −23.9 | −13 | −37 |
| 40 | −22.3 | −14.8 | −43.6 |

### *5.12.1 Other digital processing anomalous behavior*

Sadly as reliable as modern digital processing is, there are still many "crashes" that users of such processing experience. It is no different for sensor processing. SED and single-event detection (SEU/SED) protection and recovery must be incorporated into all real-time processing. SEU could be attributable to cosmic rays from outer space, power transients, bad software and user errors. Real-time systems will not tolerate the kind of nonsense most users of Microsoft Windows and Windows application software are subjected to. Real-time sensor processing systems require watchdog timers, interrupt-driven software and carefully structured programming to allow rapid recovery from crashes regardless of cause. When hardware and software loses its way, resets and reboots are required in real time to keep the process from collapsing. These processing restarts will naturally introduce noise, which can range from a momentary blind spots, false target detection or the complete loss of a whole scan. If the process is critical, then redundancy is required. Good design requires SEU/SED provisions from the outset.

## 5.13 Noise-limited tracking [11]

This section provides simple calculations for the lower bound on range, Doppler and angle tracking. The SNR sets a lower bound on tracking accuracy. Accurate tracking is dependent on the SNR but also on hardware and software inaccuracies.

### *5.13.1 Range tracking*

Figure 5.36 shows the range tracking noise situation for a split gate tracker. The matched filter for a rectangular pulse is a rectangular pulse convolved on the input. The output of the matched filter will be a triangular pulse as shown in Figure 5.36. The tracker will take samples of the range cell width apart. The samples could be anywhere relative to the output pulse in order to account for this possibility an offset variable $\alpha$ is introduced. If $\alpha$ is 1, then sample 2 will be at the peak of the output; and if $\alpha$ is 0, then sample 1 will be at the peak of the matched filter output. Each sample will have some noise voltage associated.

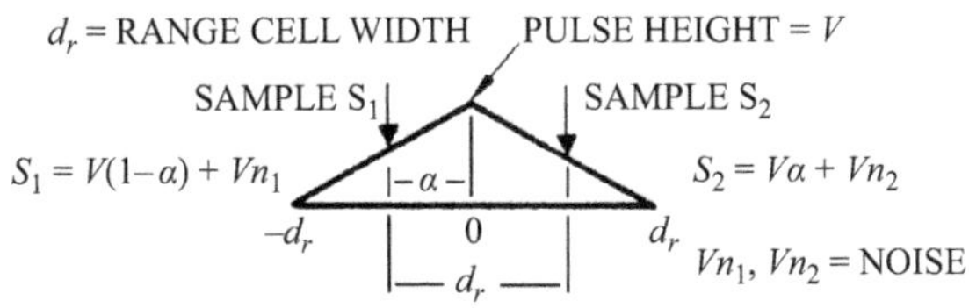


*Figure 5.36 Range track noise situation*

$$\begin{aligned}
&\text{Range gate measurements: } S_1 = V\cdot(1-\alpha)+Vn_1\ ,\ S_2 = V\cdot\alpha+Vn_2\\
&\text{Find difference: } S_1 - S_2 = V\cdot(1-2\cdot\alpha)+Vn_1 - Vn_2\\
&\text{Normalize to range units: } \delta R = \frac{d_r}{2\cdot V}\left[V\cdot(1-2\cdot\alpha)+Vn_1 - Vn_2\right]\\
&\text{Range estimate: } \delta R = \frac{d_r}{2}\cdot(1-2\cdot\alpha)+\text{ Noise error}\\
&\text{Noise variance: } \sigma_R^2 = \left(\frac{d_r}{2\cdot V}\right)^2\cdot \mathrm{E}\left\{(Vn_1 - Vn_2)^2\right\} = \frac{d_r^{\,2}}{2\cdot SNR}\\
&\text{Noise standard deviation: } \sigma_R \geq \frac{d_r}{\sqrt{2\cdot SNR}}
\end{aligned} \tag{5.35}$$

Equations (5.35) are the first-order noise model for a split gate range tracker. The two range gate measurements are subtracted from one another and then normalized by the peak voltage. Additionally, the measurement must be converted into range units by multiplying by one half the range cell size as shown in line 3 of (5.35). One can separate the noise error term from the incremental range change as in line 4. The random noise-induced range error variance is the expectation of the noise errors on each sample squared. Since they are assumed random and uncorrelated, the noise power doubles as there are two contributors. The random noise-induced lower bound on RMS range error is the range cell size divided by the square root of two times the signal to noise power ratio. The good news is that the better the SNR the smaller the range estimation error. It is usually possible to improve range accuracy by 10:1 with high SNR. Ultimately other hardware, software and environmental factors limit range accuracy no matter how good the SNR becomes.

### 5.13.2 Doppler tracking

The random noise-limited accuracy for Doppler tracking is similar to range tracking accuracy. Consider two rotating phasors from two successive measurements at the output of a filter as shown in Figure 5.37. Each vector has a small noise vector associated with it. In the time between samples, the phase may have advanced by some amount. That phase difference divided by the time between samples is an estimate of frequency.

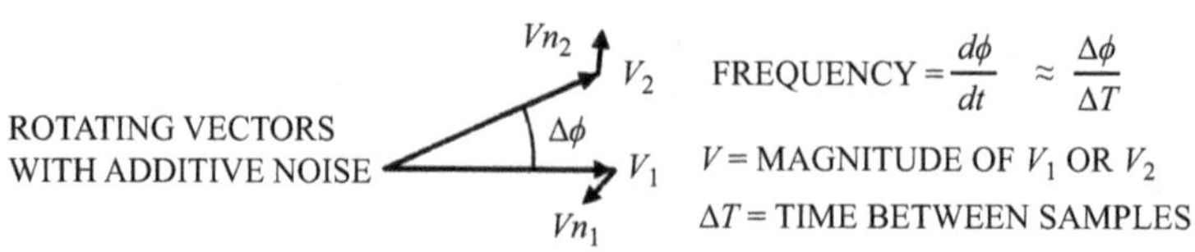


*Figure 5.37 Doppler track noise situation*

$$\text{Frequency: } \frac{d\phi}{dt} \approx \frac{\Delta\phi}{\Delta T}$$

$$\text{Small angle phase change: } \Delta\hat{\phi} \approx \frac{|V_2 + Vn_2 - V_1 - Vn_1|}{V}$$

$$\text{Frequency estimate: } \hat{f} = \frac{\Delta\hat{\phi}}{\Delta T} \approx \frac{|V_2 - V_1|}{\Delta T \cdot V} + \frac{|Vn_2 - Vn_1|}{\Delta T \cdot V}$$

$$\text{Noise variance: } \sigma_f^2 = \frac{\mathrm{E}\left\{|Vn_2 - Vn_1|^2\right\}}{(\Delta T \cdot V)^2} = \frac{2}{\Delta T^2 \cdot SNR}$$

$$\text{Noise standard deviation: } \sigma_f \geq \frac{\sqrt{2}}{\Delta T \cdot \sqrt{SNR}} \tag{5.36}$$

Equation (5.36) summarizes the random noise limit on frequency tracking. For small phase increments between samples, the normalized difference between two vectors approximates the angle between them. Then separating the noise terms in the frequency estimate, the random noise-induced frequency variance is calculated in line 4 of (5.36). Again since there are two noise contributors, the noise power is doubled. Finally, the lower bound noise-induced RMS frequency error is the square root of two divided by the sample interval and the square root of the SNR in the last line of (5.36).

## 5.13.3 Angle tracking

There are many different forms of angle tracking but almost all of them can be resolved into the form of difference to sum ratio. A typical angle track scenario is a sum to difference ratio approximated by an almost straight line between the half power points on the antenna beam. For the simplified analysis here, an azimuth beam is chosen but the elevation dimension will be similar.

Consider the geometry of Figure 5.38, a difference to sum ratio of $2/\theta_{az}$ is assumed. The estimate of an off boresight angle to the target is the difference to sum ratio times the angle discriminant slope. As before there will be a random

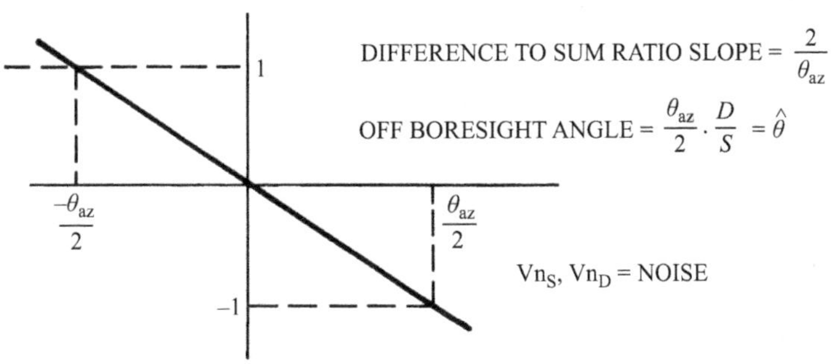


*Figure 5.38 Angle track noise situation*

uncorrelated noise term associated with both the sum and difference measurements.

$$\text{Difference to sum ratio slope: } \approx \frac{2}{\theta_{az}}$$

$$\text{Target off boresight angle: } \hat{\theta} = \frac{\theta_{az}}{2} \cdot \frac{D + Vn_D}{S + Vn_S}$$

$$\hat{\theta} \approx \frac{\theta_{az}}{2} \cdot \left[ \frac{D}{S} + \frac{Vn_D}{S} - \frac{Vn_S \cdot D}{S^2} - \frac{Vn_S \cdot Vn_D}{S^2} \right]$$

$$\text{Noise error: } \frac{\theta_{az}}{2} \cdot \left[ \frac{Vn_D}{S} - \frac{Vn_S \cdot D}{S^2} - \frac{Vn_S \cdot Vn_D}{S^2} \right]$$

$$\text{Noise variance: } \sigma_\theta^2 = \left( \frac{\theta_{az}}{2} \right)^2 \cdot \mathrm{E}\left\{ \left( \frac{Vn_D}{S} - \frac{Vn_S \cdot D}{S^2} - \frac{Vn_S \cdot Vn_D}{S^2} \right)^2 \right\}$$

$$\text{If SNR is high and off boresight is small then: } \sigma_\theta \geq \frac{\theta_{az}}{2 \cdot \sqrt{SNR}} \tag{5.37}$$

Referring to (5.37), the estimated off boresight angle to a target will be the slope times the difference to sum ratio including the noise terms as shown in the second line of (5.37). Expanding that equation using the approximation that $1/(1 + x) \approx (1 - x)$ yields the third line in (5.37). Then separating the random noise angle error terms from the desired measurement, the random noise angle variance expectation is calculated in line 5 of (5.37). If one assumes that the SNR is high and that the difference to sum ratio is small, then the lower bound on the RMS angle error is half the beamwidth divided by the square root of the SNR. Normally it will be significantly worse than this. Nonetheless, looking into space, with nitrogen-cooled front ends and long slow tracking, better than 1/100 of a beamwidth is achieved.

## References


[1] Lynch, D., *Introduction to RF Stealth*, SciTech Publishing, Raleigh, NC (2004), pp. 98–114, 446–451, 82–84, 198–221, 467–470, 492–501, 504–531.

[2] Radant, M., Lewis, D., Iglehart, S., "Radar Sensors", *UCLA Short Course Notes*, July 1973.

[3] Harris, F., D. Lynch, "Digital Signal Processing and Digital Filtering with Applications", *Evolving Technology Institute Short Course Notes*, 1971–1983 (February 1978), pp. 366, 744–748.

[4] Eliot, D., ed., *Handbook of Digital Signal Processing*, Academic Press, San Diego, CA (1987), pp. 364–464, 527–589, 590–593, 594–631.

[5] Nitzberg, R., *Radar Signal Processing and Adaptive Systems*, Artech House, Norwood, MA (1999), pp. 199–202, 207–236, 267–290.

[6] Low Phase Noise SAW VCOs, www.Synergymwave.com, HFSO1000-5 descriptive material, 2014.

[7] Ivanov, E.N., Tobar, M.E , Woods, R.A., "Advanced Phase Noise Suppression Technique for Next Generation of Ultra Low-Noise Microwave Oscillators", *1995 IEEE International Frequency Control Symposium*, pp. 314–320.

[8] Kester, W., "Intermodulation Considerations for ADC's", *Analog Devices Web Page*, October 2008, MT-012.

[9] Kester, W., "Aperture Time, Aperture Jitter, Aperture Delay Time-Removing the Confusion", *Analog Devices Web Page*, October 2008, MT-007.

[10] Brigham, E.O., *The Fast Fourier Transform*, Prentice Hall, Englewood Cliffs, NJ (1974), pp. 172–197.

[11] Aks, S., Lynch, D.D., Pearson, J.O., Kennedy, T., "Advanced Modern Radar", *Evolving Technology Institute Short Course Notes*, November 1984.

# Chapter 6
# The GMTI idea

## 6.1 Introduction

Ground or surface moving target detection (often called ground moving target indication or GMTI) consists of radar or EO detection of humans or tactical vehicles moving on or near the earth's surface. Animals and humans have little difficulty detecting movement and sorting unimportant movements from an important movement at short range. This is the result of hundreds of millions of years of selective evolution. For example, the frog's eye is especially adapted to detect and react to a fly-sized object moving through its field of vision. It is so good that its tongue is on the move to where the fly will be (not where it is) in less than 100 ms. Sadly, radars and EO sensors are not nearly that good over their ranges of operation (much longer than a frog's tongue!).

Detecting important movement at longer range by automatic means is very difficult. The principal reason is that in a larger FOR there are many things, which are moving, most of which are uninteresting. Humans do this sorting process at closer ranges with such ease that they often underestimate the complex processing that goes on. For instance, a human can tell the difference between a golf ball and a sparrow in flight at more than 100 yards even though the average human eye doesn't have the acuity to resolve the detailed features of sparrows or golf balls at that distance. We use a wide variety of cues to do this including trajectory and movement in a background, which does not change its position.

GMTI and ground moving target track (GMTT) radar modes have different challenges. First, target detection is usually the easy part; the RCS of most manmade and many natural moving targets is large (10–1000 $m^2$). Even humans have an RCS, which is between ½ and 2 $m^2$. Unfortunately, there are many moving but stationary objects such as fences, ventilators, ocean waves, watercourses and suspended objects (e.g. intersection stop lights), which lead to apparent false alarms. Often slow-moving vehicles have fast-moving parts (e.g. helicopters, agricultural irrigators). Most land areas have very large numbers of vehicles and scatterers that could be vehicles or animals (including humans). Processing capacity must be adequate to handle and discriminate thousands of high SNR threshold crossings and hundreds of moving targets of interest. In most cases, all targets must be tracked and then recognized on the basis of Doppler spectrum (helicopters vs. wheels vs. tracks), rate of measured location change (ventilator and fencing locations don't change much) and consistent trajectory (60 mph where there are no roads is improbable for a surface vehicle). In addition, vehicles of interest can have relatively low radial velocities requiring endo-clutter (inside mainlobe clutter (MLC)) processing [1,2].

The basic idea behind GMT detection is that radar targets of interest are moving at significant speed relative to the movement of the background. This naïve idea goes back to at least World War II. The idea is true for many airborne targets traveling at 50 m/s (~100 knots or ~112 miles/h) or more against a ground clutter background. Sadly, it is not nearly true for most surface targets against their backgrounds. Recalling Figure 3.31, there are common natural objects that have apparent velocities up to 40 m/s. Even trees and fences have apparent short-term velocities in the presence of wind. In addition, many manmade stationary objects have apparent velocities up to 100 m/s such as fans on ventilators, air conditioners, wind turbines, windmills, heat exchangers, etc. Furthermore, there are generally so many real GMTs that the problem is not detection but sorting and discrimination between the objects of interest and all other movers. These attributes require significant signal and data processing to detect and track just the moving targets of interest. In the following sections, the fundamental characteristics for discrimination of ground mover RCS, Doppler signature and clutter will be discussed.

## 6.2 Doppler clutter spread [3]

First, let us consider the simplest confusor for a moving platform attempting to detect surface moving targets, that is, the Doppler spread across the antenna beamwidth. Consider the geometry of Figure 6.1. The antenna azimuth 3 dB beamwidth, $\theta_{az}$, has a "footprint" (see Section 3.3) on the surface as shown at the left in Figure 6.1. The center of the beam is at some distance, $R$, and composite angle, $\eta$, made up of an elevation angle, $\varepsilon$, and azimuth angle, $\theta$, with respect to the platform composite velocity vector, $V_a$. The amplitude of the surface return in the upper center of Figure 6.1 is some bump shape in range. If one were to examine a single-range bin,

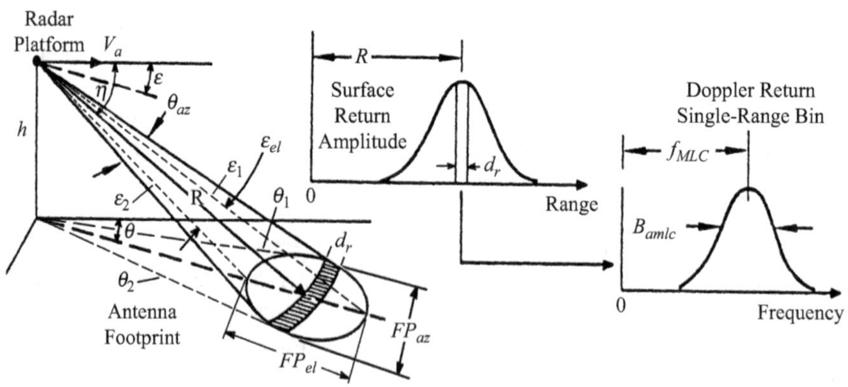


Figure 6.1 Clutter Doppler spread

$d_r$, from that return, the Doppler spectrum also would be some bump shape in frequency as suggested in the right hand of Figure 6.1. Since the Doppler returns from cells near the edges of the antenna footprint are at slightly different angles $\theta_1$ and $\theta_2$, they will have slightly different Doppler frequencies. In fact, each cell across the beam footprint will be at a slightly different Doppler. Thus one can reduce the competing clutter area in a single-range bin by breaking up the return Doppler into multiple frequency/angle bins with a filter bank as suggested by the lines in the bin, $d_r$, at the left in Figure 6.1.

The approximate center frequency of MLC, $f_{MLC}$, and half power spread of the Doppler spectrum of MLC in azimuth for a given elevation, $B_{amlc}$, and similarly the MLC spread in elevation for a given azimuth, $B_{EMLC}$, are given in (6.1). If the beamwidth is reasonably symmetrical, then either of the forms in the bottom equations (6.1) are adequate estimates of Doppler spread across the beam.

$$
\begin{aligned}
&f_{MLC} = \frac{2V_a}{\lambda}\cos\eta = \frac{2V_a}{\lambda}\cos\left(\operatorname{acos}\left(\cos\varepsilon\cos\theta\right)\right) \\
&B_{amlc} = \frac{2V_a}{\lambda}\left(\cos\eta_1 - \cos\eta_2\right) = \frac{2V_a}{\lambda}\cos\varepsilon\left(\cos\theta_1 - \cos\theta_2\right) \\
&\text{If } \theta_{az} \text{ and } \varepsilon_{el} \text{ are small,} \\
&\text{then the sine of the angle can be approximated by the angle:} \\
&B_{amlc} \approx \frac{2V_a}{\lambda}\cos\varepsilon\cdot\theta_{az}\cdot\sin\theta \\
&\text{Similarly:} \\
&B_{emlc} \approx \frac{2V_a}{\lambda}\sin\varepsilon\cdot\varepsilon_{el}\cdot\cos\theta
\end{aligned}
\tag{6.1}
$$

From here on in the text the −3 dB beamwidths in azimuth and elevation will be designated as $\theta_{az}$ and $\varepsilon_{el}$, respectively. One natural question that arises when considering not only Doppler spread across the beam but also possible ambiguities in range and Doppler is the footprint of the mainbeam shown in Figure 6.1.

Once the footprint is large enough to cover the area of interest, why should you care what the size of the mainlobe footprint may be? Sadly, the answer is that the ambiguities in range and Doppler as described in Section 2.13 create significant constraints on detection and tracking. In GMTI, the achievable azimuth accuracy is always much poorer than the achievable range resolution. This implies that minimizing Doppler ambiguities is usually more important than the number of range ambiguities. In SAR, PRF coding (sometimes called waveform diversity, see Section 4.11) can be used to partially cancel range ambiguities. For conservation of ambiguity reasons, this doesn't work well for GMTI because part of the Doppler detection space will be occupied by the coding spectrum. The time-tested alternative is the use of multiple waveform repetition frequencies whose least common multiple allows ambiguity resolution if GMTs are relatively sparse compared to the number of range-Doppler bins.

Everything in a single-range-Doppler bin competes with a possible GMT. The unresolved azimuth Dopplers as well as the antenna and range sidelobes all compete with a GMT. If there were no other way to separate these different competing returns, then they would make detection impossible for some moving targets. Fortunately, multiple phase centers and Doppler filters come to the rescue.

The number of range ambiguities, $N_{ramb}$, in the mainlobe elevation footprint is given in (6.2). A simpler version of (6.2) was given in (5.23) because the more distant ambiguities are usually not difficult to reject. This equation makes the approximation that the grazing angle is almost constant over the elevation footprint (often not true!).

$$N_{ramb} \cong \frac{2 \cdot FP_{el} \cdot \cos\psi}{c \cdot PRI} \tag{6.2}$$

For example, suppose $FP_{el} = 280$ km, which is typical of the ground ranges over 100 km at an altitude of 7.5 km as shown in Figure 3.18(b), $PRI = 200$ μs, $\psi = 0.05$ average grazing angle over a footprint centered at 150 km, then $N_{ramb} =$ 9.35. Since a radar is typically blind for part of the PRI around the main bang, probably as many as 4–8 separate waveforms will be required to resolve the several thousand GMTs in the footprint.

## 6.3 Doppler spectral situation

In Section 2.9, the pulse Doppler situation from a moving platform relative to a moving target near or on the surface was introduced. The LOS velocity for a moving target, $V_{tLOS}$, is shown in Figure 6.2 (same as Figure 2.27). That LOS velocity could be redefined as the platform velocity times the cosine of a slightly different angle $(\eta + \Delta\eta)$ as given in (6.3). The variables in (6.3) are all defined in Figure 6.2.

Apparent target velocity:

$$V_{tLOS} = V_a \cdot \cos\eta_t - V_t \cdot \cos\zeta_t \equiv V_a \cdot \cos(\eta_t + \Delta\eta)$$

Apparent clutter velocity at target angle:

$$V_{cLOS} = V_a \cdot \cos\eta_t$$

Assuming $\Delta\eta$ is small then:

$$\Delta\eta \approx \frac{-V_t \cdot \cos\zeta_t}{V_a \cdot \sin\eta_t} \quad \text{(note this approximation blows up as } \eta_t \to 0\text{).} \tag{6.3}$$

Observe from (6.3) that the apparent target velocity (and Doppler) appears to come from a different angle than the apparent clutter velocity (and Doppler). The clutter that competes with a moving target is at a different angle than the target! Figure 6.3 shows the typical ambiguous range-Doppler situation before the MLC is shifted to the PRF and its harmonics. In general, the center of MLC is not the same in each range bin. Normally, the center of MLC in the center range bin is shifted to zero frequency (often called DC). There will be some Doppler offset between the center bin and the edges of the range swath. If the range swath is small relative to the range, then the offset is small. But there are many geometries in which this will

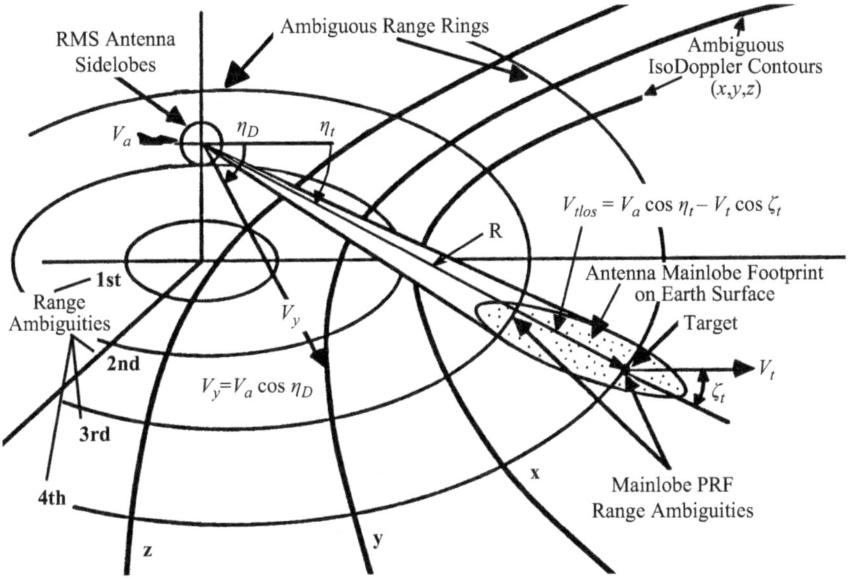


*Figure 6.2 GMTI pulse Doppler situation (repeated Figure 2.27) [4]*

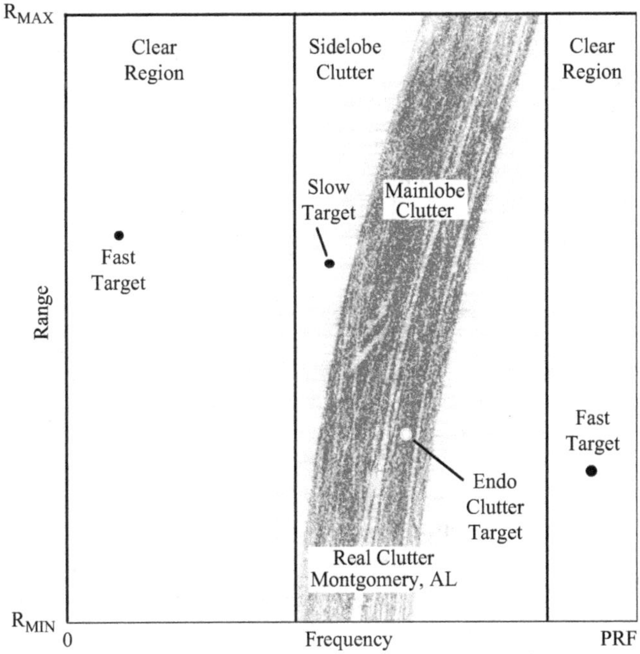


*Figure 6.3 Ambiguous range-Doppler clutter and targets shifted to 0.6 PRF (MontAlSARModel001d)*

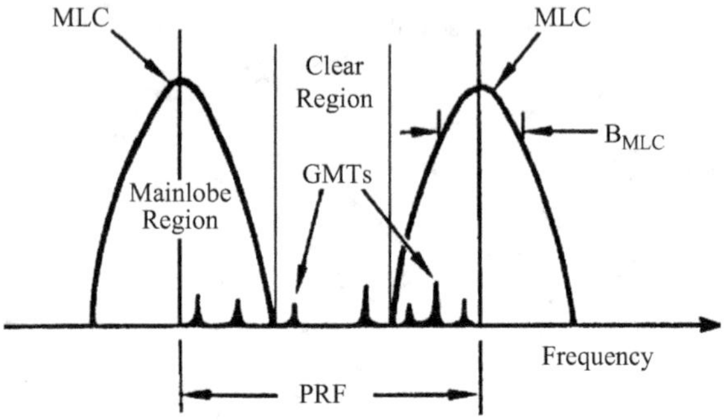


*Figure 6.4 Single-range bin Doppler GMT situation*

not be true. There are three regions in which GMTs may be detected (Figure 6.3). They are the clear region, the near sidelobe region (SLC) and the MLC region (MLC). Each requires a different strategy.

As soon as the apparent angle offset is outside the antenna mainlobe, the antenna pattern dramatically reduces the amplitude of the competing clutter and a GMT is in the clear region shown in Figures 6.3 and 6.4.

Consider a single-range bin that has had MLC shifted to zero frequency and the PRF harmonics as shown in Figure 6.4. Some GMT speeds and corresponding Dopplers are far enough away from MLC Doppler spread that they are only competing with thermal and sidelobe clutter noise. These "fast" GMTs are easily detected in what is known as the clear or exo-clutter region. Angle discrimination always is used for GMT detection with the simplest case being just the normal antenna pattern itself. GMTs moving relatively slower than the spread of MLC are competing with progressively more clutter power as shown in Figures 6.3 and 6.4. This naturally leads to the need for antenna beamwidths as narrow as possible for a given platform. Some GMTs will be in the transition or near SLC between MLC and the clear region. Some will have enough RCS to be detectable at the edge of MLC. For those, perhaps only a few filters to reduce competing clutter area will be required for detection. GMTs in the mainlobe will require very narrow filters and may require additional angle information for detection. An example of real clutter in range and Doppler from Montgomery, AL, is shown in Figure 6.5. Please note that the clutter is anything but homogeneous. It is filled with streaks from roads, waterways and other low reflectivity features interspersed with very bright scatterers. Note that the clear region contains streaks that are the result of very bright scatterers burning through the Doppler filter sidelobes. In this case, the clutter covers about 1/3 of the spectrum between the PRF ambiguities.

More elaborate angle discrimination can be used to reduce or cancel that clutter competing with a slow radial velocity GMT. This is commonly called endo-clutter detection. Dozens of schemes using this angle of arrival property have been implemented including monopulse anomaly detection (MAD), monopulse null detection (MND), displaced phase center antenna (DPCA) and space-time adaptive processing (STAP) clutter cancellation. These methods are especially important in GMT

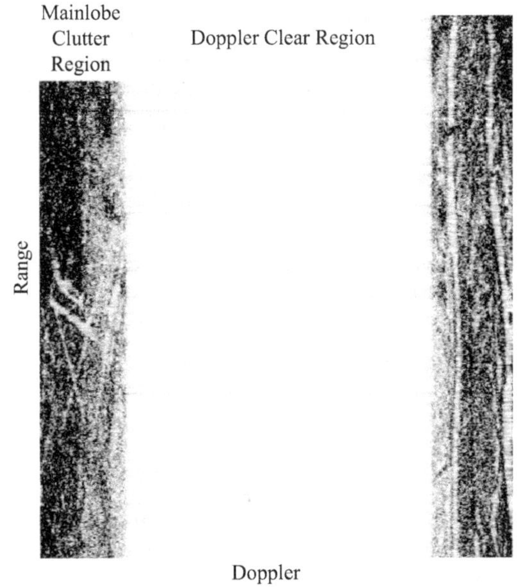


*Figure 6.5 Real clutter and the Doppler clear region, clutter at DC and PRF*

detection and track from space because spacecraft are moving fast and the fraction of GMTs that have a high radial velocity in the spacecraft direction is much smaller. Some form of each of these techniques will be discussed in a later section [5–10].

Referring to (6.3), as the target angle, $\eta_t$, approaches the platform velocity vector, the apparent target to clutter angle, $\Delta\eta$, gets very large. At that point the Doppler spread across the antenna beam is very small and in-mainlobe-clutter angle-based cancellation is usually not required. There are several other slow-moving target detection schemes such as SAR-based change detection, video SAR, synthetic-DPCA and interferometry. These will be discussed in Chapter 7.

Doppler spread across the antenna varies with the LOS relative to the velocity vector. It is the first item of consideration in what follows.

---

**Example 6.1** Suppose a radar in an aircraft has the following parameters given in Table 6.1.

Then the Doppler spread across the 3 dB azimuth beamwidth using equations (ex 6.1) would be

$$B_{amlc} = \frac{2 \cdot 300 \cdot 3° \cdot \sin(45°) \cdot \cos(6°)}{0.03 \cdot 57.3°} = 736.37 \text{ Hz}$$

and

$$B_{amlc} = \frac{2 \cdot 300 \cdot 3° \cdot \sin(20°) \cdot \cos(6°)}{0.03 \cdot 57.3°} = 356.17 \text{ Hz} \qquad \text{(ex 6.1)}$$

*Table 6.1 Example 6.1 parameters*

| Parameter | Value | Parameter | Value |
|---|---|---|---|
| $\theta_{az}$ | 3° | $\theta$ | 45°, 20° |
| $V_a$ | 300 m/s | $\varepsilon$ | 6° |
| $\lambda$ | 0.03 m | $\varepsilon_{el}$ | 6° |

A GMT traveling radially at 5.3 m/s (a little less than 12 miles/h) would be inside MLC Doppler for the 45° look angle. Although 5 m/s seems slow, many moving vehicles still might be visible inside MLC including most large trucks. Unfortunately, many others will be masked by MLC. However, if the angle off the velocity vector was 20° instead of 45°, then the spread would be one half as much and slower targets approaching 2.67 m/s could be detected at the edge of MLC.

In addition, one must remember that vehicles and people move in the local surface plane with a smaller radial component to an airborne or spaceborne observer. Depending on the grazing angle, $\psi$, and target heading, $\zeta_t$, relative to the observer, such a limitation might exclude half the targets of interest. On the other hand, a narrower beam, shallower depression angle and many narrow Doppler filters in the MLC spectrum might allow most GMTs of interest to be visible. Often, the center of MLC, $f_{MLC}$, is placed at zero frequency and cancelled. Before such cancellation, the spectral situation is similar to that shown in Figure 6.3.

## 6.4 Typical RF front end

As has been described earlier, low-level RF receiver and exciter are required to establish basic noise-limited performance in every GMTI or SAR radar. Most GMTI receivers have multiple channels. A simplified multichannel receiver block diagram is shown in Figure 6.6. The channels may come from segments of the

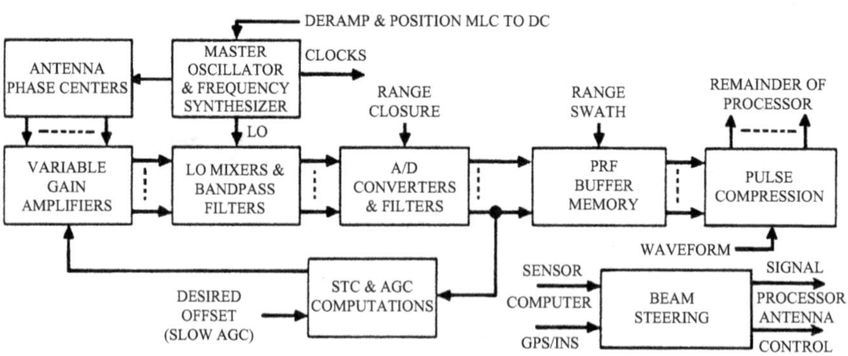


*Figure 6.6 Simplified receiver-exciter block diagram (repeated Figure 4.5)*

antenna array, the entire antenna at separate frequencies or separate antennas with different functions, for example, guard, IFF, radar, comm., etc.

The radar MO controls both the transmitted signal and the noise spectrum as well as the signal demodulation reference on receive. Unfortunately, the noise on the RMO establishes the ultimate limit on both SAR and GMTI performance not thermal noise. The exciter usually contains a frequency synthesizer driven by the MO, which creates frequency and phase references for translation mixers, range bin sampling clocks and waveform demodulators. Often there is pulse-to-pulse phase or time modulation on the transmitted waveform to suppress ambiguities, which must be compensated in the receiver demodulators. In addition, MLC is usually placed either at zero frequency and its PRF ambiguities or at PRF/2 and its ambiguities. The MLC demodulation frequency is a combination of the predicted clutter center frequency based on antenna pointing and motion compensation plus some signal-based clutter tracking in the processor and radar control computer. This signal-based clutter tracking is required because even with the best short-term stability reference oscillators, excellent GPS-inertial motion sensors and antenna pointing accuracy, there still will be uncompensated clutter centroid frequency errors.

As mentioned in Section 2.12, the amplitude range of input signals will easily be more than 10 billion to 1. In order to accommodate this large dynamic range, the receiver must have time-variable gain amplifiers that are adjusted inside a single PRI as well as PRI to PRI. The single PRI variable gain adjustment is usually called STC. The PRI to PRI adjustment is usually called AGC. Both your car radio and TV have AGC and some have fine frequency tracking.

Following variable gain, amplifiers and mixers will be A/D converters, which digitize the analog signals. Usually the conversion rate is quite high for relatively short bursts covering the range swath. A high-speed PRF buffer memory smoothes out the data transfer rate between the A/Ds and the signal processor. Also, the timing to the A/Ds is usually adjusted for range closure in the mainlobe swath. Digital signals are picked off from the A/D outputs to calculate and create the commands necessary for the RF front-end variable gain amplifiers.

A more detailed diagram of the receiver channels is shown in Figure 6.7. Each channel must have several levels of CAGC ($CAGC_{1,2}$ ), which is often common across all channels, a phase-modulated first local oscillator (1st LO), bandpass filtering for interference and cross-talk rejection, a second local oscillator (2nd LO) and gain stages with fine automatic gain control (FAGC), which is usually unique to each channel. Following establishment of the noise figure, selection of *a priori* frequency shifts and of general frequency band, the channels may be further demodulated with a variable frequency oscillator (VFO) reference locked to the MO. The VFO may be used to provide some dechirp or to position the largest amplitude portion (usually clutter) of the input spectrum to DC (zero frequency) on a range bin by range bin basis to minimize the effects of distortion and simplify adaptive thresholding.

Subsequent to VFO demodulation, the signal is bandpass filtered (BPF) in an approximation to the matched filter for the chip/bin time. The BPF outputs are converted from A/D with sample intervals chosen for acceptable performance (not necessarily the chip/bin spacing as will be seen later).

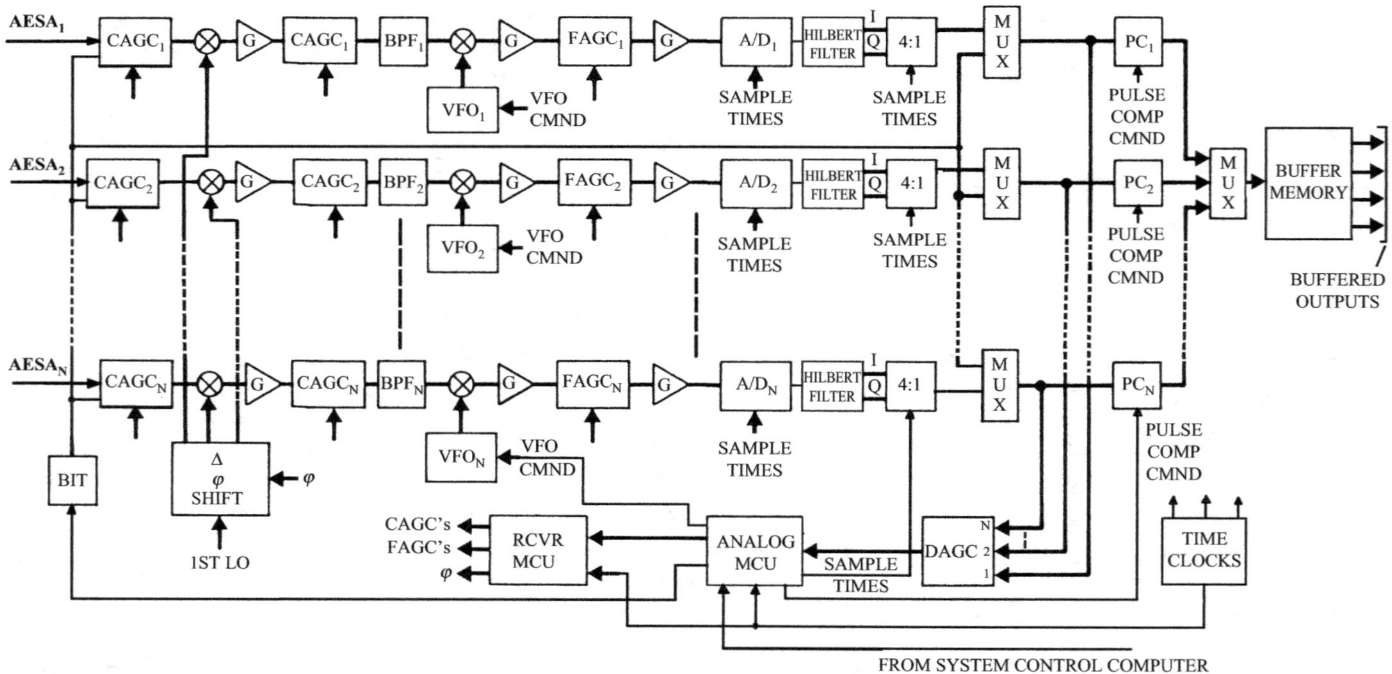


*Figure 6.7 Typical GMTI/SAR RF front end (similar to Figure 2.21)*

I/Q demodulation may also be accomplished after A/D conversion in a digital Hilbert transform filter. Often the A/D in the Hilbert transform technique has its sample frequency set close to the IF center frequency. Sometimes the sampling frequency is chosen so that a sampling ambiguity of the signal appears at one quarter of the sampling rate or its ambiguities. The advantage of I/Q conversion after digitization is that I/Q imbalance can be made much smaller, thus improving false target image rejection [11].

There are usually several places in the front end where built-in test (BIT) signals can be multiplexed into the data stream. There may be 1,000s of wavelengths between the antenna and the A/D converters. Tiny amounts of parameter drift can cause large phase and amplitude shifts. Therefore, BIT signals are usually inserted every second or two between CPIs to calibrate/compensate for these possible changes.

Prior to pulse compression (PC), if any, the sampled digital signals are delivered to a digital AGC function that buffers and averages the high-speed samples for further manipulation in the analog microprocessor control unit (MCU). The analog MCU controls all the functions in the front end. The analog MCU delivers AGC, STC and other commands to the receiver MCU (RCVR MCU). Because most RF circuits are very nonlinear, the RCVR MCU uses internal hardware calibration data as well as the analog MCU commands to set the parameters for the low-level RF circuits. Each MCU has a common interrupt-driven operating system with watchdog timing designed for embedded real-time operation systems (ERTOS). The ERTOS is usually in every location where an MCU is used even though the applications might be quite different (e.g. in the power supply).

PC may be applied to each channel. The PC is not necessarily complete or the same in each channel. Further PC may be performed after digital narrowband filtering. The outputs are multiplexed into a rate change PRF buffer memory and subsequently read out into the signal processor complex.

## 6.5 GMTI signal processing

A GMTI processor functional block diagram might look like that shown in Figure 6.8. The output bandwidth from the receiver intermediate frequency (IF) amplifiers usually is approximately matched to the transmitted bandwidth. The IF output often goes through a set of in-phase (I) and quadrature (Q) detectors (synchronous mixers) producing I/Q video as described in the previous section. Each channel is A/D converted and the range bin samples are stored in memory. As previously mentioned, there could be a single A/D that is synchronous with the IF center frequency followed by a digital Hilbert transformer, which creates digital in-phase and quadrature components. The advantage of this second method is that I/Q imbalance is much smaller, thereby suppressing the unwanted image term to −60 dB or better. For analog I/Q detection, the image term usually is only −40 dB below the peak input amplitude.

Each subsequent PRI produces another set of range bins, which may be sent to a clutter canceller (if present). The resultant is returned to memory until enough

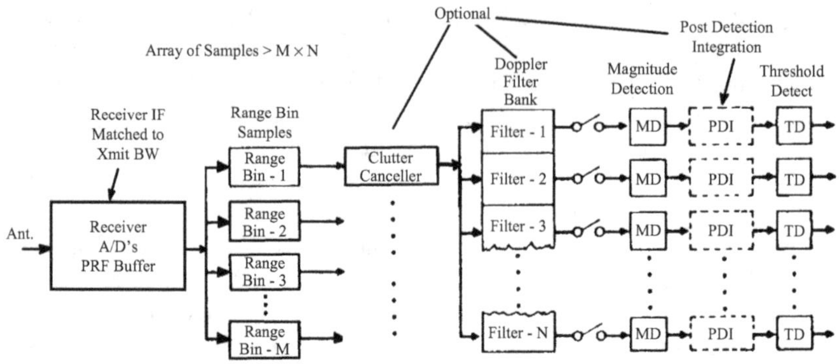


*Figure 6.8 Typical GMTI processing*

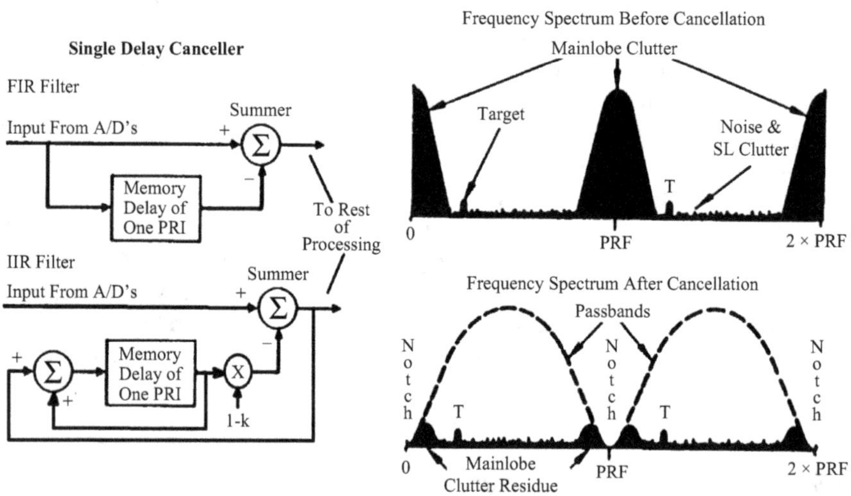


*Figure 6.9 Single-delay canceller*

samples are accumulated to allow filter formation in each range bin. The most common Doppler filter bank processing is an amplitude weighted FFT with at least N samples. The filter bank is formed for all of the M range bins. The processed data array is at least M × N but often it is made larger by filling part of the array with zeros. This process is sometimes called data turning. Subsequent to the filter bank, each filter output may be magnitude detected or monopulse processed. There may be PDI following magnitude or monopulse detection if the SNR warrants it. Following PDI there will be threshold detection, which may be quite complex as shown later.

Cancellation of the surface scatterers, which are truly stationary, is often performed by very simple filters such as single or two delay cancellers. A single-delay canceller and its effect on the GMT frequency spectrum are shown in Figure 6.9. The only reason for such a filter is to control or reduce dynamic range since most of the filtering and integration is performed by the Doppler filter bank (often an FFT). There will still be some MLC residue from a canceller filter as suggested by Figure 6.9, which will compete with some slow-moving targets. Typically the individual filter bandwidth will only be limited by expected target geometrical acceleration, internal motion (i.e. target Doppler spread, see Chapter 3) and the time that fast targets may be in a single-range bin.

Two different examples of single-delay cancellers are given in Figures 6.9 and 6.10. One is a FIR filter and one is an IIR filter. The frequency response for these filters is shown in Figure 6.10 for a PRF of 5000 and ***k*** of 0.9. As can be seen, the clutter notch is quite narrow for both of these filters. The principal advantage these filters provide is reduction of dynamic range in the remainder of the processing. The main weakness of the FIR filter, which is also true of three- and four-pulse filters, is that endo-clutter moving targets will also be attenuated.

Simple two-pulse DPCA clutter cancellation before Doppler filtering has a clutter cancellation response similar to that in Figure 6.10 single-delay FIR. It also attenuates some faster-moving surface targets just as the filters of Figure 6.9 suggests. For example, for the two-pulse DPCA case at X band with 5 kHz PRF, there will be 10 dB of attenuation for a 7.5 m/s target radial velocity (500 Hz relative Doppler).

A natural question to ask is, "What range and Doppler resolution is necessary to reliably detect tactical GMTs?" First, the cell size should not be smaller than the smallest desired tactical target. Typical dimensions for cars, medium trucks and semi-tractor trailers were given in Table 3.5. Depending on the type of target mix, the cell size might range from 2 × 2 ft for people (0.61 × 0.61 m) to 20 × 150 ft (6.1 × 45.7 m) for large vehicles. Second, since the movers will be competing with

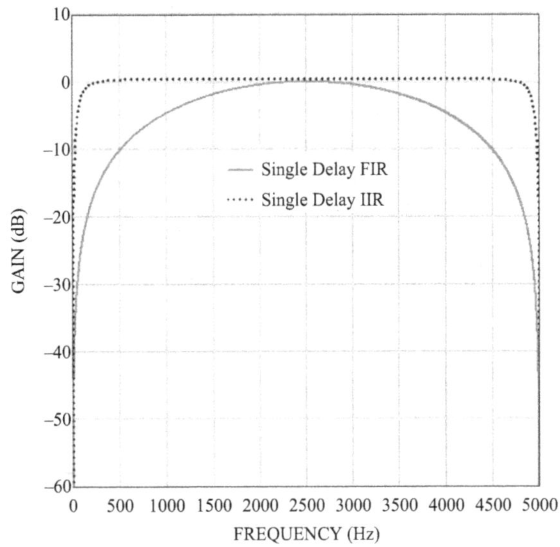


*Figure 6.10 Single-delay clutter canceller frequency responses (Appendix A.6, Cancellers, B)*

clutter in a similar size cell, the backscatter cross section of competing cells should be significantly smaller than targets of interest. Recall from Figure 3.10 that the patch competing with a GMT prior to Doppler filtering is as given in (6.4).

$$\sigma_{clutter} = \sigma_0 \cdot R \cdot \theta_{az} \cdot \frac{c \cdot \tau}{2} \cdot \sec\psi = \sigma_0 \cdot R \cdot \theta_{az} \cdot d_r \cdot \sec\psi \tag{6.4}$$

---

**Example 6.2** Suppose a radar had the parameters of Table 6.2:

Then:

$$\sigma_{clutter} = \frac{1}{200} \cdot 100000 \cdot \frac{1}{57.3} \cdot 6 \cdot \sec 6° = 52.6\ \text{m}^2 \tag{ex 6.2.1}$$

Then the Doppler spread, $B_{amlc}$, from (6.1) would be approximately

$$B_{amlc} = \frac{2 \cdot 204.27 \cdot \sin 45° \cdot 1° \cdot \cos 6°}{0.03 \cdot 57.3°} = 167\ \text{Hz} \tag{ex 6.2.2}$$

If the target of interest is a car, then from Table 3.5, a typical $-1$ sigma cross section might be 5 m$^2$. For high probability of detection with low false alarms, the SNR after Doppler processing needs to be 40 (16 dB) in a clutter background. Therefore, the 10:1 input disadvantage target-to-clutter needs to be improved by 400:1. This doesn't seem too hard but one must remember that internal motion limits the minimum Doppler cell size. GMTs might be moving at speeds between 3.5 ft/s (1.07 m/s) and 120 ft/s (36.6 m/s) with internal motion from below 2 ft/s (0.61 m/s) and up to 6 ft/s (1.83 m/s). Thus, first-level Doppler filters probably shouldn't be smaller than say 1 m/s. For X band, this corresponds to approximately a 66 Hz filter bandwidth ((2 × 1)/0.03). For slow-moving targets, filter bandwidth might be as small 3–5 Hz.

There is an interaction between the antenna patch size on the surface and the PRF caused by the sampling ambiguities as previously mentioned. The sidelobes of the range or Doppler filters, which fall outside the antenna mainlobe, are attenuated by the antenna sidelobes. The inter-pulse-period (IPP) and corresponding PRF must be adjusted relative to the desired swath (SW) so that the ambiguous responses are antenna limited not PRF limited as shown in Figure 6.11.

*Table 6.2 Example 6.2 parameters*

| Parameter | Value | Parameter | Value |
|---|---|---|---|
| Range, $R$ | 100 km | RCS, $\sigma_0$ | 1/200 |
| Range cell, $d_r$ | 6 m | Velocity, $V_a$ | 400 kts (204.27 m/s) |
| Beamwidth, $\theta_{az}$ | 1° | Azimuth angle, $\theta$ | 45° |
| Grazing angle, $\psi$ | 6° | Elevation angle, $\varepsilon$ | 6° |
| Wavelength, $\lambda$ | 0.03 m | | |

With Figure 6.11 in mind, the PRF should be higher than the combined maximum opening and closing speeds in Doppler. For X band using the maximum target speed mentioned earlier, this would be 2 × 2440 Hz or 4880 Hz ((2 × 2 × 36.6/0.03); see (2.20)). Selecting a PRF of 5000 Hz would allow a filter bank of 250 overlapping Doppler cells made up of 250 heavily amplitude weighted samples, for example, Hamming weighting (see Table 2.2). This would result in a CPI of 50 ms. CPIs should be overlapped at least 2:1 to keep up in real time. Such a strategy would break the clutter return into some number of smaller patches.

For target Dopplers, which are outside MLC, detection usually is much easier since sidelobe clutter and noise will be small, perhaps equivalent to 0.1 m$^2$ compared to a 5 m$^2$ target return. The Doppler filter spectral situation might be as shown in Figure 6.12. As the figure shows, there may be a clutter canceller notch placed in the middle of MLC to improve dynamic range. The Doppler filter banks may span all the rest of the region between zero and the PRF. Of course, because of sampling ambiguities, these filters seem to repeat throughout the sampling spectrum all the way to transmit waveform-matched filter cutoff. Some filters will be in the MLC region and some will be in the near sidelobe or clear region. As mentioned earlier, the design is usually such that detection of "fast" moving targets relative to sidelobe clutter and thermal noise in this region is relatively easy. For the parameters of Example 6.2, the residual clutter in mainlobe filters would be about 33 Hz × 52.6 m$^2$/167 Hz or about 10 m$^2$.

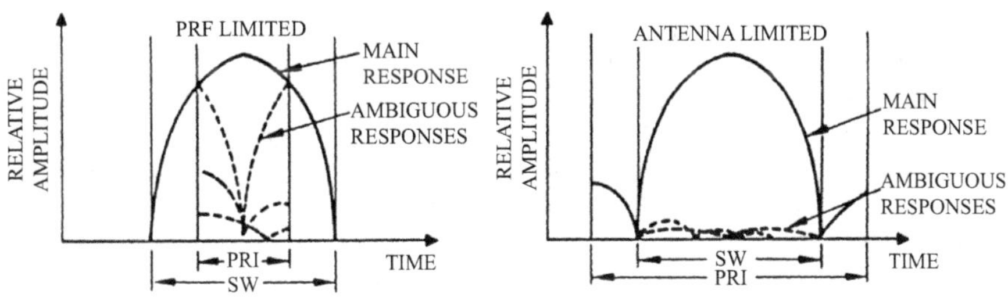


*Figure 6.11 Interaction between PRF, antenna and ambiguities [12]*

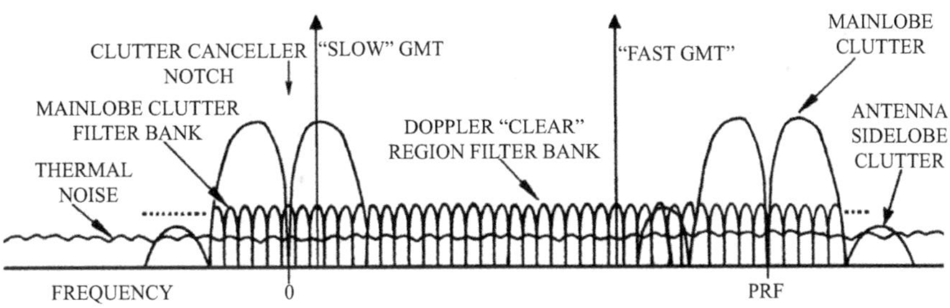


*Figure 6.12 Typical Doppler filter situation for each range bin*

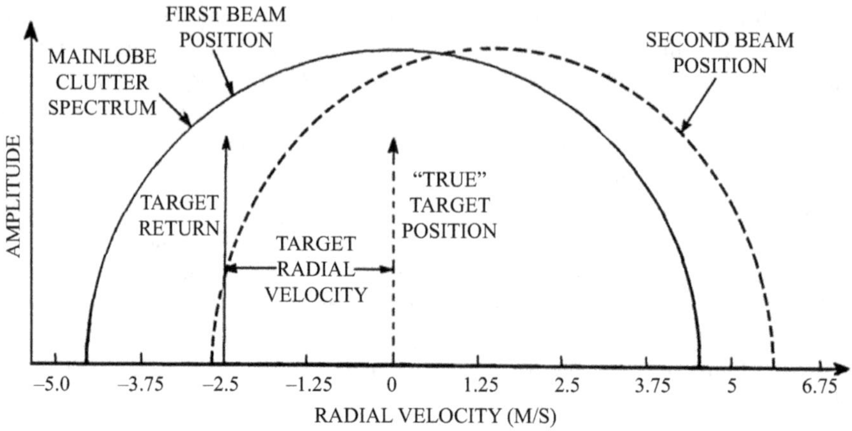


*Figure 6.13 Half beam step scan each CPI improves "semi-slow" GMT detection [13]*

After filtering, there would still be an additional net improvement required for the GMTs, which are competing with MLC. There would have been a 5:1 improvement in signal to clutter from the filter bank but that is still a long way from the desired 400:1 improvement stated for Example 6.2. Another 80:1 improvement would be required. One possibility is to use a low threshold for first detection coupled with step scanning and a second confirming detection at beam edge where the competing clutter is reduced.

An example of the notion of step scanning is shown in Figure 6.13. The idea is that a target near the center of the antenna beam but with a radial velocity competing with clutter near the −3 dB point of MLC will be at the −3 dB point with a half beam step but the competing clutter will be a full beamwidth away and at −25 dB or more for a well-designed antenna. For the extended Example 6.2 given earlier, the equivalent radial velocity across the beam is roughly 5 m/s. Using the parameters of Example 6.2, this strategy will allow detection of target radial velocities down to 2.5 m/s. The strategy achieves more than the 80:1 improvement mentioned earlier (up to 158:1) for targets down to half the beam velocity spread.

Another possibility is to use an antenna with multiple phase centers to cancel clutter based on the correlation between its Doppler and its angle of arrival. This requires multiple channels to be carried along far enough in the processing chain to establish a high-enough clutter-to-noise-ratio (CNR) that the angle of arrival cancellation of clutter has high accuracy (typical CNR should be 30 dB). As mentioned earlier, the multiple phase center technique is called by various names including monopulse clutter cancellation, adaptive array techniques and STAP. The technique relies on the fact that the Doppler in MLC arrives from slightly different angles as mentioned earlier. Multiple phase centers allow the placing of an angle of arrival null in each of the Doppler filter outputs. If the Doppler and the angle of arrival coincide, then it is cancelled. If the Doppler and angle of arrival don't correspond

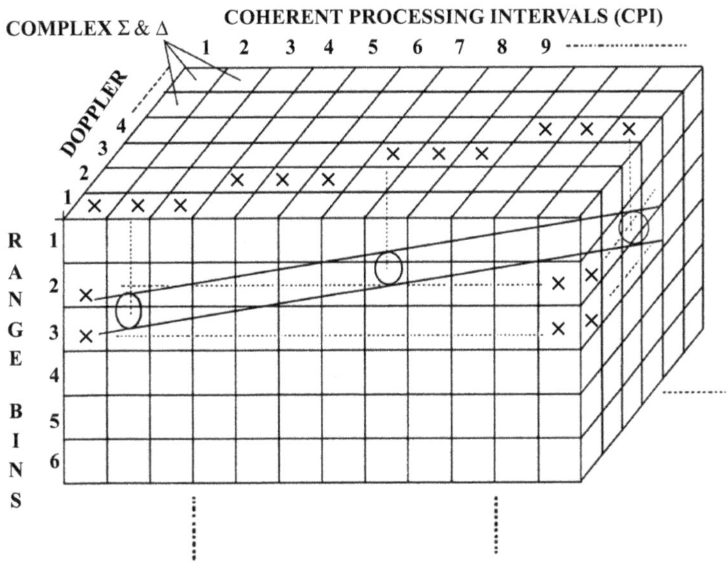


*Figure 6.14 Following a single target across multiple-range-Doppler-CPI cells*

to a stationary reflector, then the output is not cancelled and it represents a "slow" moving target. More about this in a later section.

Often a 3D array of range bin-Doppler bin-CPI sets from the complex filter outputs of a processor like Figure 6.8 before magnitude detection are created as shown in Figure 6.14. Each range bin and Doppler filter output representing single CPIs then may be coherently correlated across the range-Doppler space for various assumed trajectories. A lower threshold (7–8 dB SNIR) is used to nominate possible real targets. All of the raw data for these first threshold crossings is retained. After a sequence of these crossings (~5 in 10 observations) that represent a roughly curvilinear trajectory are associated, the corresponding raw data is integrated and subjected to a much higher threshold (16–18 SNIR). If there is a second threshold crossing, a target detection is declared and a track started. This process is sometimes called track before detect. If these low threshold detections can be associated with an existing target cluster, they are used directly in an existing tracking filter. Although the process takes a lot of number smashing, it is often required to achieve enough SNIR so that cross-range accuracy is tactically useful (geolocation to hundreds of meters is worthless).

As previously mentioned in Chapter 3, the internal motion of GMTs is correlated to ground speed. Slow-moving targets, which are in MLC, are more narrowband and so filters can be narrower. One way to accomplish this is to use the coherent outputs from each Doppler filter over multiple CPIs to form cascaded 4–16 point discrete

Fourier transforms in the MLC region. As the filter bandwidth gets narrower, the competing equivalent clutter area gets smaller. Obviously, this only goes so far since some GMTs will have significant ground speed but an unfavorable trajectory relative to the observer. It also assumes that there may be some bright individual scatterers that by themselves have very little internal motion, for example, truck bed corner reflector. One may expect to achieve a signal-to-clutter-and-noise-ratio (usually called SNIR) improvement of between 3 and 12 (4.8–11 dB) from this multirate filtering process under the right circumstances. Such a process will make a mover in the extended example above achieve over unity SNIR. At that level, noncoherent PDI can now be used to improve SNIR and corresponding angle accuracy to track and geolocate ground movers to tactically useful coordinates (<100 m CEP).

---

## 6.6 GMTI wide area surveillance [13]

The first important class of GMTI is wide area surveillance (WAS). The primary objective is the detection of concentrations of large-scale vehicle activity. This might involve an area of regard (AoR) of 40,000 km$^2$ as suggested by Figure 6.15. The figure shows a hypothetical conventional battle area in western Europe (the heavily studied so-called Fulda Gap). In addition to combatants, there might be tens of thousands of fleeing refugees. The typical AoR is greater than Los Angeles, Orange and San Diego counties in California, which have a population of over 20 million (not much different than the total populations of Iraq, Korea or Taiwan) and 20 million cars.

The tens of thousands of potential movers can only be dealt with in clusters of hundreds or thousands at a time. There might be 250–500 antenna beam steps spanning the area with 500–1000 bona fide near-surface moving targets visible per beam. Another 1000 stationary movers could be detected per beam. To sort the wheat from the chaff, all must be tracked in ground coordinates and the stationary movers and ridiculous observations removed. Second, those fleeing should be eliminated unless there is collateral intelligence that requires tracking (e.g. recognized cellphone numbers or large heat emissions). The tracking could involve clusters of targets (5–50) moving at the same speed in the same direction. For example, there are 17000 cars and trucks per hour traveling on many sections of the Southern California San Diego freeway in the same direction, at roughly the same speed. Those freeway commuters might be a single large cluster and only those that take off-ramps become other clusters.

Obviously, it is important to have baseline traffic flows in order to detect extraordinary flows of surface vehicles. How do you tell the difference between "normal" traffic and military or guerilla convoys? The baselines might include railcars/day (or per hour) on rail lines, cars/day on major traffic arteries, heavy trucks/day, etc. In this environment, the RCS of individuals is not an issue but rather knowledge of the mix of vehicle types and gross numbers. For example, Figure 6.16 shows GMTI WAS for Kuwait City in the 1990 Gulf War. There are so many movers that even individual clusters are not resolvable on a high-resolution display screen.

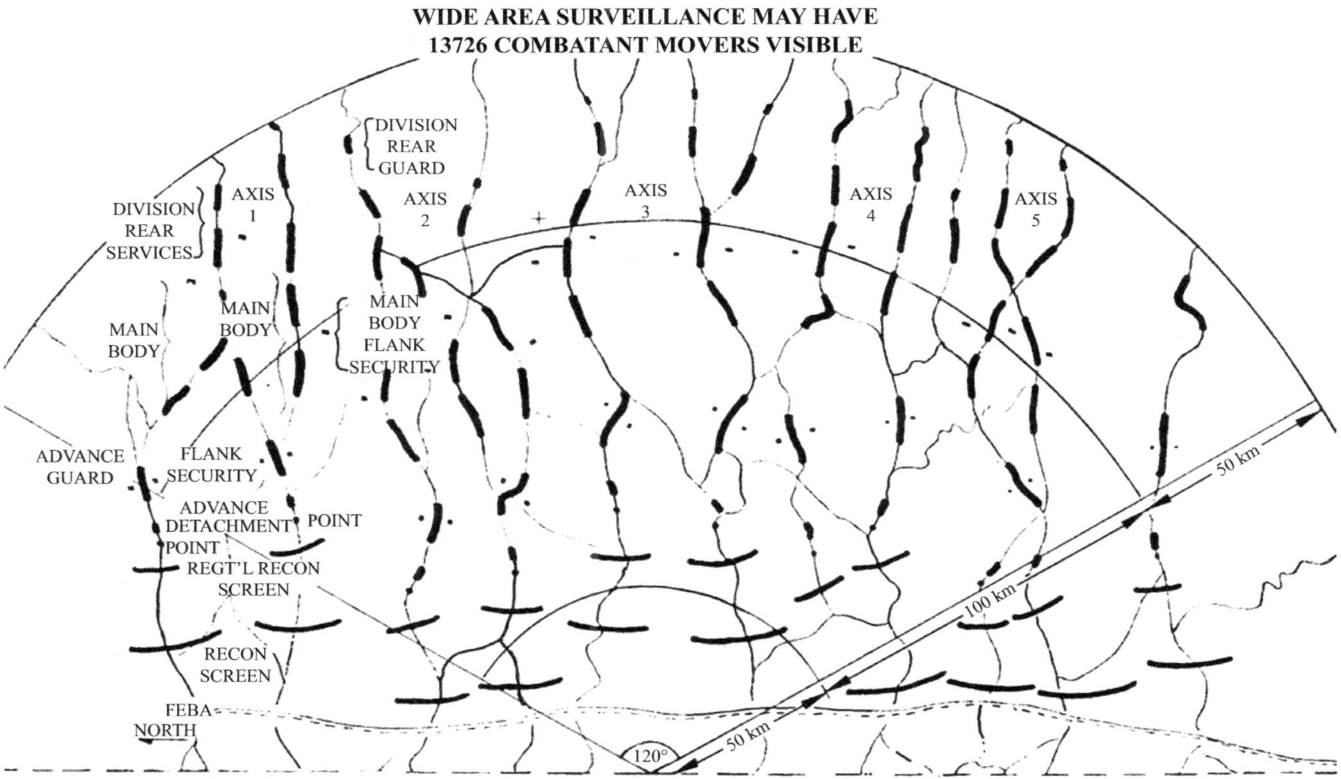


*Figure 6.15 Wide area GMTI surveillance for military targets [13]*

*Figure 6.16 JSTARS wide area surveillance GMTI; adapted from [14] (repeated Figure 1.24)*

Also note in Figure 6.16 that there is a road grid underlay onto which the GMTs often are mapped. Very few vehicles are capable of traveling over 20 mph (32.2 kmph) completely off road. All of the faster vehicles can be pushed on to the nearest roads to improve geolocation accuracy. If there are no roads within the radar error ellipse, then targets should not be relocated. It sometimes occurs that a road not in the DTED has high traffic. Then the GMTs coupled with a SAR map can be used to define a new road in the database by operator designation. Of course, very slow GMTs can be anywhere and cross-range geolocation is problematical and significantly less accurate. Usually a counterpart SAR map is created and underlaid for context and location relative to terrain features.

Many military and transport vehicles have large RCS (see Figures 3.7–3.12). The issue is discrimination not detection. Some of the discrimination can be based on Doppler spread as tabulated in Tables 3.4 and 3.5. Other discriminants include speeds, on–off road, vehicle mix and other cues such as communications emissions and higher radar resolution polarization features.

Higher accuracy in angle is essential to tracking and discrimination. Cross-track accuracy must be 10–100 m or less; otherwise tracking of clusters or recognition algorithms will be competing with too many confusors. The additional angle discrimination may come from sequential lobing, mainbeam centroiding (if the beam is narrow enough), monopulse or other multiple phase center techniques (STAP). Since update times are slow in WAS, track while scan must be used and inevitably it is inaccurate if the group ground speed is high. But since only groups of targets are being tracked and their extent may be a mile, this inaccuracy is not too detrimental to the mode objectives. Stationary movers will be accurately located over multiple scans and then easily censored.

---

**Example 6.3** Consider the following requirements for a hypothetical WAS design (Tables 6.3 and 6.4).

*Table 6.3 Example 6.3 parameters*

| Parameter | Value |
|---|---|
| Scan coverage | 120° × 50–200 km (similar to Figure 6.15) |
| GMT maximum range resolution | 6 m |
| Target velocity range | 4–132 kmph |
| Target cross section | 10–1000 m$^2$ |
| Revisit time | 1/min |
| Cross-range accuracy @ max. range | 100 m |
| Aircraft speed | 400 kts (~206 m/s) |
| Aircraft altitude | 30 kft (~9144 m) |

*Table 6.4 Example 6.3 additional parameters*

| Parameter | Comment |
|---|---|
| $R_{max} = 200$ km | Dictated maximum range |
| $\lambda = 0.03$ | Assumed above, X band |
| $A = 1.8$ m$^2$ | Assumed above, 0.3 × 6 m |
| $V_{tgtmin} = 1.1$ m/s | Dictated minimum radial velocity |
| $R_{cr} = 100$ m | Dictated, maximum allowable for tactical utility |
| $k_B = 1.38 \times 10^{-23}$ | Dictated by nature, Boltzmann's constant |
| $T_0 = 290$ K | Assumed standard temperature |
| $NF = 2$ dB | A good value for X band |
| $L = 4$ dB | A good value for all sources of loss |
| $\sigma_T = 10$ dBm$^2$ | Dictated minimum vehicle RCS (see Chapter 3) |
| $D_T = 0.2$ | Very little improvement for larger values due to eclipsing |
| $\theta_0 = 60°$ | Dictated worst-case scan angle from broadside |
| $N_{freq} = 2$ | Usually required for ambiguity and scintillation reasons |
| $c_{min} = 0.05$ | Typical antenna |
| $N_B = 1$ | All that's required in this case |
| $eff_n = 0.75$ | Optimistic performance gain for multiple CPI integration |
| $T_{GMTISNR} = 0.1091$ s | Calculated above for 550 beam positions |

A system architect might then propose a radar embedded longitudinally along the fuselage of an aircraft to obtain the aperture size necessary for the required range of velocities and cross-range accuracy. Although there are many configurations that might work, suppose the antenna was an X band electronically scanned array (ESA) gimbaled to allow looks to both sides of the aircraft with dimension 6 × 0.3 m in a radome on the belly of the aircraft.

Having settled on possibly the largest antenna that could be installed on the aircraft, the next most important decision is the required power. The radar equation form shown in (6.5) can provide an estimate of the required transmitter peak power. Note that both the azimuth footprint, $FP_{az}$, and projected antenna area, $A_e$, with electronic scan angle are significant drivers of the required power. Since the cross-range accuracy needs to be a small fraction of the beamwidth, the SNR must be high, typically 30–40 dB.

$$P_T = \frac{4\pi \cdot FP_{az}^2 \cdot \lambda^2 \cdot R^4 \cdot k_B \cdot T_0 \cdot NF \cdot N_B \cdot L}{eff_n \cdot T_{GMTISNR} \cdot D_T \cdot A_e^2 \cdot \sigma_T \cdot R_{CR}^2}$$

$$FP_{az} \cong R_{max} \cdot \theta_{az} = \frac{R_{max} \cdot BW_0 \cdot (1 + c_{min})}{(\cos(\theta_0) + c_{min})}$$

$$A_e \cong A \cdot (\cos(\theta_0) + c_{\min}) \tag{6.5.1}$$

Because there is a revisit time requirement, the required power is driven by Doppler clear region detection and cross-range accuracy performance. With the assumed aperture, the beamwidth varies between 0.3725° at broadside to 0.745° at 60° (see (2.7) and (2.11)), thus 550 half beam steps are required to cover 120°. With a required 1 min revisit time, the maximum time per beam must be less than 0.1091 s. This would allow two 54 ms. CPIs plus some time for end of scan calibration, AGC set up and time for the initial pulses to arrive and return from maximum range. Equations (6.5.1) show a version of the radar equation, which derives required peak power from allowable dwell time and other radar parameters defined in (6.5.2).

where: $\psi$ is grazing angle, $\lambda$ is operating wavelength (m),
$FP_{az}$ is single beam azimuth ground footprint (m), $R$ is range (m),
$k_B$ is Boltzman's constant, $T_0$ is standard temperature,
$NF$ is noise figure, $N_B$ is number of sub-beams,
$P_T$ is transmitted peak power (w), $D_T$ is transmit duty ratio,
$A_e$ is antenna aperture effective area ($\mathrm{m}^2$), $\sigma_T$ is target cross section ($\mathrm{m}^2$),
$R_{CR}$ is cross range accuracy (m), $L$ is aggregate loss, $N_{freq}$ is number of independent frequency looks, $N_{PRF}$ is number of different PRF's,
$\theta_0$ is azimuth scan angle off antenna normal, $eff_n$ is integration efficiency
$R_{\max}$ is maximum operating range, $BW_0$ is antenna beamwidth at zero scan angle
$A$ is antenna area, $c_{\min}$ is maximum beamwidth constant, $T_{GMTISNR}$ is smaller of: dwell time to achieve desired SNR or maximum allowed dwell time. (6.5.2)

The following parameters are assumed or dictated by the requirement stated earlier:

Then calculating the elements of (2.7), (2.11) and (6.5) for the parameters in Example 6.3 yields

$$\theta_{az0} = BW_0 \cong 1.3 \cdot 0.03/6 = 0.0065 \text{ mr}, A_e \cong 1.8 \cdot \left(\cos\left(60^o\right) + 0.05\right) = 0.99 \text{ m}^2$$

then: $FP_{az} \cong 200000 \cdot \dfrac{0.0065 \cdot (1+0.05)}{\left(\cos\left(60^o\right)+0.05\right)} = 2482 \text{ m}$

$$P_T = \frac{4 \cdot \pi \cdot 2364^2 \cdot 0.03^2 \cdot \left(2 \cdot 10^5\right)^4 \cdot 1.38 \cdot 10^{-23} \cdot 290 \cdot 1.58 \cdot 1 \cdot 2.5}{0.75 \cdot 0.1091 \cdot 0.2 \cdot 0.99^2 \cdot 10 \cdot 100^2} = 1098 \text{ w} \quad \text{(ex 6.3)}$$

Thus, approximately 1098 W peak will be required to achieve the required cross-range accuracy with a 20% duty cycle. This calculation is optimistic because it assumes high-efficiency coherent integration. It does not take into account the losses (typically 3–5 dB) associated with ambiguity resolving and multiple frequency looks necessary for real target detection.

Ambiguity resolution is an important consideration, which depends on the antenna footprint in azimuth and elevation. Often, the main issue is range ambiguities when considering the desired range of unambiguous GMT Dopplers and corresponding PRFs. Range ambiguities in the mainlobe are determined by the elevation footprint. The larger the number of range ambiguities, the larger the number of PRFs required to resolve those ambiguities. Of course, all ambiguity resolution depends on a sparse number of targets in the mainbeam. The method for calculating the elevation footprint was given in (3.4). To find the elevation footprint of Example 6.3, consider Example 6.4.

---

**Example 6.4** Substituting the values from Example 6.3 and assuming the ground range in the beam center is 81.75 km, then the elevation footprint using (3.4) would be

$$\varepsilon_{el} \cong 1.3 \cdot 0.03 / 0.3 = 0.13 \text{ radians}$$

$$R_g = 81.75 \text{ km}, h = 10 \text{ km}, R_e = 8495 \text{ km (4/3 earth)}$$

$$R = \sqrt{10^2 + 2 \cdot \left(8495^2 + 8495 \cdot 10\right) \cdot \left(1 - \cos\left(\frac{81.75}{8495}\right)\right)} = 82.41 \text{ km}$$

$$\varepsilon = \text{acos}\left(\frac{8495}{82.41} \cdot \sin\left(\frac{81.75}{8495}\right)\right) = 0.1265 \text{ radians}$$

$$\psi_{heel} = \text{acos}\left[\left(1 + \frac{10}{8495}\right) \cdot \cos\left(0.1265 + \frac{0.13}{2}\right)\right] = 0.1853$$

$$\psi_{toe} = \text{acos}\left[\left(1 + \frac{10}{8495}\right) \cdot \cos\left(0.1265 - \frac{0.13}{2}\right)\right] = 0.0378$$

$$R_{g-heel} = 8495 \cdot \left(0.1265 + \frac{0.13}{2} - 0.1853\right) = 52.67 \text{ km}$$

$$R_{g-toe} = 8495 \cdot \left(0.1265 - \frac{0.13}{2} - 0.0378\right) = 201.33 \text{ km}$$

$$FP_{el} = 201.33 - 52.67 = 148.66 \text{ km} \quad \text{(ex 6.4)}$$

which is just adequate to stretch from desired minimum ground range of 50 km to desired maximum ground range of 200 km. If one chooses to have a high enough PRF to prevent any GMT Doppler ambiguities at X band, then the PRF will be about 5 kHz. Recalling (6.2), there will be approximately five range ambiguities in the mainlobe footprint for a 5 kHz PRF (~30 km/ambiguity). Clearly multiple PRFs will be required to resolve these range ambiguities. Depending on the expected density of GMTs, at least 4–8 separate PRF or frequency "looks" will be required to resolve GMT detections to a single-range bin.

Another important consideration is range closure per CPI. When range resolutions are high, the range bin that contains a single target must open or close with the radar motion as given by (6.6).

$$\begin{aligned} \Delta R_{CPI} &= CPI \cdot V_{tLOS} = CPI \cdot (V_a \cdot \cos\eta_t - V_t \cdot \cos\varsigma_t) \\ \Delta R_{CPI} &\cong CPI \cdot (V_a \cdot \cos\theta \cdot \cos\varepsilon \pm V_{tmax} \cdot \cos\psi) \end{aligned} \quad (6.6)$$

---

---

**Example 6.5** For Example 6.5, suppose the radar-carrying platform is traveling at $V_a = 204$ m/s, the maximum target velocity is $V_{tmax} = 36.6$ m/s, grazing angle $\psi = 5.5°$ and the look angle off the velocity vector is $\theta = 30°$, $\varepsilon = 6°$ and the $CPI = 54$ ms (see earlier Example 6.2). Then

$$\begin{aligned} \Delta R_{CPI} &= 0.54 \cdot (204 \cdot 0.866 \cdot 0.9945 \pm 36.6 \cdot 0.9954) \\ \Delta R_{CPI} &= 11.45 \text{ or } 7.52 \text{ meters} \end{aligned} \quad \text{(ex 6.5)}$$

Since the assumed range resolution in Example 6.3 was 6 m, range closure compensation to keep a target return in the same range cell will improve integration efficiency and reduce detection losses. The most common way to compensate for range closure is to slightly advance or retard the range bin sampling instants at each PRI by a time amount given by (6.7). In modern digital radars, there is a PEG that creates most of the fine timing. It is easy to adjust range sample times so that each radar pixel stays on the same piece of terrain throughout the CPI even through maneuvers. Some SAR processing (Chapter 7) goes to great lengths to compensate for range closure in the matched filtering when just moving the sample instants is so much easier.

$$\Delta t_{rc} = \frac{2 \cdot PRI \cdot V_a \cdot \cos\theta \cdot \cos\varepsilon}{c} \quad (6.7)$$

---

---

**Example 6.6** Using the parameters of Example 6.5, then the range closure compensation for the look angle given earlier and a 5 kHz PRF would be

$$\Delta t_{rc} = \frac{2 \cdot 0.0002 \cdot 204 \cdot 0.866 \cdot 0.9945}{2.99 \cdot 10^8} = 0.235 \text{ nanoseconds/PRI} \quad \text{(ex 6.6)}$$

Although this seems like a small amount per PRI, if uncompensated, it will result in an additional potential range-Doppler bin straddling loss of several dB.

Each CPI interval may be at a different frequency and/or PRF with perhaps four different waveforms in two half beam steps. With half beam steps, one would expect to see most "semi-fast" and "fast" GMTs in at least three out of four CPIs as the beam crosses a target location. Thus the detection criteria is often set at 3 of 4. Subsequent to detection the targets must be range resolved and placed in their estimated true range-Doppler location based on the three or four successful detections. If there are too many targets detected relative to the total number of bins, then more PRFs or frequencies in shorter CPIs are necessary to resolve target location. There might be $10^6$ ambiguous range-Doppler bins, and if there are 2000 targets detected per full beam step, then the probability of false correlations will approach 1%. This is probably acceptable for a WAS mode but unacceptable for a target tracking mode.

## 6.7 GMTI fast target detection and track [15]

### *6.7.1 Overview*

Detection and tracking of GMTs in the near mainlobe or Doppler "clear" (antenna sidelobe) region is often called fast-moving target detection and track (FMTT). Fast target detection and track is primarily oriented toward GMTs whose speed in ground coordinates is above 30 kmph. (The GMT LOS velocity could be as low as 5–10 kmph, see Figure 3.28.) These GMTs are most likely traveling on roads and their cross-range accuracy can be significantly improved by pushing the targets onto the closest road.

Just as in WAS mentioned in the previous section, the SNIR must be high enough for cross-range accuracy to be about 100 m or less; otherwise, there will be ambiguity as to which road to push the target return onto. Furthermore, the region of interest or "patch" for tracking of target clusters usually is much more restricted – typically "patch" sizes range from $4 \times 4$ km to $20 \times 20$ km. This operating mode is often interleaved with WAS. Some number of patches, limited by power-aperture product scheduling, are periodically illuminated to maintain tracking on target clusters.

Figure 6.17 shows a typical display from FMTT for an $8 \times 8$ km patch. The radar display of Figure 6.17 shows annotation data across the top. There is a rectangular $2 \times 2$ km grid and primary/secondary road underlay with north at the top of the display. The trapezoidal $8 \times 8$ km box shows the radar illumination of the patch. In this particular example, two radar beams are required to illuminate the $8 \times 8$ km patch and the centers of those two beams are shown as cross-shaped ground stake indicators. The display shows three target clusters in track pushed onto the nearest roads. The dotted boxes show the extent of the three clusters with a circled cross showing the centroid as well as a Beaufort symbol giving the approximate direction and velocity vector. Note also that one of the three clusters

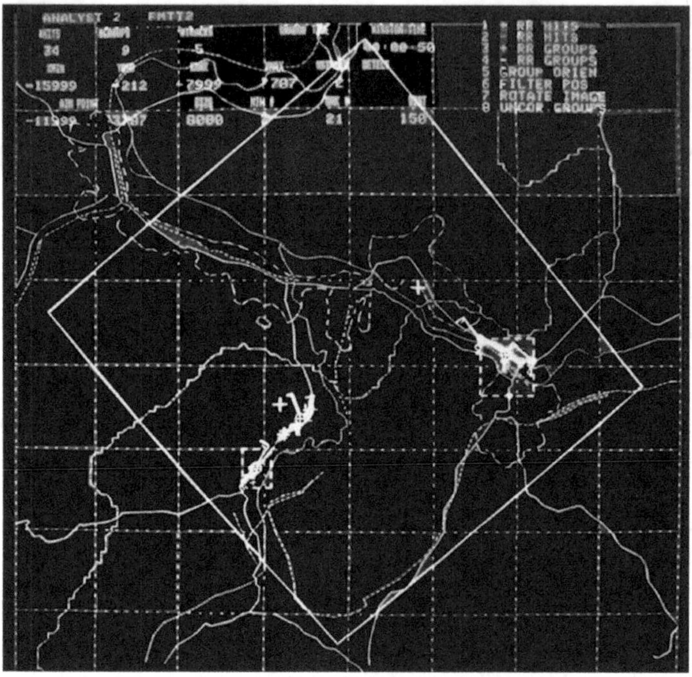


*Figure 6.17 Fast-moving target detection and track [13]*

(rightmost) has an obvious misassociated target traveling down a different road than the rest of the cluster. This happens for short periods on many clusters as a vehicle leaves a cluster or was traveling across the cluster trajectory.

When this mode is performed from space, the relative ground velocity is higher depending on orbital altitude, which necessitates a much narrower antenna beamwidth for the same Doppler spread and number of range ambiguities. Also, the requirement for 100 m cross-range accuracy to allow tracking and target discrimination requires a high gain narrow antenna beamwidth. It often turns out that multiple beams are required to cover the patch. Figure 6.18 shows an example multibeam laydown with its corresponding ground stakes.

As previously mentioned, there are several common types of GMT cross-range geolocation; many of them are based on using one or more of the following: DTED, digital elevation model (DEM) and digital line graphs (DLG). Modern GIS databases have multiple layers that contain terrain elevation on a grid of ground stakes (DEM or DTED) and cultural features. The map underlay in Figure 6.17 is DLG showing roads, political boundaries, natural features including rivers and dry waterways.

One method using such data (DLG) is shown in Figure 6.19. An error ellipse and its corresponding eccentricity are calculated for each detected target based on range, angle and Doppler centroiding and estimated SNIR. If the ellipse eccentricity is less than some arbitrary threshold (e.g. 0.866 relative), the minimum perpendicular distance is calculated for road segments inside the three times

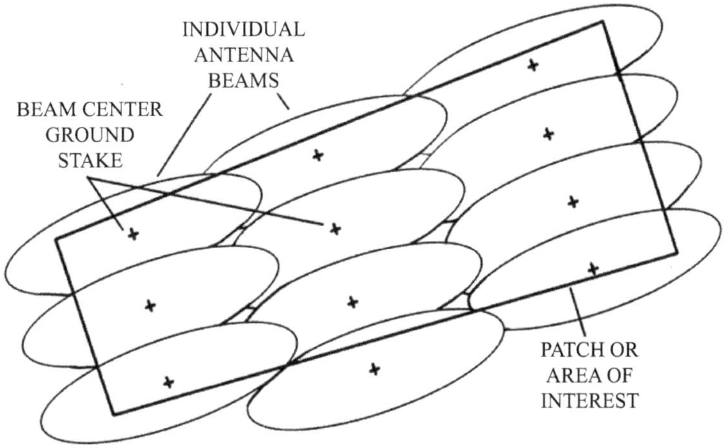


*Figure 6.18 Multibeam laydown to cover FMTT patch [16]*

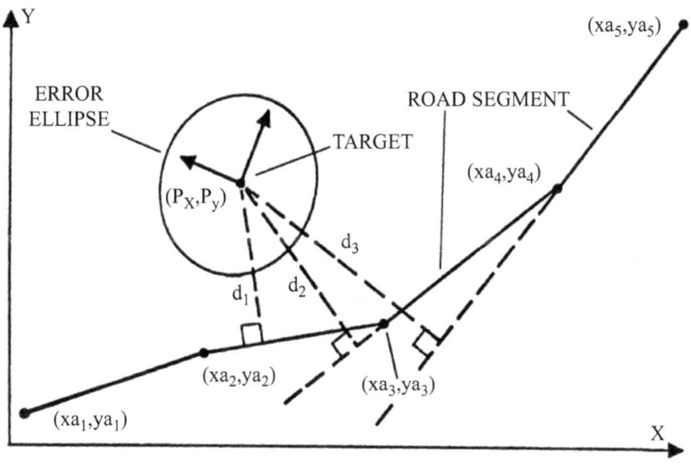


*Figure 6.19 Cartographic-assisted GMT geolocation [16]*

standard deviation $(3 - \sigma)$ ellipse. As shown in the figure, the perpendicular intercept may not lie inside the road segment (e.g. $d_2$ and $d_3$ in Figure 6.19) and will be discarded. The minimum distance for a valid road segment will be selected as the GMT location. If the eccentricity is greater than the threshold, the road segments that have a major ellipse axis intercept inside 3 sigma are compared and the minimum distance is selected. This accuracy improvement method can be used before or after measurements are entered into a tracking filter and sometimes it is made a radar operator option.

Obviously, some other screening must also be applied such as prior hits at that same location (probable fans, ventilators and wind turbines) or Doppler extent. For

*Table 6.5 Example FMTT parameters [13]*

| Parameter | Value |
|---|---|
| PRF | 3.5–7 kHz |
| Maximum average power | 1000 W |
| Pulse compression | 2048:1 P-codes |
| Range resolution | 6 m |
| Duty ratio | 20% |
| Frequency channels | 2–4 |
| Maximum range | 200 km |
| Range bins processed | 1536 |
| Doppler filters | 1024 |
| Antenna scan rate | Ground stabilized |
| Azimuth coverage | ±60° |
| Elevation coverage | 0°–70° |
| Frame time | 0.2–5 s |
| Total bandwidth | 250 MHz |
| Operating band | ~8–10 GHz |

example, some roads cannot support high speeds (fan?) and tanks (narrow spectrum) do not have to be on roads, etc.

Table 6.5 summarizes the leading parameters for a typical FMTT mode. Transmitted average power may be 1000 W. PRFs may be ambiguous in both range and Doppler over all angles but unambiguous in Doppler in the mainbeam and near sidelobes. PRFs encoded with P codes are often used because they can achieve canceling sidelobes for a few range ambiguities. A 2048:1 PC code might be used and 1536 range bins retained for tracking. A 6 m range cell size might be chosen to match the smallest vehicle of interest and break GMTs out of clutter. As usual high duty ratios and multiple frequency channels would be used. A FMTT mode may have $1536 \times 512 \times 3$ processed bins (range × Doppler × sum, delta, guard) per CPI. Antenna illumination usually is ground stabilized and may have 2–5 beam positions depending on range. Corresponding frame times might be 0.2–5 s. Total bandwidth might be 25 MHz per CPI with multiple channels available to be used for avoiding interference or ambiguity resolution.

A processing block diagram for FMTT is shown in Figure 6.20. Although there are alternate ways to perform minor edge-of-mainlobe-clutter processing, a monopulse-based processing scheme is given in Figure 6.20. Sum and difference monopulse channels for multiple frequencies are digitized and pulse compressed. Periodic calibration signals are used to create a gain and phase correction table for all frequencies, antenna beam steering and monopulse channels, which is then applied to the digitized measurements. Motion compensation to a fraction of a wavelength for platform deviations is applied to the data. High-resolution Doppler filtering is performed in a conventional FFT. Doppler filter outputs are used to form MLC discriminants for the final motion compensation. MLC is not in the same frequency location for each range bin and so filter output indices must be adjusted

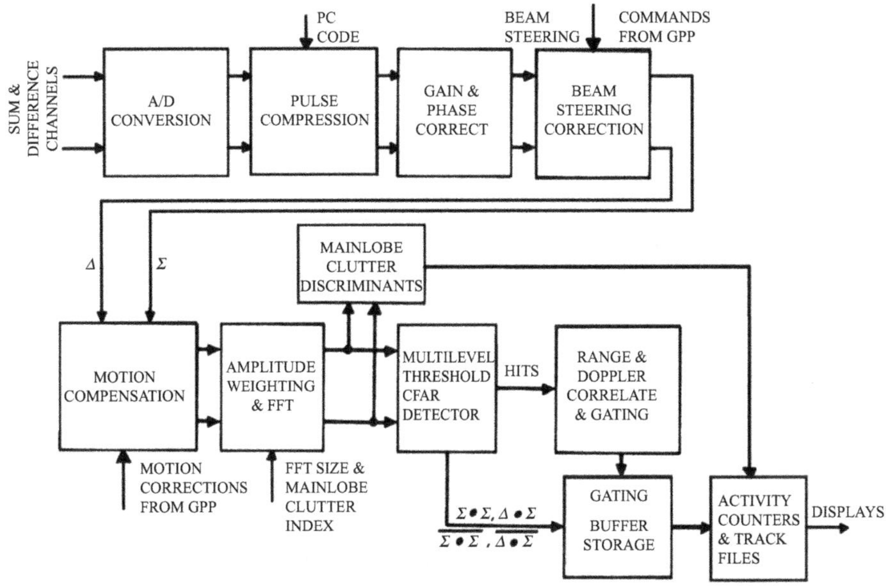


*Figure 6.20 Ground-moving target detection and track processing*

to present a common input to the threshold detector. The Doppler filter bank outputs are applied to a multilevel threshold detector similar to those described in the next section.

The multilevel threshold has several unique features. In addition to the alert-confirm properties, it also uses monopulse discriminants (or multiple phase centers if you prefer) as well as near sidelobe threshold multipliers. Threshold crossings are correlated in range and Doppler and buffered along with corresponding monopulse or phase center discriminants, which are presented to tracking filters or activity counters.

The processing functions are performed in two different architectures. A PSP usually performs PC, gain and phase correction, beam steering, motion compensation, Doppler filter bank (FFT), MLC discriminants, threshold detection, range-Doppler centroiding, correlation and buffering. A GPP calculates the required PRF and transmit frequency, PC code, beam steering commands, motion compensation, filter bank size and MLC filter output index and all the target tracking filters.

## *6.7.2 GMT thresholding*

A general concept for GMT thresholding is shown in Figure 6.21. There are three regions of thresholding: mainlobe, fringe (mainlobe edge and near sidelobe) and Doppler clear. Near-surface targets of interest will often have radial velocities of a few kilometers per hour for long periods of time, which forces the detection of GMTs well into edges or middle of MLC. Monopulse or multiple phase center

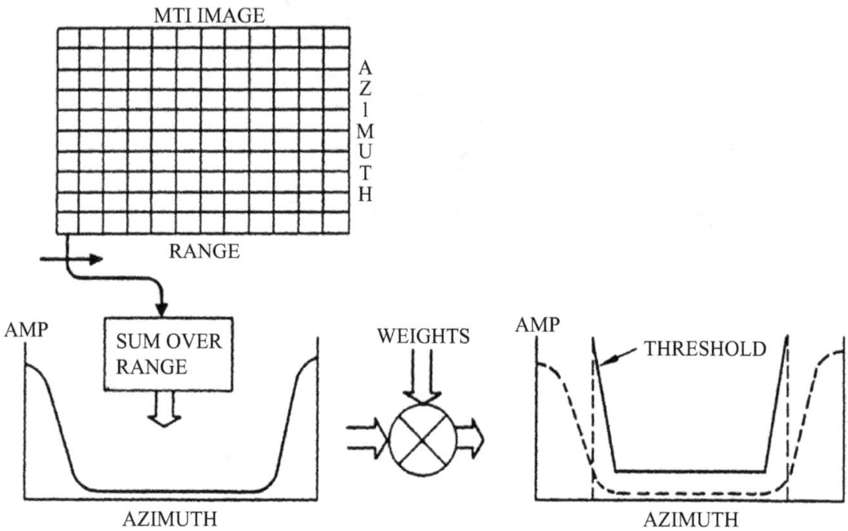


*Figure 6.21 Variable threshold concept*

processing allows the first-order cancellation of clutter for many slow-moving targets. Unfortunately, clutter does not always have well-behaved statistical tails and to maintain a constant false alarm rate the threshold must be raised for endo-clutter targets. The output of the Doppler filter bank might be thought of as a 2D array range-Doppler image. After index adjustment, the amplitude versus Doppler profile in any range bin is as shown in the upper right in Figure 6.21. By summing over segments of range, an estimate of clutter amplitude can be formed and then weighted by assumed clutter statistical behavior to form the threshold profile shown in the lower right of the figure. There may still be parts of MLC that are not used for "fast" GMT detection. As previously mentioned, MLC is used to form a coarse DBS map and to calculate Doppler centroiding for use as motion compensation position and velocity estimates.

An example actual thresholding scheme based on this concept is shown in Figure 6.22. The range-Doppler space is broken up into a grid. Each cell in the grid is perhaps 64 × 64 bins with 384 grid cells total. The grid locations spanning MLC are used for forming MLC Doppler discriminants and a coarse Doppler-beam-sharpened map to improve the accuracy of the patch ground stake. The bins in each grid cell are ensemble averaged (*EA*) in sum and difference for each frequency channel. There are usually overlapping grid cells as suggested in Figure 6.22 upper-left clear region. There may also be estimates of higher moments to allow elimination of jamming, discretes, rain and chaff. The power in each bin in a grid cell in the clear region is compared to a threshold, $P_{TH1}(EA)$, which is a function of the *EA* in that grid cell. In the sidelobe fringe region, a discriminant, $C_s$, is formed in addition. It is used to provide monopulse sidelobe/fringe clutter cancellation prior

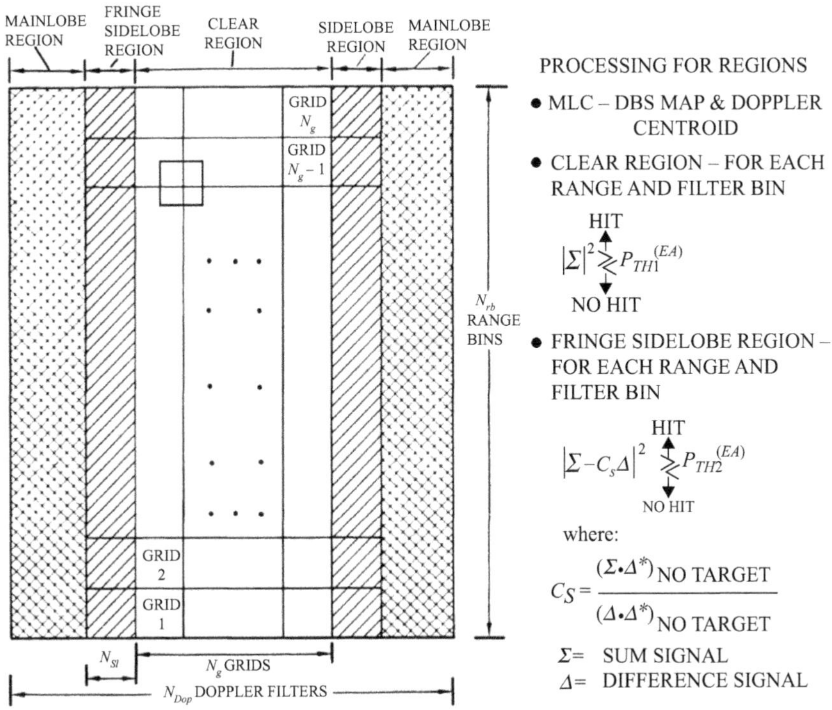


*Figure 6.22 Multiregion GMT thresholding (no STAP)*

to thresholding. Note that this formulation is different than would be used in mainlobe monopulse clutter cancellation (shown later in this chapter). In the sidelobe/fringe region, the difference, Δ, is larger than the sum, Σ, so the clutter cancellation weighting estimate, $C_S$, is applied to the difference channel. (See Section 6.8, Figures 6.34 and 6.36.) In the mainlobe slow target clutter region and after further processing, the Σ is larger than the Δ and so the clutter cancellation weighting is applied to the sum channel. (More about this in a later section.)

Again, the threshold, $P_{TH2}(EA)$, is a function of the *EA* around the cell under test and *a priori* knowledge of the clutter statistics. Clearly, if the sum SNIR is low ($<2$ typical) or the difference to sum is too high ($>1.5$ typical), the cell is not used in the *EA*. Although only one threshold is described, two are often used before hits and their counterpart monopulse discriminants are passed to the track files. All low threshold hits are passed to activity counters. The low threshold is usually set for a false alarm rate of $10^{-3}$. As complex as this thresholding scheme seems to be, it is very power efficient with low detection losses [1].

A contour plot of a simple 2D sliding window threshold scheme including real clutter and simulated moving targets in mainlobe, sidelobe and clutter-free regions in three separate range bins is given in Figure 6.23. Many additional examples are given in the Appendix A.6 DVD-R or internet download.

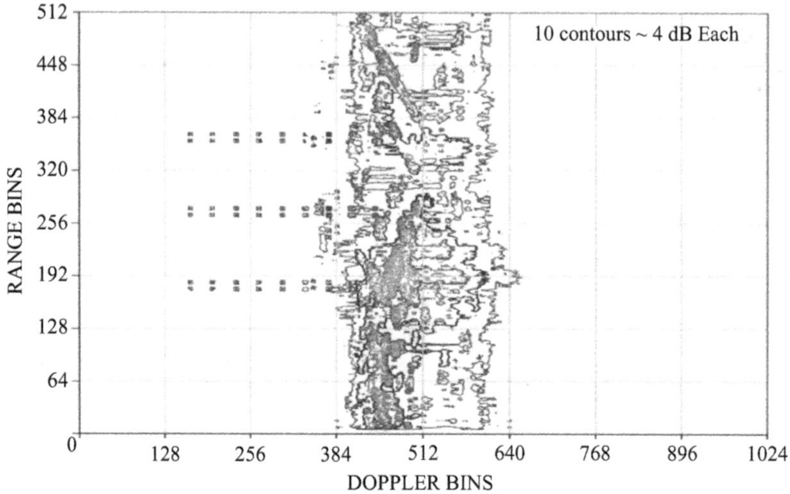


*Figure 6.23 2D sliding window thresholds shifted to PRF/2 (repeated Figure 4.20) (GMTIStLouis001)*

### *6.7.3 GMT tracking*

There normally are so many near GMTs that they all must be tracked even in search to select the interesting movers. Figure 6.20 alludes to this by showing a track function in the FMTT block diagram. Typically, the movers are sorted out by their signature and movement characteristics. For example, manmade targets are usually moving in a specific direction, whereas birds and insects tend to change directions abruptly and stay in the same area. Ventilators stay in the same geographic location even though they may be moving fast and helicopters have high subsonic moving parts but move much more slowly. It is not uncommon to have hundreds or thousands of near surface moving targets in track simultaneously.

An overview of the major functions GMT tracking is shown in Figure 6.24. Since vehicle tracks may often cross each other as well as disappear behind structures and terrain features, radar observations and correlations with prior tracks are essential. Because terrain can mask targets, they often will abruptly appear or disappear, which requires sophisticated track initiation and coasting functions.

The track gate acceptance functions are conditioned on the target SNR and similarity with current tracked target clusters. Once observations are associated with existing tracks, they are used to update individual Kalman filters for each target. The parameters of high-priority targets are provided to displays as well as used to select the transmitted waveform to maintain high-quality tracks on those targets.

Experience has shown that GMT tracking is best done in a ground-referenced coordinate system (e.g. NED). This is true for several reasons: first, most targets

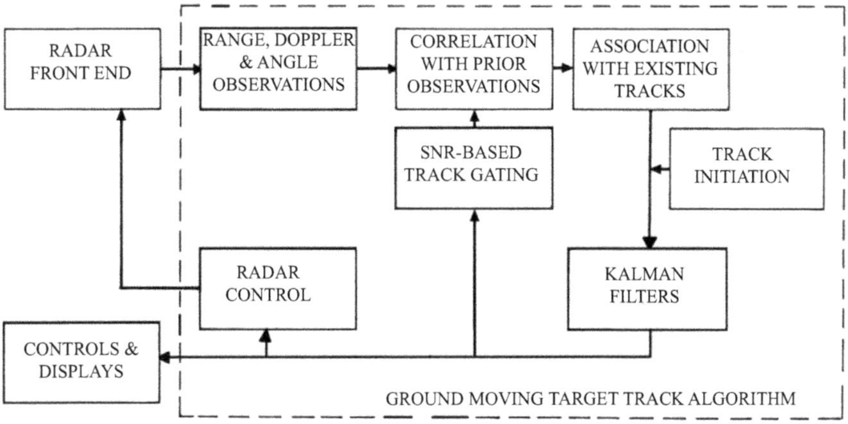


*Figure 6.24 GMT track overview [4]*

are on or so near the ground that they are ground stabilized; second, geographic (hypsographic) databases are often used to correct the location of target tracks; and third, target speed relative to ground features is used to recognize target signatures (e.g. surface vehicles cannot travel in road-less country at 70 mph but helicopters can). A block diagram of all the elements of ground-stabilized tracking is shown in Figure 6.25. First, the observing platform navigation data in all six degrees of freedom (6 DOF, 3 position and 3 attitude) must be known and extrapolated to the target measurement time for input to the Kalman filter. Next track measurements in range, range rate, altitude, azimuth and elevation as well as noise estimates are transformed into ground-stabilized coordinates. Then track coordinates and track errors are input to Kalman track filters. The track filter output is propagated to the next forecast measurement time. These estimated outputs are transformed into radar coordinates ($R$, $\acute{R}$, $AZ$, $EL$), subtracted from the next measurement and used to form error terms for the next input to track filters. Since most targets are on or near the ground, the known ground altitude ($h_G$) from the hypsographic database is subtracted from the tracker estimated altitude ($h_T$) to form an altitude error term.

The required antenna beam position in ground-stabilized coordinates for the next set of measurements as well as the apparent position, velocity and acceleration is calculated ($P_F$, $V_F$, $A_F$) for a fixed ground location at beam center (often called the ground stake). The antenna is then pointed in that direction and, since the actual measurements are in radar-based angle and range coordinates, the apparent dynamics of the fixed ground stake ($R_F$, $\acute{R}_F$, $AZ_F$, $EL_F$) are added with suitable transformations to the current measurements to provide an estimate of target location in earth-stabilized dimensions. Usually the ground stake is known with highest precision because it can be correlated with stored terrain data using DBS or SAR. There will be atmospheric refraction along the LOS, some of it will be anomalous. Without reference to a correlated ground stake in an accurate GIS database, target estimated positions may have additional errors above that caused by SNIR limits of over 100 m depending on LOS path length. Tracking everything relative to the ground stake gives best results. The radar and signal processor are provided with calculated parameters from the radar control function (usually a general purpose processor (GPP)) to allow proper waveform and processing of the expected GMTI image. The processor provides the next set of target measurements (discriminants) in radar-based coordinates to repeat the whole cycle.

### 6.7.4 GMT Kalman tracking filters

More details of the Kalman track filters and track propagation described in Figures 6.24 and 6.25 are given in Figures 6.26 and 6.27. Figure 6.26 shows the

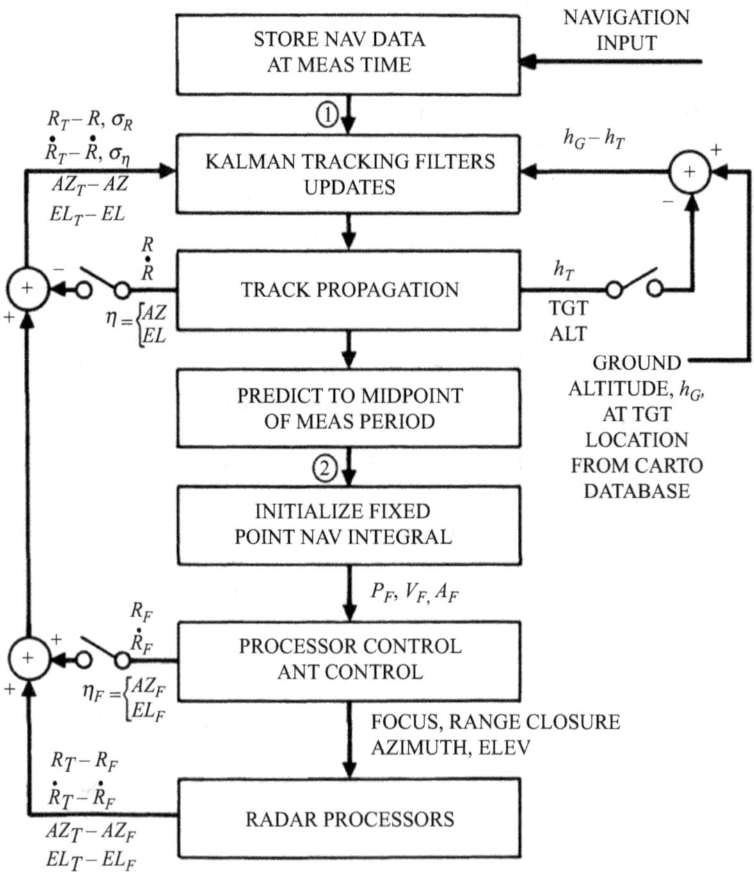


*Figure 6.25 Example GMT tracking block diagram [4]*

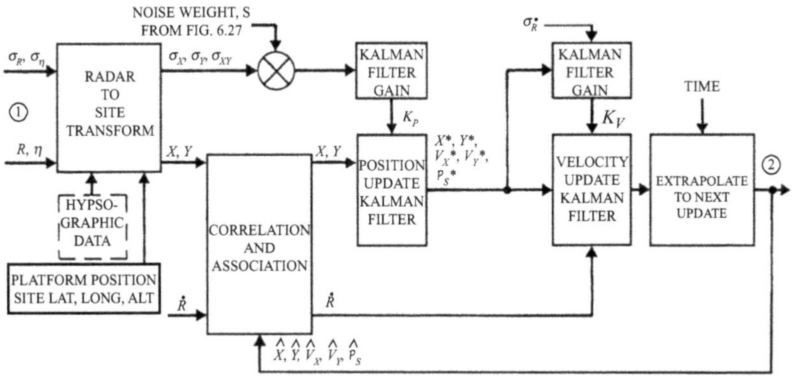


*Figure 6.26 GMTT Kalman filter diagram [4]*

details between point 1 and 2 in Figure 6.25. The natural tendency is to lump all these filter functions into a single Kalman filter with a large number of degrees of freedom (i.e. everything but tide tables and prayer times). This usually results in filters, which are very unstable. Alternatively, more stable tracking is achieved by having dedicated Kalman filters, which only track position or velocity. Often, multiple Kalman filters are used that hypothesize a specific class of movement (e.g. around a corner, straight line, sinusoidal in N-E, etc.) and the lowest variance is selected. The simple filter of Figure 6.26 is still reasonably complex considering that 1000s usually are running simultaneously. Radar range, *R*, composite angle, $\eta$, and estimated standard deviation, $\sigma_R$, $\sigma_\eta$, are transformed to ground-stabilized site coordinates, ***X***, ***Y***, conditioned on vehicle position and attitude as well as the hypsographic database. Ground coordinates are provided to a position Kalman filter after being gated by their forecast value. Their counterpart cross-correlation, $\sigma_{XY}$, and standard deviations, $\sigma_X$, $\sigma_Y$, conditioned by a noise weight based on the average of all the prior track associations of the current target. The weighted outputs are used to set the Kalman filter gain for the current measurement. The Kalman filter outputs are estimates of position, velocity and the error covariance matrix $P_S$. Smoothed position, velocity and error covariance estimates are provided to the velocity Kalman filter. Again, gain is conditioned on velocity standard deviation, range rate standard deviation and covariance estimates. The velocity filter outputs as well as propagated positions are provided to an extrapolator based on the expected time of the next radar measurement. The extrapolated position and velocity conditioned on power-based variation is used to form the correlation and association window for the next measurement. Then the cycle begins again for each of several hundred targets (or 1000s) that might be in track.

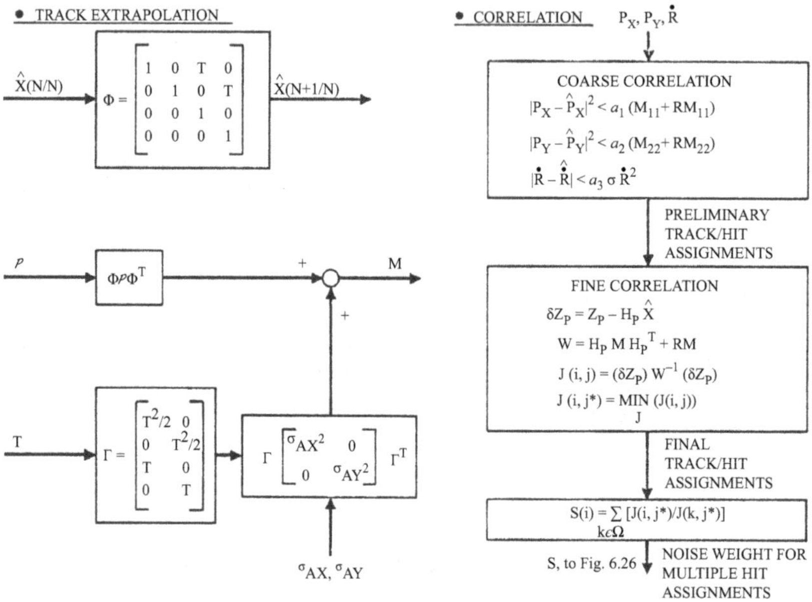


*Figure 6.27 GMTT track and correlation [4]*

The details of two functions correlation and extrapolation shown in Figure 6.26 are given in Figure 6.27. The outputs are the expected next state of **X**, the forecast measurement error matrix **M** and the track association noise weight **S**. There are several other terms that require some explanation, $\delta \mathbf{Z_P}$ is the normalized distance from existing target track to the current measurement, $R$ is the measurement covariance in the site coordinates. The $\boldsymbol{a}_i$s are threshold multipliers that reduce false correlations. The detection is assigned to the track, which has the shortest normalized distance to a current track. $\mathbf{H_P}$ is the position measurement selection matrix that selects the positions $\mathbf{P_x}$ and $\mathbf{P_y}$ from the position and velocity outputs from the first position Kalman filter. The platform x and y position measurement standard deviations are $\sigma_{AX}$ and $\sigma_{AY}$. The error covariance matrix is $P$.

## 6.7.5 GMT dwell times

Usually moving target tracking is interleaved with moving target detection in the same radar so that time management is an important issue. The time necessary to form a filter to detect the minimum required target velocity is given in the first line of (6.8). In addition, the dwell time required to achieve an adequate SNR for detection of the minimum cross-sectional target is given in the second line of (6.8). Either one of these times may be larger depending on the situation geometry.

The total dwell time ($T_{total}$) is the larger of ($T_{GMTIFilt}$) or ($T_{GMTISNR}$) times the number of independent frequency ($N_{freq}$) or PRF ($N_{PRF}$) looks necessary to resolve ambiguities, which is shown in the third line of (6.8).

$$
\begin{aligned}
&T_{GMTIFilt} = \frac{\lambda}{2 \cdot V_{tgt\min} \cdot \cos\psi} \\
&T_{GMTISNR} = \frac{4\pi \cdot FP_{az}^2 \cdot \lambda^2 \cdot R^4 \cdot k_B \cdot T_0 \cdot NF \cdot N_B \cdot L}{P_T \cdot D_T \cdot A_e^2 \cdot \sigma_T \cdot R_{CR}^2} \\
&T_{total} = (\text{The greater of } T_{GMTIFilt} \text{ or } T_{GMTISNR}) \cdot N_{freq} \cdot N_{PRF} \\
&\text{where: } V_{tgt\min} \text{ is required minimum detectable velocity(m/s),} \\
&\psi \text{ is grazing angle, } \lambda \text{ is operating wavelength(m),} \\
&FP_{az} \text{ is single beam azimuth ground footprint(m), } R \text{ is range(m),} \\
&k_B \text{ is Boltzman's constant, } T_0 \text{ is standard temperature,} \\
&NF \text{ is noise figure, } N_B \text{ is number of sub-beams,} \\
&P_T \text{ is transmitted peak power(w), } D_T \text{ is transmit duty ratio,} \\
&A_e \text{ is antenna aperture effective area(m}^2\text{), } \sigma_T \text{ is target cross section(m}^2\text{),} \\
&R_{CR} \text{ is cross range accuracy(m), } L \text{ is aggregate loss, } N_{freq} \text{ is number} \\
&\text{of independent frequency looks, } N_{PRF} \text{ is number of different PRF's.}
\end{aligned}
\tag{6.8}
$$

---

**Example 6.7** For Example 6.7, assume an airborne GMTI radar has the parameters listed in Table 6.6.

Then using (6.8) the required dwell time can be estimated. Figure 6.28 shows the dwell time required either by minimum Doppler detection or by required SNR for cross-range accuracy. The graph shows that out to about 90 km the dwell time is determined by the minimum detectable velocity. From there on, dwell time is determined by required cross-range SNR.

*Table 6.6 Aircraft Example 6.7 parameters*

| Parameter | Value | Parameter | Value |
|---|---|---|---|
| Operating altitude | 7 km | No. PRFs | 1 |
| Platform velocity | 250 m/s | Minimum grazing angle | 0.0524 rad |
| Peak power | 5 kW | Min. target radial speed | 4 km/h |
| Duty | 0.1 | Maximum range | 115 km |
| Antenna aperture | 3 $m^2$ | Azimuth footprint width | 54–856 m |
| PRF | 3600 Hz | Front-end noise power | $10^{-20}$ W |
| Cross-range accuracy | 100 m | Losses | 2.75 dB |
| Target RCS | 10 $m^2$ | No. beams | 1 |
| Wavelength | 0.03 m | No. frequencies | 2 |

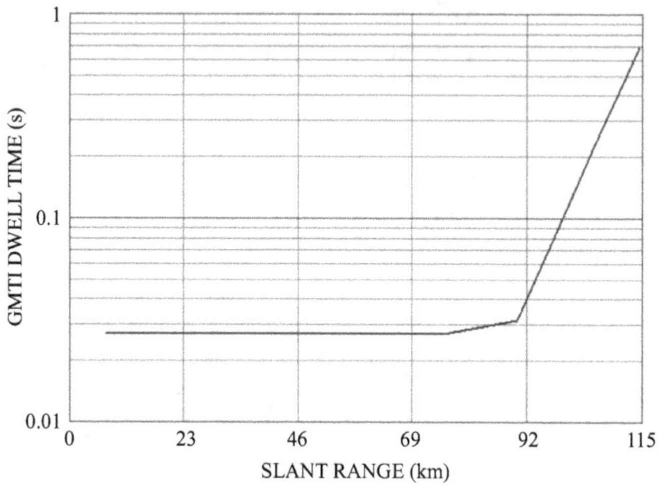


*Figure 6.28 Airborne GMTI dwell times (AircraftRdr001)*

---

**Example 6.8** In a second Example 6.8, assume a spacecraft has a GMTI radar with the following parameters (Table 6.7).

Figure 6.29 shows the dwell time required either by minimum Doppler detection or by required SNR for the cross-range accuracy. The graph shows that out to about 1500 km the dwell time is determined by the minimum detectable velocity. From there on, dwell time is determined by required cross-range SNR. If the same cross-range accuracy (100 m) required for the airborne case were required for a space platform, the maximum range dwell time would be 57.6 s. Even for slow-moving targets, this will require tracking each moving target to achieve the required cross-range accuracy. Cross-range accuracy of 800 m is just enough to assure successful placement of a 4 × 4 km box for a SAR map on the movers.

*Table 6.7 Spacecraft Example 6.8 parameters*

| **Parameter** | **Value** | **Parameter** | **Value** |
|---|---|---|---|
| Operating altitude | 1000 km | No. PRFs | 1 |
| Platform relative velocity | 7570 m/s | Minimum grazing angle | 0.105 rad |
| Peak power | 5 kW | Min. target radial speed | 4 km/h |
| Duty | 0.1 | Maximum range | 3100 km |
| Antenna aperture | 78.5 $m^2$ | Azimuth footprint width | 5.29–15.7 km |
| PRF | 3600 Hz | Front-end noise power | $10^{-20}$ W |
| Cross-range accuracy | 800 m | Losses | 2.75 dB |
| Target RCS | 10 $m^2$ | No. beams | 1 |
| Wavelength | 0.03 m | No. frequencies | 2 |

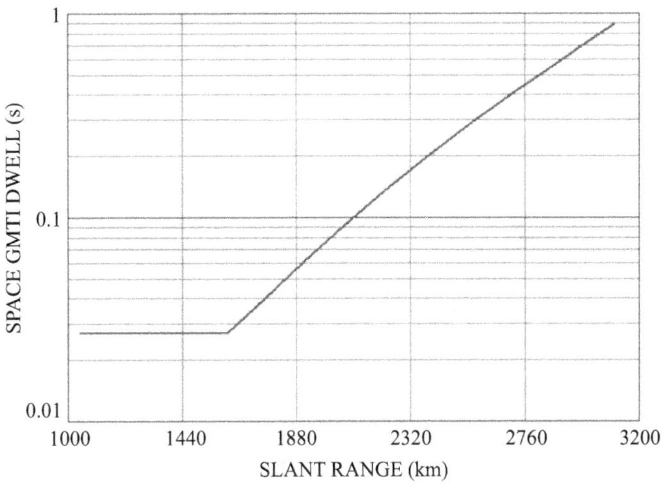


*Figure 6.29 Space GMTI dwell times (SpaceRdr001)*

## 6.8 Slow GMT detection by multiple phase centers

### 6.8.1 Slow GMT clutter phase history

Earlier in this chapter two classes of moving target detection have been described. The first is detection of GMTs in the Doppler clear region where the full effect of the antenna sidelobes suppresses most clutter and allows close to thermal noise-limited detection. The second is outside the 3 dB mainlobe into the first few antenna sidelobes where half beam stepping will allow GMTs to be well into the low sidelobes in the next half beam step. The third class is GMTs that are competing with the MLC and must be detected by multiple beams or phase centers, which allow angle discrimination coupled with filtering. One issue already mentioned with the detection of slow-moving surface targets is the fact that there are so many objects on the surface that have some movement. This fact requires tracking to suppress all of the semi-stationary movers.

First, consider Figure 6.30, at some range, R, there are some possible targets designated as 1, 3, 5 and each is at a slightly different composite angle, $\eta_n$, then, each target exhibits a total Doppler as a function of angle and time $f_d(\eta_n,t)$. Typically, a GMT is moving slowly along the LOS (even if it is a helicopter) relative to an aircraft or spacecraft. Then the Doppler target return is made up of three major components: the center frequency of the ground Doppler at the target angle, a slope term that is the rate of change of the ground geometry and finally the Doppler of the LOS velocity from the moving target. In Figure 6.30 only one of these terms is shown as time varying, which is approximately true for the typical dwell time of a scanning radar. Clearly over longer periods of time all are time varying. Each target exhibits a Doppler center frequency that is a combination of the ground LOS velocity and the moving target LOS velocity. Since the geometry for each of these returns is continually changing, the Doppler is continuously changing. Ideally, all targets should be match filtered to their true range-Doppler trajectory. Although the slope of a target at beam center can be completely compensated, the slopes of targets at 1 and 5 are much harder to compensate without significant additional complexity.

Also, as mentioned in Section 6.2, there is a Doppler ground clutter spread across the entire mainbeam $\Delta f_d$, which is also competing with the moving target returns through filter sidelobes. Examining this behavior in more detail, consider Figure 6.31. There is some range-Doppler bin at maximum range for a given surface patch under observation. Its vector is $\mathbf{R}_0(x,y,z,t)$. All the other range-Doppler bins are related to it by some offset vector $\Delta(x,y,z)$. Assume that the radar platform is traveling along the $x$-axis at an almost constant altitude above the earth of, $z = h$, and velocity $\mathbf{V_a}$, then the ground return in any cell is shown in (6.9). It is difficult to visualize what is going on with the phase and frequency equations in (6.9). Figure 6.30 suggests what the major elements of the frequency history for each range-Doppler cell will be. An approximation to (6.9) in the vicinity of a specific $R_0$ can be found using the first few terms of a Taylor series expansion of the phase with time. It is this Taylor series approximation that is alluded to in Figure 6.30.

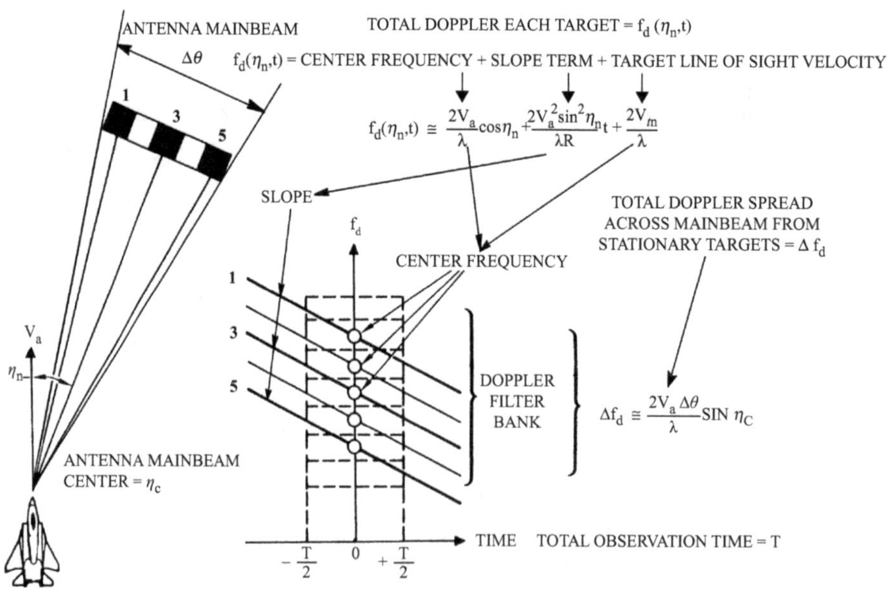


*Figure 6.30 Slow ground-moving target Doppler geometry [16]*

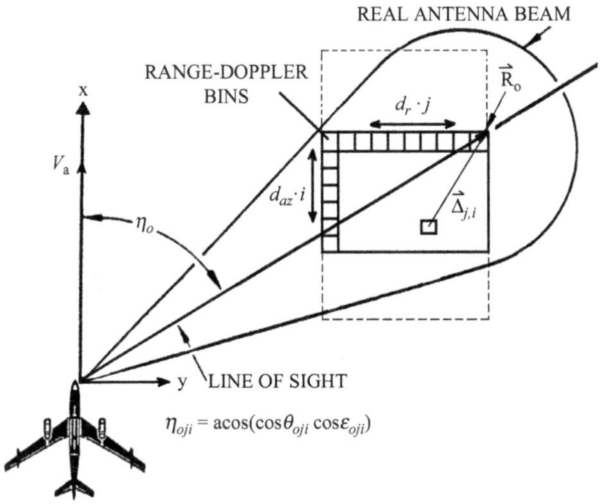


*Figure 6.31 Range-Doppler bin geometry including range closure*

Note that relative to the moving platform the position of each range-Doppler bin is changing throughout the observation time, which may require antenna scanning to stay on the required patch. There also will be amplitude modulation as a bin moves across the antenna beam as suggested by the dotted outline in Figure 6.31.

$$\mathbf{R}_0(x,y,z,t) = \mathbf{R}_0(x,y,z,0) - \mathbf{V}_a \cdot t$$

For the simple case of $\mathbf{V}_a$ in the x direction $V_{ax}$ and a flat earth then:

The initial distance to each range-Doppler bin in cartesian coordinates is:

$$R_{0\,j,i} = \sqrt{(R_0 \cdot \cos\theta_0 \cdot \cos\varepsilon_0 - i \cdot d_{az})^2 + (R_0 \cdot \sin\theta_0 \cdot \cos\varepsilon_0 - j \cdot d_r) + h^2}$$

Where: $R_0$ is the range to the farthest range bin, $h$ is the altitude,
$d_{az}$ is the azimuth (x) direction bin size, $d_r$ is the range (y) bin size,
$\theta_0$ is the initial azimuth angle, $\varepsilon_0$ is the initial elevation angle, and
$i$ is azimuth index to single cell, $j$ is range index to single cell.

The distance to each range-Doppler bin with time is approximately:

$$R_{j,i}(t) = \sqrt{(R_{0x} - V_{ax} \cdot t - i \cdot d_{az})^2 + (R_{0y} - j \cdot d_r)^2 + h^2}$$

Where: $R_{0x} = R_0 \cdot \cos\theta_0 \cdot \cos\varepsilon_0$ and $R_{0y} = R_0 \cdot \sin\theta_0 \cdot \cos\varepsilon_0$

The distance can also be expressed as:

$$R_{j,i}(t) \cong \sqrt{\left(R_{0_{j,i}} - V_{ax} \cdot t \cdot \cos\theta_{0\,j,i} \cdot \cos\varepsilon_{0\,j,i}\right)^2 + \left(V_{ax} \cdot t \cdot \sin\theta_{0\,j,i} \cdot \cos\varepsilon_{0\,j,i}\right)^2}$$

The phase with time is: $\varphi_{j,i}(t) = \dfrac{4 \cdot \pi}{\lambda} \cdot R_{j,i}(t)$ radians (6.9.1)

And the frequency with time is: $f_{d_{j,i}}(t) = \frac{2}{\lambda} \cdot \frac{dR_{j,i}(t)}{dt}$ Hertz

$$f_{d_{j,i}}(t) = \frac{2}{\lambda} \cdot \frac{-\left(R_{0x} \cdot V_{ax} - V_{ax}^2 \cdot t - i \cdot d_{az} \cdot V_{ax}\right)}{\sqrt{\left(R_{0x} - V_{ax} \cdot t - i \cdot d_{az}\right)^2 + \left(R_{0y} - j \cdot d_r\right)^2 + h^2}} \tag{6.9.2}$$

Assume that time zero occurs when the farthest range-Doppler bin is at $\eta_0$, a composite angle of azimuth $\theta_0$ and elevation $\varepsilon_0$ as shown in Figure 6.31. The phase with time can be approximated by the first few terms in a Taylor series expansion shown at the top of (6.10). By calculating the derivatives and evaluating those derivatives at time zero, one obtains an approximation of the phase history with time given in the bottom of (6.10). For MTI filtering this approximation is more than adequate. It is not close for high-resolution SAR, which requires the third-order derivative to be incorporated into the approximation. The time $t = 0$ is assumed to be in the middle of the array time for best accuracy.

Note that: $\eta_{0j,i} = \operatorname{acos}\left(\cos\theta_{0\,j,i} \cdot \cos\varepsilon_{0\,j,i}\right)$

Let $t = 0$ when $\theta = \theta_0$ and $\varepsilon = \varepsilon_0$ and expand $\varphi_{j,i}(t)$ in Taylor series

$$\varphi_{j,i}(t) = \varphi_{j,i}(0) + \left.\frac{d\varphi_{j,i}}{dt}\right|_{t=0} \cdot t + \frac{1}{2} \cdot \left.\frac{d^2\varphi_{j,i}}{dt^2}\right|_{t=0} \cdot t^2 + \frac{1}{6} \cdot \left.\frac{d^3\varphi_{j,i}}{dt^3}\right|_{t=0} \cdot t^3 + \ldots$$

$$\varphi_{j,i}(0) = \frac{4 \cdot \pi}{\lambda} \cdot R_{j,i}(0)$$

$$\left.\frac{d\varphi_{j,i}}{dt}\right|_{t=0} = \frac{4 \cdot \pi}{\lambda} \cdot \frac{-2 \cdot R_{0\,j,i} \cdot V_{ax} \cdot \cos\eta_{0\,j,i}}{2 \cdot R_{0\,j,i}} = \frac{-4 \cdot \pi}{\lambda} \cdot V_{ax} \cdot \cos\eta_{0\,j,i}$$

$$\left.\frac{d^2\varphi_{j,i}}{dt^2}\right|_{t=0} = \frac{4 \cdot \pi}{\lambda} \cdot \left[\frac{V_{ax}^2}{R_{0\,j,i}} - \frac{R_{0\,j,i}^2 \cdot V_{ax}^2 \cdot \cos^2\eta_{0\,j,i}}{R_{0\,j,i}^3}\right] = \frac{4 \cdot \pi}{\lambda} \cdot \left[\frac{V_{ax}^2 \cdot \left(\sin^2\eta_{0\,j,i}\right)}{R_{0\,j,i}}\right]$$

Then the first three terms of the Taylor series expansion of the phase are:

$$\varphi_{j,i}(t) \approx \frac{4 \cdot \pi}{\lambda} \cdot \left[R_{0\,j,i} - V_{ax} \cdot \cos\eta_{0\,j,i} \cdot t + \frac{V_{ax}^2 \cdot \sin^2\eta_{0\,j,i} \cdot t^2}{2 \cdot R_{0\,j,i}}\right] \tag{6.10}$$

---

**Example 6.9** For example, suppose that the most distant range-Doppler bin is 30 km, the platform altitude is 10 km, the platform velocity is 225 m/s, the azimuth angle is 60°, the elevation depression angle is 19°, the operating wavelength is 0.03 m and the patch size is 1024 m × 768 m. Then the phase behavior is shown in Figure 6.32 for the farthest range-Doppler bins as well as the approximation given in (6.10). (See Appendix A.6 RangeChangevsTime007d on DVD-R or internet download.)

It is obvious that not only does the frequency change for single bins but also the slope is different for the closest and farthest bins in a patch. Best performance is

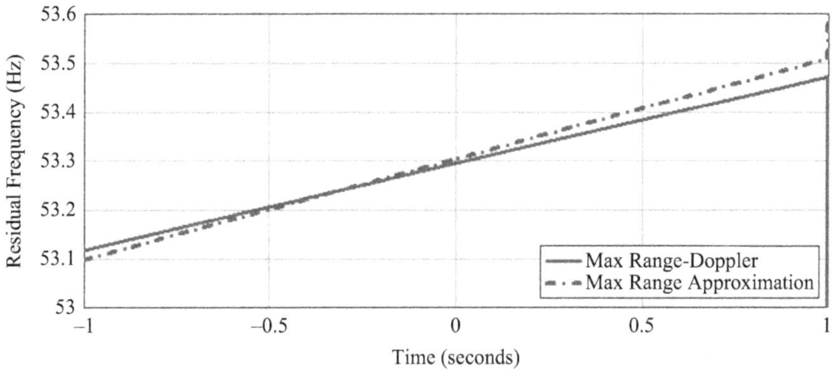


*Figure 6.32 Range-Doppler bins phase slope change (RangeChangevsTime007d)*

achieved if most of this known frequency change with range and angle is compensated. For best performance, all of the systematic Doppler from a surface clutter return should be compensated before any other Doppler signal processing.

### 6.8.2 Monopulse clutter cancellation [5,7,10,13]

The simplest multiple phase center detection scheme for slow-moving targets uses some form of monopulse or sequential lobing (pseudo-monopulse) cancellation. There are at least four schemes for canceling clutter in the mainlobe to detect slow-moving targets. Monopulse was introduced in Chapter 2. Since monopulse is already incorporated into many radar antennas (as well as in subarrays of ESAs), it is common to use it for clutter cancellation. These techniques have been in use for at least 50 years. Its performance is best if applied after Doppler processing, which requires multiple monopulse channels to be carried until the monopulse discriminants are formed. The clutter cancellation can either be done by monopulse discriminant curve fitting or by entirely signal-based adaptive cancellation. This is the earliest known (1950s) practical use of what we now call STAP. STAP uses both the time waveform and multiple phase centers to provide both angle of arrival and Doppler data for rejection of clutter and jamming. Most monopulse schemes result in almost equivalent performance. If the monopulse pattern is well characterized, a previously calculated or measured monopulse discriminant profile can be fitted in a least-square sense to the received Doppler spectrum. That resulting difference-to-sum ($\Delta/\Sigma$) value in each Doppler bin can be multiplied by the sum channel and the product subtracted from the difference channel to cancel clutter (or vice versa). Such a scheme can achieve an output from a Doppler filter bank and angle filter with greater than 20 dB improvement in GMT to noise and clutter interference (SNIR). If the input SNR is high (e.g. 18 dB or more) but the input SCR is low (e.g. $-18$ dB), then significant clutter rejection can be achieved. A well-characterized $\Delta/\Sigma$ pattern can then be used on the clutter-cancelled GMT residue to provide a more accurate estimate of GMT location in the mainbeam (~10 to 100:1 depending on SNR, more about this in what follows).

The basic idea in monopulse cancellation is that clutter in the current range/Doppler bin is correlated well enough with adjacent range/Doppler bins so that a clutter estimate, based on a number of adjacent range bins (16–64 per Doppler filter typical), can be subtracted from the current bin leaving a moving target, if any, visible. This has been demonstrated to provide 20–30 dB clutter cancellation. Obviously, since clutter is neither ergodic nor stationary, thresholds must be set higher than for pure thermal noise or more false alarms must be acceptable to the processing that follows. The other important aspect is that bright discretes usually are cancelled but at the expense of sensitivity in adjacent bins. This is also true of antenna guard channels.

One of the most important adjuncts for monopulse in the presence of ground clutter and other potential interference is a guard channel. This is an auxiliary antenna and receiver channel whose gain is above all the sidelobes in both the sum and difference antenna patterns but less than the A and B mainlobes (see Figures 2.8 and 6.34). Usually a guard channel is used in multiple beam or phase center antennas. The guard channel is provided to prevent false detections from clutter bright discretes (see Section 3.3.3) and interference in the sidelobes. The guard channel processing may be deterministic or adaptive. Many systems have multiple guard channels to adaptively null jammers and sidelobe bright discretes. In addition, returns from the skirts of the mainlobe, that is, outside the 3 dB beamwidth where the difference pattern exceeds the sum pattern but above the guard channel gain must also be suppressed by some combination of half beam stepping and monopulse.

One way to create a monopulse pattern is to design an antenna that has two or more offset beams created with separate antenna beamforming networks. For a 2D monopulse, these beams are often called *A*, *B*, *C* and *D* as shown in Figure 6.33.

These separate beams often have separate phase centers on the antenna aperture. Those beams can be used to produce either amplitude comparison or phase comparison monopulse. Either can provide an angle of arrival discriminant usable both for clutter rejection as well as angle measurement inside the mainbeam.

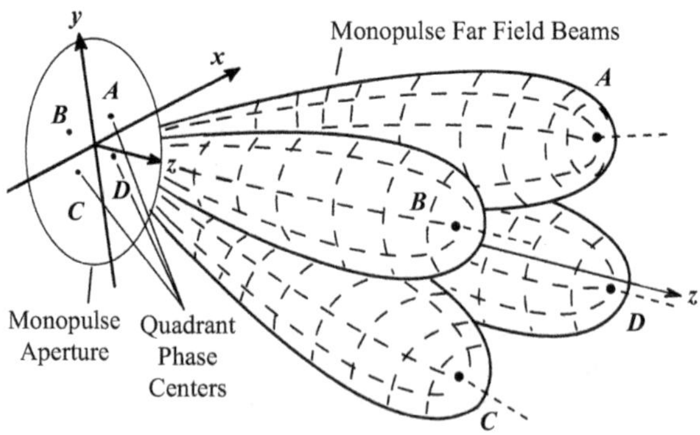


*Figure 6.33 Circular aperture 2D monopulse with separate phase centers; adapted from [17]*

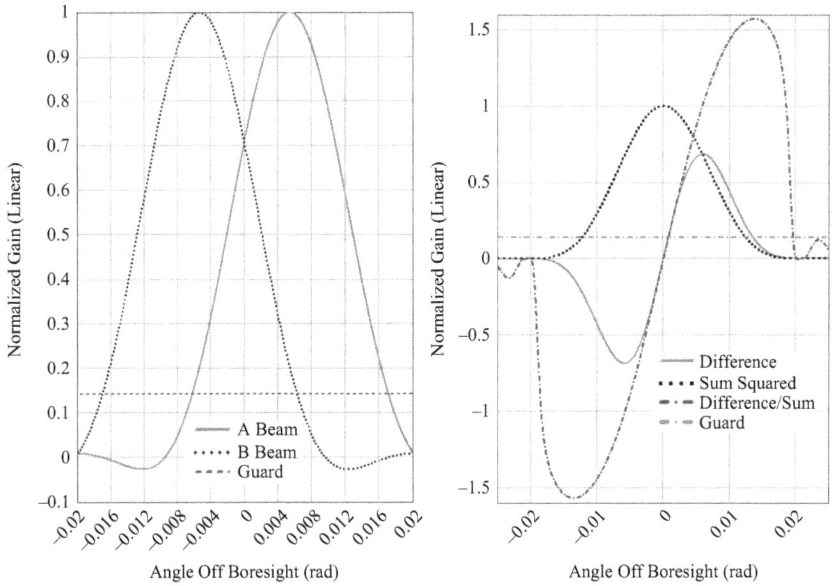


*Figure 6.34 Typical 1D amplitude monopulse pattern (Monopulse22)*

For amplitude monopulse, often the phase centers of the beams are close enough together that there is very little phase warp across the pattern except for the sign reversal going through boresight. For phase monopulse, the phase centers are placed so that a little less than 360° of phase change occurs across the 3 dB sum beamwidth.

An example of two amplitude monopulse beams in a single axis is shown on the left hand of Figure 6.34. Such a pattern is typical for reflector antennas whose feeds are less than a wavelength apart. Those two beams are added to create a sum beam and subtracted to create a difference beam as shown on the right in Figure 6.34. Both plots in Figure 6.34 also show a typical guard channel response. The right-hand plot shows the difference to sum ratio. Note that as soon as the difference/sum ratio exceeds 1–1.25 it is usually suppressed to zero somewhere in the processing. The guard channel performs the same suppression function for the sidelobe region. The guard channel angular pattern is almost omnidirectional. The guard spectrum and statistical distribution often change slowly sometimes allowing it to be multiplexed with the difference channel.

The far-field antenna patterns in Figure 6.34 use an antenna aperture amplitude distribution called Hanning weighting (raised cosine or cosine squared). Its principal feature here is that its far-field response can be written in closed form and is given in (6.11). Hanning weighting is somewhat unrealistic because the edge of the antenna has a zero amplitude, which cannot happen unless the antenna is infinitely wide. Hanning weighting causes the 3 dB beamwidth to be 1.4403 the value for an unweighted aperture. Normally, a real antenna will have edge amplitude about 5%–10% of peak.

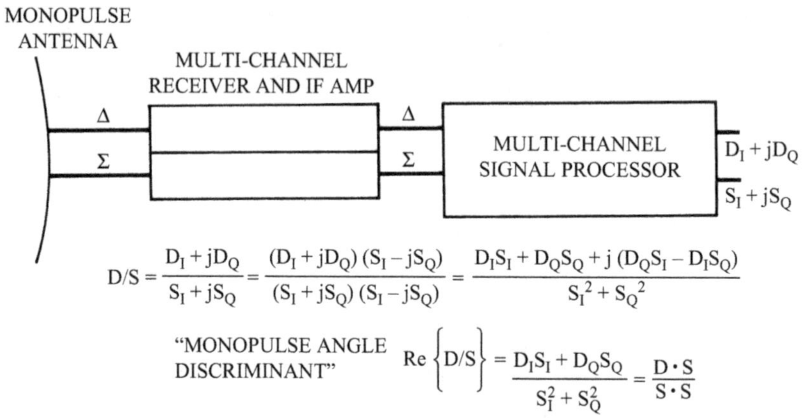


*Figure 6.35 Amplitude monopulse angle processing [3]*

The two-way sum and difference patterns given in Figure 6.34 right are processed as shown in Figure 6.35. The monopulse discriminant shown in Figure 6.35 is given in (6.11) for amplitude monopulse. The *A* and *B* beam far-field patterns are given in the first line of (6.11). The amplitude monopulse angle discriminant is the real part of the difference to sum ratio and is usually implemented as the scalar (dot) product of the sum times the difference divided by the squared magnitude of the sum.

The fine details of an actual monopulse discriminant *D/S* can be calculated from the measured antenna patterns. Often, the monopulse discriminant can be curve fitted with two offset sine curve approximations for the A and B sides. Although the actual antenna pattern may be symmetrical in angle space, the equivalent pattern in Doppler space may not be because the Doppler will be proportional to the cosine of the look angle from the velocity vector. With suitable curve fitting, the discriminant can then be applied to sum and difference Doppler filter outputs to cancel clutter filter by filter.

Typically, a planar antenna monopulse pattern has significant phase shift with angle in the mainbeam and the two or more beams, which make up the pattern are many wavelengths apart in the near field. An example pattern is shown in Figure 6.36. It is based on a 6 m aperture length, a wavelength of 0.03 m and 2.775 m spacing between the A and B beam near-field centers, sidelooking. The figure shows the sum and difference magnitude pattern but also shows the variation in phase angle in units of radians as well as the magnitude of the monopulse discriminant. Although the discriminant is quite linear in the mainlobe, it has significant phase shift, which must be taken into account for clutter cancellation. The phase warp shows up both on the clutter and any GMTIs in the mainlobe. Most modern monopulse planar or electronically scanned antennas exhibit behavior like that shown in Figure 6.36.

In the case of Figure 6.36, the receive beam patterns are as given in (6.11) including the addition of a phase term accounting for the distance between phase centers. This model assumes an electronic scan antenna with normal to the velocity vector and beam spread associated with scan angle beam broadening. The complex

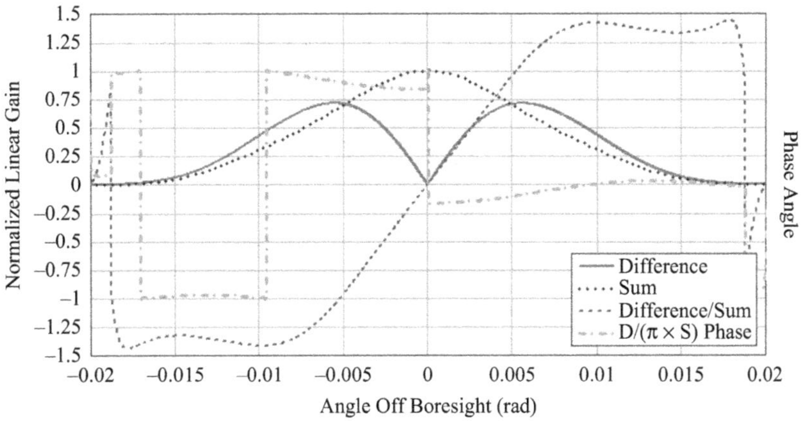


*Figure 6.36 Phase and amplitude monopulse (Monopulse28f)*

monopulse discriminant can be calculated on the fly by averaging multiple range bins around the current bin for each Doppler filter in MLC. This discriminant can then be used as a coefficient multiplier on the sum channel, which is then subtracted from the difference channel. The residue is the remaining uncancelled clutter or a moving target, which will be thresholded for a possible detection. This technique has been used for more than 50 years and has been demonstrated on USAF FLAMR, DARPA/USAF Pave Mover and many other programs. The method is shown in (6.12).

$$A \cong \left[ \frac{\sin\left(\frac{\pi}{\lambda}\cdot\left(u-D_{off}\right)\right)}{\frac{\pi}{\lambda}\cdot\left(u-D_{off}\right)\left(1-\left(u-D_{off}\right)^2\right)} \right] \cdot \frac{\exp\left[-j\cdot\frac{\pi}{\lambda^2}\cdot D_{off}\cdot\cos\left(\theta_0\right)\right]}{\sin\left(\theta_0\right)+0.15},$$

$$B \cong \left[ \frac{\sin\left(\frac{\pi}{\lambda}\cdot\left(u+D_{off}\right)\right)}{\frac{\pi}{\lambda}\cdot\left(u+D_{off}\right)\left(1-\left(u+D_{off}\right)^2\right)} \right] \cdot \frac{\exp\left[j\cdot\frac{\pi}{\lambda^2}\cdot D_{off}\cdot\cos\left(\theta_0\right)\right]}{\sin\left(\theta_0\right)+0.15},$$

$$\Sigma = A+B\ ,\ \Delta = A-B\ ,\ Guard = 0.1\cdot\Sigma_{peak}\ ,\ D/S = \left[\frac{\Delta\cdot\overline{\Sigma}}{\Sigma\cdot\overline{\Sigma}}\right] \tag{6.11.1}$$

where: $u = 2/3\cdot D_x\cdot\theta$,

$\overline{\Sigma}$ is the complex conjugate of $\Sigma$,

$A$ and $B$ beams typically use 2/3 of the antenna aperture width, $D_x$,

$D_{off}$ is the near field distance between the $A$ and $B$ beam illumination and is typically one half of the 3 dB beamwidth times $\lambda$,

$\lambda$ is the operating wavelength. (6.11.2)

All of the apparent phase shift for the center of each range cell approximated by (6.11) must be removed for monopulse or DPCA clutter cancellation to work well. Figure 6.37 shows the spectrum in the sum channel for a single-range bin. Figure 6.38 shows the average spectrum over all the range bins for the sum channel and the average clutter cancellation over those same range bins.

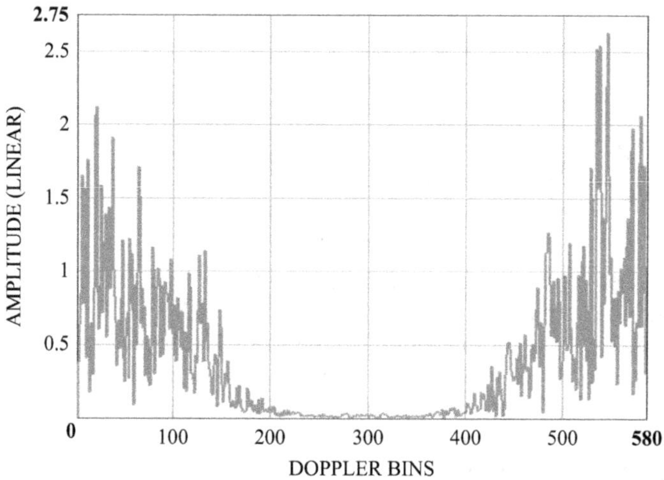


*Figure 6.37 Example received mainlobe spectrum including bright discretes*

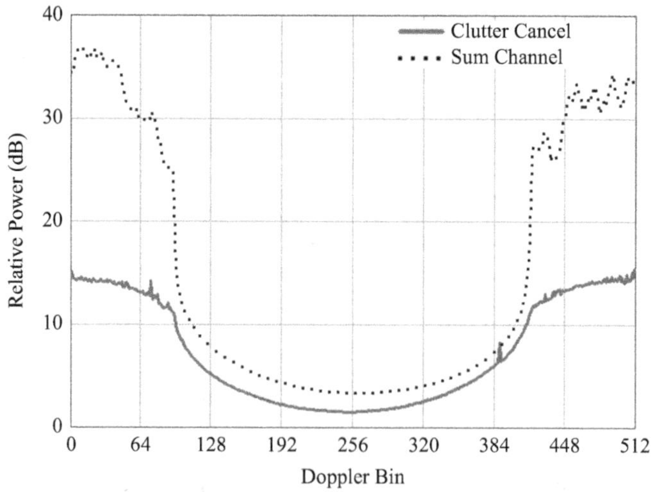


*Figure 6.38 Example monopulse average real clutter cancellation [18] (MontAlSARModel 005b)*

**Example 6.10** A clutter cancellation example for a 90° look angle with 200 m/s platform velocity, antenna aperture 6 m × 0.5 m, X band, range 78 km, competing clutter patch ~2400 $m^2$ with RCS of 72 $m^2$ competing with a 30 $m^2$ moving target is given in Figures 6.39–6.41. This particular example shows approximately 30 dB cancellation, which is better than normally achieved with real clutter and motion compensation as shown in Figures 6.39–6.41. Performance is a strong function of the number of clutter samples in range and Doppler used to calculate the adaptive weight $FDSA_m$. Often the Doppler filters are much closer in the MLC region. As mentioned earlier, a second stage of filtering can be implemented by using I/Q filter

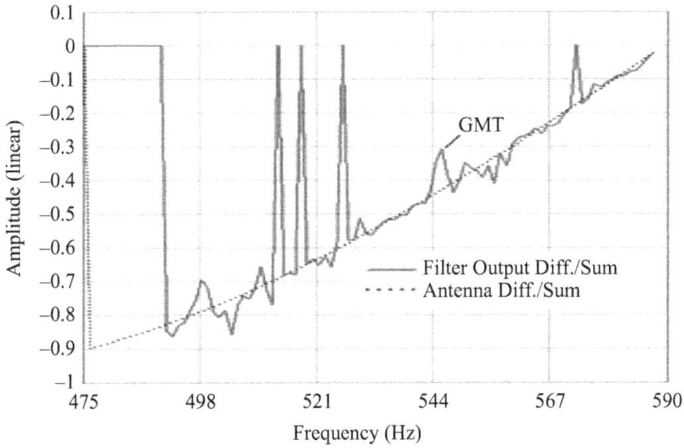


*Figure 6.39 Monopulse curve fit to D/S spectrum*

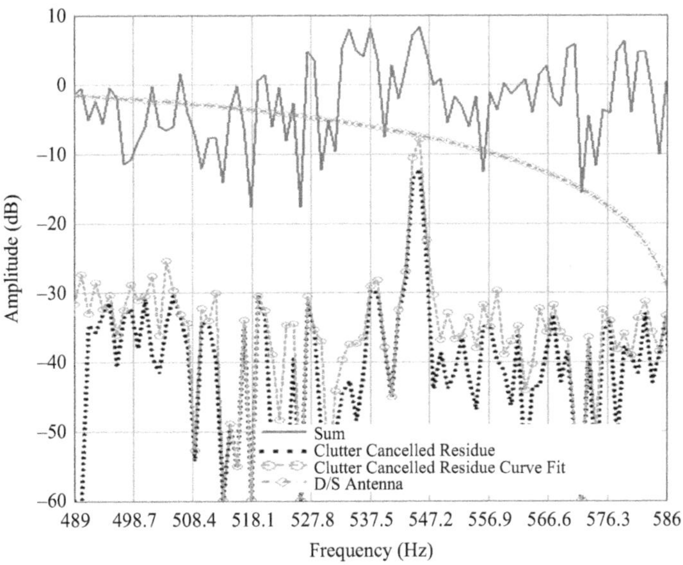


*Figure 6.40 Monopulse clutter-cancelled spectrum (two methods)*

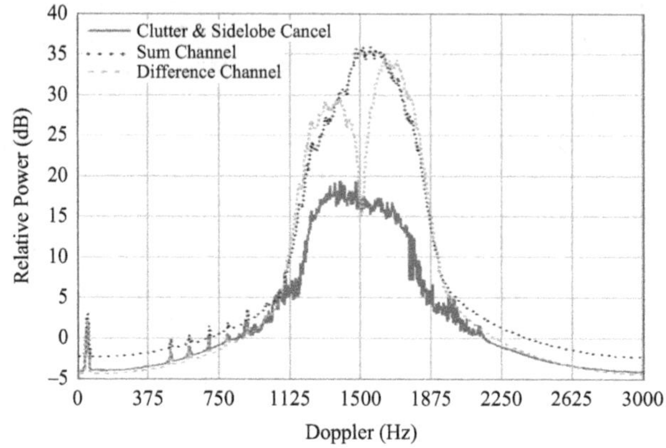


*Figure 6.41 Averaged sum, difference and clutter-cancelled spectra shifted to PRF/2 [18] (GMTIStLouisMono005)*

outputs from multiple CPIs to form narrower filters. Since GMT apparent velocities are lower endo-clutter, narrower filters can improve detection. About 25 dB of cancellation is achievable with this scheme.

Form range-Doppler filter bank for Sum ($FS_{n,m}$), Difference ($FD_{n,m}$) and Guard ($FG_{n,m}$) channels after filtering.

Calculate mean noise in Doppler filters using clear region indices ($n_c$, $m_c$).

$$FN = \frac{1}{n_c \cdot m_c} \cdot \sum_{n_c, m_c} \left| FS_{n,m} \right|$$

Calculate difference to sum ratio at filter outputs for all range-Doppler cells.

$$FDS_{n,m} = \left[ \frac{FD_{n,m} \cdot \overline{FS_{n,m}}}{FS_{n,m} \cdot \overline{FS_{n,m}}} \right]$$

Censor out-of-mainlobe & low SNR data for each $n, m$ cell.

If $\left( \left| FDS_{n,m} \right| > 1.25 \cup \left| FS_{n,m} \right| < \left| FG_{n,m} \right| \cup \left| FS_{n,m} \right| < V_{TH1} \cdot FN \right)$

Then $FDS_{n,m} = 0$, otherwise $FDS_{n,m} = FDS_{n,m}$

Find complex difference to sum average for each Doppler bin.

$$FDSA_m = \frac{1}{N} \cdot \sum_n FDS_{n,m}$$

Form sliding window threshold in range where $\underline{n}$ is current cell.

$$FT_{\underline{n},m} = \frac{V_{TH2}}{n-3} \cdot \left[ \sum_n \left| FS_{n,m} - FS_{\underline{n}-1,m} - FS_{\underline{n},m} - FS_{\underline{n}+1,m} \right| \right]$$

Form clutter cancellation and threshold for each $\underline{n}$ .

If $\left( \left| -FDSA_m \cdot FS_{\underline{n},m} + FD_{\underline{n},m} \right| > FT_{\underline{n},m} \right)$

Then $FT_{\underline{n},m} = 1$, otherwise $FT_{\underline{n},m} = 0$ (6.12)

One method used for monopulse clutter cancellation is to fit the known beamshape to clutter output spectrum to improve cancellation since the raw clutter data is often very nonuniform bin to bin. This is shown in Figure 6.39.

Figure 6.40 shows the detection of a single target by both curve fitting and bin averaging. Curve fitting is slightly better in this example. The overall average clutter cancellation is shown in Figure 6.41; note that the cancellation is worse when the signal strength is lower on the left side of the antenna pattern.

---

## *6.8.3 DPCA clutter cancellation*

Another method for detecting slow-moving targets is called DPCA. This is also a subset of STAP. The earliest versions (1950s) of DPCA were based on monopulse antennas. Two-pulse DPCA processing usually has been implemented with monopulse antennas. The fundamental notion of this scheme is to cause the total path length between the transmitting antenna phase center and the receiving antenna phase center to be constant for two or more pulses. Then, the clutter can be cancelled using typical ground-based one, two or more pulse cancellers. It works best if the antenna is pointed normal to the velocity vector and has no sidelobe weighting so that manipulating the phase center is easiest. Although the earliest versions relied on analog cancellation at RF or IF, it also works best if performed after Doppler processing, which is similar to monopulse cancellation. One embodiment of DPCA is given in Figure 6.42, showing successive positions of the antenna along the flight path. The figure is obviously exaggerated but the idea is to move the phase center along the receive aperture so that the total path length to a single scatterer is exactly the same distance pulse to pulse while the radar platform and antenna continue to move. Figure 6.42 shows separate transmit and receive

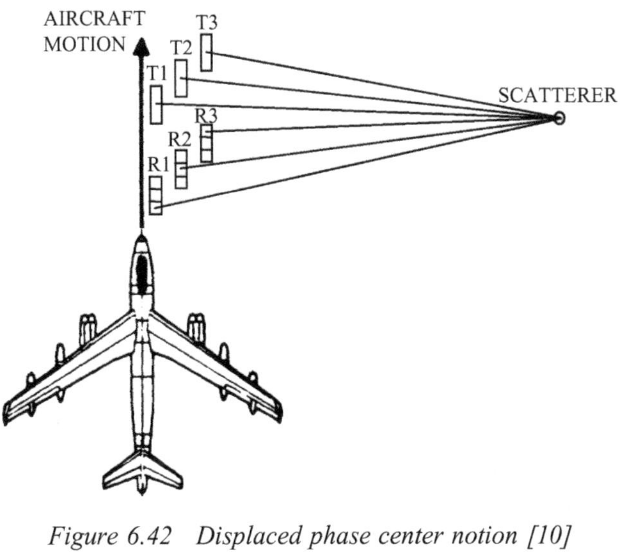


*Figure 6.42 Displaced phase center notion [10]*

apertures but obviously the transmit and receive apertures can be the same aperture with adjusted phase centers for transmit and receive. Transmission 1 (T1) would go to receive phase center 1 (R1) and transmission 2 (T2) would go to receive phase center 2 (R2), etc. This requires that the PRF is "magic," that is, directly related to platform speed, look angle and the dimensions of the possible distances between phase centers. Equation (6.13) shows the constraint.

$$PRF = \frac{2 \cdot V_a \cdot n}{D_\phi}$$

Where $V_a$ is platform velocity,

$D_\phi$ is the distance between phase centers,and

$n$ is the index of individual phase centers. (6.13)

Figure 6.42 depicts three successive phase centers and is typical of practical implementations. Researchers have hypothesized up to 10 phase centers but even so the limitations imposed by (6.13) are obvious, that is, for some combination of velocities, look angles and ranges, there may not be an acceptable PRF. Furthermore, as the look angle moves away from the normal to the velocity vector, the PRF selection gets progressively more difficult. The PRF must often be adjusted during the CPI due to aircraft or spacecraft motion variations.

The DPCA clutter rejection performance is the same as available from a one- or two-pulse canceller. Its principal advantage is that processing is conceptually simple, assuming the hardware implementation is not too difficult. Most modern DPCA and monopulse systems use adaptive weighting for cancellation. The weaknesses of DPCA are due to antenna phase center mismatches, timing errors, atmospheric refraction variations and uncompensated motion errors.

Target detection losses for both monopulse and DPCA clutter cancellation are similar and to a first order are given in (6.14). The loss is inversely proportional to two important elements: the ratio of target velocity to platform velocity (not surprising) and the distance between phase centers relative to a wavelength. Once adaptive weighting is used for either DPCA or monopulse cancellation, the techniques are just a step away from full STAP with a small number of phase centers.

$$L_t \cong \left( \sin\left( \frac{2 \cdot \pi \cdot V_t \cdot D_\phi}{V_a \cdot \lambda} \right) \right)^{-2} \quad \text{where } V_t \text{ is target velocity.} \tag{6.14}$$

Three phase centers in the horizontal direction can be achieved with a 2D monopulse antenna by rotating it through 45°. This is shown in Figure 6.43 design study.

In this design study, four possible configurations of a 2D monopulse post-Doppler filtering DPCA were examined for a nose-mounted fighter radar. The assumed aircraft velocity is high but typical of what a fighter attacking a ground target might actually do. One configuration, D, was just rotation with no special phase center processing. Two configurations, B and C, used the existing feed structure in various combinations. The last configuration, A, used an additional

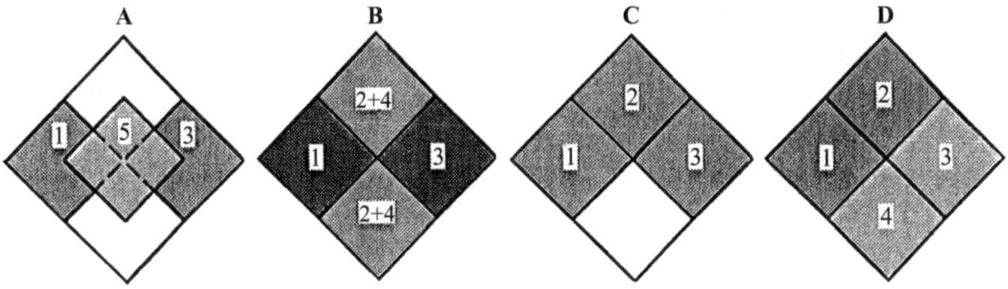


*Figure 6.43 Design study of four possible ADPCA configurations*

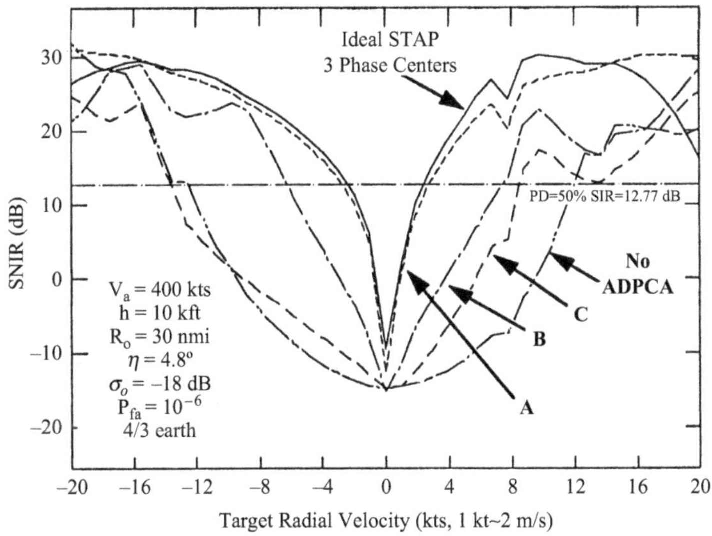


*Figure 6.44 ADPCA design study clutter rejection performance*

feed structure, 5, to form three identical overlapping beams. The principal weakness with B and C was that the beams were unsymmetrical and dissimilar in shape. What is disappointing but probably not surprising is that many of the adaptive degrees of freedom were consumed "fixing" the beamshape differences rather than canceling clutter. The clutter rejection performance is shown in Figure 6.44. The A configuration approaches the best available performance but none is very good relative to slower-moving platforms with larger antennas.

### 6.8.4 STAP clutter cancellation

A third method for detecting slow-moving targets is STAP. It is more generalized in nature and follows the same basic notion as monopulse and ADPCA clutter cancellation. STAP cannot do anything that couldn't be done with conventional techniques with perfect knowledge of the external environment. Since the external environment contains many unknowns, STAP performance can often be much better than existing *a priori* knowledge. The concept is just an extension of monopulse and ADPCA to more phase centers, range and Doppler bins, which provide more degrees of freedom for adaptive cancellation. Much of the literature is devoted to unweighted individual phase centers in which all the cancellation of both jammers and clutter is provided by adaptive weighting. Often, the performance of an alternate method to full STAP is compared to an ideal 3D cancellation scheme in angle, Doppler (slow time) and range (fast time). The main weakness of this approach is that perhaps hundreds of thousands of very large matrix inversions are required, which results in poor performance in real life. This is analogous to what happened in the 1960s when attempts to apply adaptive array techniques to the stock market were abject failures. The best actual performance is obtained when the possible degrees of freedom are dramatically reduced, separated into range, Doppler and angle domains as well as restricted by *a priori* information (sometimes called Cognitive Radar [2]). That is, aperture-weighted phase centers are steered toward the area of interest and a number of adjacent angle, range and Doppler bins are averaged (called modified sample matrix inversion (MSMI)) for calculating adaptive weighting using known information about types of jamming, terrain, noise, beamshape, system shortcomings, etc.

The adaptive cancellation of clutter or interference in DPCA or monopulse basically attempts to reduce the clutter background or bright interferors to the level of noise. They are "whitening" filters. Similarly, STAP carries this one step further to multiple phase centers and range-Doppler bins. Adaptive array techniques have been used since the late 1950s and early 1960s [1,5–10]. The author witnessed a real-time adaptive array radar field test against jammers and clutter in fall of 1970 at Hughes Aircraft using eight phase centers and processing hosted on an HP 2114 mini-computer. It used 2048 angle-range-Doppler degrees of freedom (8 angle $\times$ 16 Doppler $\times$ 16 range bins) to "whiten" interference. Later, real-time radar systems have used at least 8,388,608 degrees of freedom (16 angle $\times$ 512 Doppler $\times$ 1024 range) to cancel jammers and clutter while still detecting low observables targets. The object in this section is to show with a small number of degrees of freedom what processing is going on with STAP.

STAP bears many similarities with Kalman filtering introduced in Chapter 4. The fundamental idea is to multiply and sum measured data by a set of adaptive weights that maximizes the SINR in each range-Doppler-angle bin.

In the most general case, all the dimensions are weighted. The range dimension may need adaptive weighting to counter a type of jamming called "hot clutter" or ground bounce jamming in which clutter is not well behaved with range and angle. A single repeater jammer can produce a significantly large set of jammed ranges and angles in the Doppler clear region. For conventional slow-moving surface target detection, range is only adaptively weighted by such things as sliding window AGC/STC/CFAR. Figure 6.45 shows the completely general case in which adaptive weights, $\mathbf{W}(n,l,j)^{\mathbf{H}}$, are calculated and applied to the raw data $\mathbf{Z}$ for every phase center, PRI and range bin. The weights are most commonly calculated using most or all of a CPI data array. In the most general case, the correlation matrix of the entire data array except for the bins just around a specific range-Doppler-angle bin is used to calculate the optimum weight for the subject bin. Just as in the Kalman filter case (see Section 4.12), that weight calculation requires inversion of the correlation matrix. It is one thing to invert a $4 \times 4$ or $6 \times 6$ matrix for 2000 MTI

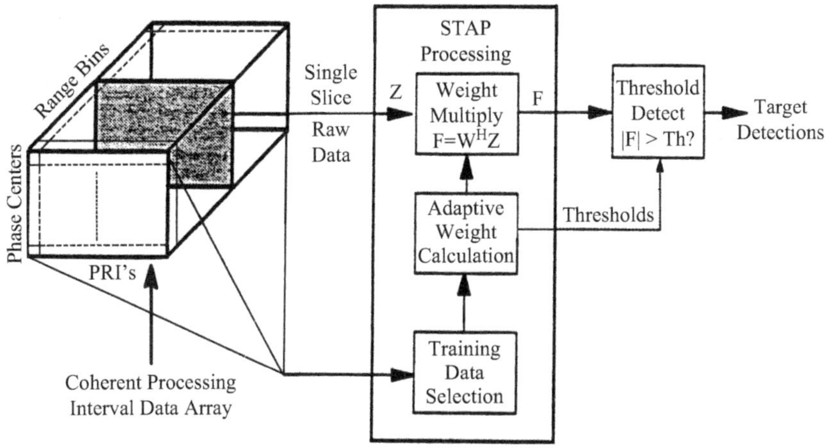


*Figure 6.45 Elements of STAP; adapted from [6]*

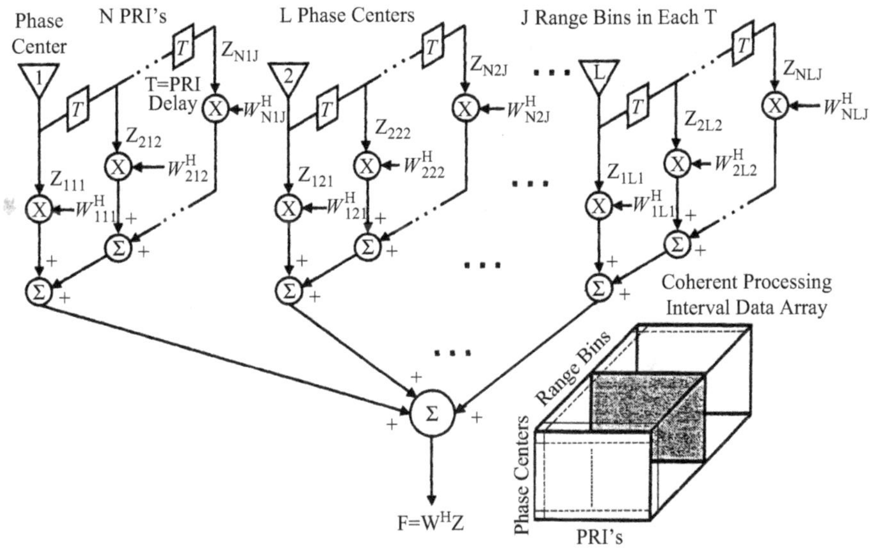


*Figure 6.46 STAP application of adaptive weights; adapted from [7]*

targets in track; it is quite another to invert a unique matrix for each of 8 million bins of potentially a little less than $16 \times 512 \times 1024$ matrix elements! That naturally leads to a search for a reduced dimension, partially adaptive STAP. Reduced dimension components of the total data field might be in range, Doppler or angle. More often STAP is applied to just the angle and Doppler dimensions as shown in Figure 6.46. The figure assumes that PC has already been performed and adaptive weighting in range is not required. Typically this is because large bandwidth PC often requires very fast number smashing. There are many nonlinearities in the broadband RF front end of most radars. It often turns out that Doppler filtering before adaptive filtering allows easier compensation of those nonlinear effects saving the adaptive degrees of freedom for external interferors.

The optimum weight matrix is given by (6.15). The correlation matrix is the expectation or mean for each range-Doppler-angle bin. If too few or too many cells are averaged, it won't be representative enough to properly cancel noise and interference in an individual bin.

$$\mathbf{W} = \mathbf{WN}^{-1} \cdot \mathbf{WA}^{-1} \cdot \mathbf{S}$$

where: **WN** is a normalizing vector having a single value for each bin,

**WA** is an $N \times L \times J$ correlation matrix averaged over

adjacent range-Doppler-angle bins, **S** is a steering vector in range-Doppler-angle, which may include sidelobe weighting. (6.15)

The problem with this idea is to select all of the cross-correlation terms, which might actually apply to a specific range-Doppler-angle bin but not terms that have no real significance to it. This is the problem with life, medicine, engineering, etc., that is, what is correlated but not causative. The classic of this dilemma was the discovery that children in Norway within one mile of a high-voltage electrical transmission line had a statistically higher incidence of leukemia. There are so many other correlations not eliminated that optimizing on distance from transmission lines provides no improvement in incidence of leukemia. (Perhaps, one must live in the frozen wilderness in northern Norway herding reindeer without electricity to escape the one in a million odds of leukemia?) Usually the general problem is reduced to a simpler problem by a transformation to another domain, which requires a simpler correlation matrix with fewer but with adequate degrees of freedom for reasonable performance. Sometimes weighting windows are applied to improve general sidelobes. The steering matrix or vector in (6.15) for angle and Doppler only is given in (6.16).

$$\mathbf{S} = (\mathbf{B_D} \otimes \mathbf{A_\theta}) \odot (\mathbf{W_D} \otimes \mathbf{W_\theta})$$

where the Doppler steering vector $\mathbf{B_D}$ is:

$$\mathbf{B_D} = \begin{bmatrix} 1 & \exp(j2\pi \cdot f_{DN} \cdot 1) & \exp(j2\pi \cdot f_{DN} \cdot 2) & \cdots & \exp(j2\pi \cdot f_{DN} \cdot (M-1)) \end{bmatrix}^{\mathrm{T}},$$

and an angle (azimuth) steering vector $\mathbf{A_\theta}$ is:

$$\mathbf{A_\theta} = \begin{bmatrix} 1 & \exp(j \cdot \beta_0 \cdot 1) & \exp(j \cdot \beta_0 \cdot 2) & \cdots & \exp(j \cdot \beta_0 \cdot (N-1)) \end{bmatrix}^{\mathrm{T}},$$

$\mathbf{W_D}$ = weighting window in Doppler, $\mathbf{W_\theta}$ = weighting window in angle,

$N, M$ = Doppler and angle sample indices, $\beta_0 = 2 \cdot \pi \cdot d \cdot \sin\theta_0 \cdot \cos\phi_0 / \lambda$,

$f_{DN}$ = Doppler frequency normalized by the PRF, $d$ = spacing between phase centers,

$\odot$ = Hadamard product i.e. each element in the first matrix is multiplied by the counterpart element in the second matrix,

$\otimes$ = Kroneker product i.e. each element in first vector or matrix is multiplied by all the elements of the second vector or matrix. (6.16)

The generalized STAP idea usually must be tailored to the specific situation geometry and hardware realization for it to work well. Many field and simulation experiments have been conducted to reduce the dimension of STAP correlation matrices as well as make experimental matrices well conditioned. Some of the dimension reduction happens naturally when filter banks and beam steering are formed *a priori* and only subsequent weighting is performed. Nevertheless, when proper tailoring is applied STAP can work very well as will be shown in examples that follow.

The Hermitian (conjugate transpose) of the optimum weighting matrix should be multiplied by the input data vector in a specific range-Doppler-angle bin and normalized to achieve desired clutter and jamming cancellation as given in (6.17).

### *6.8.5 STAP examples*

Some examples will help to understand what is actually going on. Three simple examples will be presented. The first example uses post-Doppler STAP for monopulse channels averaged in range using single Doppler bins. Phase only monopulse STAP doesn't seem to work as well as monopulse in which amplitude and phase are used. The second example uses post-Doppler STAP averaged in range using monopulse and three subarray monopulse channels. The third example uses post-Doppler STAP averaged in range and three adjacent Doppler bins. These examples were chosen because there was existing post-PC I/Q clutter data for sum, difference, guard and also for three subapertures that could be used from several flight test programs. The I/Q data is in Chapter 6 Appendix on DVD-R or internet.

$$F_{canc}(n,l,j) = \mathbf{W}^H \cdot Z(n,l,j) = \left[\left(\mathbf{WN}^{-1} \cdot \left[\mathbf{WA}^{-1} \cdot \mathbf{S}\right]^T\right)\right] \cdot Z(n,l,j)$$

where: $F_{canc}(n,l,j)$ is desired cancellation output bin,

$\mathbf{W}^H$ is Hermitian of optimum weighting matrix,

$\mathbf{WN}$ is a normalizing number for each output bin,

$Z(n,l,j)$ is the input data bin requiring SNIR improvement. (6.17)

---

**STAP Example 6.11** First, Example 6.11 is the simplest. It is a STAP-based 1D azimuth monopulse clutter cancellation using averaging in range for single Doppler filters with clutter from the St. Louis metropolitan area. The ESA antenna was oriented along the velocity vector and looked normal to the velocity vector ±60°. The original I/Q data was lowpass filtered previously and so moving targets were removed from the original data. Ten moving targets from angles all across the mainbeam and at Dopplers well inside MLC to well into the clutter-free region were inserted into three range bins. Input moving target signal to clutter ratios (SCRs) were from −40 to −50 dB and the thresholds were set to limit false alarm probability to less than $0.5 \times 10^{-6}$. Table 6.8 summarizes the parameters.

The same two-way antenna pattern that the clutter embodied was applied to the simulated targets. The starting phase of each target was random and independent of the other simulated targets. Since the targets were spread over the two-way mainbeam, there naturally occurred a 6 dB variation in individual target amplitudes. The composite target and clutter frequency at beam center is shifted to the PRF/2 for each range

*Table 6.8 Example 6.11 simple monopulse STAP*

| **Parameter** | **Value** | **Parameter** | **Value** |
|---|---|---|---|
| Operating altitude | 10 km | Coherent dwell time | 0.34 s |
| Platform relative velocity | 225 m/s | Minimum grazing angle | 0.05 rad |
| Peak power | 5 kW | Min. target radial speed | 4 km/h |
| Duty | 0.2 | Maximum range | 101.6 km |
| Antenna aperture | 3 × 0.5 m | Azimuth footprint width | 1.95 km |
| PRF | 3000 Hz | Front-end noise power | $10^{-20}$ W |
| Cross-range accuracy | 100 m | Losses | 2.75 dB |
| Target RCS | 10 m$^2$ | RMS clutter/range bin | 55,300 |
| Wavelength | 0.03 m | FFT filter bandwidth | 4.22 Hz |
| Velocity vector off angle | 60° | Range resolution | 13 m |
| No. time samples | 1024 | No. range bins | 512 |

bin (bin by bin frequency differences were slight). Amplitude weighting is applied. The frequency shift and amplitude weighting are given in (6.18).

Amplitude Weighting:

$$AW(n) = \left[a - (1-a)\cdot\cos\left[2\cdot\pi\cdot\left(\frac{n}{N}\right)\right]\right]$$

Beam Center Phase/Frequency Compensation:

$$PD(n,j) = \exp\left[\sqrt{-1}\cdot\left[2\cdot\pi\cdot\left[-\left(f\left(\frac{i\max}{2},j\right)-\frac{PRF}{2}\right)\cdot t(n) + \Delta f\left(\frac{i\max}{2},j\right)\cdot t^2(n)\right]\right]\right] \tag{6.18.1}$$

Where: time index starting at zero = $n$, total number of time samples $=N$,

Doppler frequency: $f\left(\frac{i\max}{2},j\right) = \frac{2\cdot V\cdot\cos(\theta_0)\cdot\cos(\varepsilon_j+\varepsilon_0)}{\lambda}$,

amplitude weighting coefficient = $a$, range bin index = $j$,

angle index representing beam center = $\frac{i\max}{2}$,

pulse repetition frequency = PRF, discrete sampling time $t(n) = \frac{\mathrm{n}}{\mathrm{PRF}}$,

Doppler rate of change: $\Delta f\left(\frac{i\max}{2},j\right) = \frac{V^2\cdot\left(1-\cos^2(\theta_0)\cdot\cos^2(\varepsilon_j+\varepsilon_0)\right)}{\lambda\cdot R(j)}$,

beam center azimuth look angle = $\theta_0$, elevation look angle = $(\varepsilon_j+\varepsilon_0)$,

slant range to the $j$th range bin = $R(j)$. (6.18.2)

The amplitude-weighted and frequency shifted I/Q time samples from both the sum and difference channels are FFTed. The guard channel data was not used in this example. The FFT filter outputs for both sum and difference were cross-correlated for every range-Doppler bin as shown in (6.19). Thus there were N × J two by two correlation matrices, that is, 512 × 1024 = 524,288 individual matrices of four values each. Those matrices were averaged in range as shown in the second equation line yielding 1024 two-by-two matrices. Also in that line was a definition of the steering vector, which in this case picks out the first column in the optimum weighting matrix. An estimated noise floor is added along the diagonal to prevent a singularity (really quite common otherwise) when the matrices are inverted. This method is called the MSMI. The third equation line shows the 1024 matrix inversions, **WAI(n)**, as well as the calculation of the 1024 normalizing numbers $WN(n)$, one for each frequency bin. Again to ensure good behavior for the normalizing number, the noise minimum times the threshold multipliers in range and frequency is added. Finally, the complex conjugate transpose is multiplied times each range-Doppler bin for sum and difference channels.

Cross correlation of sum and difference range-Doppler bins:

$$\mathbf{W}(\mathbf{n},\mathbf{j}) = \begin{bmatrix} FS(n,j)\cdot\overline{FS(n,j)} & FS(n,j)\cdot\overline{FD(n,j)} \\ FD(n,j)\cdot\overline{FS(n,j)} & FD(n,j)\cdot\overline{FD(n,j)} \end{bmatrix}$$

Averaging of correlation in range including diagonal loading, define steering vector:

$$\mathbf{WA}(\mathbf{n}) = \left[\left(\frac{1}{J}\right)\cdot\sum_{j}\mathbf{W}(\mathbf{n},\mathbf{j})\right] + \begin{bmatrix} Noise & 0 \\ 0 & Noise \end{bmatrix}, \quad \mathbf{S} = \begin{bmatrix} 1 \\ 0 \end{bmatrix}$$

Find matrix inverse and normalizing term:

$$\mathbf{WAI}(\mathbf{n}) = \mathbf{WA}(\mathbf{n})^{-1}, \quad WN(n) = \mathbf{S}^T\cdot\mathbf{WAI}(\mathbf{n})\cdot\mathbf{S} + Noise\cdot TMR\cdot TMF$$

Form clutter cancellation:

$$Fcanc(n,j) = \overline{\left[\left[\mathbf{WAI}(\mathbf{n})\cdot\mathbf{S}\right]^T\cdot\left(WN(n)\right)^{-1}\right]}\cdot\begin{bmatrix} FS(n,j) \\ FD(n,j) \end{bmatrix} \tag{6.19.1}$$

Where: $FS(n,j)$ is sum channel FFT output in bin $n, j$,

$FD(n,j)$ is difference channel FFT output in bin $n, j$,

$TMR$ is sliding window range multiplier, Noise is expected noise floor,

$TMF$ is frequency sliding window multiplier. (6.19.2)

A plot of the average clutter cancellation with frequency is given in Figure 6.47. It can be compared with the average sum and average difference outputs from the FFT also shown in Figure 6.47. One can also see that targets influence the clutter cancellation especially in the clutter-free region. The clutter reduction is very dramatic but one must also realize that there is some signal reduction as well because STAP is a “whitening” filter. Figure 6.48 shows the range-Doppler plot of MLC and the simulated MTI targets at the input to the STAP canceller shifted to the PRF/2.

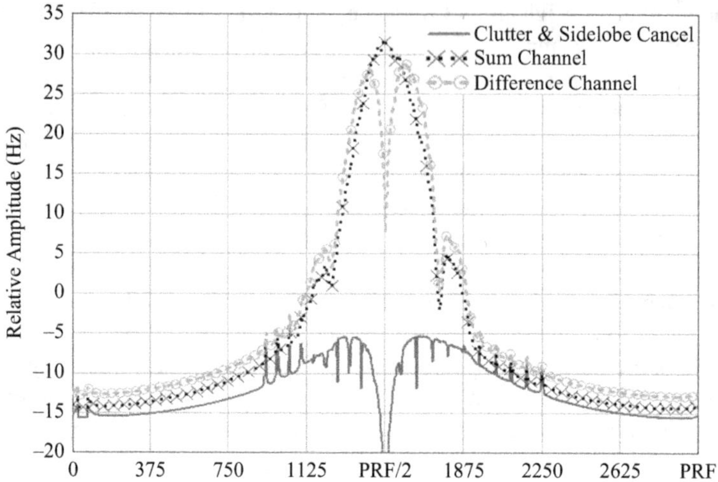


*Figure 6.47 Example 6.11 STAP-based monopulse clutter cancellation (GMTIStLouisSTAP005x2)*

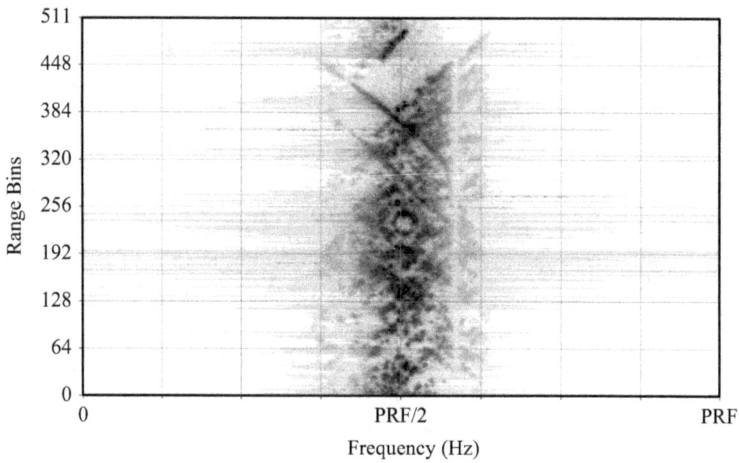


*Figure 6.48 Example 6.11 range-Doppler plot – targets and clutter @ PRF/2 (GMTIStLouisSTAP005x2)*

One can see that the clutter is not uniform in the mainlobe and near sidelobes. In some range bins, sidelobe clutter almost covers the entire ambiguous Doppler spectrum in Figure 6.48. The slowest target simulated is approximately 1 m/s radial, which is equivalent to just under 4 km/h.

Figure 6.49 shows a single-range bin with cancelled clutter. One can see that targets are detected well into MLC as well as the fact that target amplitudes will be affected by the STAP processing even with normalization. Also sliding window

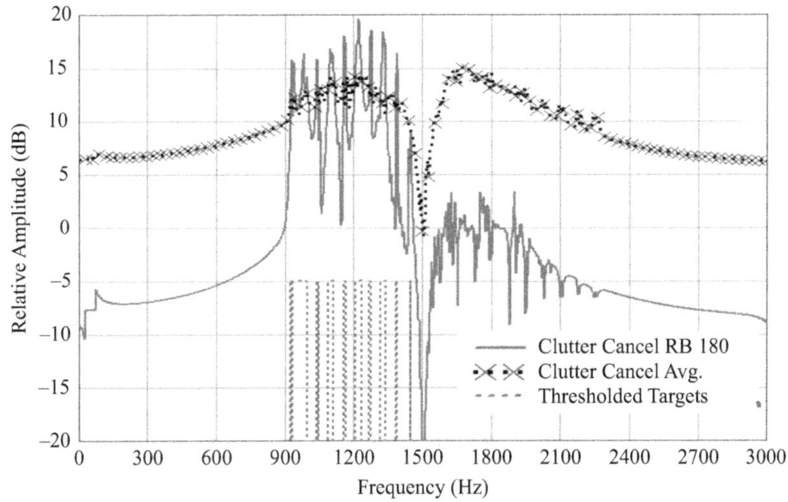


*Figure 6.49 Range bin 180 clutter cancellation and detected targets (GMTIStLouisSTAP005x2)*

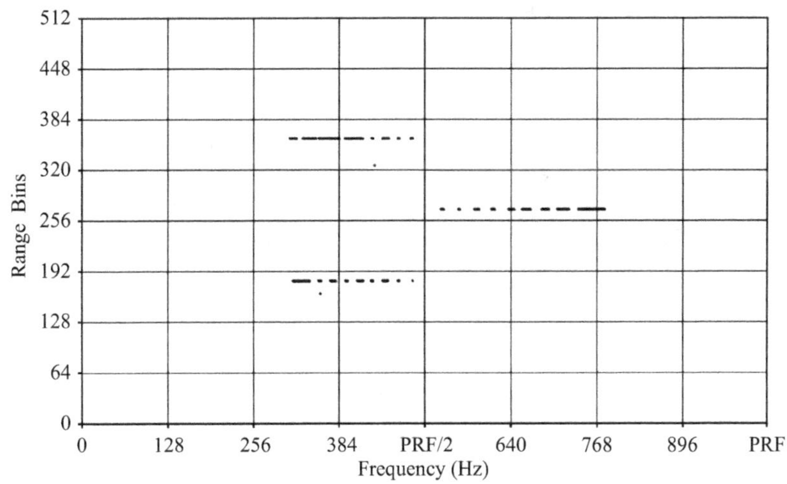


*Figure 6.50 Monopulse STAP GMTI targets: two false alarms (GMTIStLouisSTAP005x2)*

thresholding is required because of the variation in output amplitude with frequency and range.

The output detected target range-Doppler plot in Figure 6.50 shows all 30 simulated targets detected and 2 false alarms. Because the threshold is set at an SNR of 13.4 dB and input SCRs −40 to −50 dB, STAP cancels as much as 63 dB of clutter.

All of the details of this example are in Appendix A.6 on the DVD-R or internet in a file called GMTIStLouisSTAP005x2. Unfortunately the raw clutter data contains a known birdie at about 50 Hz, which appears in each range bin. Rather than use up degrees of freedom and thresholding to remove the birdie, those Doppler bins were skipped and a surrounding average was substituted for those filters. Close examination of Figure 6.47 shows the result of this substitution. The substitution results in a small Doppler blind region near maximum GMTI opening Doppler at about 78 km/h.

**STAP Example 6.12** Second, Example 6.12 applies STAP to three overlapping subapertures formed into three monopulse channels in subsequent processing. (There is a three-phase center monopulse cancellation in Appendix A.6 as well as GMTIStLouisMono006.) When the desire is to cancel clutter in the mainlobe, part of the application of STAP is the requirement to form a blocking matrix in the direction of clutter. Monopulse fills that requirement very well since the difference beams already have a null along boresight. As was mentioned previously, it is common to combine either active or passive subarrays in electronically scanned antennas into sum and difference channels to improve dynamic range and minimize the effects of nonlinearities. This example is thus representative of some modern systems. Example 6.12 also uses clutter from St. Louis with a different beam configuration, a higher PRF, a slightly different mainbeam pointing angle and simulated GMTI targets as before. The slowest targets detected are 4 km/h [8,9].

The basic antenna configuration is shown in Figure 6.51 in which three overlapping subarrays A, B, C each have a receive phase taper in the look direction, $\theta_0$. Each subarray has a different phase center. The three beams not only have a phase offset but also the A and B beams are squinted to provide both phase and amplitude angle discrimination. There are similar modern systems that have as many as 10 or more overlapping subarrays with monopulse in each. Since the monopulse difference channel is already a clutter blocking matrix, it is perfect for clutter rejection. (Sidelobe interference is a different problem and may still require a blocking matrix with the guard channels depending on their antenna patterns.)

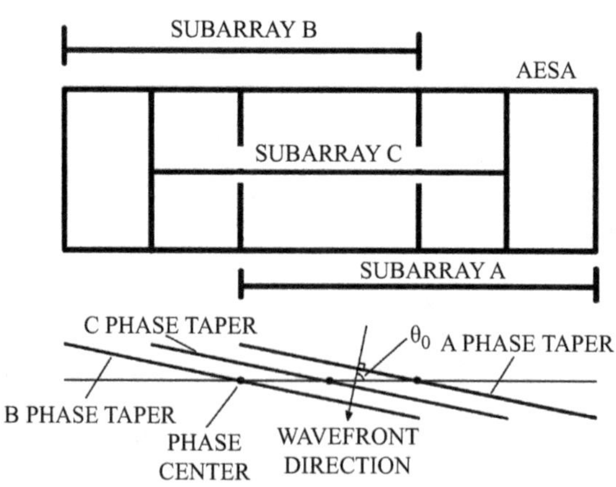


*Figure 6.51 Example 6.12 three overlapping subarrays*

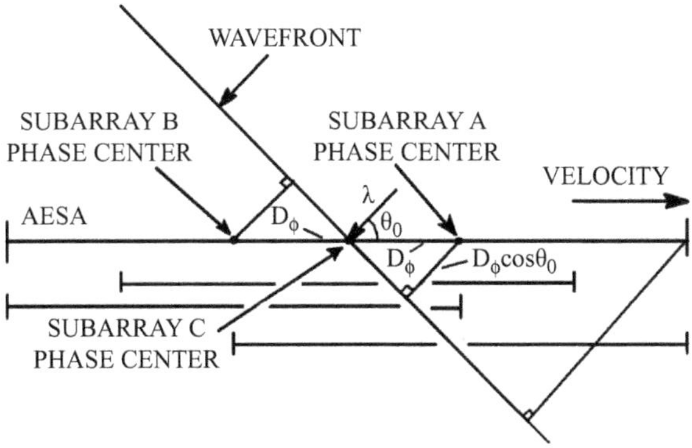


*Figure 6.52 Subarray geometry*

*Table 6.9 Example 6.12 radar parameters*

| Parameter | Value | Parameter | Value |
|---|---|---|---|
| Operating altitude | 10 km | Coherent dwell time | 0.28 s |
| Platform relative velocity | 225 m/s | Minimum grazing angle | 0.05 rad |
| Peak power | 5 kW | Min. target radial speed | 4 km/h |
| Duty | 0.2 | Maximum range | 101.6 km |
| Antenna aperture | 3 × 0.5 m | Azimuth footprint width | 1.95 km |
| PRF | 3600 Hz | Front-end noise power | $10^{-20}$ W |
| Cross-range accuracy | 100 m | Losses | 2.75 dB |
| Target RCS | 10 $m^2$ | RMS clutter/range bin | 55,300 |
| Wavelength | 0.03 m | FFT filter bandwidth | 4.22 Hz |
| Velocity vector off angle | 1.05 rad | Range resolution | 13 m |
| No. time samples | 1024 | No. range bins | 512 |
| Parabolic transmit beam | 25 mr | Hanning receive beams | 22 mr |

A more detailed view of the phase geometry of the three subarrays is given in Figure 6.52. One of the requirements of the configuration of Figure 6.51 for angles away from array normal is that the phase offset subarray to subarray must be compensated when processing all of them for STAP clutter cancellation. The required phase compensation is the distance to the phase center in wavelengths from the center of the array times the cosine of the look angle, $\theta_0$ relative to the velocity vector. The required phase center distance compensation is shown in Figure 6.52. The transmit beam must cover most of the subarray beams so it should be slightly larger than the receive beams. Each subarray is not only phase tapered but also Hanning amplitude weighted to improve sidelobes. Monopulse beams are created by adding and subtracting the A, B, C beams after Doppler processing to create one sum and three difference channels.

The radar parameters for Example 6.12 are summarized in Table 6.9. They are very similar to Example 6.8 except for the very different transmit and receive

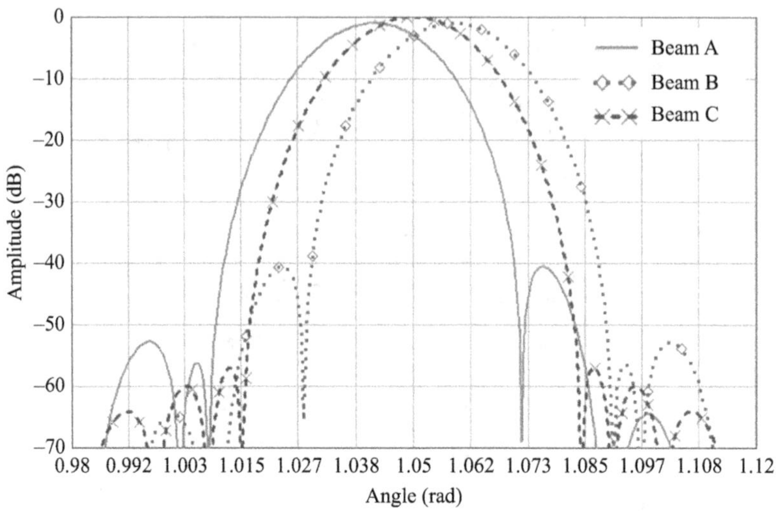


*Figure 6.53 Example 6.12 two-way antenna patterns*

patterns. This set of array patterns could almost be called amplitude-weighted ADPCA but STAP provides more freedom in selecting parameters.

The two-way antenna beam patterns are given in Figure 6.53. The sum beam uses parabolic weighting, which covers about 1.25 m in the center of the 3 m aperture. The A, B and C receive beams cover an overlapping 2 m each with phase centers 0.5 m apart. The A and B beams cross at the 3 dB point and have amplitudes somewhat smaller than the C beam due to their offset inside the transmit beam.

The A, B, C beams subsequent to Doppler filtering in an FFT can be combined to form multichannel monopulse sum and difference signals as shown in (6.20). Those signals can be cancelled pairwise and used to estimate angle of arrival using the phase differences between them. They all can be combined for a single cancelled output for the range-Doppler cells. Using three apertures doesn't improve cancellation null depth, which seems driven primarily by the greatest distance between phase centers. The normalization terms are at transmit beam center in azimuth and elevation. (This is *not* the clutter Doppler center.) Fortunately, there is no birdie in this raw data.

Calculate normalizations for each beam: $\mathrm{AN}=\left|\mathrm{A}_{\mathrm{imax/2,J2}}\right|$ , $\mathrm{BN}=\left|B_{\mathrm{imax/2,J2}}\right|$ , $\mathrm{CN}=\left|C_{\mathrm{imax/2,J2}}\right|$

Calculate sum channel: $FS(n,j)=\left[\dfrac{FA(n,j)+FB(n,j)+FC(n,j)}{AN+BN+CN}\right]$

Calculate difference channels: $FD1(n,j)=\left[\dfrac{FA(n,j)-FC(n,j)}{AN+CN}\right]$,

$$FD2(n,j)=\left[\frac{FB(n,j)-FC(n,j)}{BN+CN}\right],\ FD3(n,j)=\left[\frac{FA(n,j)-FB(n,j)}{AN+BN}\right] \tag{6.20}$$

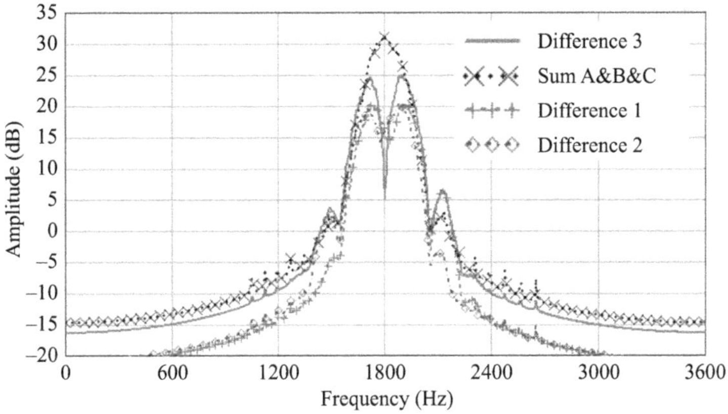


*Figure 6.54 Example 6.12 sum and difference channels (GMTIStLouisSTAP10xd)*

Figure 6.54 shows the sum and the three difference channels containing both clutter and simulated GMTI targets. Because the difference 1 and 2 channels cross over at roughly the 1 dB amplitude point and the phase centers are only 0.5 m apart, the difference patterns are significantly lower than difference 3 channel. Thus individual moving target angle of arrival estimates will be less accurate depending on SNIR. Another weakness in angle of arrival estimates is that the target angle is derived from phase comparison only, which usually is a little poorer than phase and amplitude combined.

The STAP cross-correlation is shown in (6.21) for this example. It is very similar to Example 6.11 except that it uses a $4 \times 4$ matrix for each of 524,288 range-Doppler bins. Similar to Example 6.11 the matrix is averaged in range to yield 1024 $4 \times 4$ matrices, one for each Doppler bin. Noise loading is added to the averaged cross-correlation (a number that is an estimate of RMS noise at the filter outputs). The steering vector picks out the sum channel similar to Example 6.11. The cancelled output $F_{canc}$ for each range-Doppler bin is the product of the normalized conjugate transpose of the optimum weighting matrix times the four terms from each range bin. A single-range bin is shown in Figure 6.55 including the uncancelled sum channel, clutter cancelled output and thresholded targets. The three traces in 6.55 have been offset to allow separation between signals.

Form cross correlation matrix for each range-Doppler bin:

$$\mathbf{W(n,j)} = \begin{bmatrix} FS(n,j)\cdot\overline{FS(n,j)} & FS(n,j)\cdot\overline{FD1(n,j)} & FS(n,j)\cdot\overline{FD2(n,j)} & FS(n,j)\cdot\overline{FD3(n,j)} \\ FD1(n,j)\cdot\overline{FS(n,j)} & FD1(n,j)\cdot\overline{FD1(n,j)} & FD1(n,j)\cdot\overline{FD2(n,j)} & FD1(n,j)\cdot\overline{FD3(n,j)} \\ FD2(n,j)\cdot\overline{FS(n,j)} & FD2(n,j)\cdot\overline{FD1(n,j)} & FD2(n,j)\cdot\overline{FD2(n,j)} & FD2(n,j)\cdot\overline{FD3(n,j)} \\ FD3(n,j)\cdot\overline{FS(n,j)} & FD3(n,j)\cdot\overline{FD1(n,j)} & FD3(n,j)\cdot\overline{FD2(n,j)} & FD3(n,j)\cdot\overline{FD3(n,j)} \end{bmatrix} \tag{6.21.1}$$

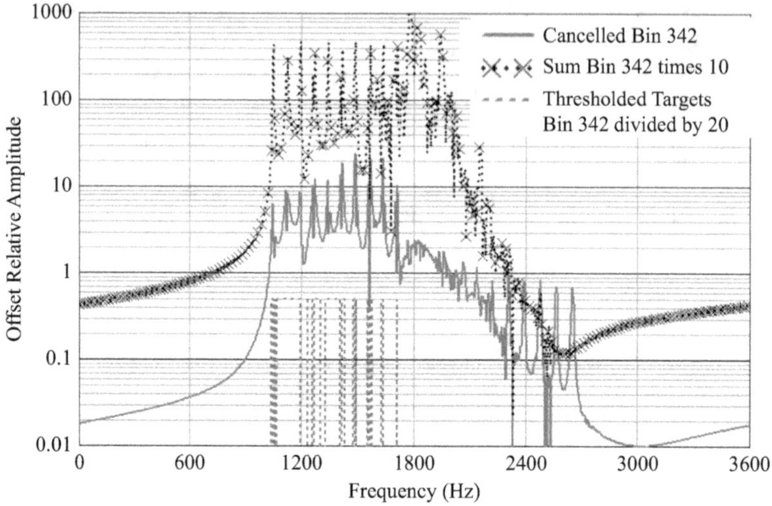


*Figure 6.55 Range bin 342 with 10 GMTI targets detected*

Averaging of correlation in range including diagonal loading, define steering vector:

$$\mathbf{WA(n)} = \left[\left(\frac{1}{j_{max}}\right)\cdot\sum_j \mathbf{W(n,j)}\right] + \begin{bmatrix} Noise & 0 & 0 & 0 \\ 0 & Noise & 0 & 0 \\ 0 & 0 & Noise & 0 \\ 0 & 0 & 0 & Noise \end{bmatrix}, \quad \mathbf{S} = \begin{bmatrix} 1 \\ 0 \\ 0 \\ 0 \end{bmatrix}$$

Find matrix inverse and normilization coefficient for each Doppler bin.

$$\mathbf{WAI(n)} = \mathbf{WA(n)}^{-1}, \quad WN(n) = \mathbf{S}^T\cdot\mathbf{WAI(n)}\cdot\mathbf{S} + Noise$$

Form clutter cancellation.

$$F_{canc}(n,j) = \left[\left(\left[\mathbf{WAI(n)}\cdot\mathbf{S}\right]^T\cdot WN(n)^{-1}\right)\right]\cdot\begin{bmatrix} FS(n,j) \\ FD1(n,j) \\ FD2(n,j) \\ FD3(n,j) \end{bmatrix} \tag{6.21.2}$$

The cancellation still must be followed by a 2D sliding window threshold scheme that uses three bins on both sides of three bins potentially containing a target in the range and Doppler dimensions. The overall sliding window is shown in Figure 6.56. The sliding window threshold equations are shown in (6.22) and are similar to those in (4.6).

The initial calculation of the early-range bins average also contains two other terms: an RMS term over all range bins for each Doppler bin divided by the range and Doppler threshold multipliers. This is to compensate for subsequent multiplication of the sliding window by these multipliers. The second term is the random noise estimate at the filter outputs. That term is usually de-minimus because for STAP to work well CNR must be high. The primary purpose of these terms is to

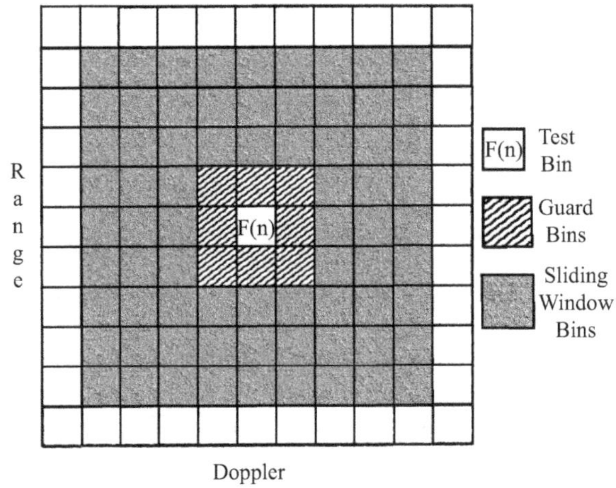


*Figure 6.56 Example 6.12 sliding window*

prevent very low thresholds that cause false alarms.

Calculate range early and late windows:

$$ERA_{out}(n, j+2) = \left[\frac{1}{J} \cdot \sum_{j2=0}^{J-1} \left|E_{in}(n,(j-j2))\right|\right] + \frac{RMS(n)}{TMR \cdot TMF} + Noise,$$

$$LRA_{out}(n, j) = ERA_{out}(n,(j+J+3)) \ ,$$

$$TH1(n, j) = 0.5 \cdot \left[ERA_{out}(n, j) + LRA_{out}(n, j)\right]$$

Apply range dimension tilt logic:

$$TH2 = if\left(ERA_{out}(n, j) \geq 2 \cdot LRA_{out}(n, j), ERA_{out}(n, j), TH1(n, j)\right),$$

$$TH3 = if\left(LRA_{out}(n, j) \geq 2 \cdot ERA_{out}(n, j), LRA_{out}(n, j), TH2(n, j)\right),$$

$$THR(n, j) = TMR \cdot TH3(n, j)$$

Calculate Doppler early and late windows:

$$EFA_{out}((n+2), j) = \left[\frac{1}{N} \cdot \sum_{n2=0}^{N-1} THR(n-n2, j)\right],$$

$$LFA_{out}(n, j) = EFA_{out}((n+N+3), j),$$

$$TH4(n, j) = 0.5 \cdot \left[EFA_{out}(n, j) + LFA_{out}(n, j)\right],$$

Apply frequency dimension tilt logic:

$$TH5 = if\left(EFA_{out}(n, j) \geq 2 \cdot LFA_{out}(n, j), EFA_{out}(n, j), TH4(n, j)\right),$$

$$TH6 = if\left(LFA_{out}(n, j) \geq 2 \cdot EFA_{out}(n, j), LFA_{out}(n, j), TH5(n, j)\right),$$

$$THF(n, j) = TMF \cdot TH6(n, j) \tag{6.22.1}$$

where: $n$ is the frequency sample index, $j$ is the range sample index,
$j2$ and $n2$ are averaging indices, $J$ is number of range bins averaged
on each side of the target bin, $N$ is number of Doppler bins averaged
on each side of the target bin, $TMR$ and $TMF$ are threshold multipliers
to control false alarms, their product is typically 1 to 5,
$THF(n, j)$ is the final 2 D sliding window threshold. (6.22.2)

Just as in the sliding window in Figure 4.17, tilt logic is used in case the window is riding up on a large bright discrete or clutter blob. The impulse response of (6.22) is given in Figure 4.19. One will notice in Figure 6.57, which is the threshold contour plot for Example 6.12, that the characteristic four lobes around each target as well as target "ghosts" appear in many places in the plot. In addition, large uncancelled clutter sidelobes can raise the threshold in some range bins for the entire spectrum.

Finally, the target detection plot in both range and Doppler is shown in Figure 6.58. All but one of the simulated moving targets are detected with no false alarms. The average simulated target amplitudes are 2 dB higher than in Example 6.11, which was required to achieve the same detection probability and false alarm rate.

All of the details of this example are on Appendix A.6 DVD-R or internet in a file called GMTIStLouis10xd.

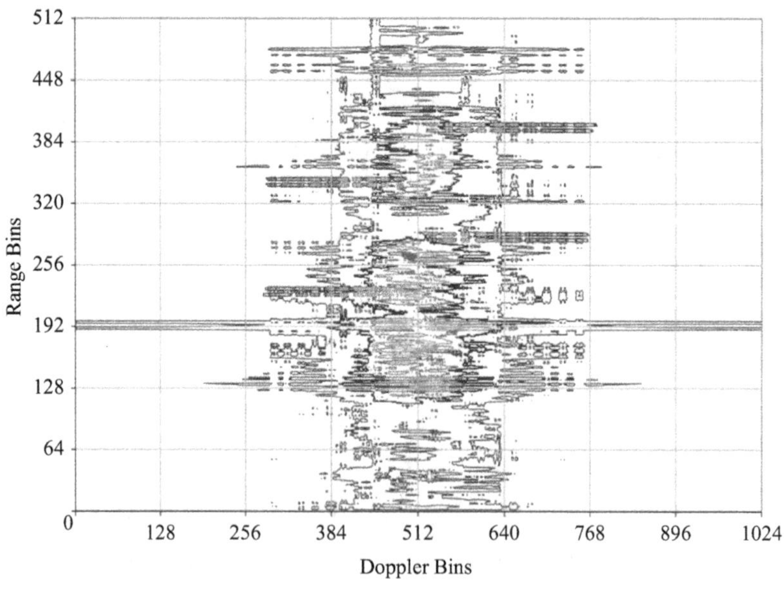


*Figure 6.57 Example 6.12 2D threshold contour plot*

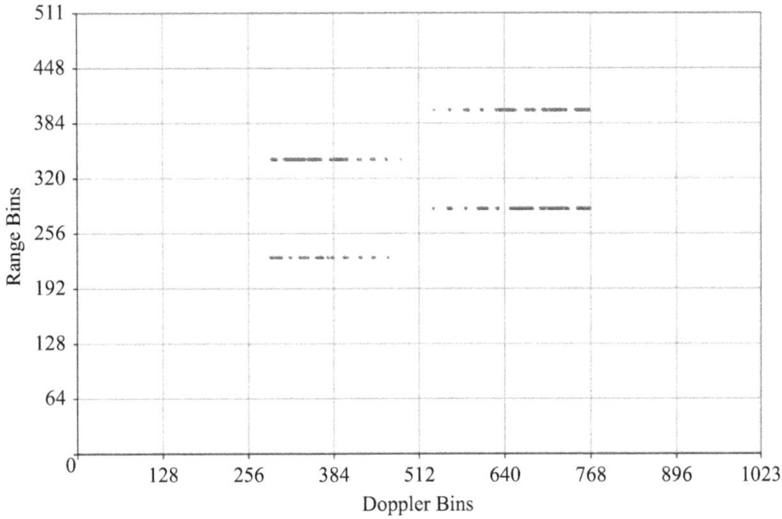


*Figure 6.58 39 out of 40 targets detected, no false alarms*

---

**STAP Example 6.13** The third example is 1D azimuth monopulse using three adjacent Doppler filters to improve the clutter cancellation. Again it uses St. Louis clutter and simulated MTI targets are inserted [1,5,10]. Table 6.10 summarizes the leading parameters. The clutter cancellation performance of this example is really excellent. In addition, it actually works better with fewer range bins averaged to derive the optimum weighting (as few as 128).

Figure 6.59 shows the uncancelled sum and difference channels as well as the average clutter and sidelobe cancellation across the range bins. Not only is clutter

*Table 6.10 STAP Example 6.13*

| Parameter | Value | Parameter | Value |
|---|---|---|---|
| Operating altitude | 10 km | Coherent dwell time | 0.17 s |
| Platform relative velocity | 225 m/s | Minimum grazing angle | 0.05 rad |
| Peak power | 5 kW | Min. target radial speed | 1 km/h |
| Duty | 0.2 | Maximum range | 101.6 km |
| Antenna aperture | $3 \times 0.5$ m | Azimuth footprint width | 1.95 km |
| PRF | 3000 Hz | Front-end noise power | $10^{-20}$ W |
| Cross-range accuracy | 100 m | Losses | 2.75 dB |
| Target RCS | 10 $m^2$ | RMS clutter/range bin | 55,300 |
| Wavelength | 0.03 m | FFT filter bandwidth | 8.44 Hz |
| Velocity vector off angle | 60° | Range resolution | 13 m |
| No. time samples | 512 | No. range bins | 256 |

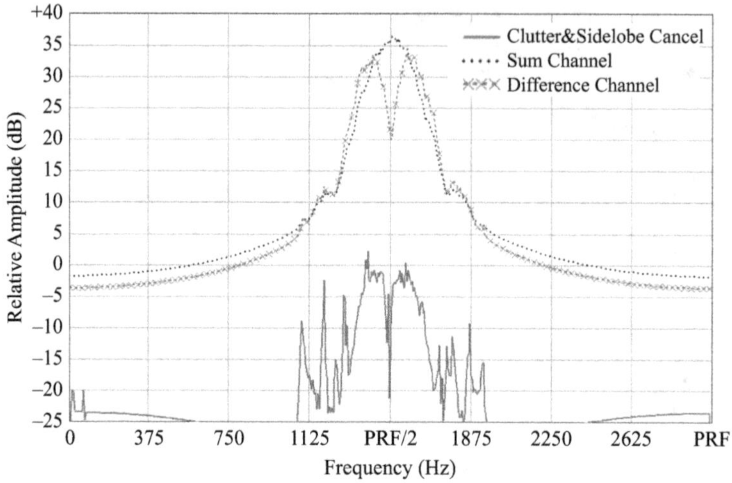


*Figure 6.59 Example 6.13 average clutter cancellation (GMTIStLouis20x2g)*

cancelled by 37 dB but the signal is increased by the FFT by 35 dB relative to clutter. The correlation and cancellation equations are shown in (6.23).

Define diagonal loading and steering vector:

$$\mathbf{N01}(\mathbf{n},\mathbf{j}) = \begin{bmatrix} 0.25 & 0 & 0 & 0 & 0 & 0 \\ 0 & 0.25 & 0 & 0 & 0 & 0 \\ 0 & 0 & 0.25 & 0 & 0 & 0 \\ 0 & 0 & 0 & 0.25 & 0 & 0 \\ 0 & 0 & 0 & 0 & 0.25 & 0 \\ 0 & 0 & 0 & 0 & 0 & 0.25 \end{bmatrix}, \ \mathrm{S}=\begin{bmatrix} 0 \\ 1 \\ 0 \\ 0 \\ 0 \\ 0 \end{bmatrix} \tag{6.23.1}$$

Calculate cross correlation for sum and difference for three adjacent Doppler bins:

$$\mathbf{W}(\mathbf{n+1},\mathbf{j}) = \left[\begin{matrix} FS(n,j)\cdot\overline{FS(n,j)} & FS(n,j)\cdot\overline{FS(n+1,j)} & FS(n,j)\cdot\overline{FS(n+2,j)} \\ FS(n+1,j)\cdot\overline{FS(n,j)} & FS(n+1,j)\cdot\overline{FS(n+1,j)} & FS(n+1,j)\cdot\overline{FS(n+2,j)} \\ FS(n+2,j)\cdot\overline{FS(n,j)} & FS(n+2,j)\cdot\overline{FS(n+1,j)} & FS(n+2,j)\cdot\overline{FS(n+2,j)} \\ FD(n,j)\cdot\overline{FS(n,j)} & FD(n,j)\cdot\overline{FS(n+1,j)} & FD(n,j)\cdot\overline{FS(n+2,j)} \\ FD(n+1,j)\cdot\overline{FS(n,j)} & FD(n+1,j)\cdot\overline{FS(n+1,j)} & FD(n+1,j)\cdot\overline{FS(n+2,j)} \\ FD(n+2,j)\cdot\overline{FS(n,j)} & FD(n+2,j)\cdot\overline{FS(n+1,j)} & FD(n+2,j)\cdot\overline{FS(n+2,j)} \end{matrix}\right. \cdots$$

$$\cdots \left.\begin{matrix} FS(n,j)\cdot\overline{FD(n,j)} & FS(n,j)\cdot\overline{FD(n+1,j)} & FS(n,j)\cdot\overline{FD(n+2,j)} \\ FS(n+1,j)\cdot\overline{FD(n,j)} & FS(n+1,j)\cdot\overline{FD(n+1,j)} & FS(n+1,j)\cdot\overline{FD(n+2,j)} \\ FS(n+2,j)\cdot\overline{FD(n,j)} & FS(n+2,j)\cdot\overline{FD(n+1,j)} & FS(n+2,j)\cdot\overline{FD(n+2,j)} \\ FD(n,j)\cdot\overline{FD(n,j)} & FD(n,j)\cdot\overline{FD(n+1,j)} & FD(n,j)\cdot\overline{FD(n+2,j)} \\ FD(n+1,j)\cdot\overline{FD(n,j)} & FD(n+1,j)\cdot\overline{FD(n+1,j)} & FD(n+1,j)\cdot\overline{FD(n+2,j)} \\ FD(n+2,j)\cdot\overline{FD(n,j)} & FD(n+2,j)\cdot\overline{FD(n+1,j)} & FD(n+2,j)\cdot\overline{FD(n+2,j)} \end{matrix}\right] \tag{6.23.2}$$

Find weighting average in range, find inverse and normalizing factor:

$$\mathbf{WA}(\mathbf{n}) = \left[\left(\frac{1}{J}\right) \cdot \sum_{j} \left(\mathbf{W}(\mathbf{n}+\mathbf{1},\mathbf{j}) + \mathbf{N01}(\mathbf{n},\mathbf{j})\right)\right], \mathbf{WAI}(\mathbf{n}) = \left(\mathbf{WA}(\mathbf{n})\right)^{-1},$$

$$WN(n) = S^T \cdot \mathbf{WAI}(\mathbf{n}) \cdot S + (TMR \cdot TMF) \cdot Noise, \; WNI(n) = \left(WN(n)\right)^{-1}$$

Form clutter cancellation in each range-Doppler bin:

$$F_{canc}(n,j) = \overline{\left[\left[\mathbf{WAI}(\mathbf{n}) \cdot S\right]^T \cdot WNI(n)\right]} \cdot \begin{bmatrix} FS(n,j) \\ FS(n+1,j) \\ FS(n+2,j) \\ FD(n,j) \\ FD(n+1,j) \\ FD(n+2,j) \end{bmatrix} \quad (6.23.3)$$

As before diagonal loading is required to improve the behavior of the matrix inversion as well as the normalizing factor. The steering vector has six entries but only the middle sum channel of the three Doppler filters is selected. The cross-correlation is a 6 × 6 matrix of three adjacent Doppler sum channels and three adjacent Doppler difference channels. The literature suggests that it is best to use unweighted FFT output filters for the correlation because there is less SNR loss. However, the author's experience is that unweighted filters lead to more false alarms with real clutter forcing higher thresholds. There is no net gain by foregoing sidelobe weighting.

One of the important attributes of using three adjacent filters for monopulse cancellation is the narrowness of the clutter null at beam center as indicated in Figure 6.59. This allows detecting moving targets endo-clutter at very low Doppler offsets. Under some clutter circumstances, targets as low as 0.2 m/s can be detected. Combined with FFT coherent integration, SCRs can be improved by 54–60 dB for Dopplers well below walking speeds.

In this example, there are 256 range bins by 512 filters for 131,072 6 × 6 correlation matrices. After the averaging, **WA(n)**, shown in the middle equations of (6.23), there are 512 6 × 6 matrices to be inverted one for each center Doppler bin. The normalizing factor calculation as before must contain a noise-related term to improve behavior and reduce false alarms. Finally, the vector of three adjacent sum channels and three adjacent difference channels is multiplied by the optimum weighting factor to produce the clutter-cancelled output $F_{canc}$ for each range-Doppler bin.

The average simulated target signal to real clutter is shown in Figure 6.60. Even though at input the simulated targets are all the same amplitude, since they are in different angular positions inside the mainbeam their amplitudes before filtering vary by as much as 10 dB. In addition, the real clutter varies greatly range bin to bin, so the input SCR varies all over the map (real life!). It is amazing how well very simple STAP as in Examples 6.11, 6.12 and 6.13 works in real clutter (with suitable limits).

Figure 6.61 shows the sum input and STAP output for a single-range bin 180. Also shown are thresholded targets in bin 180. The average target SCR in bin 180 is

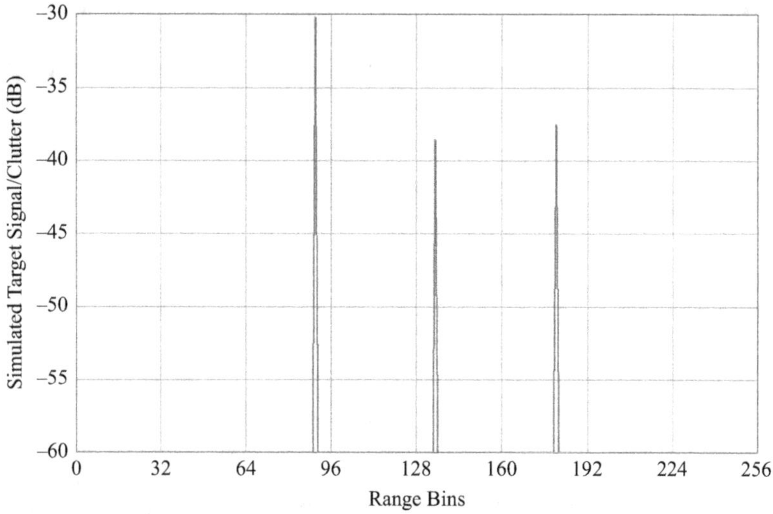


*Figure 6.60 Simulated GMTI targets average signal to clutter ratio*

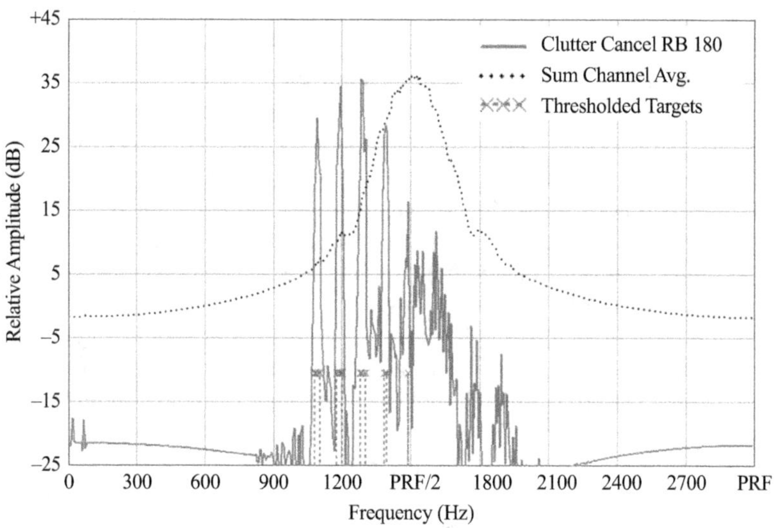


*Figure 6.61 Bin 180 targets detected to 0.2 m/s radial velocity (GMTIStLouis20x2g)*

−37 dB based on Figure 6.60. Even the target at 12 Hz Doppler offset from beam center has a positive SCR of 13.5 dB at the output and is detected.

In summary for this example, Figure 6.62 shows the overall target detection performance in a standard ambiguous range-Doppler plot. There is one false alarm and all 15 simulated targets detected including those near 0.2 m/s are detected.

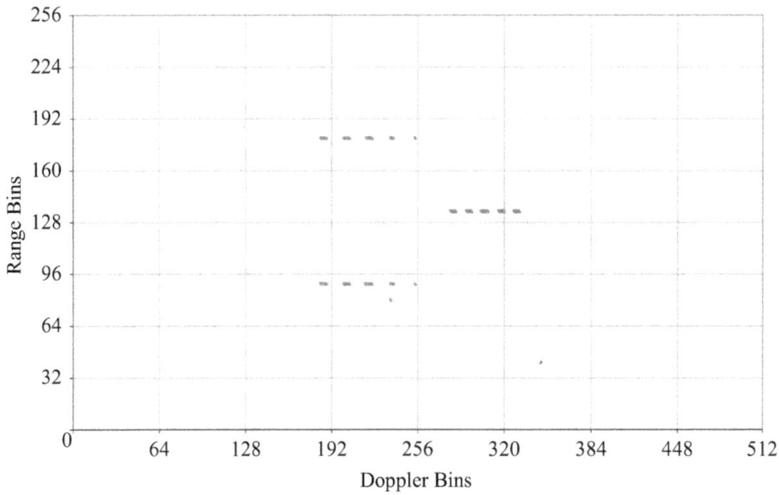


*Figure 6.62 All 15 GMTI targets detected – one false alarm*

However, as mentioned several times earlier these detections are worthless unless they can be geolocated and put in a proper context.

For a large Doppler filter bank (FFT), null depth/width is closely matched to individual filter width and seems to do best with just 2–3 phase centers and multiple Doppler channels. In order to see the slowest targets, one must use long array times, which will not be appropriate for faster targets. Octave decimation used in sonar signal processing is appropriate for seeing fast and slow targets with the same waveform set. Recall Figure 6.14, tracking across coherent arrays may require small FFTs on the very low frequency I/Q filter outputs in which the targets may move very little in range or Doppler.

All of the details of this example are on Appendix A.6 DVD-R or internet download in a file called GMTIStLouis20x2g. There are also several raw I/Q data files for Montgomery, AL, Andalusia Airport, AL, Scott AFB, IL, St. Louis, MO, and Southern California at various resolutions, which can be used in most of these programs.

## 6.9 Integrated MTI and SAR processing

Sometimes a single waveform is used to acquire and track fast, slow and stationary targets. Such waveforms are usually filtered with respect to Doppler and separated into three groups: stationary targets, slow mover but accelerating targets and "fast" targets. The waveform is selected to provide the desired SAR resolution but oversampled enough to provide resolvable fast target Dopplers and range collapsed to allow fast target detections. The lower Dopplers are presummed to provide higher-resolution SAR maps to allow slower target detection and focused SAR map underlays. An example block diagram is shown in Figure 6.63.

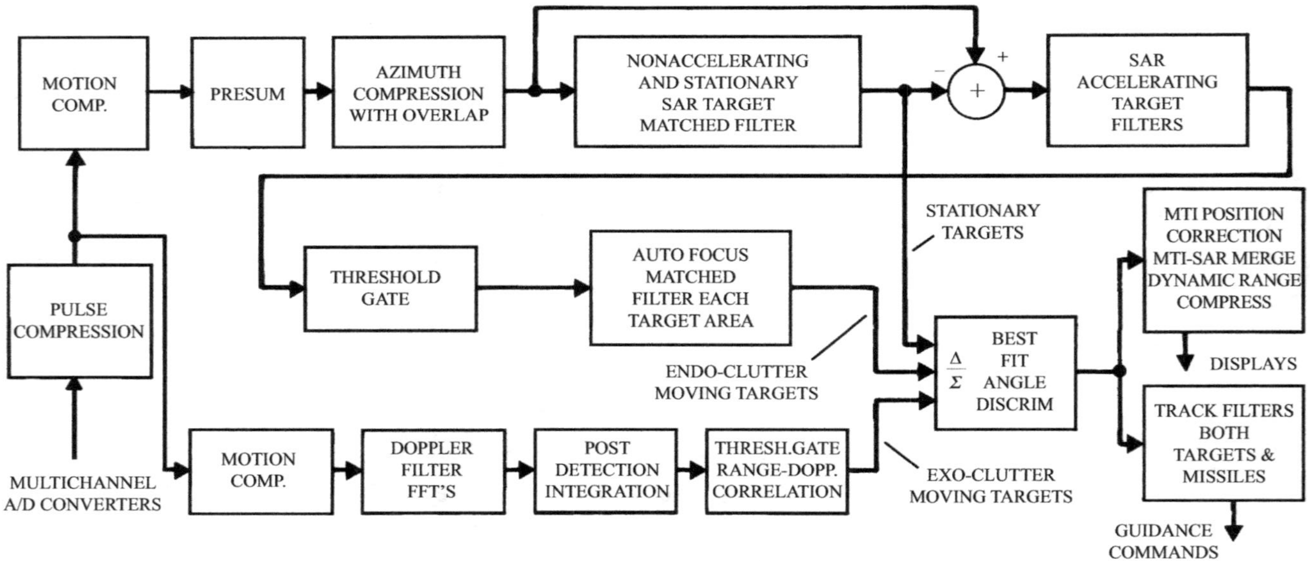


*Figure 6.63   Integrated SAR and GMTI [15]*

## 6.10 Synthetic monopulse MTI

Synthetic monopulse MTI uses interleaved SAR images to create two or more synthetic phase centers, which can then be used to detect and locate moving targets that would otherwise not focus or be detected. This does not work as well as real monopulse because it requires a better SCR. However, if a radar only has one real phase center, this method allows detection of moving targets in clutter. The method presented here creates synthetic monopulse in the time domain. This general technique can also be performed in the frequency domain. The author presented these methods to Boeing in 2004 and Hank McCord presented a different but similar method to Boeing in 2006 [19,20].

---

**Example 6.14** Example 6.14 uses IQ data from a SAR map of Andalusia Airport, AL, with resolution of about 5 m after presumming and before azimuth-matched filtering. The basic idea is to create two (or three) beams pointed in the direction of the target patch using adjacent PRI time samples. In this example, the A and B beams use two time samples each and overlap by one time sample. The basic beamforming geometry is given in Figure 6.64. The A beam uses samples at time $t_n$ and $t_{n-1}$ summed together with appropriate phase shift so that the two samples are in the plane normal to the look direction. Similarly, the B beam uses time samples $t_{n-1}$ and $t_{n-2}$ appropriately phased and summed. Equations (6.24) show the processing. Those beams are combined into sum (Σ) and difference (Δ) patterns to allow formation of an angle discriminant. Once sum and difference beams are formed, they are transformed into the frequency domain and conventional monopulse STAP processing is used to cancel clutter. Subsequent to clutter cancellation, a normal 2D sliding window threshold is applied for target detection. The spatial distance between sample times $t_n$ is $D_{PRF}$ shown in Table 6.11 as 0.45 m. Thus the synthetic phase

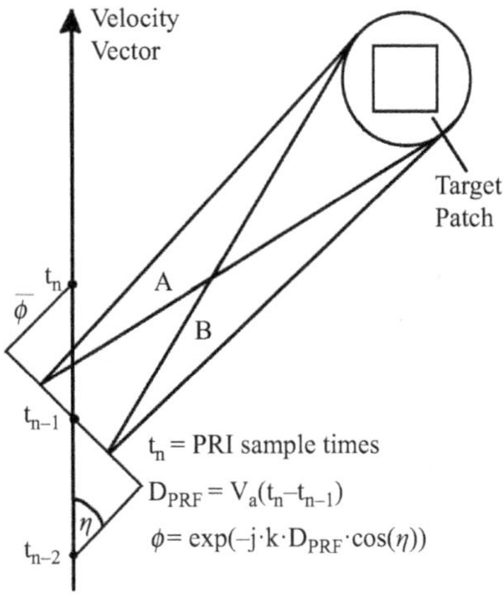


*Figure 6.64 Synthetic monopulse geometry*

*Table 6.11 Example 6.14 synthetic monopulse*

| Parameter | Value | Parameter | Value |
|---|---|---|---|
| Radar altitude, km | 10 | PRF, Hz | 3000 |
| Radar velocity, m/s | 225 | Presum sample rate, Hz | 500 |
| Antenna width, m | 1 | Array time, s | ~2 |
| Antenna height, m | 0.8 | Initial look angle, ° | 60 |
| Antenna orientation | Sidelook | Wavelength, m | 0.03 |
| Max. ground range, km | 95 | Raw IQ samples | 1024 × 512 |
| PRI sample distance, m | 0.45 | Resolution, m | 5 |

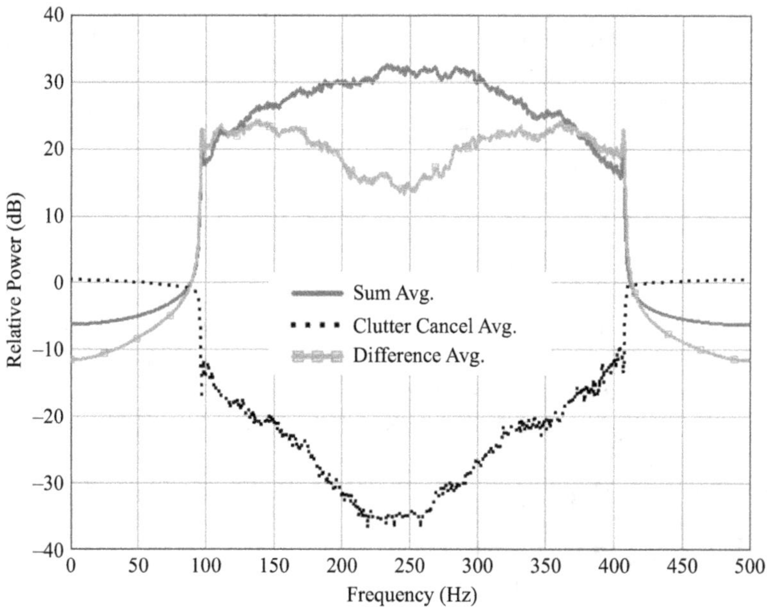


*Figure 6.65 Average sum, difference and clutter cancellation (GMTIAnApSynMono006h1)*

centers are $D_{PRF}\sin(\eta)$ apart, which is not the same as the amount of phase shift that must be added or subtracted to place them all on the same focal plane.

The average sum and difference power as well as the clutter cancellation power is shown in Figure 6.65. The difference pattern is not all that sharp or deep but it still is adequate for this example.

Obviously, the method is not limited to just a few beams or just a few time samples per beam. The more samples per beam the narrower the beamwidth, which results in improved SCR and STAP degrees of freedom but at the expense of more processing to cover the clutter spectrum and to resolve ambiguities in mover tracking. The cancellation does cause some reduction in the mover power as well. Figure 6.66 shows a single-range bin, 180, for sum, difference, cancellation and

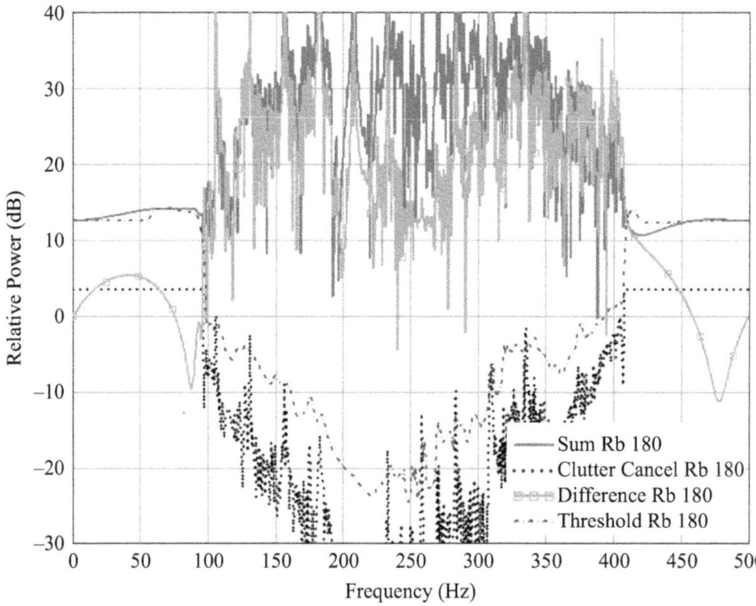


*Figure 6.66 Example 6.14 range bin 180 with 10 movers (GMTIAnApSynMono006h1)*

sliding window threshold. The Mathcad software is given in the Appendices Chapter 6 DVD-R or internet GMTIAnApSynMono006h1.

Equations (6.24) provide the required processing for synthetic monopulse. First step is to form synthetic beams A and B. Then combine those beams into sum, Σ, and difference, Δ. The next step is to perform Fourier transforms on Σ and Δ. Then bin-by-bin cross-correlation must be performed on all the range-Doppler bins, **WP**. The bins are averaged in range with noise loading, **WPA**. Finally, normalized cancellation is performed on each range-Doppler bin for the entire *N x J* array. All the definitions necessary for (6.24.1) and (6.24.2) are given in (6.24.3).

$\mathrm{F1\Sigma}_{n,j} = \text{Pulse Compressed Raw IQ Data}$ ,

Shift to baseband: $\mathrm{F2\Sigma}_{n,j} = \mathrm{F1\Sigma}_{n,j} \cdot \mathrm{PD}_{n,j}$

Form A and B beams:

$\mathrm{A} = \mathrm{F2\Sigma}_{n,j} \cdot \mathrm{PDp}_j + \mathrm{F2\Sigma}_{(n-1),j}$ , $\mathrm{B} = \mathrm{F2\Sigma}_{(n-1),j} + \mathrm{F2\Sigma}_{(n-2),j} \cdot \mathrm{PDm}_j$

Form Sum and Difference Beams and Amplitude Weight:

$$PF\Sigma_{n,j} = \left[\mathrm{F2\Sigma}_{n,j} \cdot \mathrm{PDp}_j + \mathrm{F2\Sigma}_{(n-2),j} \cdot \mathrm{PDm}_j + 2 \cdot \mathrm{F2\Sigma}_{(n-1),j}\right] \cdot \mathrm{PD2}_{n,j} \cdot \mathrm{AW}_n$$

$$PF\Delta_{n,j} = \left[\mathrm{F2\Sigma}_{n,j} \cdot \mathrm{PDp}_j - \mathrm{F2\Sigma}_{(n-2),j} \cdot \mathrm{PDm}_j\right] \cdot \mathrm{PD2}_{n,j} \cdot \mathrm{AW}_n$$

Find Fourier transforms of PFΣ and PFΔ in $n$ for each range bin $j$ :

$$PFS(n,j) = \mathrm{FT}\left(PF\Sigma_n^j\right), PFD(n,j) = \mathrm{FT}\left(PF\Delta_n^j\right) \tag{6.24.1}$$

Form cross correlation of synthetic sum and difference range-Doppler bins:

$$\mathbf{WP(n,j)}=\begin{bmatrix} PFS(n,j)\cdot\overline{PFS(n,j)} & PFS(n,j)\cdot\overline{PFD(n,j)} \\ PFD(n,j)\cdot\overline{PFS(n,j)} & PFD(n,j)\cdot\overline{PFD(n,j)} \end{bmatrix}$$

Average correlation matrix in range including diagonal loading:

$$\mathbf{WPA(n)}=\left[\left(\frac{1}{j_{\max}}\right)\cdot\sum_j \mathbf{WP(n,j)}\right]+\begin{bmatrix} Noise & 0 \\ 0 & Noise \end{bmatrix}$$

Define steering vector: $\mathbf{S}=\begin{bmatrix} 1 \\ 0 \end{bmatrix}$

Find matrix inverse and normalizing term:

$$\mathbf{WPAI(n)}=\mathbf{WPA(n)}^{-1} \quad , \; WPN(n)=\mathbf{S}^T\cdot\mathbf{WPAI(n)}\cdot\mathbf{S}+Noise\cdot TMR\cdot TMF$$

$$\text{Cancel: } FPcanc(n,j)=\overline{\left[\left[\mathbf{WPAI(n)}\cdot\mathbf{S}\right]^T\cdot\left(WPN(n)\right)^{-1}\right]}\cdot\begin{bmatrix} PFS(n,j) \\ PFD(n,j) \end{bmatrix}$$

(6.24.2)

Where: $\mathrm{AW}_n=a-(1-a)\cdot\cos\left(\frac{2\cdot\pi\cdot n}{n_{\max}}\right)$ and

$$\mathrm{PD}_{n,j}=\exp\left(\sqrt{-1}\cdot 2\cdot\pi\left(\begin{array}{l}\left(V_{ax}\cdot\cos\eta_{0imax2,j}-f_{off}\right)\cdot t_n-\dfrac{V_{ax}^2\cdot\sin^2\eta_{0imax2,j}\cdot t_n{}^2}{2\cdot R_{0imax2,j}}\ldots \\ -\dfrac{V_{ax}^3\cdot\cos\eta_{0imax2,j}\cdot\sin^2\eta_{0imax2,j}\cdot t_n{}^3}{2\cdot R_{0imax2,j}^2}\end{array}\right)\right)$$

and $imax2=\frac{imax}{2}$ , $f_{off}$ = frequency necessary to bring F1Σ to IQ baseband,

$\mathrm{PDp}_j=\exp\left(\sqrt{-1}\cdot \mathrm{k}\cdot D_{PRF}\cdot\cos(\eta_j)\right)$ , $\mathrm{PDm}_j=\exp\left(-\sqrt{-1}\cdot \mathrm{k}\cdot D_{PRF}\cdot\cos(\eta_j)\right)$ ,

$\lambda$= wavelength, $D_{PRF}$ = distance between PRI's, $R_0$= back of patch range @ $t_n=0$

$\mathrm{PD2}_{n,j}=\exp\left(-\sqrt{-1}\cdot\pi\cdot PRF\cdot t_n\right)$ , $a=0.54$, $V_{ax}$ = radar velocity, $\mathrm{k}=\frac{2\cdot\pi}{\lambda}$

$\eta_j$ = angle to center of processed array, $TMR$ = sliding window range multiplier,

Noise = expected noise floor, $TMF$ = frequency sliding window multiplier.

(6.24.3)

Figure 6.67 shows the thresholded output of 30 simulated moving targets after synthetic monopulse processing. Using more samples and forming more beams can improve performance. This example is to allow an understanding of the basic process.

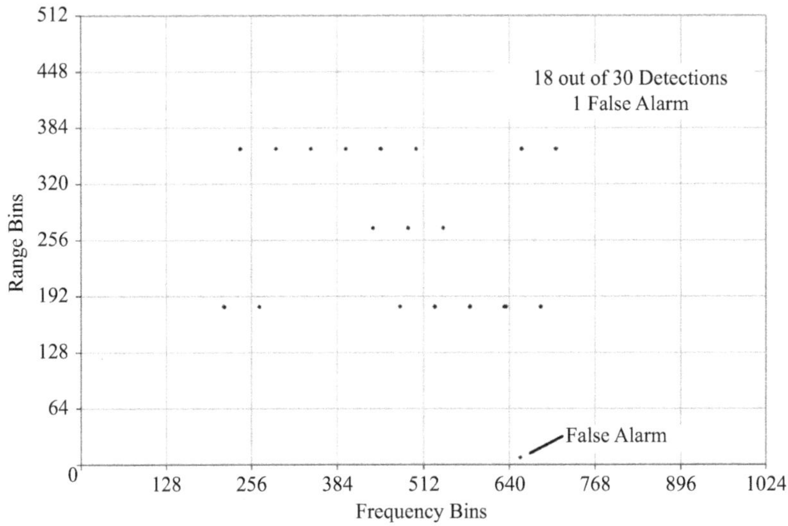


*Figure 6.67 Example 6.14 synthetic monopulse target detections (GMTIAnApSynMono006h1)*

## References


[1] Brown, R., Wicks, M., Zhang, Y., Zhang, Q., Wang, H., "A Space-Time Adaptive Processing Approach for Improved Performance and Affordability", *IEEE National Radar Conference*, May 1996, pp. 321–326.

[2] Guerci, J., *Cognitive Radar*, Artech House, Norwood, MA (2010), pp. 23–26, 135–153.

[3] Lynch, D., Mutton, J., Pearson, J., Shamash, E., Williams, F., "Advanced Radar", *Evolving Technology Course Notes*, June 1982.

[4] Bogdanovic, P., Graham, G., Pearson, J., Lynch, D., "Radar Data Processing", *Evolving Technology Course Notes*, 1986.

[5] Brown, R., Schneible, R., Wicks, M., Wang, H., Zhang, Y., "STAP for Clutter Suppression with Sum and Difference Beams", *IEEE Transactions*, Vol. 36, No. 2, April 2000, pp. 634–646.

[6] Ward, J., "Space-Time Adaptive Processing for Airborne Radar", *MIT Lincoln Laboratory Technical Report 1015*, December 13, 1994.

[7] Guerci, J., *Space-Time Adaptive Processing for Radar*, Artech House, Norwood, MA, (2003), pp. 4–66.

[8] Nohara, T., "Derivation of a 3-Channel DPCA/Monopulse Radar Using Phased Arrays", *1994 IEEE National Telesystems Conference*, pp. 243–246.

[9] Nohara, T., "Comparison of DPCA and STAP for Space Based Radar", *IEEE International Radar Conference 1996*, pp. 113–119.

[10] Graham, G., Lynch, D., Murray, J., Pearson, J., "Air to Ground Multimode Radar", *Evolving Technology Course Notes*, 1987.
[11] Harris, F.J., *Multirate Signal Processing for Communication Systems*, Prentice Hall, Englewood Cliffs, NJ (2004), pp. 210–214, 396–409, 446–458.
[12] Tylka, J., Swiger, J.M., Pearson, J., Lynch, D., "Synthetic Aperture Radar", *ETS Short Course Notes*, 1977.
[13] "Pave Mover Final Report", *Hughes Aircraft Company*, December 1984.
[14] Northrop Grumman Inc., JSTARS Marketing Brochure, 2006.
[15] Lynch, D., Kopp, C., "Multifunctional Radar Systems for Fighter Aircraft", Chapter 5, *Radar Handbook*, 3rd ed., M. Skolnik, Ed., McGraw Hill, New York, NY (2008), pp. 5.1–5.46.
[16] Lynch, D., *Introduction to RF Stealth*, SciTech Publishing Inc. (2004), pp. 492–544.
[17] Peebles, P., *Radar Principles*, John Wiley & Sons, New York, NY (1998), pp. 288–318, 557–614.
[18] Williams, R., "AFCD Final Report", *Hughes Aircraft Company*, October 1982.
[19] Lynch, D., "GMTI III", *Boeing Study Contract*, 2004.
[20] McCord, H.L., "Single Aperture DPCA Design Study", *Boeing Study Contract*, 2006.

# Chapter 7
# The synthetic aperture radar idea

## 7.1 SAR notions

Another way to detect slow-moving tactical targets is by means of HRM coupled with some form of change detection between multiple images. The change detection methods are coherent change detection (CCD) and shadow detection, non-coherent change detection (NCD), video SAR, inverse SAR (ISAR) and synthetic monopulse. Each of these will be discussed but first an understanding of conventional SAR is required. Often, fast(er) moving targets and slow-moving targets and SAR maps are acquired by the same waveform as described at the end of Chapter 6. The separate target types are usually separated by filtering.

HRM of the ground involves breaking up the antenna beam into fine resolution cells in azimuth and range as shown in Figure 7.1. The map is then made by forming those cells by some means and then measuring the signal intensity in each cell. The differing contrast in cells allows one to easily discern terrain features. One possible technique is tomography in which multiple observations at different angles are overlaid on one another and the slight contrast changes with angle ultimately allow a build up of an object map. This technique has been used with radars but it is often too slow in resolving the azimuth cell dimension.

SAR depends on relative motion of objects with respect to an observer within the beamwidth and resolving that motion into individual cells by Doppler or phase history. This can be done either actively or passively at any wavelength.

Every short-term motion of a point on a rigid body can be transformed into a translation along some LOS and a rotation about that LOS. If the extent of the body is small relative to an antenna beamwidth, the translation motion causes so little phase change that it usually can't be seen during the observation time. The rotation motion on the other hand creates enough differential phase change over reasonable observation times to resolve individual scatterers on an object. Figure 7.2 summarizes the SAR notion. There are three major applications of the SAR idea. First is classic SAR imaging in which the observer moves relative to a stationary object. Second is SAR imaging of an object on a turntable using a stationary radar. This is very common on radar cross-section ranges. Third is what is called inverse SAR imaging in which both the radar observer and the target are moving but the dominant motion is the target motion once the target centroid is tracked and motion compensated. This is common for ship and aircraft recognition as well as surface

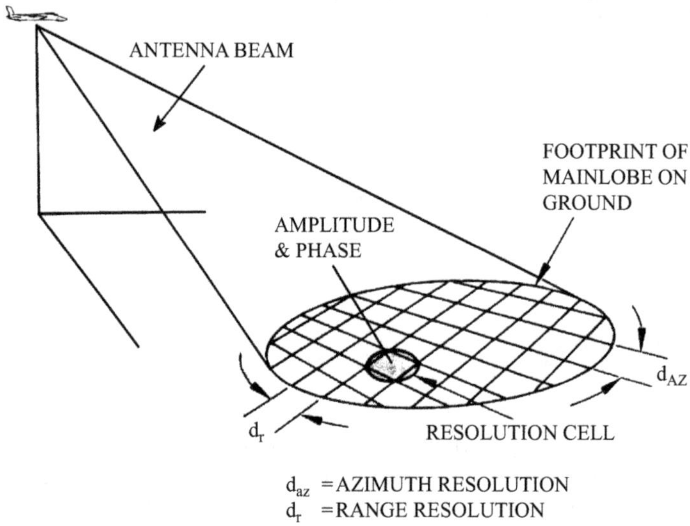


*Figure 7.1 High-resolution mapping (repeated Figure 2.39)*

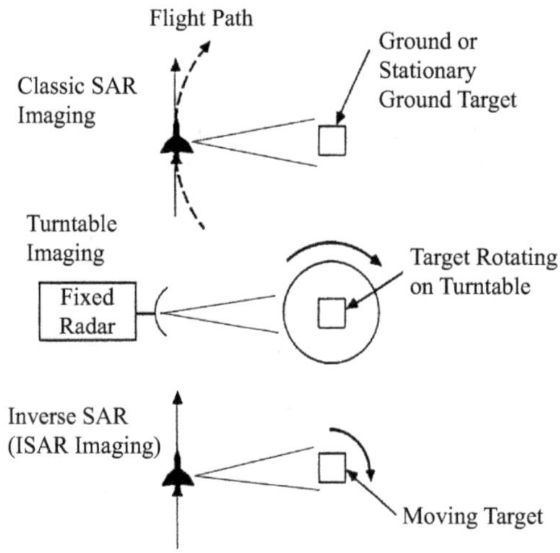


*Figure 7.2 Synthetic aperture radar notion*

vehicle imaging in turns. It will later be explained that all of these types of SAR arise in a naturally spherical or cylindrical coordinate space. The individual radar samples are not uniform in rectilinear coordinates.

The natural question to ask is, "what turn angle is required to achieve a certain resolution on the target?" Azimuth resolution only requires that there is an angular change between the image scene and the radar LOS. Either the target or the radar or

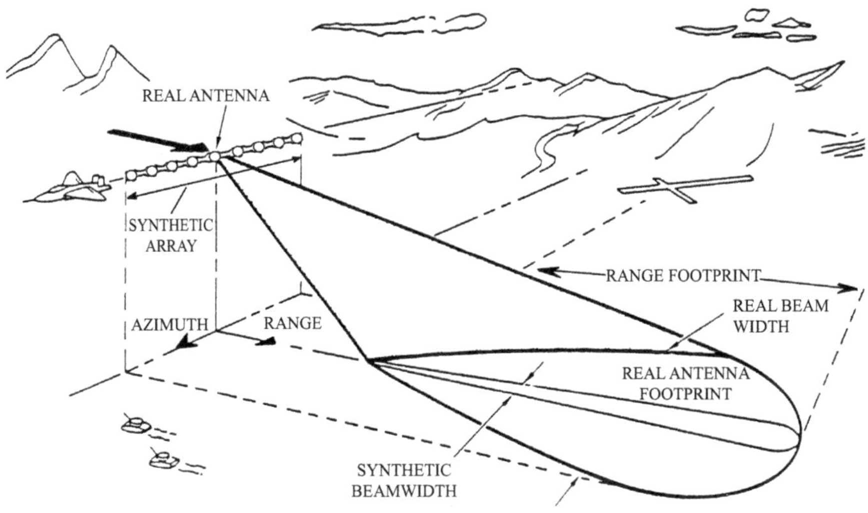


*Figure 7.3 SAR concept from an antenna point of view [1]*

both may move to achieve the required angle change. The angle changes required are surprisingly small. Equations (7.1) summarize this viewpoint.

$$d_{az} = \frac{k_w \cdot R \cdot \lambda}{2 \cdot D_S \cdot \sin\eta} = \frac{\lambda}{2 \cdot \alpha} \text{ achievable azimuth resolution}$$

$$\alpha = \frac{\lambda}{2 \cdot d_{az}} \text{ turn angle}$$

Where: $R$ = range to target, $\lambda$ = wavelength, $\alpha$ = turn angle,

$D_S$ is the SAR array length, $k_w$ = sidelobe weighting factor,

$\eta$ = compound look angle from velocity vector.

For example: $d_{az} = 0.3\text{m}$, $\lambda = 0.03\text{m}$, $\alpha = \frac{0.03}{2 \cdot 0.3} = 0.05$ rad. or $2.86^\circ$ turn angle.

(7.1)

The second way to think about SAR mapping is from an antenna aperture point of view. A real antenna on an airborne or spaceborne platform "flies-out" a longer synthetic aperture as shown in Figure 7.3. The real antenna takes on successive positions of the elements of a larger phased array. If the radar receiver can accurately measure the phase return from each object in the real antenna field of view, then phased array beamforming techniques can be used to form a synthetic beam inside the real antenna. Since the synthetic aperture is much longer than the real aperture, the synthetic beamwidth can be much narrower in the direction normal to the flight path.

**Example 7.1** Suppose a real array radar had the parameters in Table 7.1. It would have a cross-range resolution shown in the first line of example equations (ex 7.1). But if the same antenna flew out a synthetic array, then the cross-range resolution would be as shown in the second line of equations (ex 7.1). That is, a 100:1 improvement in resolution.

$$\text{Real array cross range resolution: } d_{az} = \frac{2\cdot k_w \cdot R \cdot \lambda}{D_x} = \frac{1.3\cdot 50000\cdot 0.03}{6} = 325\text{ m}$$

$$\text{Synthetic array cross range resolution: } d_{az} = \frac{k_w \cdot R \cdot \lambda}{2\cdot D_S \cdot \sin\eta} = \frac{1.3\cdot 50000\cdot 0.03}{2\cdot 300\cdot 1} = 3.25\text{ m} \quad \text{(ex 7.1)}$$

Using (7.1) it is easy to calculate the required synthetic array lengths necessary for various azimuth resolutions at broadside relative to the velocity vector. Figure 7.4 plots array lengths for an L band radar for resolutions between

*Table 7.1 Example 7.1*

| Parameter | Value |
|---|---|
| Antenna length, $D_x$, m | 6 |
| Wavelength, $\lambda$, m | 0.03 |
| Range, $R$, m | 50,000 |
| Synthetic array, $D_S$, m | 300 |
| Look angle, $\eta$, rad | $\pi/2$ |
| Sidelobe weighting factor, $k_w$ | 1.3 |

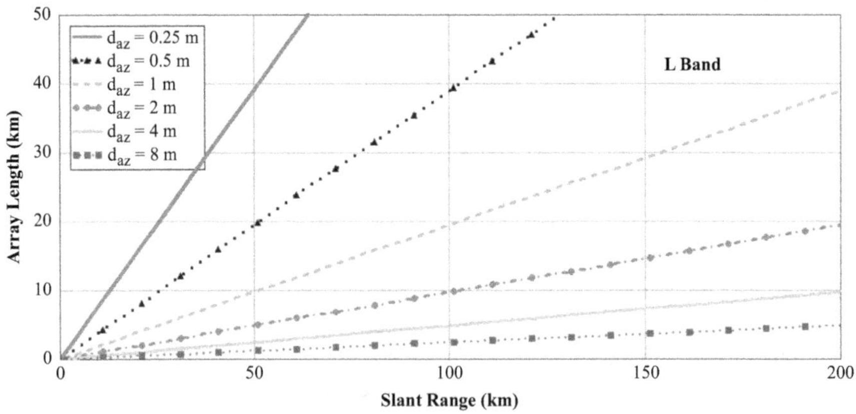


*Figure 7.4 Aircraft L band SAR array lengths (SARArrayLengthvsResolution)*

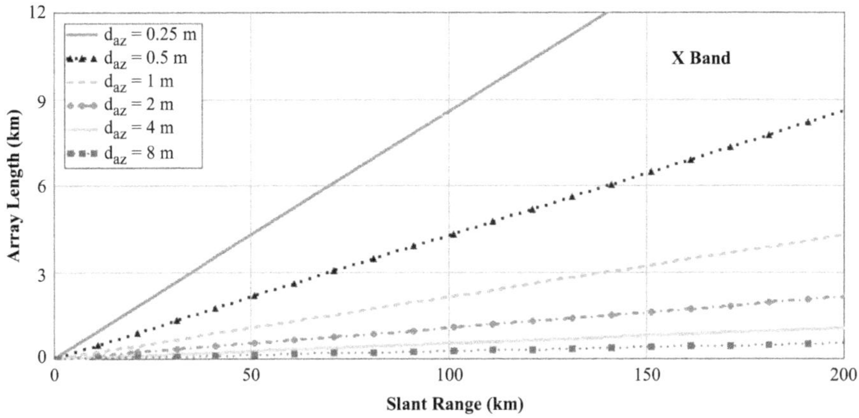


*Figure 7.5 Aircraft X band SAR array lengths (SARArrayLengthvsResolution)*

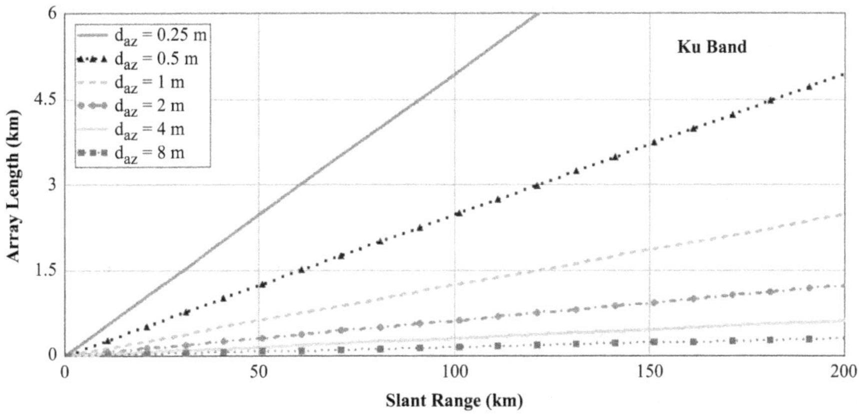


*Figure 7.6 Aircraft Ku band SAR array lengths (SARArrayLengthvsResolution)*

0.25 and 8 m. Figures 7.5 and 7.6 show similar array lengths for radars at X and Ku bands.

Figure 7.7 shows X band array lengths for the same resolutions as for aircraft, but assumes a spacecraft in a 1000 km orbit so the slant ranges extend to 3000 km. Because spacecraft travel so much faster than aircraft, it takes less than 30 s to fly out a 225 km synthetic array to achieve 0.25 m resolution at 2500 km slant range. Referring to Figure 7.5 an aircraft traveling at 250 m/s flies out an X band 12 km synthetic array in 48 s to achieve 0.25 m resolution at 140 km. They are roughly the same for comparable resolution.

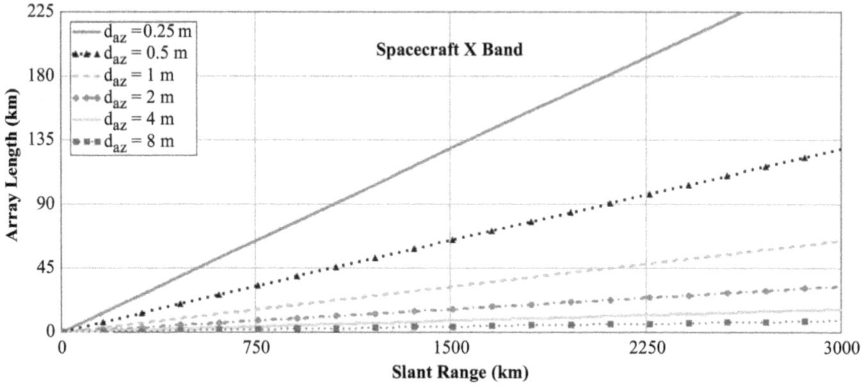


*Figure 7.7 Spacecraft X band SAR array lengths (SARArrayLengthvsResolution)*

## 7.2 Synthetic array geometry

Recall from Chapter 6 Figure 6.31 and (6.9) that the distance with time for each range-Doppler bin was derived for a flat earth model assuming without loss of generality that the velocity is along the $x$ axis. At many ranges and altitudes, this approximation is still valid as long as the SAR mapped patch is small relative to the range and altitude.

Figure 6.31 and (6.9) are repeated here as Figure 7.8 and (7.2), respectively. If the local terrain or earth curvature is to be taken into account, the height of the terrain for each bin in $x$ and $y$ must be added or subtracted from the mean height of the mapped site, $h$. This can cause some bins to have the same phase history, which is often called layover. Usually it is enough for adequate imaging to map $x$ and $y$ onto a plane through the mean height of map center canted so that the RMS error between the true height in each radar pixel and the approximating plane is minimum. This last method is commonly called conversion to site coordinates. Figure 7.8 shows an angle of $\eta_0$ to the $R_0$ range, which is a composite of the azimuth $\theta_0$ and elevation $\varepsilon_0$, that is, $\eta_0 = \text{acos}(\cos(\theta_0)\cos(\varepsilon_0))$. Obviously, the 4D vector equation stated at the beginning of (7.2) is correct everywhere. As mentioned in Chapter 6, often the range and range rate history are approximated by a Taylor series expanded around $R_0$. This is only accurate in the immediate vicinity of $R_0$ but often more than adequate for GMTI. In summary, the vector range history with time is equal to the square root of the vector range squared at $R_0$ plus the squared quantity of the vector velocity multiplied by time as shown in (7.2) for $R_{i,j}(t)$. From these equations, the range rate history, phase history and Doppler history are all derived. As mentioned in Section 2.9, constant range contours are arcs of ellipses or circles on the surface of the earth. Contours of constant Doppler are arcs of hyperbolas or catenaries. For small areas, these contours are often approximated in range by circular arcs and in Doppler by constant angle lines.

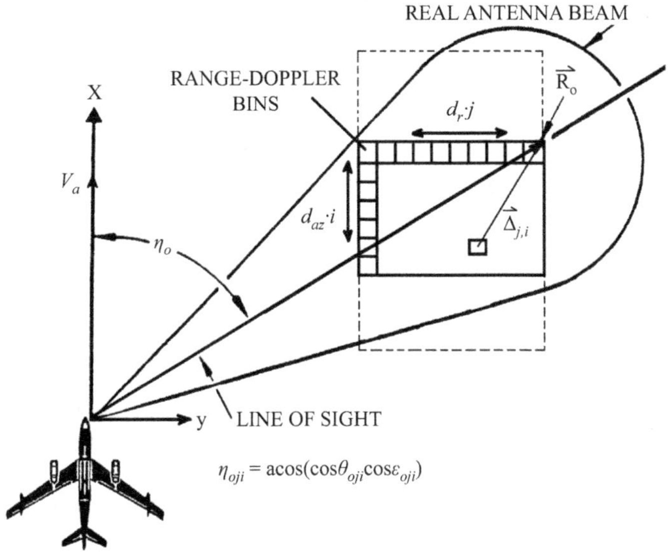


*Figure 7.8 Range-Doppler bin geometry including range closure (repeated Figure 6.31)*

$\mathbf{R}_0(x,y,z,t) = \mathbf{R}_0(x,y,z,0) - \mathbf{V}_a \cdot t$

For the simple case of $\mathbf{V}_a$ in the x direction $V_{ax}$ and a flat earth then:

The initial distance to each range-Doppler bin in cartesian coordinates is:

$R_{0i,j} = \sqrt{(R_0 \cdot \cos\eta_0 - i \cdot d_{az})^2 + (R_0 \cdot \sin\eta_0 - j \cdot d_r) + h^2}$

Where: $R_0$ is the range to the farthest range bin at initial angle $\eta_0$ ,

$h$ is the altitude, $d_{az}$ is the azimuth (x) bin size, $d_r$ is the range (y) bin size,

$i$ is azimuth (x) index to a cell, $j$ is range (y) index to a cell.

Let: $R_{0x} = R_0 \cdot \cos\eta_0$ and $R_{0y} = R_0 \cdot \sin\eta_0$

Then the distance to each range-Doppler bin with time is:

$R_{i,j}(t) = \sqrt{(R_{0x} - V_{ax} \cdot t - i \cdot d_{az})^2 + (R_{0y} - j \cdot d_r)^2 + h^2}$

Let: $R_{0xi} = R_{0x} - i \cdot d_{az}$ and $R_{0y\,j} = R_{0y} - j \cdot d_r$, $R_{0gi,j} = \left(R_{0xi}{}^2 + R_{0y\,j}{}^2\right)^{0.5}$

And the frequency with time is: $f_{D_{i,j}}(t) = \dfrac{2}{\lambda} \cdot \dfrac{dR_{i,j}(t)}{dt}$ Hertz

$$f_{D_{i,j}}(t) = \frac{2}{\lambda} \cdot \frac{-\left(R_{0x} \cdot V_{ax} - V_{ax}^2 \cdot t - i \cdot d_{az} \cdot V_{ax}\right)}{\sqrt{(R_{0x} - V_{ax} \cdot t - i \cdot d_{az})^2 + (R_{0y} - j \cdot d_r)^2 + h^2}} \tag{7.2.1}$$

The distance with time can also be expressed as:

$$R_{i,j}(t) \cong \sqrt{\left(R_{0_{i,j}} - V_{ax} \cdot t \cdot \cos\theta_{0i,j} \cdot \cos\varepsilon_{0i,j}\right)^2 + \left(V_{ax} \cdot t \cdot \sin\theta_{0i,j} \cdot \cos\varepsilon_{0i,j}\right)^2}$$

$$\text{Where: } \theta_{0i,j} = \operatorname{atan}\left(\frac{R_{0yj}}{R_{0xi}}\right) \text{ and } \varepsilon_{0i,j} = \operatorname{atan}\left(\frac{h}{R_{0gi,j}}\right) \tag{7.2.2}$$

Equations (7.2) can be approximated with the first few terms in a Taylor series but when the approximations are projected onto the earth surface, the resulting SAR image can be distorted if the resolution is less than about 1 m. In recognition of this fact for higher resolutions, the polar format often is used. Polar format consists of interpolating between the range-Doppler samples, which are spaced nonlinearly in the sense that their projected positions on the earth are not uniform but through transcendental equations containing sines and cosines of grazing and look angles. Once these equations are interpolated on to the surface, a uniform filtering using 1D or 2D Fourier transforms can be applied to the data, which greatly simplifies the map processing (more about this later in the chapter).

Also shown in Figure 7.8 is a dotted outline of where the range bins would appear to the sensor if the sensor was stationary and the map surface was moving. This is to remind the reader that the relative location of each bin is moving with time. Often the antenna beam must be moved during the array dwell time to keep the mapped patch in the center of the illumination. The antenna must be scanned while maintaining phase coherence from beam position to beam position. This is another constraint for electronically scanned arrays used for SAR mapping.

Figure 7.9 summarizes the two ways to think about the SAR principle. The basic properties important to SAR imaging are given in Figure 7.9 for each

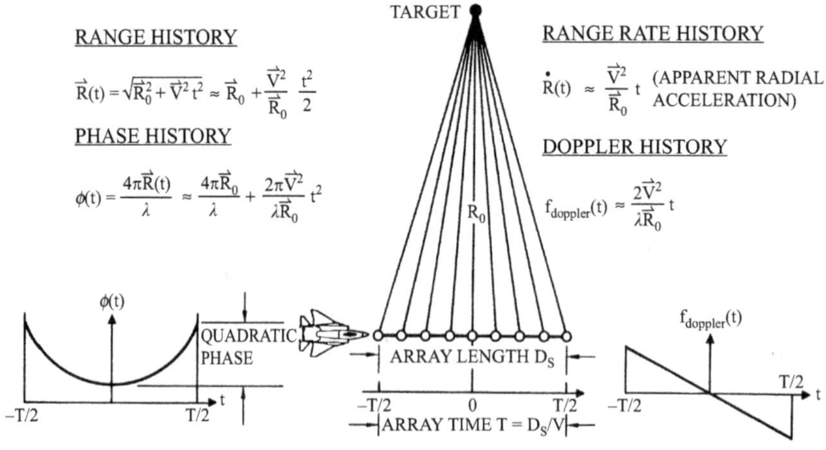


*Figure 7.9 SAR approximation summary*

target point. Note there are approximations for each of these attributes, which are in common use.

## 7.3 Doppler histories of point scatterers

Recall from Chapter 6 that the approximate phase history of a stationary object on the surface from a spaceborne or airborne platform was given in (6.10) and the Doppler frequency history is the derivative of the phase history as given in (7.3). For short segments, the frequency return from a single-range Doppler bin is approximately a linear FM or chirp. For MTI clutter cancellation, the approximations of (6.10) are adequate but for high-resolution SAR the approximations in Figure 7.9 may not be adequate. Usually the Doppler centroid is derived by tracking the mainlobe clutter signal and then using that frequency to partially demodulate the swath to be mapped. This notion is shown in Figure 7.10. SAR autofocus, which will be treated later in Chapter 7, is really a signal tracking method.

The exact Doppler history of each range-Doppler bin is:

$$f_{D_{i,j}}(t) = \frac{2}{\lambda} \cdot \frac{-\left(R_{0x} \cdot V_{ax} - V_{ax}^2 \cdot t - i \cdot d_{az} \cdot V_{ax}\right)}{\sqrt{\left(R_{0x} - V_{ax} \cdot t - i \cdot d_{az}\right)^2 + \left(R_{0y} - j \cdot d_r\right)^2 + h^2}} \text{ Hertz}$$

The highest Doppler frequency comes from the $i = 0$, $j = j_{\max}$ cell:

$$f_{D_{0,j_{\max}}}(t) = \frac{2}{\lambda} \cdot \frac{-\left(R_{0x} \cdot V_{ax} - V_{ax}^2 \cdot t\right)}{\sqrt{\left(R_{0x} - V_{ax} \cdot t\right)^2 + \left(R_{0y} - j_{\max} \cdot d_r\right)^2 + h^2}} \text{ Hertz} \tag{7.3.1}$$

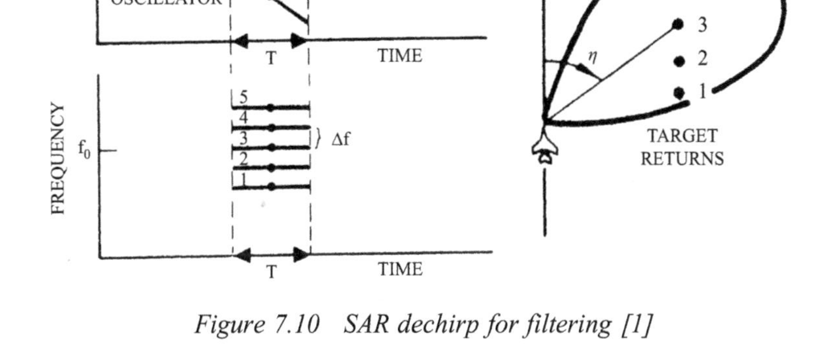


*Figure 7.10 SAR dechirp for filtering [1]*

The lowest Doppler frequency comes from the $i = i_{\max}$, $j = 0$ cell:

$$f_{D_{i_{\max},0}}(t) = \frac{2}{\lambda} \cdot \frac{-\left(R_{0x} \cdot V_{ax} - V_{ax}^2 \cdot t - i_{\max} \cdot d_{az} \cdot V_{ax}\right)}{\sqrt{\left(R_{0x} - V_{ax} \cdot t - i_{\max} \cdot d_{az}\right)^2 + \left(R_{0y}\right)^2 + h^2}} \text{ Hertz}$$

The frequency centroid of the range-Doppler bins is approximately at cell $i = {i_{\max}}/{2}$, $j = {j_{\max}}/{2}$:

$$f_{Dcentroid}(t) \cong \frac{2}{\lambda} \cdot \frac{-\left(R_{0x} \cdot V_{ax} - V_{ax}^2 \cdot t - {i_{\max}}/{2} \cdot d_{az} \cdot V_{ax}\right)}{\sqrt{\left(R_{0x} - V_{ax} \cdot t - {i_{\max}}/{2} \cdot d_{az}\right)^2 + \left(R_{0y} - {j_{\max}}/{2} \cdot d_r\right)^2 + h^2}} \text{ Hertz} \tag{7.3.2}$$

Returning to the phase history and its Taylor series approximation, one can derive an approximate time history of each point in a radar map as depicted in Figure 7.8 and shown in (7.4). The approximate phase history given in (6.10) can be improved by adding an additional term in the Taylor series approximation. This third-order term moderates the chirp slope. The time history has some complex amplitude, $A_{j,i}$, times the complex phase history in an exponential as shown at the bottom of (7.4). The third-order term in the exponential complicates the Fourier transform computation and is often left out for focusing after coarse azimuth compression. By splitting the presum output into multiple subbands and coarse compression using the third-order phase term at input to the FFT, a simple linear chirp may be all that is required at FFT output.

Note that: $\eta_{0i,j} = \operatorname{acos}\left(\cos\theta_{0i,j} \cdot \cos\varepsilon_{0i,j}\right)$

Let $t = 0$ when $\theta = \theta_0$ and $\varepsilon = \varepsilon_0$ and expand $\varphi_{i,j}(t)$ in Taylor series

$$\varphi_{i,j}(t) = \varphi_{i,j}(0) + \left.\frac{d\varphi_{i,j}}{dt}\right|_{t=0} \cdot t + \frac{1}{2} \cdot \left.\frac{d^2\varphi_{i,j}}{dt^2}\right|_{t=0} \cdot t^2 + \frac{1}{6} \cdot \left.\frac{d^3\varphi_{i,j}}{dt^3}\right|_{t=0} \cdot t^3 + \ldots$$

$$\varphi_{i,j}(0) = \frac{4 \cdot \pi}{\lambda} \cdot R_{i,j}(0)$$

$$\left.\frac{d\varphi_{i,j}}{dt}\right|_{t=0} = \frac{4 \cdot \pi}{\lambda} \cdot \frac{-2 \cdot R_{0i,j} \cdot V_{ax} \cdot \cos\eta_{0i,j}}{2 \cdot R_{0j,i}} = \frac{-4 \cdot \pi}{\lambda} \cdot V_{ax} \cdot \cos\eta_{0i,j}$$

$$\left.\frac{d^2\varphi_{i,j}}{dt^2}\right|_{t=0} = \frac{4 \cdot \pi}{\lambda} \cdot \left[\frac{V_{ax}^2}{R_{0i,j}} - \frac{R_{0i,j}^2 \cdot V_{ax}^2 \cdot \cos^2\eta_{0i,j}}{R_{0i,j}^3}\right] = \frac{4 \cdot \pi}{\lambda} \cdot \left[\frac{V_{ax}^2 \cdot \left(\sin^2\eta_{0i,j}\right)}{R_{0i,j}}\right]$$

$$\left.\frac{d^3\varphi_{i,j}}{dt^3}\right|_{t=0} = \frac{4 \cdot \pi}{\lambda} \cdot \left[\frac{-3 \cdot V_{ax}^3 \cdot R_{0i,j}^3 \cdot \cos^3\eta_{0i,j}}{R_{0i,j}^5} + \frac{2 \cdot R_{0i,j} \cdot V_{ax}^3 \cdot \cos\eta_{0i,j}}{R_{0i,j}^3} + \frac{R_{0i,j} \cdot V_{ax}^3 \cdot \cos\eta_{0i,j}}{R_{0i,j}^3}\right]$$

Simplfying the third derivative: $\left.\dfrac{d^3\varphi_{i,j}}{dt^3}\right|_{t=0} = \dfrac{4 \cdot \pi}{\lambda} \cdot \left[\dfrac{3 \cdot V_{ax}^3 \cdot \cos\eta_{0i,j} \cdot \left(\sin^2\eta_{0i,j}\right)}{R_{0i,j}^2}\right]$

Then the first four terms of the Taylor series expansion of the phase are:

$$\varphi_{i,j}(t) \approx \frac{4 \cdot \pi}{\lambda} \cdot \left[R_{0i,j} - V_{ax} \cdot \cos\eta_{0i,j} \cdot t + \frac{V_{ax}^2 \cdot \sin^2\eta_{0i,j} \cdot t^2}{2 \cdot R_{0i,j}} + \frac{V_{ax}^3 \cdot \cos\eta_{0i,j} \cdot \sin^2\eta_{0i,j} \cdot t^3}{2 \cdot R_{0i,j}^2}\right] \text{ rad.} \tag{7.4.1}$$

The approximate Doppler History for the $j,i$ bin is:

$$f_{D\,i,j} = \frac{2}{\lambda}\cdot\left[-V_{ax}\cdot\cos\eta_{0i,j} + \frac{V_{ax}^2\cdot\sin^2\eta_{0i,j}\cdot t}{R_{0\,i,j}} + \frac{3\cdot V_{ax}^3\cdot\cos\eta_{0i,j}\cdot\sin^2\eta_{0i,j}\cdot t^2}{2\cdot R_{0i,j}^2}\right]\ \text{Hz}$$

The time waveform is: $h_{i,j}(t) = A_{i,j}\cdot\exp\left(\mathrm{j}\cdot\varphi_{i,j}(t)\right)$ where $\mathrm{j} = \sqrt{-1}$

The approximate time waveform is:

$$h_{i,j}(t) \approx A_{i,j}\cdot\exp\left(\mathrm{j}\cdot\frac{4\cdot\pi}{\lambda}\cdot\left[\begin{array}{l} R_{0i,j} - V_{ax}\cdot\cos\eta_{0i,j}\cdot t + \dfrac{V_{ax}^2\cdot\sin^2\eta_{0i,j}\cdot t^2}{2\cdot R_{0i,j}}\ldots \\ +\dfrac{V_{ax}^3\cdot\cos\eta_{0i,j}\cdot\sin^2\eta_{0i,j}\cdot t^3}{2\cdot R_{0i,j}^2}+\cdots \end{array}\right]\right) \tag{7.4.2}$$

Usually, the fixed phase and amplitude factors shown in (7.4) are ignored when implementing the frequency domain-matched filter suggested in (7.5). The frequency chirp acceleration slope and starting frequency factors can't be ignored for high-resolution imaging. A coarse azimuth compression is usually performed by multiplying the incoming range compressed data by the complex conjugate of the $h_{j,i}(t)$ waveform at the middle angle $\mathrm{i}_{max}/2$ for each range bin. Also it is often assumed that the frequency extent of the matched filter is limited to a small number of Doppler bins around the starting frequency. Unfortunately, the chirp acceleration associated with angle is often larger than the variation with range and so matched filtering with azimuth Doppler is often required.

Using the approximate phase history in $h_{j,i}(t)$, finding its Fourier transform and grouping factors yields a frequency domain spectrum made up of four terms: an amplitude factor, a fixed phase factor, which is a function of range, a starting frequency divided by frequency chirp slope, and finally by a frequency chirp as shown in (7.5). The chirp frequency acceleration term can be approximated in various ways including using the principle of stationary phase (SP) and the first two terms of the generating series for an exponential. Both of these methods require coarse range or azimuth compression before application. It is not given here in (7.5).

The Fourier transform of the $h(t)$ for the terms up to quadratic is:

$$H_{j,i}(f) = A_{j,i}\cdot\sqrt{\frac{1}{2\cdot\Delta f_{j,i}}}\exp\left(\mathrm{j}\cdot\pi\left(\frac{4\cdot R_{0\,j,i}}{\lambda}+\frac{1}{4}\right)\right)\cdot\exp\left(-\mathrm{j}\cdot\pi\left(\frac{\left(f-f_{0\,j,i}\right)^2}{2\cdot\Delta f_{j,i}}\right)\right)$$

Expanding and grouping factors:

$$H_{j,i}(f) = A_{j,i}\cdot\left(2\cdot\Delta f_{j,i}\right)^{-1/2}\cdot\exp\left(\mathrm{j}\cdot\pi\left(\frac{4\cdot R_{0\,j,i}}{\lambda}+\frac{1}{4}-\frac{f_{0\,j,i}{}^2}{2\cdot\Delta f_{j,i}}\right)\right)\ldots$$

$$\cdot\exp\left(-\mathrm{j}\cdot\pi\left(\frac{f^2}{2\cdot\Delta f_{j,i}}\right)\right)\cdot\exp\left(\mathrm{j}\cdot\pi\left(\frac{f\cdot f_{0\,j,i}}{\Delta f_{j,i}}\right)\right)$$

$$\text{Where: } \Delta f_{j,i} = \frac{V_{ax}^2\cdot\sin^2\eta_{0\,j,i}}{\lambda\cdot R_{0\,j,i}},\ f_{0\,j,i} = \frac{2\cdot V_{ax}\cdot\cos\eta_{0\,j,i}}{\lambda}\ \text{and}\ \mathrm{j}=\sqrt{-1} \tag{7.5}$$

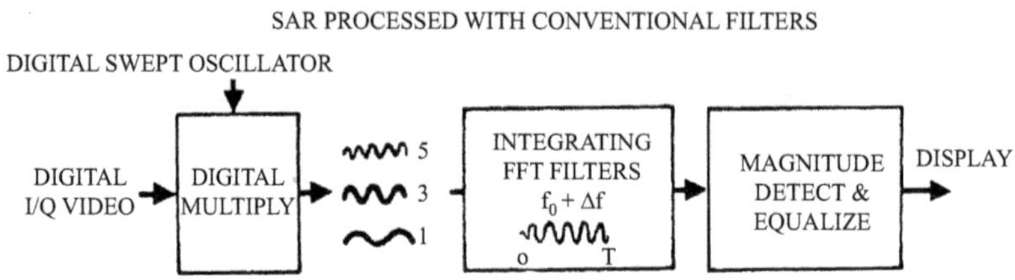


*Figure 7.11 Doppler filter bank for DBS/SAR imaging [1]*

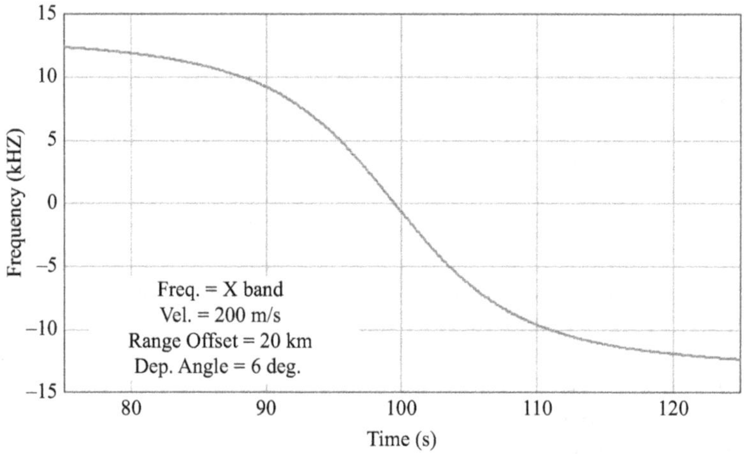


*Figure 7.12 Single-point target Doppler history 180° azimuth (RangeHistory002)*

Multiplying the range compressed raw azimuth data by the complex conjugate of the $h_{j,i}(t)$ waveform makes each range-Doppler return almost a "constant" frequency as shown in Figure 7.10. The SAR dechirp represented by Figure 7.10 in a gross view seems to provide a single tone for each azimuth return. However, close examination shows that these tones still have some residual chirp as a function of the azimuth angle and possibly Doppler shown in Figure 7.15. When applied to a subsequent FFT as shown in Figure 7.11, the output is an approximately focused SAR map, which may require some local additional focusing across a few frequency bins. When range closure (mentioned in Chapter 6) is used to compensate sample times for the target cell, the arc of competing constant range cells but with different Dopplers decorrelate and show up as a noise floor. If range closure compensation is done in a 2D matched filter as in polar processing, then that desired constant range bin must compete with all the returns from other angles. This usually requires much higher matched filter amplitude weighting for high contrast.

Figure 7.12 shows a more accurate Doppler history for a single-point target offset from the ground track by 20 km with a platform velocity vector of 200 m/s at a depression angle of 6° and a wavelength of 0.03 m. As one can see the central segment of the Doppler history slope is reasonably linear from around 45° to 135°

from the velocity vector. Figure 7.16 also shows that near the velocity vector changes in frequency are very small, thus requiring very long dwell times to achieve a SAR image. If motion compensation is very good, images of 10′ resolution can be made down to within 3° of the velocity vector. This almost linear behavior allows a swept local oscillator to dechirp the returns from individual targets and to separate and integrate returns with a Doppler filter bank. This method is suggested in Figure 7.11. If each range bin is range closure compensated as depicted by Figures 7.8 and 7.10, then each Doppler bin after dechirp will fall into a single filter as suggested by Figure 7.11. Thus single Doppler filter outputs for each range bin will represent the integrated return from a single-range-angle bin.

---

**Example 7.2** Using the equations of (7.3), the highest, lowest and approximate centroid Doppler histories can be calculated. This will determine required filter bandwidths, allowable resampling rates and final focusing matched filters. Figure 7.13 shows a range of Doppler histories across a swath to be mapped. The specific parameters are given in Table 7.2. The actual calculations are in DVD-R Appendix A.7 (RangeChangevsTime006).

Figure 7.13 shows that the instant bandwidth across the swath is a little less than 583 Hz but over the dwell time the excursion is roughly 900 Hz. With a PRF of 3498, the spectrum will be folded but will not be compromised across the swath. It will be approximately between 384 and 1221 Hz before demodulation.

Using the mid-range swath centroid frequency sweep to perform first stage demodulation will result in a spectrum extent for the swath as in Figure 7.14. Although the maximum and minimum range spectral lines appear flat in Figure 7.14, they actually have some residual slope that must be matched for a focused image. Figure 7.15 shows the maximum range Doppler with time expanded in amplitude from Figure 7.14. A similar curve with the opposite slope exists for the minimum range. Only the swath center will not require additional correction.

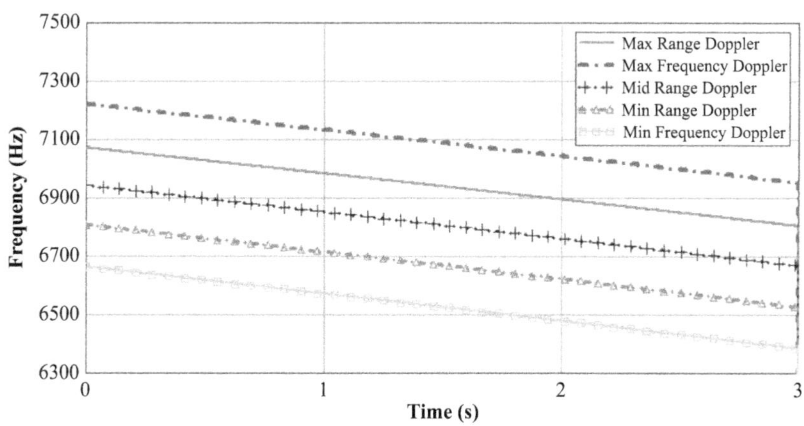


*Figure 7.13 SAR Doppler example across swath (RangeChangevsTime006)*

*Table 7.2 Parameters for SAR Doppler Example 7.2*

| Parameter | Value | Parameter | Value |
|---|---|---|---|
| Max. range (km) | 30 | PRF (Hz) | 3498 |
| Altitude (km) | 10 | Dwell time (s) | 3 |
| Wavelength (m) | 0.03 | Range resolution (m) | 1 |
| Azimuth angle (°) | 60 | Azimuth resolution (m) | 1 |
| Platform velocity (m/s) | 225 | Range bins | 768 |
| Elevation angle (°) | 19.5 | Azimuth bins | 1024 |

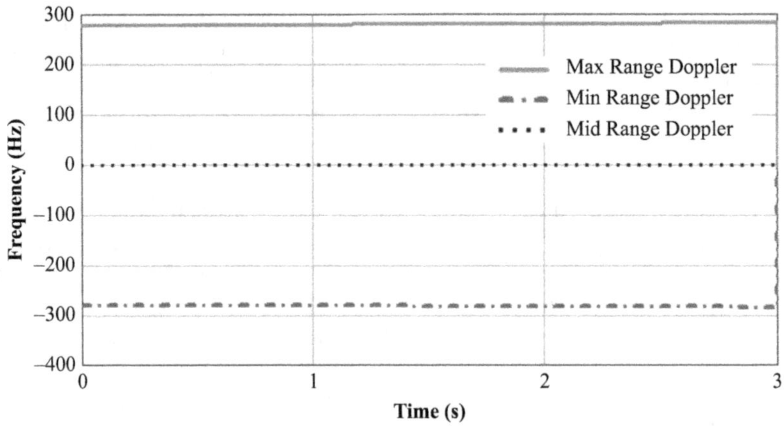


*Figure 7.14 Swath spectrum extent after centroid demodulation (RangeChangevsTime006)*

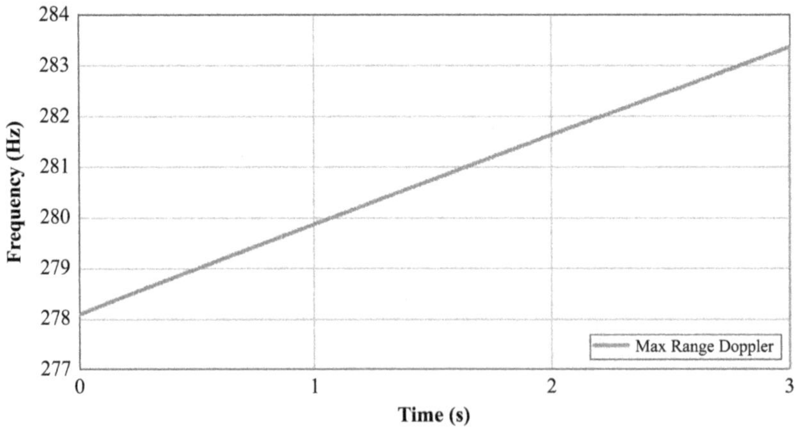


*Figure 7.15 Maximum range uncompensated chirp (RangeChangevsTime006)*

## 7.4 Doppler beam sharpening [2,3,4]

The intent of DBS is to provide a plan position indication (PPI) sector map with reasonable resolution over 90°. DBS is the first of the higher angular resolution ground mapping modes. It uses the Doppler spread of ground return across the beam to split the antenna beam into 10–100 segments. This mode is best done with 2D electronic scanning since the scan rate is nonuniform (sometimes called "funny" scan) but has been implemented with a combination of mechanical and frequency scanning as well.

DBS is very similar to SAR since both use the Doppler spread across the antenna mainlobe to create a higher resolution in the cross-beam direction [2,3,4]. The principal difference is the amount of angular coverage, beam scanning, resolution, data gathering time and accuracy of matched filtering in each range-Doppler cell. A DBS map may take a second to gather over an angle of 70°. In contrast, depending on angle from the platform velocity vector, a SAR map of a few feet resolution may take tens of seconds to gather at X band. DBS and SAR are compared in a qualitative way in Figure 7.16. Just as passive intercept of emitters is used ultimately to cue a high-resolution SAR map or electro-optical image, DBS is used to cue higher-resolution SAR maps.

As the beam is positioned closer to the velocity vector, the Doppler spread is smaller and so coherent dwell times must increase for the same resolution. Usually there is a transition from shorter CPIs and longer PDIs to longer CPIs and shorter PDIs as the beam approaches the aircraft velocity vector. Near nose-on dwell times become prohibitive and scan center is filled with real beam mapping. The real beam

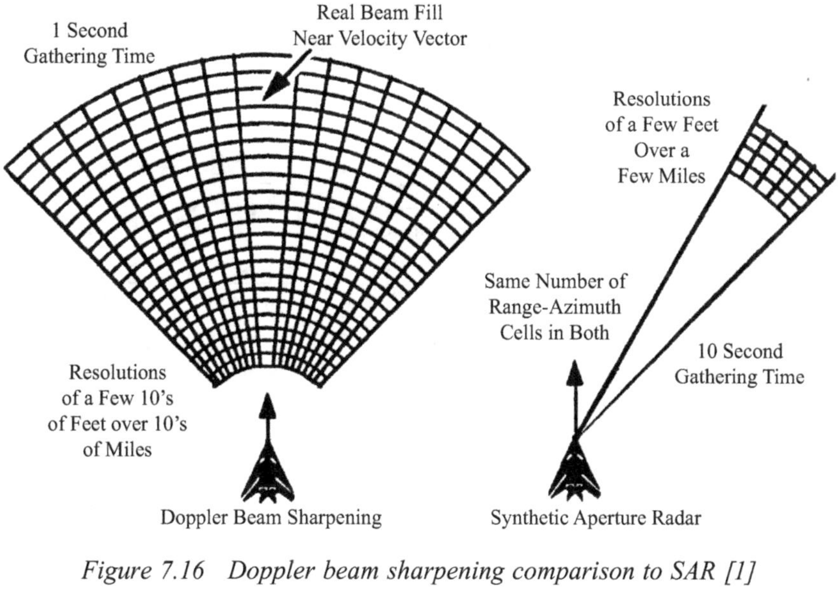


*Figure 7.16 Doppler beam sharpening comparison to SAR [1]*

uses the same range resolution, but since returns from the entire beam are used, some amplitude equalization is required to provide uniform contrast and brightness across the whole map. Some effort is made to match filter (both in range closure and phase history) the Doppler spread since iso-range and iso-Doppler's are not close to orthogonal near the aircraft velocity vector. SAR on the other hand usually is fully matched (relative to the desired resolution and phase history) in every range-Doppler cell.

### *7.4.1 DBS signal processing*

Pulse compression is assumed to be performed before the DBS processing described here. Because DBS range and cross-range resolution is usually 6–45 m, pulse compression is usually some type of phase code not chirp. Usually presumming is performed before image formation. Presummation (PRESUM) is a process in which only a fraction of the Doppler return in the mainbeam is used to form the image as shown in Figure 7.17. Presumming forms an unfocussed beam (i.e., there is little or no attempt to match the exact phase history of surface points) inside the real antenna beam by what is essentially a lowpass filter. Such filtering at the center of the mainbeam Doppler forms a sub-beam, which improves SNR and angle sidelobes.

Presumming also removes most moving targets from the DBS/SAR image formation, which may be or may not be an advantage depending on the desired functionality. Each presummed frequency channel is frequency shifted to match the clutter spectra and resampled at a lower rate for FFT filter formation. The resampling does fold the output spectrum so the noise floor does rise a little in the pass band.

Figure 7.18 shows the signal processing that might be found in a typical DBS mode. It consists of multiple time around echo (MTAE) suppression because the PRF may be ambiguous in range. Depending on grazing angle, ambiguous returns

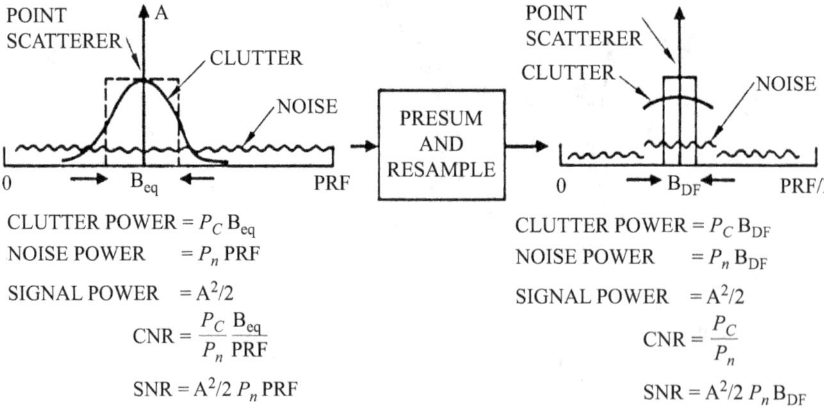


*Figure 7.17 Advantage to presumming for DBS/SAR [1]*

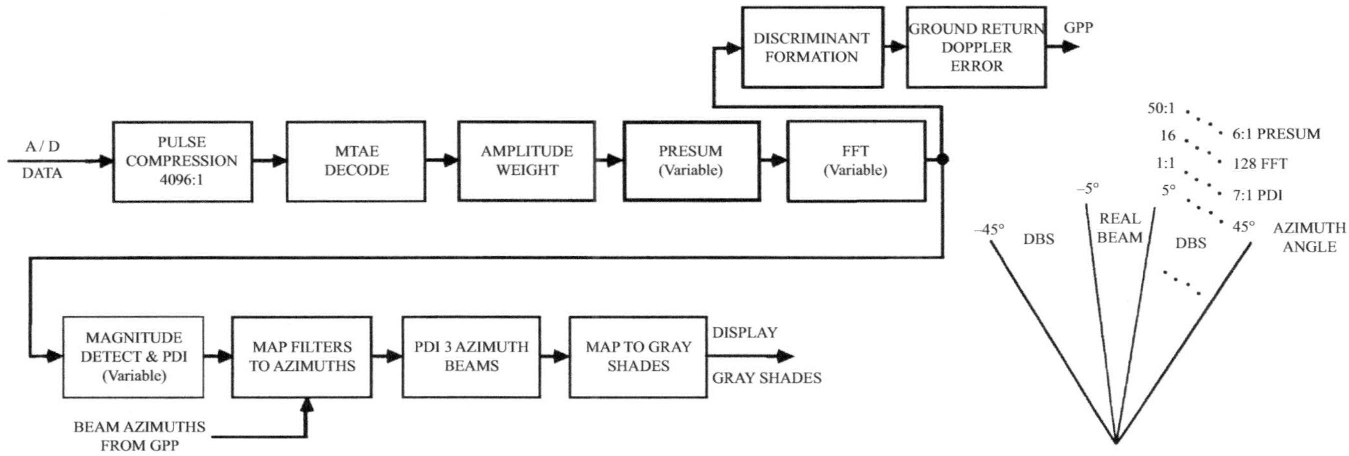


*Figure 7.18 Typical DBS processing [5]*

may compete with the region to be imaged. Often a combination of STC and pulse-to-pulse phase coding is used to reject MTAE [5]. The ground return in each frequency channel is positioned by a VFO and mixer before the A/D as well as resampling after the A/D filter to position the Doppler for most convenient processing (usually at either zero or PRF/2). There is enough range closure over the prefiltering and FFT process that range closure compensation must be performed before prefiltering. This is followed by amplitude weighting to improve sidelobes.

The bulk of the azimuth image formation is usually done with an FFT filter bank as shown in Figure 7.18. The FFT size ranges from 16 points at 5° off the velocity vector to 128 points at 45° off the velocity vector. The FFT array is filled with zeros to make the transform sizes round numbers at each beam position. The FFT outputs are used to form discriminants used for AGC and clutter Doppler error since it is not possible to predict Doppler or set the front-end VFOs accurately enough to move the center of the ground return to exactly zero or the PRF/2 frequency. Each filter output is magnitude detected and adjacent filters are collapsed and PDI to form 16 outputs across the azimuth beam. This forms an intensity map of the terrain. With a long frame time, the platform and hence the beam pointing direction can easily deviate from predicted position and so each beam actual angle must be used to integrate (PDI) the outputs from the separate synthetic beams to form the final map.

Normal movement by the platform during the gathering and processing time requires the placement of each filter output into the correct ground stabilized location. Also each CPI may have minor changes in the PRF to simplify processing and compensate for platform maneuvers. The platform may travel 1000 ft during the gathering time. There is considerable transport delay in most SAR and DBS processing; as a result, processed returns must be rectified (i.e., compensated for geometric distortion), motion compensated and mapped into the proper space angle and range position. Once radar pixels are properly located, each CPI is followed by PDI and scaling so that the display exhibits constant brightness and dynamic range.

The amount of PRESUM and PDI as a function of beam position off the velocity vector is shown in the lower right of Figure 7.18. For each different angle, there is a different Doppler spread across the beam. Therefore, in order to maintain a constant beam-sharpening ratio, different amounts of presumming must be used for each beam position. This would result in different target brightness and contrast if it were not compensated by applying a corresponding PDI for each angle as shown in Figure 7.18.

Multiple frequency looks are used to reduce speckle in the image and so several different frequencies are PDIed. The CPI is the presum ratio times the number of filter samples (128–800 typical). Since DBS usually maps a large area to provide overall ground situational awareness, the total range coverage is often completed with multiple elevation beams and range swaths. This is transparent to the operator but requires different PRFs, pulsewidths, filter shapes and dwell times.

An example DBS sector map without real beam map infill in the ±5° region around the velocity vector is shown in Figure 7.19 on the right. An accompanying printed map covering the same region is shown on the left. The outlined county

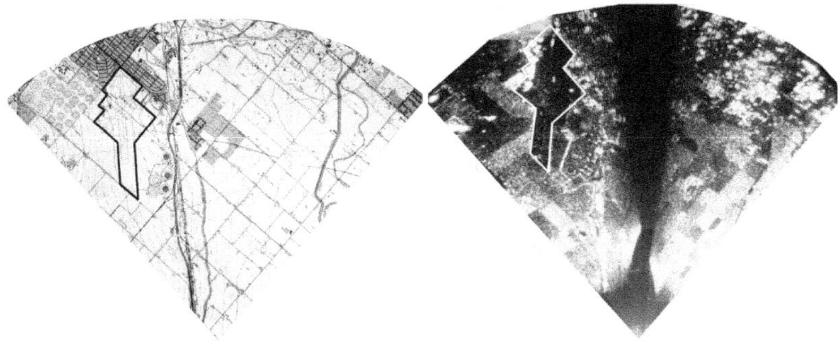

*Figure 7.19 Example DBS map without real beam map infill [3]*

*Table 7.3 Typical Doppler beam sharpening parameters*

| Parameter | Value |
|---|---|
| Waveform | 1–6 kHz PRF |
| Maximum peak power | 1 kW |
| Pulse compression | 128–4096:1 |
| Range resolution | 9–30 ft |
| Duty ratio | 5% |
| Frequency channels | 32 |
| Frequency looks/beam | 1–4 |
| Maximum range | 300 mi |
| Range bins processed | 10,000 |
| Doppler filters | 16–128 |
| Antenna scan rate | Funny |
| Azimuth coverage | 90° |
| Elevation coverage | Ground stabilized |
| Frame time | 6⅔ s |
| Total bandwidth | 50 MHz |
| Operating band | ~8–16 GHz |

airport on the map is clearly visible in the original DBS image. The airport is outlined in white in the DBS map. It should also be obvious from the DBS map that it is primarily for finding gross features for navigation or subsequent point designation for a higher-resolution spotlight SAR map. Very often hypsographic (terrain and cultural) features are added to the DBS map to facilitate recognition.

Example DBS parameters are given in Table 7.3. For the example here, the beam is split into a constant 16 segments but because the Doppler spread changes dramatically with off angle, which requires a variable presum filter, FFT size and post FFT filter collapsing and PDI. As shown in Figure 7.18 angle diagram, the changes in processing parameters used in the DBS scan over $\pm$ 45° about the velocity vector. There are three azimuth sectors: $-45°$ to $-5°$, $-5°$ to $+5°$ and 5° to 45°. The central sector about the velocity vector is often filled with real beam map. The two other sectors are DBS and have a different presum ratio, FFT size

*Figure 7.20 Alternate radar image compression clipped linear versus log*

and corresponding filter collapsing and PDI ratio for each angle. The range resolution is way out of proportion to azimuth resolution and so multiple range bins (16–32) are integrated after detection, which improves fading effects. The frame time is long but fortunately the required power is usually small because the area is big. The maximum range can be up to 300 miles and the PRF ambiguity is chosen to be just beyond the last range cell so MTAE rejection is required. Although it is not obvious from Table 7.3, the dwell time is 0.25 s per beam dwell.

The dynamic range from any mapping process can easily be 60 dB, whereas the typical display can only generate a dynamic range of 25 dB. Significant dynamic range compression often is required. Usually a nonlinear mapping of the amplitude range in any angle bin is mapped into eight shades of gray (alternatively into yellow-green on the display where the human eye is most sensitive). Sometimes the amplitude mapping is logarithmic and sometimes it is linear with clipping at maximum and minimum brightness. An example of an aircraft landing strip shows the two alternative compression schemes each with compression into 8 bits gray scale in Figure 7.20. It would seem that log compression is better in Figure 7.20 but it is a low dynamic range image. Often linear compression gives the operator or image interpreter better recognition data because shadows are darker and more defined.

## 7.5 SAR mapping

Another commonly used radar mode is SAR mapping. Any time two objects have relative rotation to one another a SAR map can be formed in the plane of rotation. This is the basis of space object imaging, inverse SAR and SAR imaging. This is another technique that has been around for 40 years. SAR imagery of your house with 3 ft resolution can be purchased over the internet if you know your latitude and longitude [6–16].

Typical examples of SAR imagery from the early 1970s are shown in Figure 7.21 for strip mapping and spotlight modes. Modern SAR imagery can be much better than this but most of the really good stuff is classified. Line-by-line strip

*Figure 7.21 Strip map and spotlight SAR imagery early 1970s [1]*

maps (bottom Figure 7.21) usually show major cultural features and can be used for cueing of subsequent spotlight SAR images. Strip map resolutions can be as low as 0.1 m resolution but usually are 1–3 m to allow coverage of large areas in a single pass.

Higher-resolution spotlight modes as shown in the top right of Figure 7.21 allow the estimation of height from shadows as obvious from the Salton Sea SAR image. The batch strip map top left in Figure 7.18 allows the differentiation between raw cotton bales in the center of the Calico Junction SAR map and vehicles in the lower right in the same image.

Table 7.4 summarizes the leading parameters for a typical SAR mode. This mode often requires the highest transmit power even though the mode has the highest level of integration because the cell RCS is so small due to a combination of grazing angle and absolute cell area, for example, $0.1 \times 0.1$ m $\times$ 0.01 m$^2$/m$^2$ = 0.0001 m$^2$ RCS. In spite of this limitation, multiple frequency looks and output dynamic range compression usually make SAR map outputs look almost photographic even with low SNR at the input.

*Table 7.4 Typical SAR parameters*

| Parameter | Value |
|---|---|
| Waveform | 2.5–6 kHz PRF |
| Maximum peak power | 1 kW |
| Pulse compression | 8192–16384:1 |
| Range resolution | ¼–3 ft |
| Duty ratio | 5%–20% |
| Frequency channels | 10 |
| Frequency looks/beam | 1–4 |
| Maximum range | 100–2000 mi |
| Range bins processed | 4096 |
| Doppler filters | 4096 |
| Antenna scan rate | Ground stabilized |
| Azimuth coverage | 90° |
| Elevation coverage | Ground stabilized |
| Frame time | Up to 60 s |
| Total bandwidth | 2 GHz |
| Operating band | ~8–16 GHz |

## 7.6 Strip mapping

The simplest SAR mode is strip mapping shown in Figure 7.21. The simplest strip mapping is fixed squint line-by-line processing, which was used in many of the early digital SAR mapping systems. Most of California was digitally mapped to a resolution of 11′ in the early 1970s by Hughes Aircraft under contract to the US Navy using line-by-line SAR. This doesn't sound like much, but there are roughly 36.6 GB 11 ft radar pixels in California, which strained the storage and analysis capacity of the Navy at the time. In line-by-line processing, each time a new azimuth element is acquired, then a new focused azimuth element-range line is produced. The second strip mapping method is batch processing in which an entire array of new data is acquired, then the entire azimuth-range array is produced. This allows the sharing of some radar functions. Batch allows the use of a 2D FFT for range, range closure and azimuth filtering. However, there will be keystoning and rotation distortion, which must be compensated if the resolution cell size is small. The batch update rate will be dependent on radar shared assets, array time and number of required looks.

In line-by-line, there is no sharing of radar functions but it allows maximum presumming and the arithmetic can be spread in time. There is minimum keystone distortion, amplitude scalloping and map updates are platform velocity dependent, which may be good or bad depending how the maps will be used. The Doppler filter design, range closure and range filter design are independent, which again can be good or bad depending on application.

Figure 7.22 shows the geometry situation for line-by-line. Line-by-line has an azimuth resolution limit, which is roughly ½ the azimuth antenna length times the

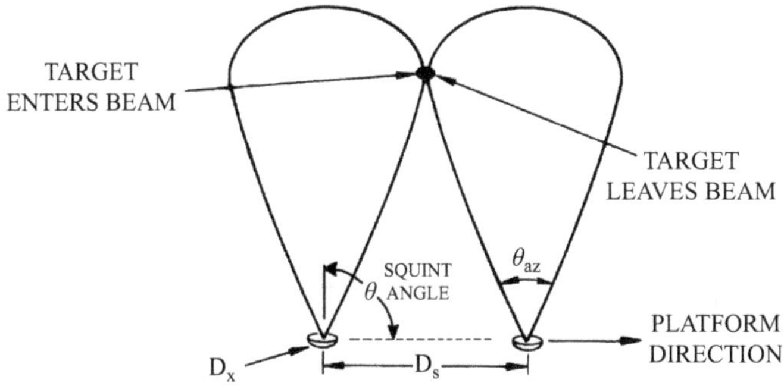


$$d_{az} = \frac{R\lambda}{2D_s} = \text{SAR RESOLUTION} \qquad \theta_{az} = \text{REAL ANTENNA BEAM}$$

$$D_{smax} = \frac{R\theta_{az}}{\sin\theta} = \frac{R\lambda}{D_x \sin\theta} \quad \therefore\ d_{azMIN} \cong \frac{D_x \sin\theta}{2}$$

*Figure 7.22 Fixed squint strip map limitations [1]*

sine of the squint angle, $D_x \sin\theta$ . The resolution is limited by the time that a single point remains in the antenna real beam as suggested in Figure 7.22, which for a fixed squint is fixed by the platform velocity. The more the beam is squinted toward the velocity vector, the longer the synthetic array length, $D_s$, can be. Of course, there is a practical limit to this approximation because as beam squint approaches the velocity vector the array times get so long that motion errors overwhelm matched filtering plus iso-range and iso-Doppler contours are no longer close to orthogonal. Typical squint angles are between 30° and 90° off the velocity vector.

### 7.6.1 Line-by-line processing

Line-by-line processing has often been done with two transversal filter shift registers, one for the range dimension and one for the azimuth dimension as shown in Figure 7.23. There have even been purpose-built cascade-able integrated circuits to implement the configuration shown in Figure 7.23. This configuration was especially advantageous in the days when large semiconductor memories were primarily serial. The range bin shift register implements a matched filter to the transmitted waveform usually a polyphase code.

Range closure compensation may be done before A/D conversion or before the azimuth shift register. In between the range compressor and the azimuth compressor is a simple serial presummer followed by a resampler to lower the rate for the azimuth shift register. The azimuth compressor implements a focusing matched filter in azimuth. This was sometimes done with simple and ingenious coefficient approximations. This configuration is often called a pipeline processor. Each time the platform travels one presum output sample or one azimuth resolution element a new map line is output.

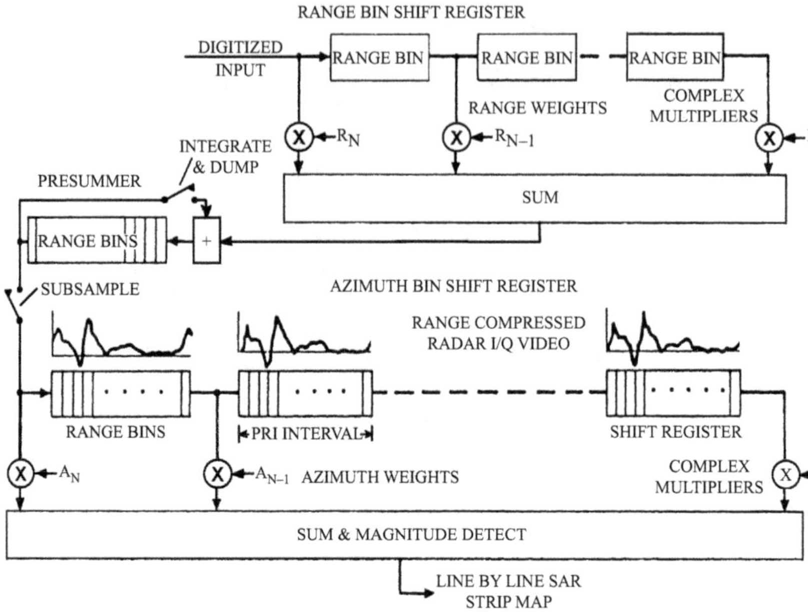


*Figure 7.23 Line-by-line strip map processing [1]*

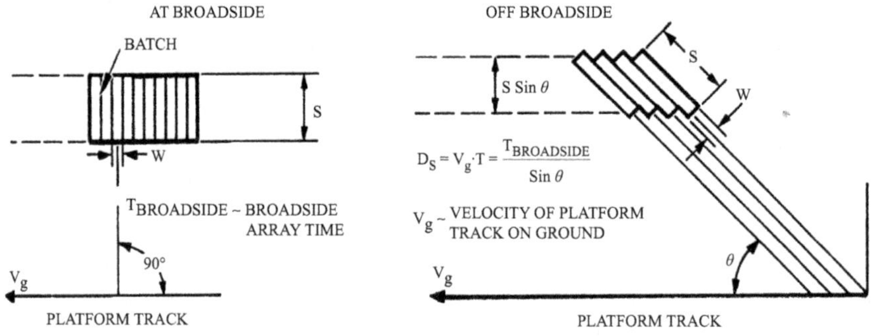


*Figure 7.24 Batch strip map [1]*

## *7.6.2 Batch strip map [1]*

The main weakness in line-by-line processing is that some of the dedicated hardware is working way below its maximum speed capacity. With the advent of large random access memories and programmable signal processors, hardware can be dramatically reduced by working all of the components near their maximum capacity. This economic fact has led to batch processing of almost all signal processing in every field – radar, EW, EOIR, acoustics, communications, TVs, CDs, DVDs, cell phones, drone autopilots, you name it. Most modern strip map processing uses batch processing. The notion is shown in Figure 7.24. There are

*Figure 7.25 Typical batch scalloping before compensation (courtesy of General Atomics) [17]*

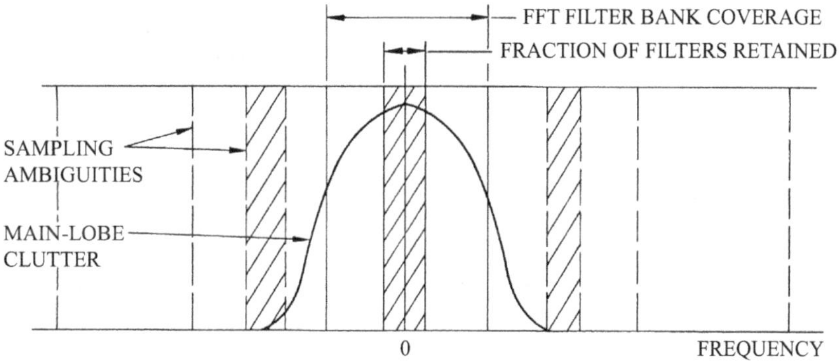


*Figure 7.26 Batch processing for strip mapping [1]*

several obvious problems with batch processing shown in Figure 7.24. First, batch uses more of the main beam so there will be amplitude scalloping across each batch shown in Figure 7.25. Second, stitching the edges of each batch together requires not only accurate motion compensation but also some interpolation at the edges to form a smooth continuous image.

There is a third challenge, which is shown in Figure 7.26. There will be an additional set of sampling ambiguities closer to the edge of the processed swath width, W. Some of the additional number smashing in the processors must be used to fix these three problems.

Fortunately, with modern signal processing if you can accurately state the problem, you can fix it with ever more complicated algorithms. Figure 7.27 shows the trade space between batch and line-by-line for strip mapping. Recall from (7.1) that $\alpha$ is the angle that the target or the observer must turn through in order to

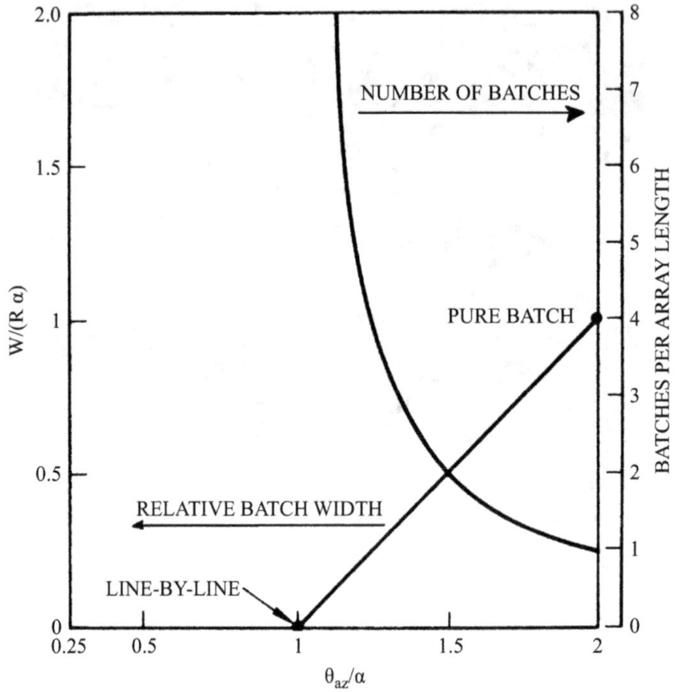


*Figure 7.27 Batch versus line-by-line tradeoffs [1]*

achieve a desired azimuth resolution. By definition in line-by-line the azimuth beamwidth and $\alpha$ must be the same. Furthermore, the desired batch width, W, divided by the range, R, to the beginning of the batch clearly must be larger than the required turn angle, $\alpha$.

Figure 7.27 shows that if the azimuth beamwidth and the turn angle are the same then all that can be achieved is a batch of one, that is, line-by-line. If the beamwidth is twice the required turn angle, then the batch angle can be the same as the turn angle and one batch can be made per array length. As the ratio of beamwidth to turn angle approaches 1, the number of batches required to make a continuous map gets very large approaching line-by-line. The message is that for batch to work well the beamwidth should be at least twice the required turn angle.

Figure 7.28 shows a block diagram of a typical strip map batch process. The figure shows an abbreviated front end including signal processing through partial pulse compression and presummation (if any). Often in both aircraft and spacecraft only a part of the processing is done on the platform and the remainder is performed in quasi-real-time in a ground station. Those front-end output samples may be encrypted and data linked or transferred by other means for the remainder of the processing at a ground processing station. This is especially true if the array size is very large, that is, over 4000 × 4000 radar or EO pixels.

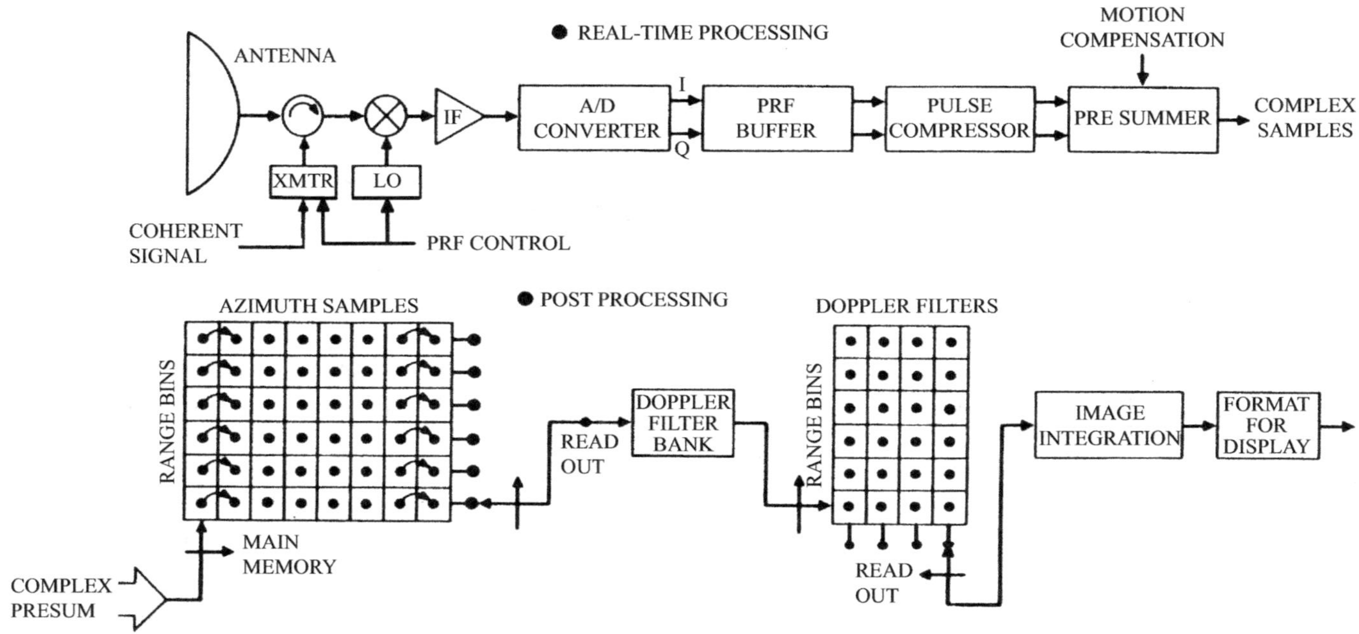


*Figure 7.28 Typical strip map batch processing [2]*

Ground station processing usually includes azimuth image formation, image integration with other data such as known terrain features, maps, cultural boundaries, recent collateral intelligence, prior images, change detection, etc. Automatic screening usually comes next (not shown in the figure) because there is just too much data to be examined by humans. Consider a 3 m resolution strip map of roughly 4000 range bins of the Los Angeles basin, the swath width would be 12 km. In a 12 × 12 km patch, there would be 144,000 houses and perhaps 25,000 vehicles. What could you do with that? Subsequent to screening the total data package may be provided to human analysts for exploitation. None of the exploitation is covered in this text but it must be and is extensive.

The subsequent postprocessing may even be done in multiple locations around the world. Usually, strip mapping even if it uses change detection does not track targets in real time but rather cues spotlight modes to specific areas for real-time observation and tracking. Clearly strip mapping can also be used for agriculture, terrain feature tracking, development surveys, water surveys, earth movement, etc.

## 7.7 Tracking telescope/spotlight SAR

The next most common SAR mode is commonly called tracking telescope or spotlight SAR. Usually spotlight will include the highest resolution mode but again lower-resolution spotlight will cue higher-resolution spotlight. Basically, the radar illuminates a single patch continuously long enough to achieve the desired resolution. For an aircraft this could even be flying in a circle.

Figure 7.29 shows one aspect of the spotlight resolution limit. Recall from (7.1) that the azimuth resolution, $d_{az}$, is related to the synthetic array length, $D_s$. The synthetic array length is the same as the velocity, $V$, times the array time, $T$, so $VT$

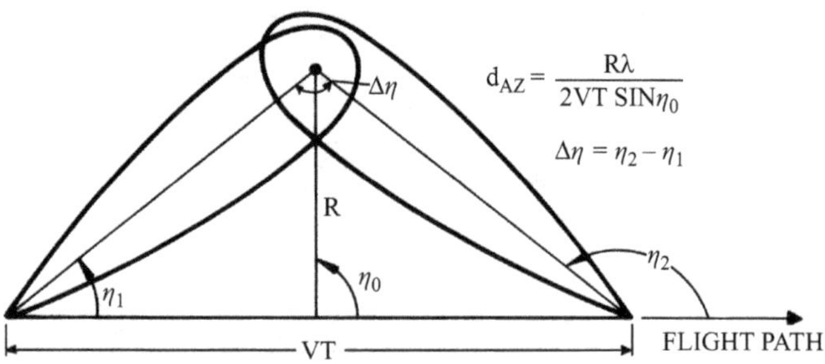


$$\text{FREQUENCY EXCURSION} = \Delta f_D \cong \frac{2V}{\lambda} \cdot SIN\,\eta_0 \cdot \Delta\eta \cong \frac{2V}{\lambda} \cdot SIN\eta_0 \cdot \frac{VT\,SIN\eta_0}{R} = \frac{2V^2T\,SIN^2\eta_0}{R\lambda}$$

$$d_{AZ} = \frac{V\,SIN\eta_0}{\Delta f_D}, \quad \Delta f_{DMAX} = \frac{4V}{\lambda} \quad \text{AND } \eta_0 = 90° \quad \text{THEN: } d_{AZMIN} = \frac{\lambda}{4}$$

*Figure 7.29 Spotlight resolution limit [1]*

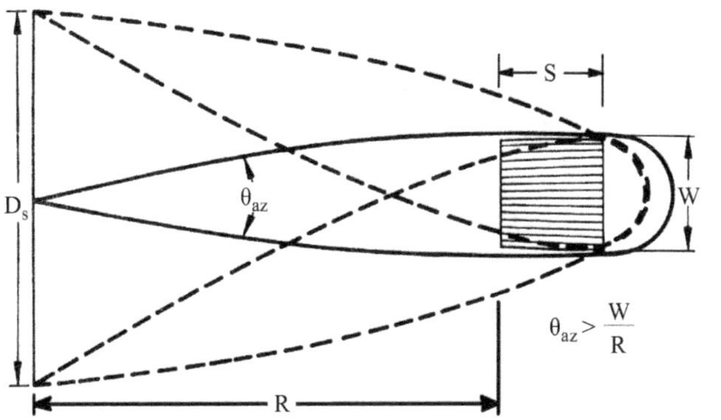


*Figure 7.30 Real beam antenna spotlight patch size limit [1]*

can be substituted for $D_s$ as shown in (7.4). If we assume a straight flight path for the platform as shown in Figure 7.29, then the maximum turn angle, $\alpha$, is 180° and the maximum Doppler change is $4V/\lambda$ (see Figure 7.11). This lower bound approximation is shown in (7.6) and Figure 7.29 and leads to a minimum resolution of $\lambda/4$. Obviously, a longer array could be achieved with a curved path but remember from Figures 3.4 and 3.6 that radar targets and terrain don't look the same from greatly different look angles. Does the side of the barn look the same as the front of the barn? No. Such an observation means that this resolution minimum is never achieved. In fact, even one wavelength resolution is very difficult to achieve in practice.

Another spotlight limitation is patch size. The real beamwidth must be greater than the angular width of the patch at the patch minimum range over the entire look angle. Figure 7.30 describes the problem. Since the look angle changes throughout the synthetic aperture, the azimuth real beamwidth must be large enough that the minimum range/azimuth bins continue to be illuminated by the radar. The real beamwidth must be greater than the angle, W/R. Although usually less of a problem, the maximum range/azimuth bins must also stay continuously illuminated.

$$d_{az} = \frac{R \cdot \lambda}{2 \cdot D_s \cdot \sin\eta_0} = \frac{R \cdot \lambda}{2 \cdot V \cdot T \cdot \sin\eta_0} \text{ and } \Delta\eta \approx \frac{V \cdot T \cdot \sin\eta_0}{R}$$

Spotlight straight path total frequency excursion

$$\Delta f_D \approx \frac{2 \cdot V \cdot \sin\eta_0 \cdot \Delta\eta}{\lambda} = \frac{2 \cdot V^2 \cdot T \cdot \sin^2\eta_0}{\lambda \cdot R} = \frac{1}{d_{az}} \cdot V \cdot \sin\eta_0$$

Solving the above equation for $d_{az}$ then: $d_{az} = \dfrac{V \cdot \sin\eta_0}{\Delta f_D}$

If $\Delta f_{DMAX} = \dfrac{4 \cdot V}{\lambda}$ and $\eta_0 = 90^o$, minimum azimuth resolution is: $d_{az} = \dfrac{\lambda}{4}$

(7.6)

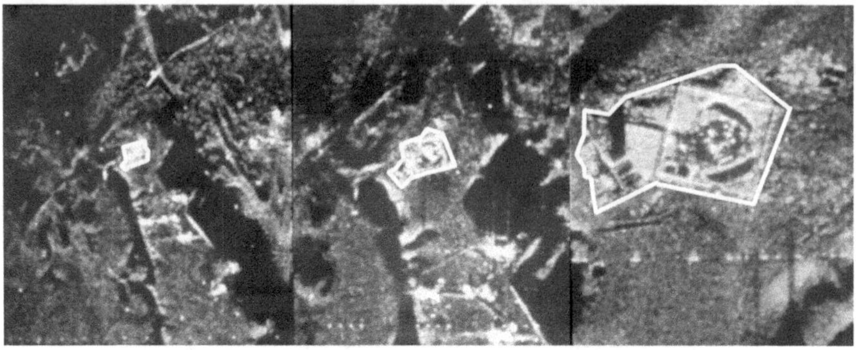

*Figure 7.31 Progressive spotlights of missile launch site at three resolutions [3]*

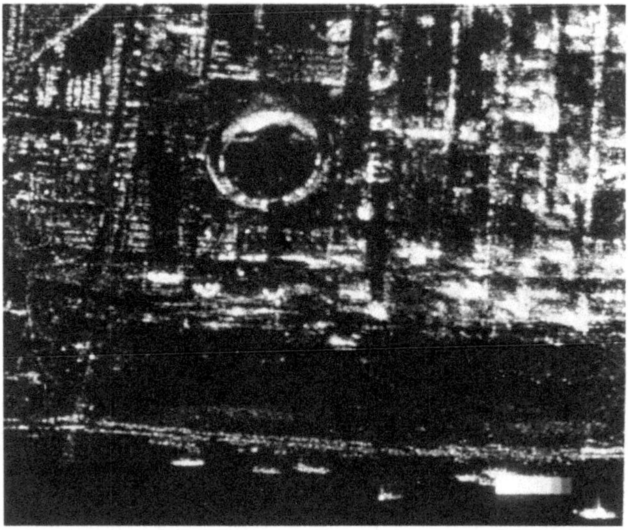

*Figure 7.32 Downtown St. Louis spotlight [15]*

Figure 7.31 shows the typical progression to higher resolution of a missile launch site in which each image cues the next usually through an operator redesignation for each. The missile launch site is outlined in white in this sequence of images.

Figure 7.32 shows a 3 m resolution spotlight map of downtown St. Louis. The dark areas at the bottom of the map are the sloping levy and the Mississippi river, only the marge of the river where there is turbulence along the levy is bright.

A top-level functional diagram of SAR processing is shown in Figure 7.33. Partially processed data is returned to memory between each major signal processing function. If the data arrays are very large, memory storage formatting is essential to maintain high throughput. Although many authors describe SAR

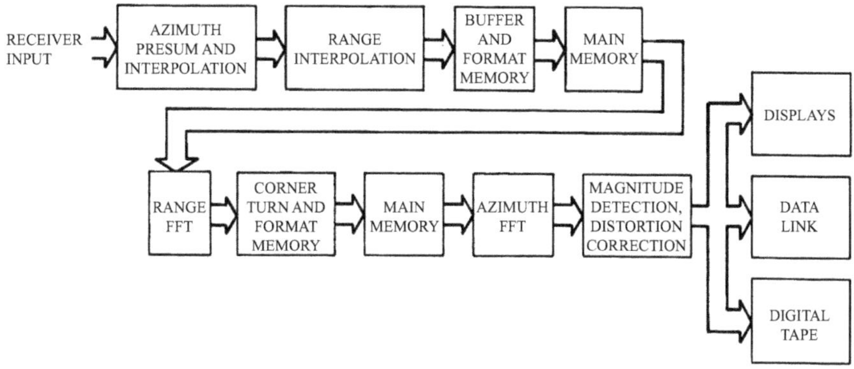


*Figure 7.33 Typical spotlight SAR processing [1]*

compression when a chirp waveform is transmitted as a 2D FFT process in which both range and azimuth (Doppler) are match filtered in the transform domain followed by inverse transform. Some of these functions may be on the platform but others may be performed at a ground station and not necessarily in real time. There are several reasons for this: first, pulse compression may not be a chirp for LPIR/LO/jamming; second, super resolution and autofocus techniques can be employed after careful analysis of the I/Q data; and on-board processing may not be adequate for all the processing that may need to be employed.

The particular spotlight or tracking telescope example mode described provides a high-resolution image of a relatively small patch on the ground. The resolution is high enough that recognition of vehicles, most manmade objects, tire tracks, buildings, fences, piping, etc., is quite reliable. Such a mode allows targeting and high-accuracy weapon delivery for virtually all military targets. The principal challenge for a radar is the fact that the cell size is quite small and hence its radar cross section is tiny. The power required may be the largest or the smallest of any radar mode depending on geometry. Large PCRs are commonly used but PRFs are restricted as will be shown later. Studies have shown that per pixel SNRs of 4 dB are adequate [6,16]. Motion compensation is essential in SAR and often autofocus techniques are required to obtain crisp maps.

Figure 7.34 provides a concept-oriented block diagram of the type of processing, which is accomplished at each stage in Figure 7.33. The digital A/D output after going through an Hilbert transform filter is phase shifted to a convenient dechirped central reference point (CRP). For high-resolution SAR images below 1 m, geometric distortion becomes a significant problem. As has been suggested elsewhere and shown in Figure 7.34, the range and azimuth samples must be interpolated and a new set of samples created, which map nonuniform spatial samples into a geometrically rectified rectangular grid. The first step is interpolating the range samples and creating output samples in the range dimension. In some cases, the range dimension is oversampled and mild presummation is performed. Range samples are skewed and so the incremental difference bin to bin is

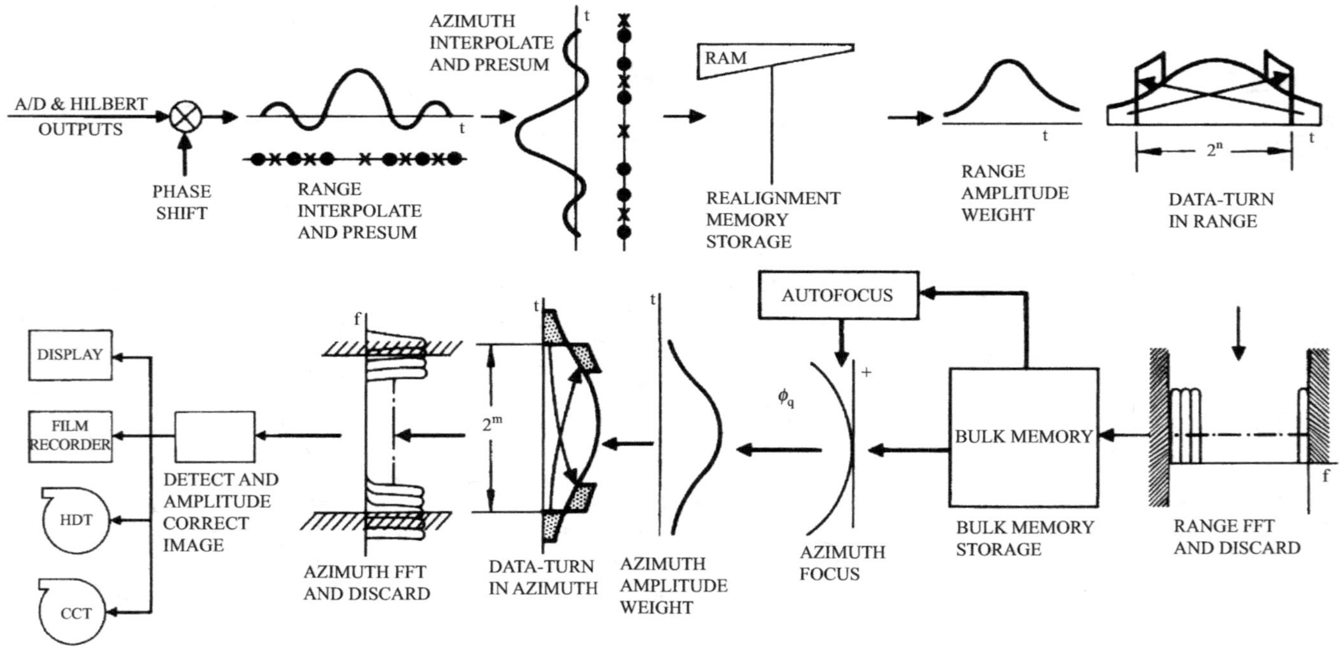


*Figure 7.34 Qualitative spotlight SAR processing [1]*

realigned as part of raw data storage. Subsequent to realignment, range dimension amplitude weighting is applied to improve range peak and integrated sidelobes (see Table 4.1 and Figures 4.12, 4.13 and 4.14). Following amplitude weighting zeros are added to the range bin record to create a convenient FFT size, typically $2^n$ (see Figure 4.15). In one embodiment of the FFT, the first pass adds and subtracts data whose indices are ½ the record length apart (see Equation (4.3)). This operation plus resampling has been called data-turning and it has often been combined with amplitude weighting (see Figure 4.11). It folds the spectrum about the middle and the tails of the amplitude record may be aliased as suggested in the upper right in Figure 7.34. The remainder of the range FFT passes are performed and the FFT outputs for the edge filters are discarded. Each PRI range swath is stored in a bulk memory.

During storage of the whole SAR array of data, smaller subarrays are retrieved from memory and focused to create subspace maps, which are used to provide autofocus for the full SAR array. These subspace outputs are just lower-resolution SAR maps covering a shorter gathering time and a shorter-range swath. Subspace maps can also be used to equalize amplitudes across the whole array, to detect moving targets, to set thresholds and to correct geometric distortion. After autofocus is applied to the phase focus terms across the array in azimuth, phase and amplitude weighting in the azimuth dimension is performed. There may be some zero fill (see Figure 4.15) to improve straddling loss as well as to optimize FFT size. Following data-turning, a second FFT is performed. Again, edge filters are discarded and the central filters are magnitude detected. Local histogram averaging as well as subspace maps are employed to normalize brightness across the map as well as to improve geometric matching to earlier maps for change detection.

The output images are then sent to operator displays, film recorders, high-density tapes (HDT) or computer compatible tapes (CCT). Why all these different formats for exploitation or storage? Displays have very limited dynamic range but allow near real-time analysis by operators. Film is the cheapest and most long-lived analog storage medium but it has very limited dynamic range. HDT is the cheapest digital storage medium but it requires maintenance to preserve its integrity. CCT or DVD/CDs are the most convenient digital formats but for large quantities of data indexing and archiving they are bulky and slow to access.

---

**Example 7.3** Referring to Figure 7.34, it often turns out that a single-phase demodulation (PD) and amplitude weight as suggested at the output of bulk memory in the lower line of the figure is adequate to focus a SAR image. For example, consider Table 7.5 Example 7.3 spotlight SAR.

Using (7.4) and its complement, PD in equations (ex 7.3), the output map cropped, amplitude detected and normalized to fit in a 256 bit dynamic range is shown in Figure 7.35. The Mathcad program SoCalSAR030a10 and it counterpart Adobe Acrobat file are in the Appendix Chapter 7 DVD-R or internet.

*Table 7.5 Example 7.3 spotlight SAR*

| Parameter | Value | Parameter | Value |
|---|---|---|---|
| Range (km) | 20 | PRF (Hz) | 3985 |
| Altitude (km) | 4.5 | Dwell time (s) | 4 |
| Wavelength (m) | 0.015 | Range resolution (m) | 0.3 |
| Azimuth angle (°) | 75 | Azimuth resolution (m) | 0.3 |
| Platform velocity (m/s) | 100 | Range bins | 1024 |
| Presum sample rate (Hz) | 398.5 | Azimuth bins | 1594 |
| Antenna width (m) | 1 | Frequency offset (Hz) | 3400 |
| Antenna height (m) | 0.8 | Hamming amplitude Wt | 0.54 |

*Figure 7.35 Cropped output from raw IQ data Example 7.3 (SoCalSar030a10)*

$\mathrm{F1\Sigma}_{n,j}$ = Pulse Compressed Raw IQ Data

$\mathrm{F2\Sigma}_{n,j} = \mathrm{F1\Sigma}_{n,j} \cdot \mathrm{AW}_n \cdot \mathrm{PD}_{n,j}$

Where: $\mathrm{AW}_n = a - (1-a)\cdot\cos\left(\frac{2\cdot\pi\cdot n}{n_{\max}}\right)$ and

$$\mathrm{PD}_{n,j} = \exp\left(j\cdot 2\cdot\pi\left(\begin{array}{l}\left(V_{ax}\cdot\cos\eta_{0imax2,j} - f_{off}\right)\cdot t_n - \dfrac{V_{ax}^2\cdot\sin^2\eta_{0imax2,j}\cdot t_n^{\,2}}{2\cdot R_{0imax2,j}}\cdots \\ -\dfrac{V_{ax}^3\cdot\cos\eta_{0imax2,j}\cdot\sin^2\eta_{0imax2,j}\cdot t_n^{\,3}}{2\cdot R_{0imax2,j}^2}\end{array}\right)\right)$$

and $imax2 = \frac{imax}{2}$, $f_{off}$ = frequency necessary to bring $\mathrm{F1\Sigma}$ to IQ baseband,

then Fourier transform in $n$ for each $j$: $\mathrm{FS}_{n,j} = \left(\mathrm{FFT}\left(\mathrm{F2\Sigma}_n\right)\right)_j$ (ex 7.3)

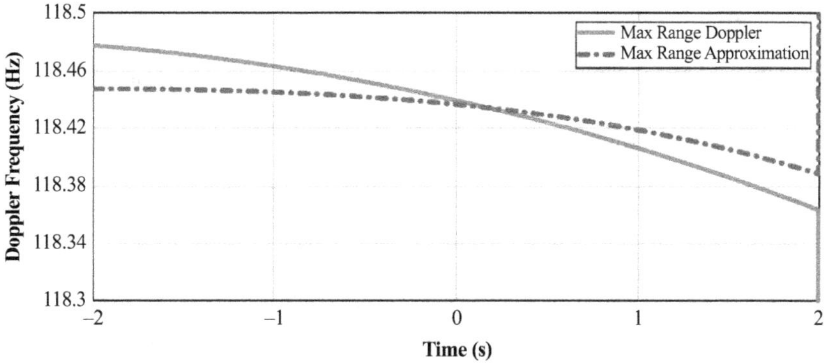


*Figure 7.36 Maximum range uncompensated Doppler (RangeHistoryvsTime007e)*

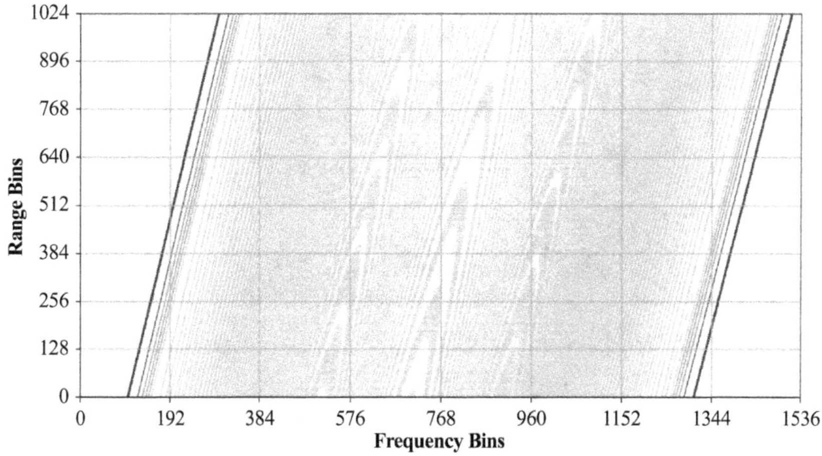


*Figure 7.37 Spectrum for range bin by range bin phase demodulation (RangeHistoryvsTime007e)*

After a single-phase demodulation, PD, shown in example equations (ex 7.3), the output residual phase difference between true frequency and the bin-by-bin approximate PD at the array edge is given in Figure 7.36. PD only compensates for the center angle in the mapped patch. Since there is a small difference between the angular center and the Doppler center, a small additional offset is required to center the map, in this case 0.08 Hz. The array filter bandwidth is approximately 0.36 Hz so the residual signal stays in a single filter for the full array time.

Figure 7.37 shows the spectrum of the PD of (ex 7.3) equations for each range-frequency bin. As can be seen, the range bin to range bin frequency varies significantly with time. After PD, the residual error for one edge of the mapped patch is shown in Figure 7.36. Because of minor flight path or orbit, a small amount of compensation is usually required by tracking the Doppler centroid to position the spectrum in the middle of the image.

Figures 7.38 and 7.39 provide other examples of high-resolution spotlight SAR maps. One is of a baseball diamond in San Diego county and another is of an array of tanks at the Barstow-Nebo, CA, tactical target array.

A careful observer will notice the shadow cast by the exterior baseball park signs as well as by several trees that allow estimation of the illumination direction

*Figure 7.38 High-resolution SAR image (courtesy General Atomics) [17]*

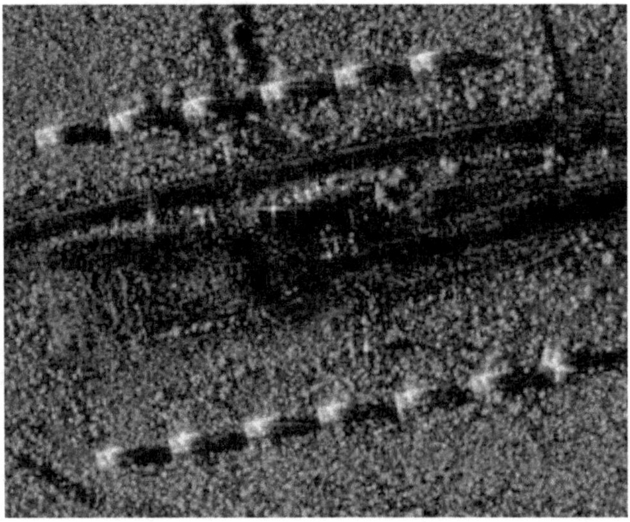

*Figure 7.39 Tanks at Nebo array (courtesy General Atomics) [17] (repeated Figure 1.27)*

as well as grazing angle in Figure 7.38. Obviously, range illumination is from the lower-left of the figure.

Similarly, Figure 7.39 shows that each tank in the Barstow-Nebo array casts a shadow, which allows an estimate of size. Some of the tanks have gun barrel reflections that allow estimates of length. In many cases, this is enough for recognition of tanks as to type. Illumination is from the left in Figure 7.39.

Figures 7.40 and 7.41 show comparisons between other types of images and SAR spotlight. Figure 7.40 compares a topographic map with a corresponding SAR

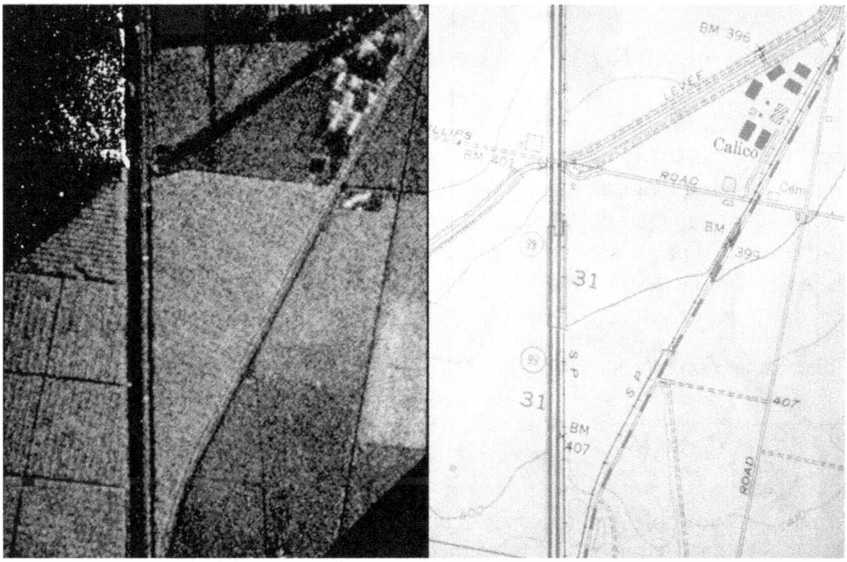


*Figure 7.40 SAR and roadmap comparison [3]*

*Figure 7.41 SAR and photo comparison [3]*

map of the same area. The area is Calico Junction just above a road junction between US Highway 99 and Famoso road as well as a railroad switch for a branch line in the central California valley. In the upper part of the map is the Friant irrigation canal as well as numerous buildings and a railroad siding serving the buildings. Figure 7.41 is an aerial photo and a 2 m resolution SAR map of Earlimart, CA, also in the San Joaquin Valley along US Highway 99. It is easy to see the highway overpass, the fence and oleanders separating the two directions of travel, the school and its ball diamond as well as fenced areas, houses, etc. One will also notice bright sidelobes, most likely from towers made of right angle bracing.

## 7.8 DBS or SAR PRF, pulse length and compression selection

For each SAR or DBS geometry, the transmitted pulse width (PW), PRI and PCR must be calculated. One possible set of selection criteria is given in (7.7) [18]. Usually, the last range ambiguity before the range swath is chosen to be outside the mainlobe far enough to be at least 20 dB down including $R^4$ effects. Often in SAR, the transmitted pulse is much larger than the range swath, $R_{swath}$. Clearly in each of the cases, the nearest integral clock interval and nearest convenient PCR is selected, since the values in (7.7) will be clock integers only by coincidence.

Pulse Repetition Interval (*PRI*):

$$\frac{\lambda}{2\cdot V_a\cdot U_0\cdot\theta_{az}\cdot\sin(\theta)}\geq PRI\geq\frac{2\cdot\left(R_1-R_{\min}+R_{swath}+R_p\right)}{\mathrm{c}}$$

Pulse Width: $R_p\leq Duty_{max}\cdot PRI\cdot\mathrm{c}$

Minimum Allowable Ambiguous Range: $R_{min}\approx h\cdot\csc\left(\varepsilon+U_1\cdot\varepsilon_{el}/2\right)$

Range Swath is Geometry and Instrumentation Dependent:

$$R_{swath}\leq h\cdot\left(\csc\left(\varepsilon-\varepsilon_{el}/2\right)-\csc\left(\varepsilon+\varepsilon_{el}/2\right)\right)\text{ and }R_{swath}\leq R_{maxswath}\qquad(7.7.1)$$

Where: $\lambda$ is transmitted wavelength, $h$ is radar platform altitude,
$\theta_{az}$ & $\varepsilon_{el}$ are the azimuth & elevation half power beamwidths,
$\theta$ & $\varepsilon$ angles between velocity vector and antenna beam center,
$R_1$ is the distance to the first range bin,
$R_{swath}$ is range swath length, $V_a$ is radar platform velocity,
$R_{maxswath}$ is maximum instrumented range swath,
$R_{min}$ is the range to the closest allowable ambiguity,
$Duty_{max}$ is allowable duty ratio, c is the velocity of light,
$R_p$ is the transmitted pulse length in distance units,
$U_0$ , $U_1$ are beamwidth multipliers at predefined power rolloff. (7.7.2)

**Example 7.4** Assume the parameters in Table 7.6; then using (7.7) a PRI range can be chosen as shown in the example equations (ex 7.4). Probably the 400 μs PRI would be chosen because it puts the first ambiguity at a longer range where it will have less return power.

*Table 7.6 Example 7.4 PRF selection*

| Parameter | Value |
|---|---|
| Platform velocity, $V_a$, m/s | 300 |
| Altitude, h, km | 5 |
| Wavelength, $\lambda$, m | 0.03 |
| Azimuth angle, $\theta$, rad | 0.5 |
| Elevation angle, $\varepsilon$, rad | 0.1 |
| Azimuth beamwidth, $\theta_{az}$, rad | 0.05 |
| Elevation beamwidth, $\varepsilon_{el}$, rad | 0.05 |
| Beam fraction, $U_0$, $U_1$ | 2.3 |
| Range swath, $R_{swath}$, km | 2 |
| First-range ambiguity, $R_{min}$, km | 32 |
| First-range bin, $R_1$, km | 50 |
| Max. duty ratio, $Duty_{max}$ | 0.25 |
| Pulse length guess, $R_p$. km | 8 |

$$\frac{0.03}{2\cdot 300\cdot 2.3\cdot 0.05\cdot \sin(0.5)} \geq PRI \geq \frac{2\cdot(50000-32000+2000+8000)}{2.9979\cdot 10^8}$$

$906 \geq PRI \geq 186$ but $\mathrm{R_{min}} \approx 32000$ or 213 $\mu$sec.

Next allowable ambiguity is just past swath.

50000+2000+8000=60000 or 400 $\mu$sec.

PRI could be 213 $\mu$sec and pulse width of 53$\mu$sec.

or 400 $\mu$sec and pulse width of 100$\mu$sec. (ex 7.4)

### 7.8.1 Ambiguities, grating and sidelobes

There are three potential sources of competing ground return associated with the choice of PRF. They are synthetic array grating lobes, Doppler sampling PRF ambiguities (spectrum folding) and range PRF sampling ambiguities (range folding). Figure 7.42 shows the general situation for SAR grating lobes. PRF sampling ambiguities were described in Section 2.13. SAR grating lobes are analogous to real antenna grating lobes where the separation between individual radiators is large enough relative to operating wavelength that the grating lobes enter real space as mentioned in Section 2.6. The initial distance between samples, that is the PRI times the velocity, $S = PRI$ x $V$, must be less than half the width, $D_x$, of the real antenna. Often the required PRI is quite short for spacecraft with electronically scanned antennas due to the combination beam spreading with off normal scan and high orbital velocity. PRIs of 300 μs may be required, which limits range swath width.

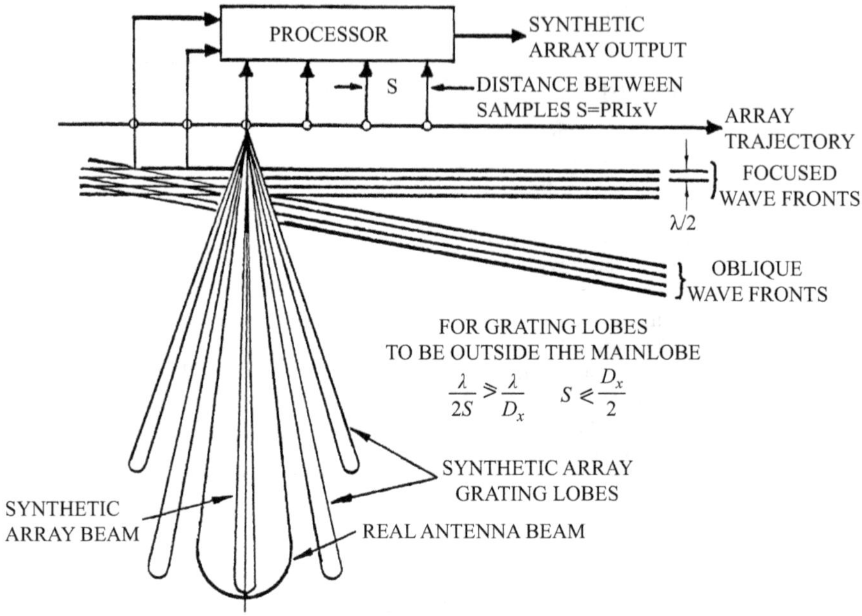


*Figure 7.42 Azimuth grating lobes [1]*

# 7.9 SAR imaging

## *7.9.1 Image quality measures*

The most important single quality of SAR imaging is the 2D impulse response of the complete SAR system including all the processing and image presentation. The impulse response determines dynamic range, shadow recognition, bright discrete tolerance, resolution and small signal suppression. Figure 7.43 highlights the elements of the impulse response in a single SAR filter output. They are the 3 dB beamwidth (BW), PSLR, RMS sidelobe ratio (RMSLR), ISLR and multilooks (m:1). All of these determine what can be detected, discriminated and recognized in SAR images. These attributes are also important in recognizing slow-moving targets with change detection, video SAR and SAR MTI.

Table 7.7 tabulates the image quality effects from each of the impulse response attributes. Resolution is important but background interference often masks recognizable features. Humans are used to seeing multifrequency optical signatures (0.45–0.65 μm wavelength, a little over ½ octave). Their recognition capabilities are strongly biased toward continuous tone images. Very few radars cover ½ octave and so interpretation of even the best radar images requires training.

Image quality attributes have been evaluated in human factors studies going back to the 1960s. This has led to the establishment of the National Image Interpretability Rating Scales (NIIRS). There are ratings for visible, radar, IR and

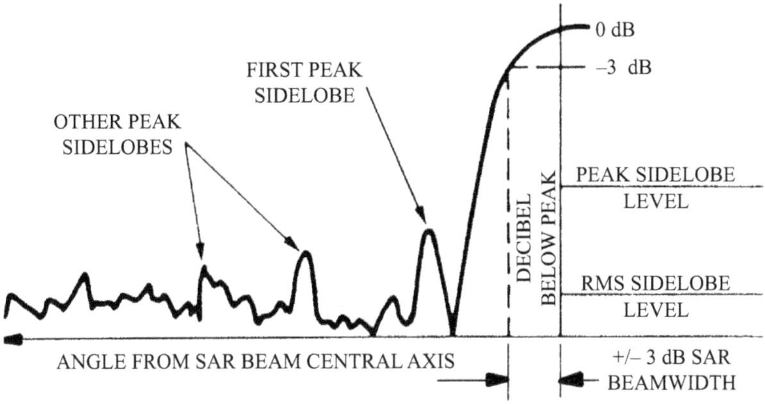


*Figure 7.43 SAR image quality considerations*

*Table 7.7 SAR image quality effects*

| Characteristic | Effect |
|---|---|
| 3 dB resolution | Minimum distance between equal intensity scatterers |
| First near sidelobe | Ratio of unequal adjacent scatterers that can be seen |
| Peak sidelobe | Large scatterers generate false returns |
| Integrated sidelobe level | Minimum distinguishable shadow |
| RMS sidelobe level | Maximum dynamic range of system |
| Multiple looks | Image smoothing improves apparent dynamic range |

multispectral imagery based on resolution. Radar NIIRS (RNIIRS) detection is usually one digit poorer than visible NIIRS for similar objects. Each system type has attributes that under special circumstances are superior to the others. For example, camouflage is easily penetrated by radar, which detects whatever equipment is under it. An artillery piece looks just like an artillery piece under a thin sheet to a radar. People under camouflage look just like hot blobs under a sheet to IR. An artillery piece or rocket launcher under a tarp covered with a few inches dirt is almost invisible to IR but is visible to radar. Radar and EW systems with large antennas exhibit above operating band resonances easily detectable in high clutter backgrounds by radars. Table 7.8 shows RNIIRS ratings versus resolution and counterpart tactical target detectability.

### 7.9.2 SAR resolution

SAR resolution is often defined as the separation between two equal amplitude radar targets each with a Gaussian amplitude response, which allows them to be "just" resolved, that is, there is a dip between them of 0.88 below their respective peaks. Each resolution cell is just a blob and it takes many resolution cells to recognize most objects of interest. This is shown in Figure 7.44 where the value D

*Table 7.8 Radar national imagery interpretability rating*

| RNIIRS | Resolution (m) | Detection or interpretation ability |
|---|---|---|
| 0 | Obscure | Can't see the baloney floating in Ballona Ck.* |
| 1 | >9 | Large paved areas, large buildings, seaports |
| 2 | 4.5–9 | Large aircraft, ships, roads, missile sites |
| 3 | 2.5–4.5 | RR yards, medium aircraft, revetments, ship features, SAM equipment |
| 4 | 1.2–2.5 | Most vehicles, bridges, dirt roads, discriminate types: ships and aircraft |
| 5 | 0.75–1.2 | Count railcars, aircraft, radar types, support equipment |
| 6 | 0.4–0.75 | Discriminate: aircraft, radars, tanks, vehicles, SS-24s |
| 7 | 0.2–0.4 | Recognize: aircraft, vehicles, fences, disturbed earth, SA-xxs |
| 8 | 0.1–0.2 | Recognize: RR cars, fighters, radars, ships, helicopters, missiles |
| 9 | 0.1< | Recognize: aircraft modifications, antennas, types of APCs and missiles |
| 10 | Prescient | Reads minds, finds good investments* |

*RNIIRS 0 and 10: usually said in ironic jest on bad data days.

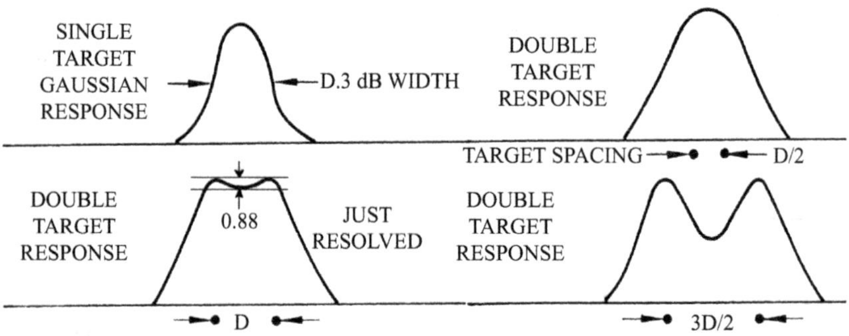


*Figure 7.44 Targets are just resolved at resolution spacing [1]*

is the stated resolution. Targets, which are closer than the resolution, just appear to be a single higher amplitude blob. Targets, which are more than the resolution apart, will have a larger and significant dip in amplitude between them. However, they each are still blobs. It may take dozens of pixels to recognize an object unless there is something very unique about its shape. For instance, it doesn't take many pixels to recognize a fence, a pond or boats on a body of water. Unfortunately, many manmade objects have a few large speculars at almost any aspect, which may make recognition difficult (see Figures 3.3 and 3.5).

### *7.9.3 SAR pulse compression*

There are three types of pulse compression used in SAR systems. First are biphase or polyphase codes, which provide superior sidelobe performance (−50 to −60 dB), with very predictable and repeatable performance (see Section 4.11). Phase codes allow their use in many different radar modes such as AMTI, GMTI, Terrain Following (TF), SAR, ISAR, data links, missile guidance, etc. They are easy to generate and can maintain excellent spectral purity. However, very high-resolution phase codes have a large instant bandwidth, which greatly complicates electronic scanned array and AESA beamforming. It usually requires real-time delays between antenna subarrays as well as element-by-element phasors. Each subarray must be no larger than the desired range resolution, that is, 0.1 m resolution requires subarrays no larger than 0.1 m. That means that in a 1 $m^2$ AESA 100 subarrays and 100 true time delay elements would be required.

The second is chirp, which is some form of frequency modulation (FM) (see Section 4.10). It is most often linear FM. With proper amplitude weighting, high FM linearity and spectral purity, it can achieve good ISLR. It is also good for simple SAR systems that use analog pulse compression. Chirp also allows simple 2D simultaneous compression in range and azimuth because both dimensions can be dechirped and processed in a 2D FFT. Originally from the 1950s to the 1970s chirp was compressed in purpose-built hardware, which involved some form of a frequency-versus-time dispersive delay line. Chirp has two weaknesses: first is range-Doppler coupling in which a moving target will appear in the wrong range-Doppler bin; second is multiple phase centers are still required unless the AESA can spin the phase shifters on transmit and then again on receive. A compromise is usually a smaller number of phase centers with true time delay coupled with limited spinning of T/R module phase shifters.

The third pulse compression variation, stretch, is based on a range dechirp of the received signal and an FFT filter bank (see Section 4.10). Stretch reduces the speed requirements on the A/D and follow-on filtering. The range compression is usually a separate FFT performed after presummation and before azimuth phase correction. The range offset between returns from the first range bin to the last can be compensated with a simple storage sequence ordering at the output of every FFT filter bin which simplifies the azimuth compression. Following azimuth phase correction, the azimuth compression is performed by FFT.

When the desired SAR resolution is high, the slight angular differences between return samples from the terrain before any compression would result in geometric distortion if not compensated. This leads to processing called "polar format" in which sample grid points are mapped onto a rectangular grid.

### *7.9.4 SAR signal processing*

Figure 7.45 shows the processing block diagram for the example SAR mode. Depending on the range to the mapped area, antenna illumination, isoDoppler curvature, Doppler/PRF ratio, phase motion correction, depth of focus, etc., may not be uniform. Although the figure and table show fixed values for many parameters, the actual PRF, presum ratio, phase corrections, PCR, FFT size, number of illuminating beams and multilook overlays are all geometry dependent.

Figure 7.46 shows representative spectra at points in the block diagram. This spectrum is shown as centered around zero frequency and the PRF; however for

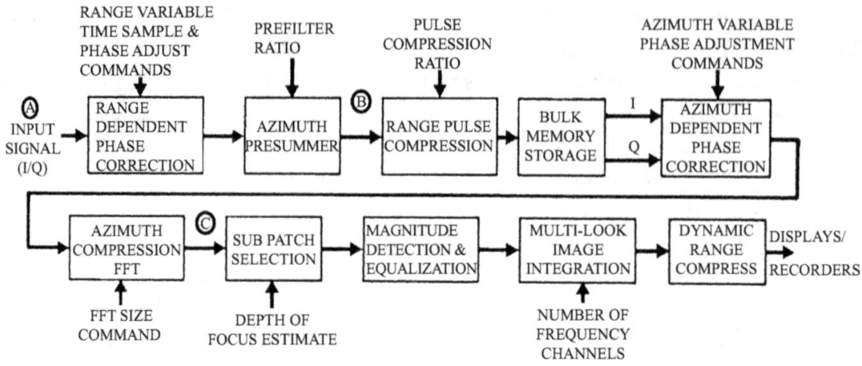


*Figure 7.45 SAR processing; adapted from [5]*

convenience, sometimes the spectrum is moved to the PRF/2. Nothing has really changed when that is done but the filtering may be slightly easier to program or simulate.

As is the case for DBS, SAR is a multirate-filtering problem, that is, a cascade of filters in which the input sampling rate is higher than the output sampling rate as shown in Figure 7.46, which requires very careful attention to range and azimuth filter sidelobes. After each filtering process the outputs are resampled to reduce throughput demands. As usual filter aliasing caused by resampling forces higher processing rates than one would hope based on the 3 dB filter width. The dynamic range relative to noise grows dramatically with each step in the process as shown for the three example spectra. Inevitably, filters at band edge, as shown in the lower right portion of Figure 7.46, will have higher noise levels and thus lower dynamic range. There is one fundamental difference between ISLR in angle or range space and filter processing space: the resample rate is variable all the way to the input sample rate. This allows the designer to be subject to fewer integrated sidelobes by using a higher resample rate and restricting the extent of the processed window. Recall that almost all weightings cited in Table 4.1 had ISLRs of approximately −22 to −25 dB. Those same weighting windows, if close in sidelobes can be excluded, can produce ISLRs lower than −40 dB when the sidelobes are defined as outside $3 \cdot U_{-3dB}$ [2].

The input signal, point A in Figure 7.45, is shown as a spectrum at A, folded about the PRF at left in Figure 7.46. Subsequently, presummation is applied, which forms an unfocussed synthetic beam or filter inside the mainlobe ground return, point B in Figure 7.45, which improves azimuth sidelobes and narrows the spectrum as suggested in the center of Figure 7.46 and in Figure 7.17. The presummer output is resampled at a lower rate, $f_S$, consistent with acceptable filter aliasing. Then, range pulse compression is performed assuming the transmitted pulse is very long compared to the range swath. If chirp (linear FM) is used, part of the "stretch" pulse compression processing is performed in the range compression function with the remainder performed in polar format processing. The dechirped and partially filtered or compressed output, at point C in Figure 7.45, may be resampled again at a new $f_S$, as indicated in the right of Figure 7.46 point C.

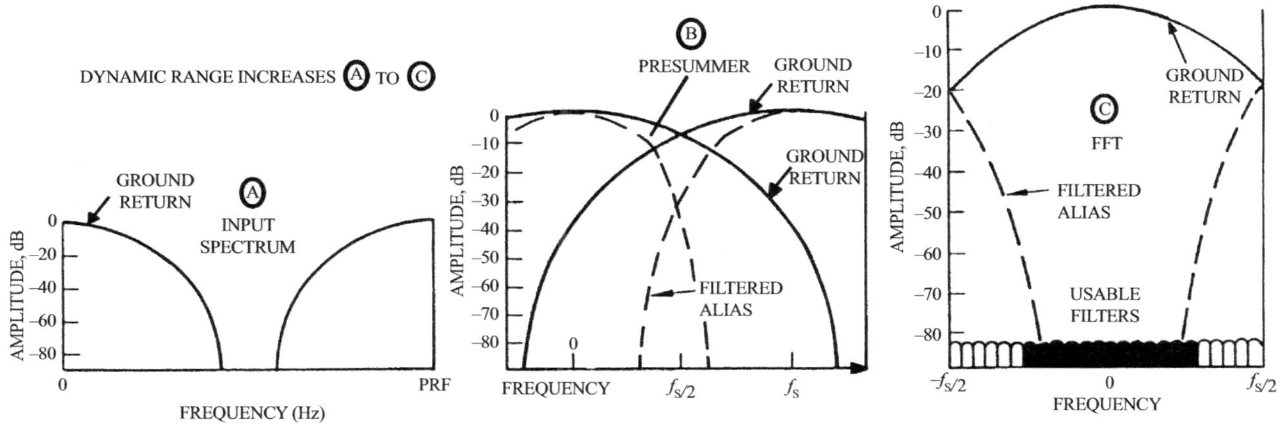


*Figure 7.46 SAR multirate filtering; adapted from [5]*

In any case, azimuth variable phase adjustment and bin mapping (which compensates for changes in measurement space angles and range closure since significant motion occurs during the data gathering time) must be performed before azimuth filtering, sometimes called compression since it is similar to phase-matched pulse compression. The output of azimuth compression is shown at point C. The complex SAR output map must be checked for depth of focus and usually requires autofocus since both atmospheric effects and locally rising or falling terrain may cause defocusing. Subsequent to refocusing, the map is magnitude detected and histogram averaged to maintain uniform brightness. The map is integrated with other looks, which requires geometrical correction and motion compensation. The total map dynamic range can easily be greater than 60 dB. The typical cockpit display is limited to 15–25 dB and dynamic-range compression, such as converting map amplitudes into their logarithms, often is performed. Typically, the spacing of individual pulses on the ground is chosen to be much closer than the desired ultimate resolution. This allows linear range closure and phase correction since each point on the surface moves a significant fraction of a range cell pulse to pulse [7–11].

### *7.9.5 Signal-to-noise ratio*

SAR SNR is governed by two alternative requirements. First, dwell time must be long enough based on platform speed to form filters whose resolution corresponds to the desired azimuth (cross-range) cell size, $d_{az}$ (see Figures 7.9, 7.22, 7.29). Second, dwell time must be long enough based on integration of enough pulses to produce a positive SNR adequate for detection.

Define: $\Delta\eta = \eta_2 - \eta_1$ , $\Delta\theta = \theta_2 - \theta_1$ , $\theta = \dfrac{\theta_1 + \theta_2}{2}$ , $\Delta\theta = \dfrac{d_{az}}{R}$

$\eta_1 = \text{acos}(\cos\theta_1 \cdot \cos\varepsilon_1)$ , $\eta_2 = \text{acos}(\cos\theta_2 \cdot \cos\varepsilon_2)$

On a spherical earth constant range, $R$, is also constant $\varepsilon$, let $\varepsilon_1 = \varepsilon_2 = \varepsilon$

$$f_{D1} = \frac{2 \cdot V_a}{\lambda}\cos(\eta_1) = \frac{2 \cdot V_a}{\lambda}\cos(\text{acos}(\cos\theta_1 \cdot \cos\varepsilon)) = \frac{2 \cdot V_a}{\lambda}\cos\theta_1 \cdot \cos\varepsilon$$

$$f_{D2} = \frac{2 \cdot V_a}{\lambda}\cos(\eta_2) = \frac{2 \cdot V_a}{\lambda}\cos(\text{acos}(\cos\theta_2 \cdot \cos\varepsilon)) = \frac{2 \cdot V_a}{\lambda}\cos\theta_2 \cdot \cos\varepsilon$$

Let $\Delta f_D = f_{D2} - f_{D1} = \dfrac{2 \cdot V_a}{\lambda}(\cos\theta_2 - \cos\theta_1) \cdot \cos\varepsilon$

$$\Delta f_D = \frac{2 \cdot V_a \cdot \cos\varepsilon}{\lambda}\left(-2 \cdot \sin\left(\frac{\theta_1 + \theta_2}{2}\right) \cdot \sin\left(\frac{\theta_2 - \theta_1}{2}\right)\right)$$

If $\Delta\theta$ is small, then:

$$\Delta f_D = \frac{-2 \cdot V_a \cdot \cos\varepsilon \cdot \Delta\theta \cdot \sin\theta}{\lambda} = \frac{-2 \cdot V_a \cdot \cos\varepsilon \cdot \sin\theta \cdot d_{az}}{\lambda \cdot R}$$

Then the minimum time to achieve resolution $d_{az}$ is:

$$T_{SARFilt} = \frac{1}{|\Delta f_D|} = \frac{\lambda \cdot R}{|2 \cdot V_a \cdot \cos\varepsilon \cdot \sin\theta \cdot d_{az}|} \tag{7.8}$$

Suppose two stationary targets at slightly different angles, $\eta$, have an azimuth angular separation of $d_{az}/R$, then from a moving platform they will each have a slightly different Doppler. In order to resolve the distance between these two targets, one must have Doppler filters, which are narrow enough to distinguish their respective frequency difference, $\Delta f_D$. Equations (7.8) derive $\Delta f_D$ in terms of platform velocity, look angle, wavelength, range and azimuth resolution. Taking the inverse of $\Delta f_D$ provides the minimum time necessary to form a filter with the desired resolution. Thus (7.8) calculates the time ($T_{SARFilt}$) to form a SAR filter adequate to achieve the desired cross-range resolution. It could take longer than this because of other implementation requirements but it cannot take less time.

Equations (7.9) calculate a version of the radar equation using the SAR resolution, range, wavelength, backscatter coefficient, transmit power, duty ratio, noise figure, antenna area, losses, minimum SNR, typically 4–7 dB and number of simultaneous beams. Equations (7.9) calculate the dwell time ($T_{SARSNR}$) to achieve a certain minimum desired SNR and then selects the greater of the times ($T_{SARSNR}$) or ($T_{SARFilt}$) multiplied by the number of diversity looks in either frequency or angle.

$$T_{SARFilt} = \frac{R \cdot \lambda}{2 \cdot V_a \cdot d_{az} \cdot \sin\theta \cdot \cos\varepsilon}$$

$$T_{SARSNR} = \frac{4\pi \cdot SNR_{\min} \cdot \lambda^2 \cdot R^4 \cdot k_B \cdot T_0 \cdot NF \cdot N_B}{P_T \cdot D_T \cdot A_e^2 \cdot \sigma_0 \cdot d_{az}^2 \cdot L}$$

$$T_{totalSAR} = (\text{The greater of } T_{SARFilt} \text{ or } T_{SARSNR}) \cdot N_{freq}$$

where: $V_a$ is the total apparent ground velocity,
$d_{az}$ is the SAR azimuth resolution, $\sigma_0$ is the assumed ground cross section, $SNR_{\min}$ is minimum signal to noise ratio per resolution cell. (7.9)

---

**Example 7.5** Consider a SAR mapping satellite with the parameters of Table 7.9. Then using (7.8) and (7.9), the required SAR dwell time per map is given in Figure 7.47. The details are given in the appendices DVD-R in a program called SpaceRdr001.

As is obvious from Figure 7.47 the closer to broadside (90°) from the velocity vector, the shorter the time to form a SAR map. At very short ranges nearly directly below the spacecraft, the radial velocity is also very low and so map imaging time goes up. At very long ranges the radar equation SNR for the required cross-range accuracy as well as the much longer array length necessary to achieve the required resolution forces the dwell times to more than a minute.

*Table 7.9 Example 7.5 space dwell times*

| Parameter | Value |
|---|---|
| Altitude, $h$, km | 1000 |
| Antenna area, $A_e$, m$^2$ | 78.5 |
| Average power, $P_T$, W | 500 |
| Frequency looks | 2 |
| Azimuth looks | 2 |
| Map resolution, $d_{az}$, m | 1 |
| Req. cross-range accuracy, $R_{cr}$, m | 800 |
| Min. grazing angle, $\psi$, ° | 6 |
| Apparent ground velocity, $V_a$, m/s | 7570 |
| RCS, $\sigma_0$, m$^2$/m$^2$ | 0.05 |
| Max. range, $R_{max}$, km | 2800 |
| Wavelength, $\lambda$, m | 0.03 |
| PRF, Hz | 3600 |
| Noise figure, $NF$, dB | 4 |
| Losses, $L$, dB | −2.75 |

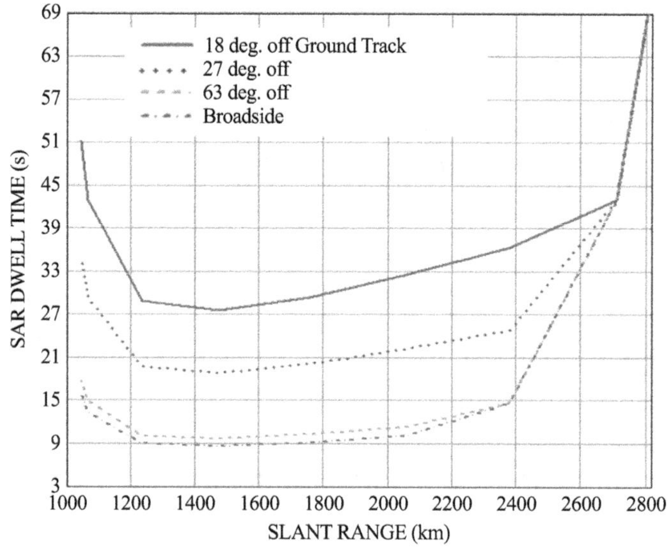


*Figure 7.47 Spacecraft SAR dwell time versus slant range – X band (SpaceRdr001)*

---

**Example 7.6** Consider an aircraft radar operating at X band with the parameters in Table 7.10. Then using (7.8) and (7.9), the dwell times are as shown in Figure 7.48.

Surprisingly, in the mid ranges the dwell times for airborne and space are similar; this is not only because antennas can be so much larger on a spacecraft but

*Table 7.10 Example 7.6 airborne dwell times*

| Parameter | Value |
|---|---|
| Altitude, *h*, km | 7 |
| Antenna area, $A_e$ m$^2$ | 3 |
| Average power, $P_T$, W | 500 |
| Frequency looks | 2 |
| Azimuth looks | 2 |
| Map resolution, $d_{az}$, m | 1 |
| Req. cross-range accuracy, $R_{cr}$, m | 100 |
| Min. grazing angle, $\psi$, ° | 6 |
| Apparent ground velocity, $V_a$, m/s | 250 |
| RCS, $\sigma_0$, m$^2$/m$^2$ | 0.05 |
| Max. range, $R_{max}$, km | 115 |
| Wavelength, $\lambda$, m | 0.03 |
| PRF, Hz | 3600 |
| Noise figure, *NF*, dB | 4 |
| Losses, *L*, dB | −2.75 |

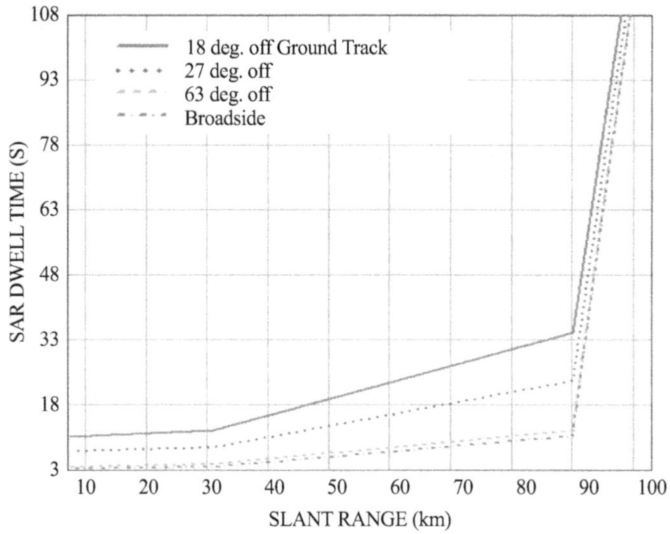


*Figure 7.48 Aircraft SAR dwell time versus slant range – X band (AircraftRdr001)*

also because spacecraft orbital velocities are so much higher. Nonetheless, when near maximum range dwell times must be very long to achieve cross-range accuracy and resolution.

### 7.9.6 Contrast ratio

In general, all map targets whether real beam, DBS or SAR are area targets in which multiple cells are involved. The target contrast ratio with respect to the background defines detection and recognition probability. There are several different measures of contrast ratio depending on the objective. One measure is the expectation of the SAR map pixel magnitude squared divided by the square of the expectation of the map pixel magnitude given in (7.10). This measure is used primarily to aid in the autofocus of the image.

$$CR = \frac{E\left(|P_{cell}|^2\right)}{E\left(|P_{cell}|\right)^2} \tag{7.10}$$

where $P_{cell}$ is the intensity in each cell.

A second contrast ratio measure is used to estimate a specific SAR design capability to detect different types of terrain in the presence of design objective SNR, ISLR and rain degradation. Its general form is given in (7.11).

$$CR = 1 + \frac{1}{\dfrac{N}{GR_{mean}} + ISLR + \dfrac{Rain}{GR_{mean}}}$$

Where: $N$ is predicted noise power at map filter outputs,

$GR_{mean}$ is the mean ground return predicted over whole map,

$Rain$ is the power returned from predicted rain, and

$ISLR$ is the all source predicted integrated sidelobe ratio. (7.11)

The third type of contrast ratio is used to predict the capacity of a SAR image to support the detection of tactical targets in a SAR map given in (7.12). It is normalized by the expected ground return mean over typical scenes.

$$CR_{i,j} \approx \frac{P_{Cell\,i,j} + N + ISLR \cdot GR_{mean} - GR_{mean}}{GR_{mean}}$$

$$CR_{i,j} \approx CR_{Cell\,i,j} + (ISLR - 1) + CNR^{-1} \tag{7.12.1}$$

Where: $P_{Cell\,i,j}$ is the intensity in the $i, j$ cell,

$GR_{mean}$ = ground return over the scene,

$N$ = receiver noise power, $CNR$ = clutter to noise ratio,

$ISLR$ = integrated sidelobe ratio for antenna and processing,

$CR_{Cell\,i,j}$ = contrast ratio in the $i, j$ cell. (7.12.2)

The maximum power from a target can be represented by its autocorrelation function. The maximum detectability relative to its background for a pixelized target can be represented by the sum of the contrast ratios in all the pixels on the target. That sum can then be compared to a conventional threshold normalized to the background power level. Although automatic screening of SAR images for tactical targets is routine to reduce image specialist workload, final detections and designations are usually done by humans. Screened images are presented to a human for final shape, shadow and contrast ratio based target detection and recognition. Thus the normalized contrast ratio for some target shape is given in (7.13). Human factors studies done in the 1970s showed that an additional 3 dB or more over that implied by a pure Gaussian statistic detection or false alarm threshold (i.e., the required $CR_{shape}$ is $13 + 3 = 16$ dB or 40) was required for humans to detect area targets in a map background [7].

$$CR_{shape} = \sum_{j=0}^{j_{\max}} \sum_{i=0}^{i_{\max}} \left| CR_{i,j} \right| \geq Th \cdot C = 20 \cdot 2$$

Where: *Th* is a conventional threshold e.g. 20, and

*C* is multiplier for human detection performance e.g. 2. (7.13)

The contrast ratio is the difference between the ground return mean as measured in the radar and the power in the *i, j* cell normalized by the ground return mean. The mean ground return as measured in the radar contains the noise power and the power from true ground return times the composite ISLR. The contrast ratio can be either positive or negative. A dark spot such as a paved road or lake can be just as useful for recognition as a bright building or stationary vehicle target.

---

**Example 7.7** Contrast ratio Example 7.7, assume the parameters given in Table 7.11. Then using (7.13), the required contrast ratio per cell is given in (ex 7.7)

$$CR_{cell} \geq \frac{40}{50} = 0.8 / cell \qquad \text{(ex 7.7)}$$

*Table 7.11 Example 7.7 contrast ratio*

| Parameter | Value |
|---|---|
| Vehicle size, $ft^2$ | $5 \times 10 = 50$ |
| Detection probability, $P_D$ | 0.9 |
| False alarm probability, $P_{FA}$ | 0.1 |
| Resolution cell, $d_{az,}$ ft | 1 |
| Required contrast ratio, $CR_{shape}$ | 40 |

In other words, the average contrast ratio per cell must be either 80% smaller or 80% greater than the surrounding background for reliable recognition.

---

**Example 7.8** Consider the real case shown in Figure 7.49, which is a SAR image of cars in a parking lot at an airport. The context gives the cues that these are indeed cars. One can also see a lamp standard in the lot (five dots in a cross). Some of the apparent smears in the image are reflections off the pavement underneath the cars, which because of multiple reflections appear beyond the cars. An image of a reflection outside the real extent of the object is very common and may be either a clue or a confusor as to recognition. Multiple reflections off roads, wheels, etc., are very common for surface vehicles. The totally black spot between some of the cars is most likely a puddle from a recent rain. The experimental contrast ratio in each cell is given in Example 7.8.

Figure 7.50 shows the contrast ratio contour map of the parked car image. Note the dark boundaries around each of the cars. Even though the resolution is 0.1 m, the cars are still just blobs in the image. Using (7.12) and (7.13) and calculating the contrast ratio per cell as well as the mean yields 0.239 per cell. Then observing that one car shape in the image is about 6.75 by 15.75 feet, the total contrast ratio for a car shape is about 1744. The car shape spans 59 by 27 radar pixels. So normalizing the car shape contrast ratio by the number of pixels yields a car shape contrast ratio average of 1.095 versus a total scene average of 0.239. Thus the car shape has an overall contrast ratio compared to the background of 4.58, more than enough to detect cars in the radar image background (obvious from the image in Figure 7.49).

*Figure 7.49 Parked cars near an airport [17]*

Contrast ratio is used to specify a new system to insure that the system will meet the requirement to successfully detect a certain type of target.

$$CR_{i,j} = \frac{\left(P_{celli,j} - GR_{mean}\right)}{GR_{mean}}$$

$$GR_{mean} = 77.545,$$

$$CR_{mean} = (974 \cdot 1241)^{-1} \cdot \sum_i \sum_j \left|CR_{i,j}\right| = 0.239,$$

Select one car shape 6.75' by 15.75' in image:

$$CR_{shape} = \sum_{i=659}^{718} \sum_{j=386}^{413} \left|CR_{i,j}\right| = 1744.335,$$

$$CR_{car-average} = (59 \cdot 27)^{-1} \cdot 1744.335 = 1.095$$

Contrast advantage car to background:

$$\frac{CR_{car-average}}{CR_{mean}} = \frac{1.095}{0.239} = 4.58 \qquad \text{(ex 7.8)}$$

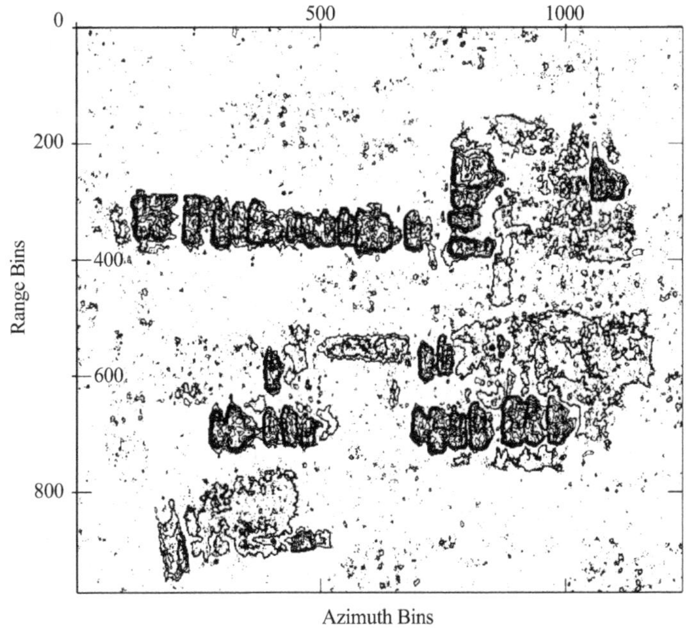


*Figure 7.50 Example 7.8 contrast ratio contour map of parked cars (ParkedCars5)*

## 7.10 Motion compensation for mapping and STAP MTI

It was pointed out in Section 2.8 that radars and passive interceptors measure distance and angle to a fraction of a wavelength. For an emitter at L band that will be less than 3″ and for a radar at X band that will be less than ¼″. Although that measurement error may be ambiguous, it still must be accurate pulse to pulse to the tolerances just stated. Thus the position pulse to pulse must be known to a small fraction of a wavelength. Even in space and high altitude, the "lumpy" gravity field on Earth will cause changes in altitude and velocity. At lower altitudes changes in local air density and velocity will also cause "lumpy" flight motion. Inaccurately known velocity and acceleration will lead to poorly focused maps and poorly cancelled clutter. Position changes inch-by-inch and millisecond-by-millisecond must be estimated. The shorter-time changes can only be estimated by inertial sensors.

The longer-time changes may be estimated by signal-based motion compensation. The antenna beam composite look angle, $\eta$, must be accurately known as well as the LOS velocity ($V_{LOS}$), the cross LOS velocity ($V_{XLOS}$) and the LOS acceleration ($\Delta V_{LOS}$). $V_{LOS}$ can be estimated with a clutter Doppler centroid tracker using differential angle (monopulse) and Doppler frequency discriminator filters. $V_{XLOS}$ can be estimated by using the two beams of the monopulse. This tracking usually results in a phase lock of the return to a local sensor reference. Figure 7.51 suggests the essential elements of motion sensing. Although these are essential for motion compensation, most of the data to arrive at these estimates may come from other sources not the radar.

If the SAR resolution or required GMTI radial detection velocities are not too stringent, this motion sensing can even be manual. However, in most modern cases, automatic motion compensation is required. If the inertial navigation system (INS) is not on the antenna, then a set of accelerometers and rate gyros must be mounted on the antenna. The antenna-mounted sensors must be sampled at the PRF rate. If the INS is on the antenna, then the motion outputs must be at the PRF rate.

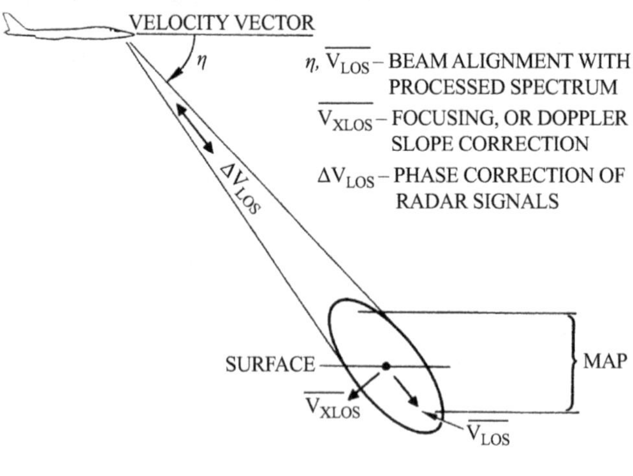


*Figure 7.51 Essential elements of motion compensation*

Figure 7.52 shows a typical motion sensing suite for an off antenna-mounted INS. Acceleration and rate sensors sampled at the PRF rate are usually noisy and must be inserted into a Kalman filter for smoothing and prediction.

The block diagram of a self-contained INS is shown in Figure 7.53. In addition to the IMU, there are usually batteries for operation through power transients,

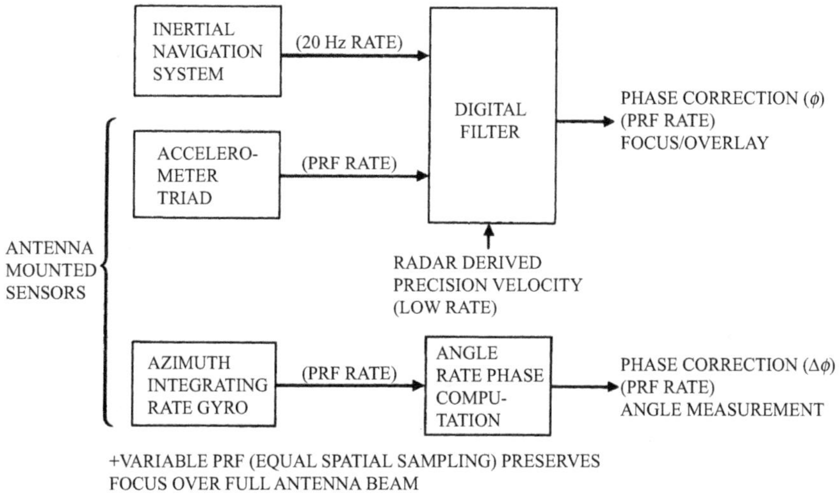


*Figure 7.52 Combined antenna sensor and INS Kalman filter inputs [16]*

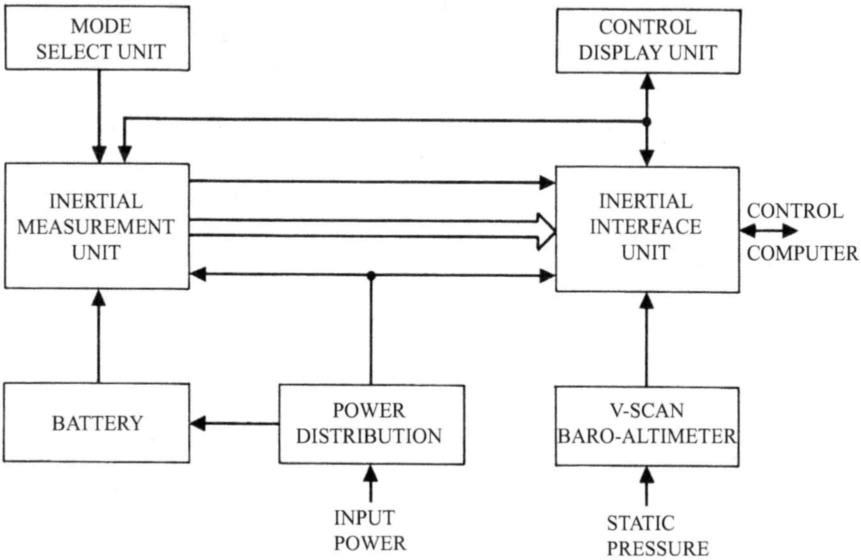


*Figure 7.53 Inertial navigation system components [16]*

power supply, controls and displays as well as a baro-altimeter. As you may be aware, inertial systems are subject to Scheuler errors. The baro-altimeter damps those errors, which then allows stable long-term operation. The INS outputs required are: pitch, roll; platform heading; velocities in Northings and Eastings; integrated acceleration in x and y; true heading with respect to North because the platform may have a yaw angle, vertical acceleration which is a combination of gravity and platform movement; elapsed time since turn on; as well as latitude, longitude and barometric altitude.

Even the best inertial sensors require periodic position and velocity updates from some other sensors such as radar, star trackers, GPS, radio navigation, etc. The INS is free running but its outputs are corrected by other sensors. Usually those updates are combined external to the inertial sensors. Figure 7.54 shows the common way that augmenting sources are used to improve an overall motion compensation for Comint/Elint/Sigint, radar and EO sensors. All the augmenting sensors have much better long-term stability but inertial is the best short-term sensor. If these other sources are continuously available, the INS can be relatively inaccurate on a long-term basis, for example, inexpensive fiber optic gyro-based inertial systems.

An example of a radar-aided navigation system is shown in Figure 7.55. It uses a relatively inexpensive commercial INS coupled with 20 s periodic radar updates. One can see that all three axes settle down over the first 10 min of Kalman filtering.

There are many sources of error but velocity errors are usually the most important. The principal effect of constant bias velocity errors is to shift the impulse response in the frequency domain as shown in Figure 7.56. For STAP this error source either uses up more degrees of freedom or results in poorer clutter cancellation. For SAR this causes the map to shift relative to the main beam and may result poorer contrast as the map is shifted near one edge of the real beam. Also the designation accuracy of an object in a map will be poorer.

The second velocity error effect arises from periodic velocity errors. These are "birdies" that cause harmonic amplitude errors that show up as sidelobes or ghost images. That effect is shown in Figure 7.57. Low-amplitude replicas of either MTI targets or maps are offset by the birdie modulation frequency. The most common source of these "birdies" is mechanical vibrations resonating in some part of the platform structure. Even spacecraft have such vibrations induced by attitude control

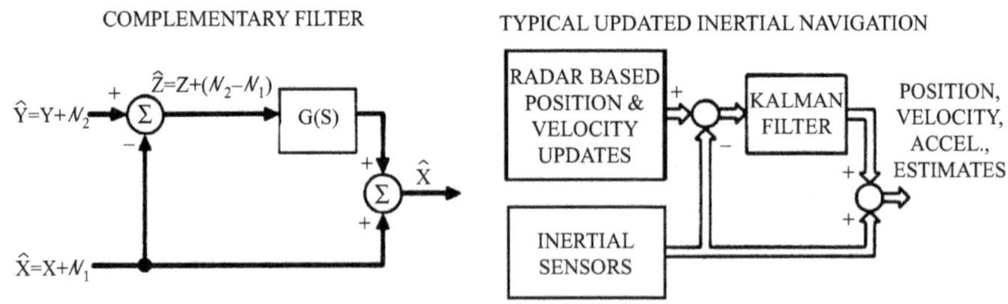


*Figure 7.54 Transfer alignment of INS with radar sources (repeated Figure 1.22)*

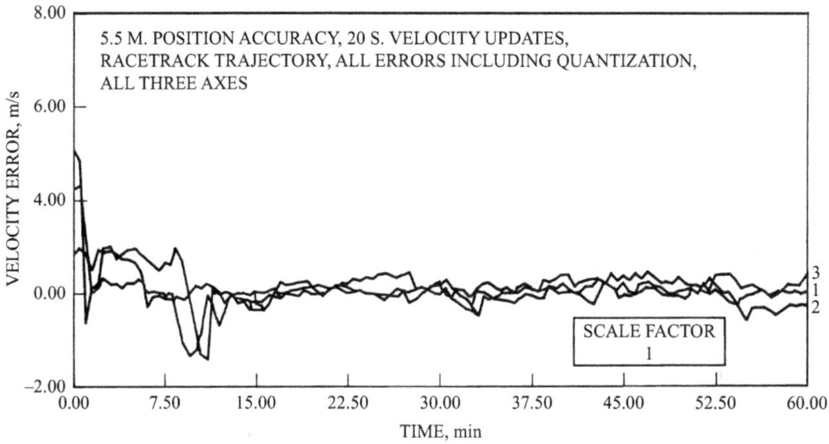


*Figure 7.55 Typical radar-aided INS velocity accuracy [3]*

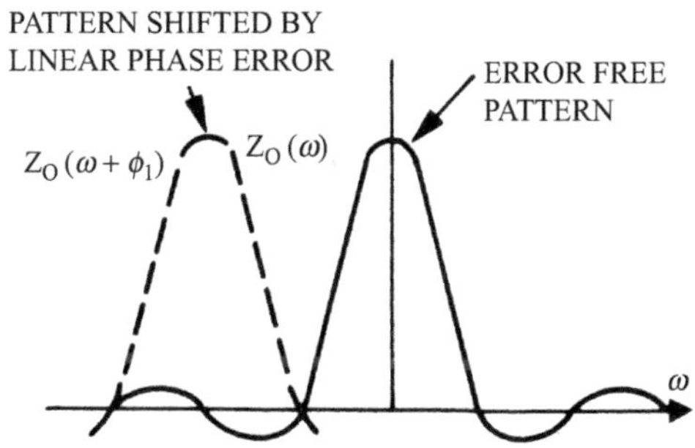


*Figure 7.56 General notion of velocity (linear phase) error [1]*

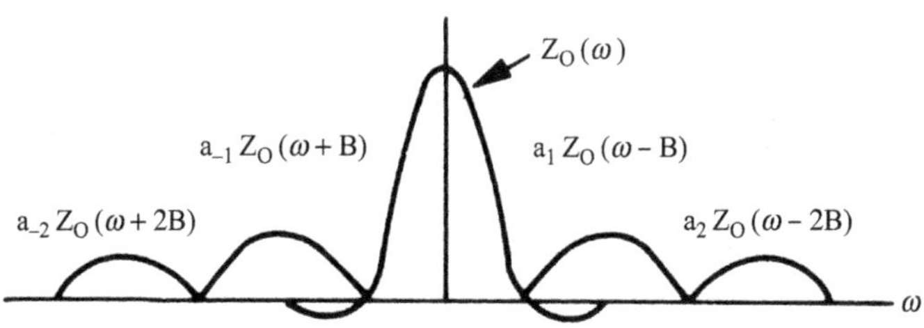


*Figure 7.57 General notion of periodic (vibration birdies) phase errors [1]*

and sunlight variations on solar arrays. For low-frequency vibration, the criteria usually is $\pi/2 \geq \phi_q = \left(\omega_{vib}^2 \cdot T^2\right)/8$. For high-frequency vibrations, the modulation results in a Bessel function where the terms are offset by the vibration frequency, $B = \omega_{vib}$ and with amplitude approximately equal to ½ the phase deviation amplitude $a_1 \approx (\phi_0 \cdot \sin \eta)/2$.

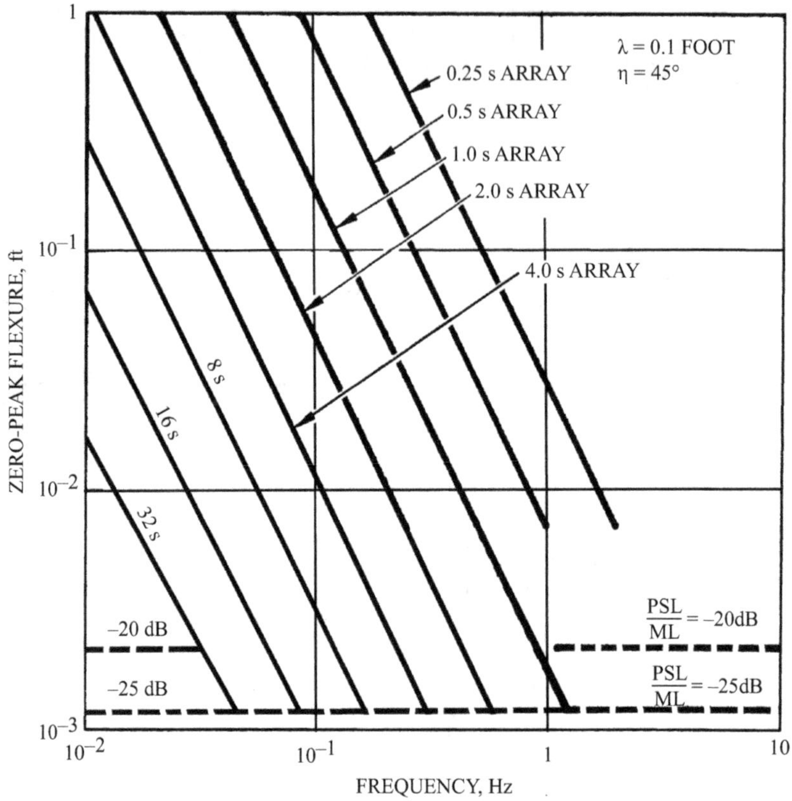


*Figure 7.58 Allowable peak vibration flexure versus frequency*

For example, combining the low- and high-frequency vibration effects and requiring them to be under $\pi/2$ requires a 20 dB per decade roll off for low frequency and a flat upper bound for high-frequency set at the required PSL/ML as shown in Figure 7.58. Figure 7.58 shows typical requirements as a function of coherent array times for an X band system at an angle of 45° off the velocity vector.

There is one more factor in motion compensation, which is the distance between the INS and the antenna phase centers. Many times the INS is mounted on an AESA within 1–2 ft of the phase center, but more often the INS is more than 3 ft from the antenna phase center. In the latter case, the moment arm between the antenna phase center and the INS is of critical importance especially for aircraft, which may undergo significant short-term acceleration. The geometry is shown in Figure 7.59. The vector equations in position, velocity and acceleration are shown in (7.14). Clearly these equations can be solved and compensation calculated, but more often the INS or augmenting motion sensors are located as close as possible to the antenna phase centers. Even 1 g acceleration can diminish performance unless the moment arm is short. This is the reason for the motion sensing configuration

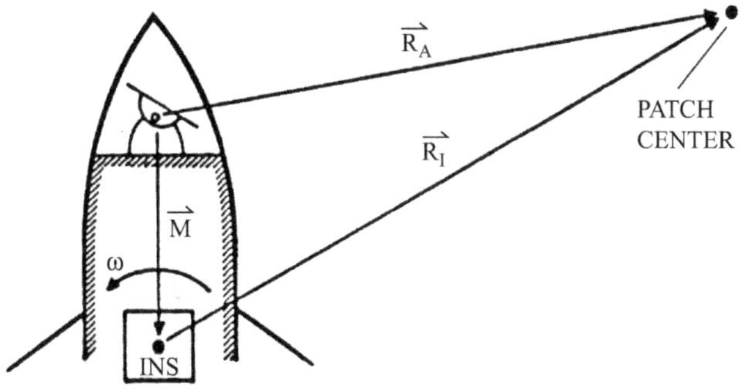


*Figure 7.59 Moment arm compensation INS to antenna phase center*

and data rates shown in Figure 7.52. In spacecraft the moment arm is a significant problem, even though accelerations are usually benign, because the antenna phase center is usually a long way from the INS (~10 m).

Position Vector Equation:

$$\mathbf{R_a}(x,y,z,t) = \mathbf{R_I}(x,y,z,t) + \mathbf{M}(x,y,z,t)$$

Velocity Vector Equation:

$$\frac{d\mathbf{R_a}}{dt} = \mathbf{V_{INS}}(x,y,z,t) + \boldsymbol{\omega}(x,y,z,t)\cdot\mathbf{M} + \mathbf{V_M}(x,y,z,t)$$

Acceleration Vector Equation:

$$\frac{d^2\mathbf{R_a}}{dt^2} = \mathbf{A_{INS}}(x,y,z,t) + \left(\mathbf{A_a}(x,y,z,t) + \boldsymbol{\omega}^2\right)\cdot\mathbf{M}$$
$$+\boldsymbol{\omega}\cdot\mathbf{V_M} + \mathbf{A_M}(x,y,z,t)$$

Where: $\mathbf{R_a}$, $\mathbf{R_I}$ are the vectors from the antenna and the INS to the patch of interest respectively;
$\mathbf{M}$ is the vector from the antenna to the INS;
$\mathbf{V_{INS}}$, $\boldsymbol{\omega}$, and $\mathbf{V_M}$ are INS angular rate and moment arm velocity vectors; $\mathbf{A_{INS}}$, $\mathbf{A_a}$, $\mathbf{A_M}$, are the INS, antenna and moment arm acceleration vectors. (7.14)

Often these moment arm errors are corrected with autofocus if the errors are small. Laser measurements can be used to correct moment-arm errors between the INS and antenna phase centers for high-resolution imaging for spacecraft. Otherwise, there are "good days" and "bad days" when detections and imaging are great on some days and are disappointingly bad on other days, which is common for most long-range systems. Anyone who has experience with real systems in real environments has seen it.

The basic requirement for total phase error across an array is that it must be at least less than $\pi/2$, that is, ¼ wavelength. Consider Figure 7.60, suppose the estimated velocity is wrong by $\Delta V$, then the coherent array will be different at each end by $(\Delta V \times T/2)$. The phase deviation at the each end of a coherent array will be $(\Delta V \times T/2) \tan \alpha$, where T is the total coherent array time. But $\tan \alpha$ is approximately $(V \times T)/(2 \times R)$, where R is the range to the maximum range bin of interest. Since $d_{az}$ is approximately $(\lambda \times R)/(2 \times V \times T)$ the phase error is approximately $(\pi \times \Delta V \times T)/(2\, d_{az})$ as in Figure 7.60. Thus $\Delta V$ must be less than or equal to $d_{az}/T$ as shown in Figure 7.60.

Although velocity errors primarily result in the impulse response shifting, acceleration errors result in broadening of the impulse response, which reduces its peak amplitude and spreads its width and sidelobes. This defocusing results in degraded SAR maps and poorer detection of STAP MTI targets. The defocusing notion is suggested by Figure 7.61. Acceleration errors are errors in the quadratic phase term of the power series expansion of the observed motion between the platform and a target whether it is a moving target or a segment of imaged terrain.

If the velocity isn't quantized accurately enough, then there could be a velocity and acceleration error associated with velocity changes during a quantum step as

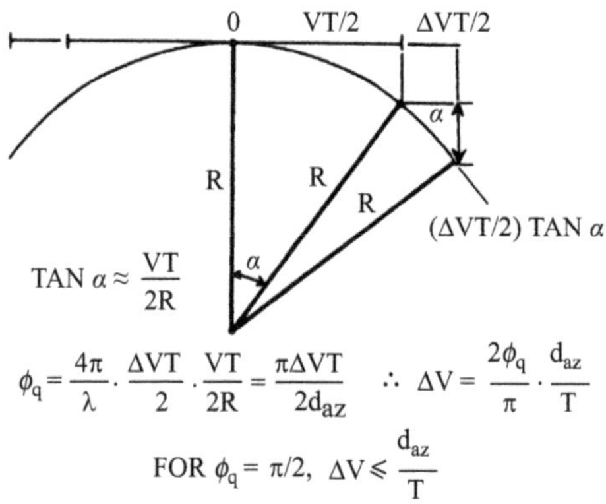


*Figure 7.60 Velocity bias error model [1]*

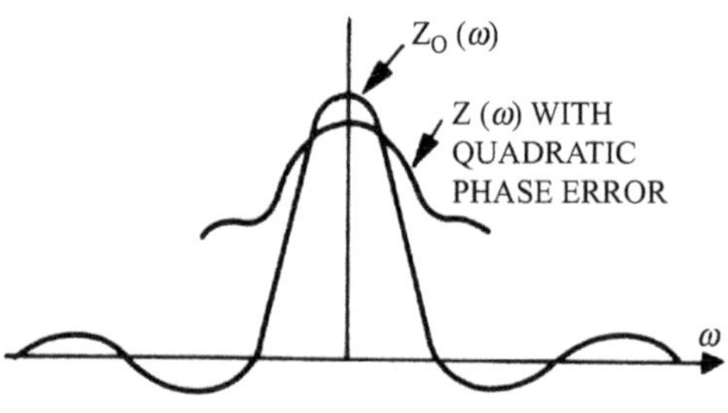


*Figure 7.61 General notion of the effect of quadratic phase error [1]*

shown in Figure 7.62. The worst case would give rise to a potential undetected acceleration of as much as the quantum step size divided by the CPI or (q/T). Recall that the integrated distance for an acceleration equals ($\frac{1}{2} \times a_q \times t^2$). The integration of that undetected error from center to edge using the model of Figure 7.62 would result in a maximum distance error of $\frac{1}{2}(q/T)(T/2)^2 = (qT)/8$. As before calculating the equivalent round trip total phase error and requiring that error to be less than $\pi/2$, requires in turn that the quantum step size must be less than $\lambda/T$ as shown in Figure 7.62. Combining this requirement with an ISLR of −20 dB yields the summary in Figure 7.63.

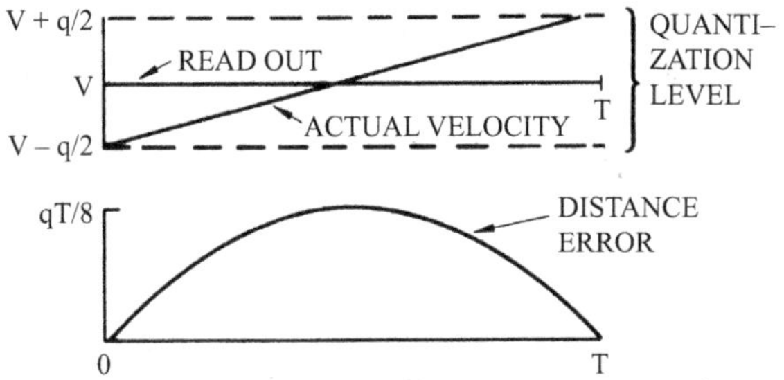


MAXIMUM UNDETECTED ACCELERATION $= \frac{q}{T}$

$$\phi_q = \frac{4\pi}{\lambda} \cdot \frac{qT}{8} = \frac{\pi q T}{2\lambda} \qquad q \leqslant \frac{2\phi_q}{\pi}\frac{\lambda}{T} \qquad \text{FOR } \phi_q = \pi/2,\ q \leqslant \frac{\lambda}{T}$$

*Figure 7.62 Velocity quantization error model [1]*

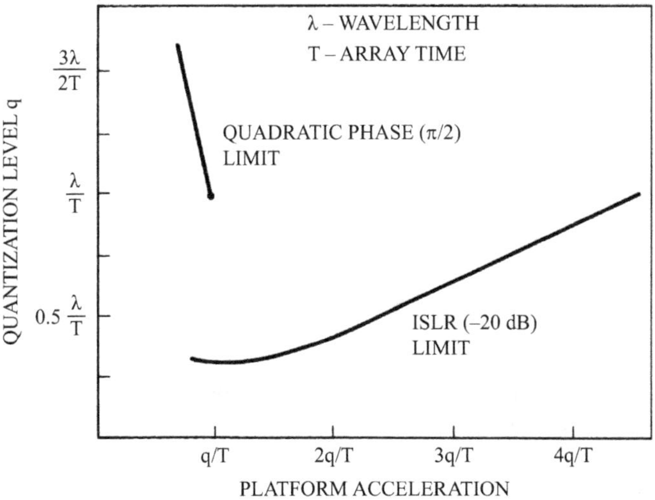


*Figure 7.63 SAR allowable velocity quantization error [1]*

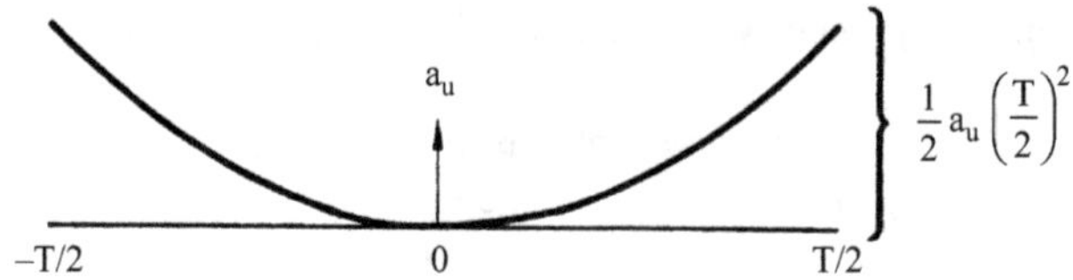


$$\phi_q = \frac{4\pi}{\lambda} \cdot \frac{a_u T^2}{8} \qquad a_u \leqslant \frac{2\phi_q}{\pi} \cdot \frac{\lambda}{T^2} \qquad \text{FOR } \phi_q = \pi/2, \quad a_u \leqslant \frac{\lambda}{T^2}$$

*Figure 7.64 Acceleration bias error model [1]*

Similarly, the effect of uncompensated acceleration error independent of velocity quantization is shown in Figure 7.64. The acceleration bias from center to edge is ½ $a_u$ $(T/2)^2$, which when solving for the allowable acceleration error so that phase error is less than $\pi/2$ requires that $a_u$ must be less than $\lambda/T^2$.

Acceleration bias error usually comes from the accelerometers and the inertial software's inability to completely eliminate the force of gravity. If the platform's attitude is not known well enough, then a fraction of the acceleration of gravity will cause an acceleration error. Accelerometers have, in addition to bias, a quasi-random component and a scale factor error component of acceleration. These three sources of acceleration error are summarized in (7.15). All of the acceleration terms must be RSS and the combination must be less than $\pi/2$.

Accelerometer Bias Error:

$$\phi_b = \frac{\pi \cdot a_b \cdot T^2}{2 \cdot \lambda} \qquad a_b \text{ is the bias error m/sec}^2,$$

Accelerometer Scale Factor:

$$\phi_{sf} = \frac{\pi \cdot sf \cdot g \cdot T^2}{2 \cdot \lambda} \qquad sf \text{ is scale factor error relative to full scale,}$$

Platform Attitude Error:

$$\phi_{att} \cong \frac{\pi \cdot \kappa_{att} \cdot g \cdot T^2}{2 \cdot \lambda} \qquad \kappa_{att} \text{ is attitude error in radians,}$$

$g$ is the acceleration of gravity, near earth $g \cong 9.8017$ m/sec$^2$

and at 1000 km altitude $g \cong 7.323$ m/sec$^2$. (7.15)

Accelerometer errors are typically measured in parts per million; however, because they must be integrated twice over long periods of time to obtain position, small numbers become large errors if not compensated. There are ways to correct a part of these errors by a known platform maneuver in 2D or 3D to allow their separation and measurement. Usually those separately identified components are then incorporated into the overall motion compensation Kalman filter and output motion estimates.

As before combining the Figure 7.61 requirement with a –20 dB ISLR results in a summary allowable acceleration error shown in Figure 7.65. Not surprisingly

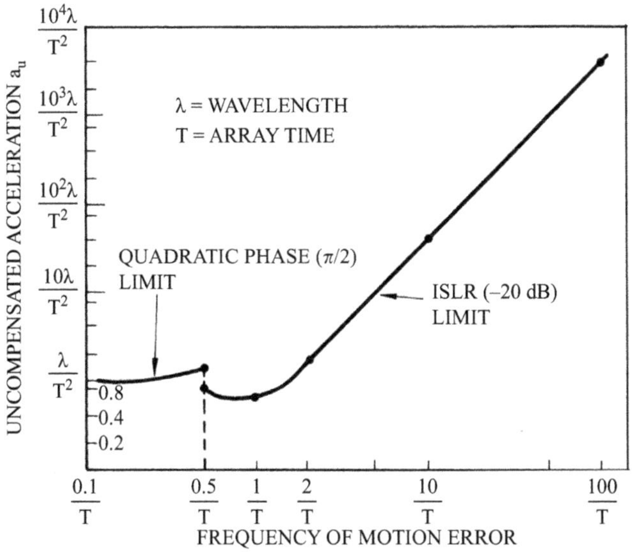


*Figure 7.65 Allowable SAR acceleration error [1]*

the high-frequency acceleration components that are a small fraction of a coherent array time are smoothed over the array and thus can be allowed to be much larger.

## *7.10.1 Motion-adaptive methods to reduce phase errors*

It has been shown that adjusting the PRF during an array time so that the spatial azimuth samples along a line through the map CRP are uniform dramatically reduces quadratic phase error [19]. This observation states that the PRF must be proportional to the angle to the CRP and the derivatives of those angles. Equation (7.16) summarizes the requirement. The phase error reduction can be as much as 100:1. This applies both to normal platform motion as well as to maneuvers during a coherent array.

$$PRF \propto \left[\cos\phi(t)\cdot\cos\theta(t)\right]\cdot\frac{d\theta(t)}{dt}+\left[\sin\phi(t)\cdot\sin\theta(t)\right]\cdot\frac{d\phi(t)}{dt}$$

Where $\theta(t)$ is the azimuth angle with time relative to map center

and $\phi(t)$ is the elevation angle with time relative to map center. (7.16)

Obviously, the requirement results in a continuously varying or "funny" PRF, which is suited to batch processing. Both "funny" PRF and similarly "funny" antenna scan have been in use in some systems since the 1960s. The most common "funny" scan is used for DBS maps.

Another motion-adaptive method in common use is antenna beam pointing angle stabilization pulse to pulse by adjusting the phase difference between two (or more) antenna phase centers based on antenna-mounted motion sensors. The measured beam error in suitable units is multiplied by the effective aperture in wavelength units to calculate a phase shift to be applied to the phase centers to compensate the short-term movement of the antenna face. An example of this technique is given in Figure 7.66 for a monopulse antenna. Both of these techniques just make subsequent processing more accurate and simple. In a STAP system, these potential error sources could be cancelled using some of the available degrees of freedom, but then those degrees of freedom wouldn't be available for other uses such as canceling jammers, clutter, etc.

### *7.10.2 Depth of focus*

The power inside the main beam, which doesn't fall on the mapped patch is wasted. This is especially a significant problem for spacecraft where the power per pass is limited by the solar arrays and battery storage. Therefore, another consideration is associated with mapping the largest possible patch of terrain within the real beam of the antenna. That inevitably leads to limitations on depth of focus. Figure 7.67 describes the problem. As the platform sweeps out a look angle large enough to

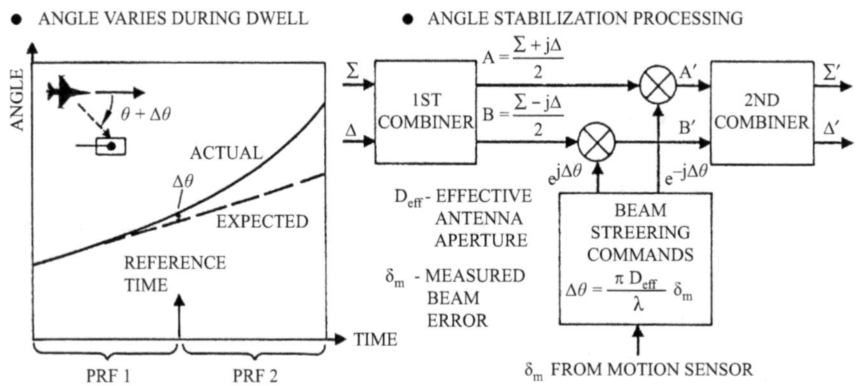


*Figure 7.66 Angle stabilization by pulse-to-pulse beam steering [19]*

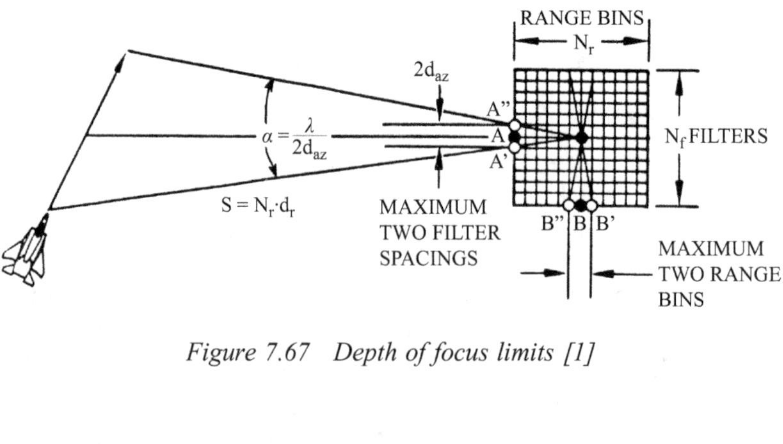


*Figure 7.67 Depth of focus limits [1]*

achieve the desired resolution, there is a natural change in aspect angle on each range-Doppler bin. This will cause walk in both range and azimuth at the edges of the mapped patch. As the resolution gets better, the change relative to a resolution cell gets much greater. This results in poorer focusing at the edges of the patch. The general geometry is described in Figure 7.67. It has been shown that for chirp-like waveforms allowing the edge range or azimuth (Doppler) cells to have an uncompensated walk of less than two cells results in a minimal degradation in focus [1]. The moment arm of the rotation for the edge cells in azimuth is half the range swath, S/2, assuming the CRP is in the middle of the swath. That length times $\alpha$, assuming the angle is small, results in the azimuth walk from A′ to A″ as shown in Figure 7.67. There is a counterpart range walk from B′ to B″.

The derivation of the depth of focus requirement is given in (7.17). The allowable azimuth walk as shown in Figure 7.67 should be less than 2 $d_{az}$ . Equations (7.1) showed the minimum rotation angle, $\alpha$, necessary to achieve a required resolution in azimuth. As stated earlier, the allowable *Walk* is given in the first line of (7.17). The range swath length is defined in the second line of (7.17). Substituting and solving for the number of range bins in focus using the unmodified conjugate of the phase coefficients defined in (7.4) is given in the second to last line in (7.17). If the range and azimuth resolutions are roughly equal, then the last line in (7.17) gives the number of range bins that can use unmodified focusing coefficients.

For example, suppose the range and azimuth resolution is 0.1 m and the wavelength is 0.05 m, then $8 \times 0.1/0.05 = 16$, so each block of 16 range bins will require a phase correction at the edges for complete focus. These phase correction terms may be formed by interpolation of phase terms in adjacent range bins. Another way to accommodate this constraint is with the use of polar format (more in a later section).

$$Walk = \frac{\alpha \cdot S}{2} \leq 2 \cdot d_{az}$$

Substituting the range swath length: $S = N_r \cdot d_r$

and the required resolution turn angle: $\alpha = \dfrac{\lambda}{2 \cdot d_{az}}$

Then: $\dfrac{\lambda}{2 \cdot d_{az}} \cdot \dfrac{N_r \cdot d_r}{2} \leq 2 \cdot d_{az}$ or $N_r \leq \dfrac{8 \cdot d_{az}^2}{\lambda \cdot d_r}$

If $d_{az} \cong d_r$ then: $N_r \leq \dfrac{8 \cdot d_{az}}{\lambda}$

Where: $N_r$ is the number of range bins

and all other variables previously defined. (7.17)

The introduction of this motion compensation section alluded to the necessity for clutter phase lock. When the loss of phase lock occurs, then both acceleration and velocity will be in error. Figure 7.68 shows an example of such an event. The

*Figure 7.68 Effect of velocity and acceleration error on an agricultural image [17]*

*Table 7.12 Motion compensation summary*

| **Error source** | **Equation must be less than 1** | **Typical motion accuracy value*** |
|---|---|---|
| Velocity | $(\Delta V T) / d_{az}$ | $\Delta V = 3 \times 10^{-3}$ |
| Velocity quantization | $(qT) / \lambda$ | $q = 10^{-3}$ |
| Acceleration bias | $(a_b T^2) / \lambda$ | $a_b = 1 \times 10^{-4}$ |
| Acceleration scale factor | $(sf g T^2) / \lambda$ | $sf = 3 \times 10^{-5}$ |
| Acceleration random | $(a_u T^2) / \lambda$ | $a_u = 1 \times 10^{-4}$ |
| Attitude | $(\kappa_{att} g T^2) / \lambda$ | $\kappa_{att} = 3 \times 10^{-5}$ |

*Table 7.12 assumes $T = 10$ s, $d_{az} = 0.3$ m and $\lambda = 0.03$ m.

velocity error results in a displacement of the image in the beam. This displacement is obvious on the left side of the figure. The acceleration error shows up as blurry image formation, which is easily seen as wider bright streaks in azimuth on the left side of Figure 7.65.

In summary, Table 7.12 shows the general requirements for each motion error source. Velocity in x and y as integrated from accelerometers usually should have an overall relative one sigma accuracy 0.003 m/s (0.2 Hz at X band). Vertical acceleration usually should have an overall relative one sigma accuracy of 0.01 m/s$^2$ (0.03 Hz/s$^2$ at X band). Overall one sigma acceleration accuracy in x and y should be 0.003 m/s$^2$ for 0.3 m resolution with ~10 s array times. Obviously, some of this performance can only be accomplished by a combination of inertial, external updates and signal-based motion compensation.

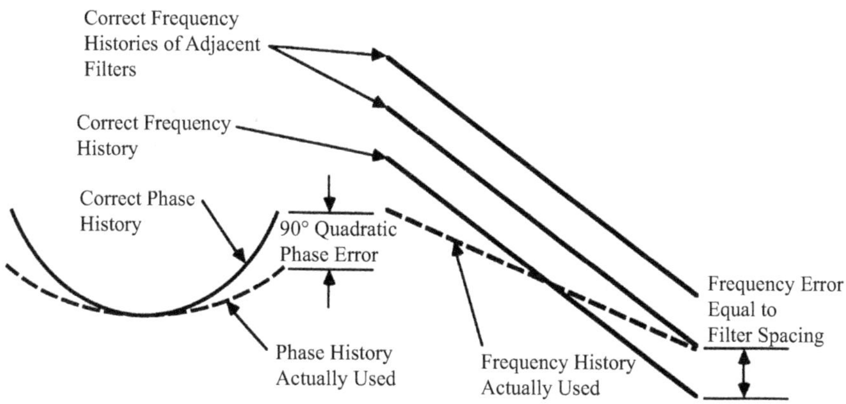


*Figure 7.69 Phase history slope error causes defocusing [1]*

## 7.11 Autofocus

As mentioned earlier, it is not possible to remove all the phase errors over the entire SAR image. Usually an automatic technique called autofocus is used to remove the last of the phase errors across the image. Figure 7.69 shows a way of visualizing the effect of phase history slope errors (acceleration error). The objective is to have less than 90° phase error at the array edges, which is to say that the accumulated frequency error at the array edge just arrives at the center of the next filter as described in Figure 7.69. This is another way of describing the depth of focus limit described in Figure 7.67. There is a classic technique for improving focus across a map called autofocus. Autofocus techniques have been known since before anything was dreamed about microwaves. The idea is to sense less than optimum focus and adjust the optics to improve it. In optical cameras, this is usually done by adjusting the focus of a set of lenses until a peak occurs in the autocorrelation function of the image. This occurs at the focal point not at the image plane. An approximation to the optical method is used by correlating sections of the SAR map to improve focus.

One technique is the "map drift" method as shown in Figure 7.70. As deviations occur from the desired or predicted flight path, coarse subarray maps will exhibit phase differences to the same patch of mapped surface. Cross correlating these subarray images will yield a phase difference, which can be used to estimate the true phase slope. This detected map phase difference may not be constant over the image due to atmospheric effects or differences in altitude to parts of a patch. There may be multiple subarrays in both range and azimuth to improve image focus.

Figure 7.71 shows the general concept and various embodiments of autofocus processing. The top two in Figure 7.71 use subarray maps. The first uses magnitude detected subarrays, which it cross-correlates to estimate the apparent movement between them to estimate a short-term velocity error, that is, an acceleration approximation. The second method uses multiple Fourier transforms on estimated autocorrelation images (not maps just bright central spots) to estimate a short-term velocity error. Both of these methods have a transport delay in phase that must be compensated at the output. In the second and third methods, phase is preserved in

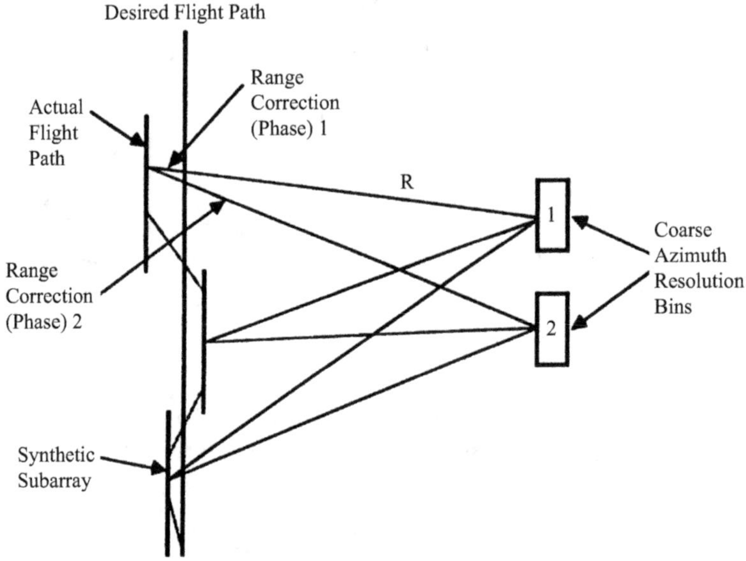


*Figure 7.70 SAR map drift autofocus sensing [1]*

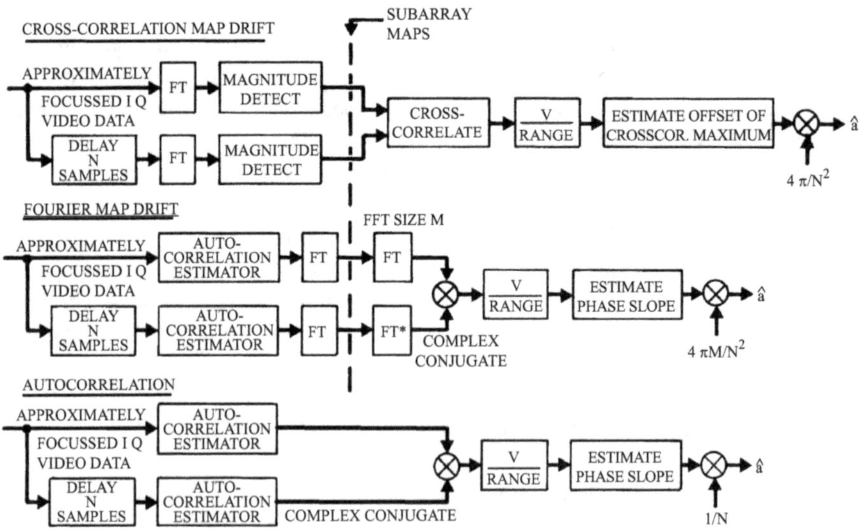


*Figure 7.71 Alternative map drift autofocus methods [1]*

the subarrays so the second image must be a complex conjugate. The third method does not form maps but rather just uses input streams with an autocorrelation estimate of subarrays delayed by some number of PRIs, which it then correlates to estimate velocity drift (approximate acceleration). The last method obviously has a different transport delay that must be phase compensated when applied to the azimuth focusing. Each of these techniques can produce eye-watering image quality under some circumstances. The most spectacular images have been processed with multiple algorithms or subarray sizes to obtain the best output. They don't all look like that!

## 7.12 Image smoothing: multilook

Forming multiple images either at different frequencies or different angles and then adding the magnitudes of each corresponding pixel can reduce image "speckle" or scintillation thereby improving target detection and recognition. The basic idea of multifrequency overlays is shown in Figure 7.72. The individual image offsets shown in Figure 7.72 are exaggerated for clarity. Usually the antenna is scanned a small amount between each image so that the number of nonoverlapping pixels is small. After image rectification each image is magnitude detected and added to produce a single image as shown in Figure 7.74.

Another method for multilook overlay is shown in Figure 7.73. This method is especially useful in strip mapping. A single frequency is used but array resolution and coverage is selected such that a single SAR array takes less time than the time

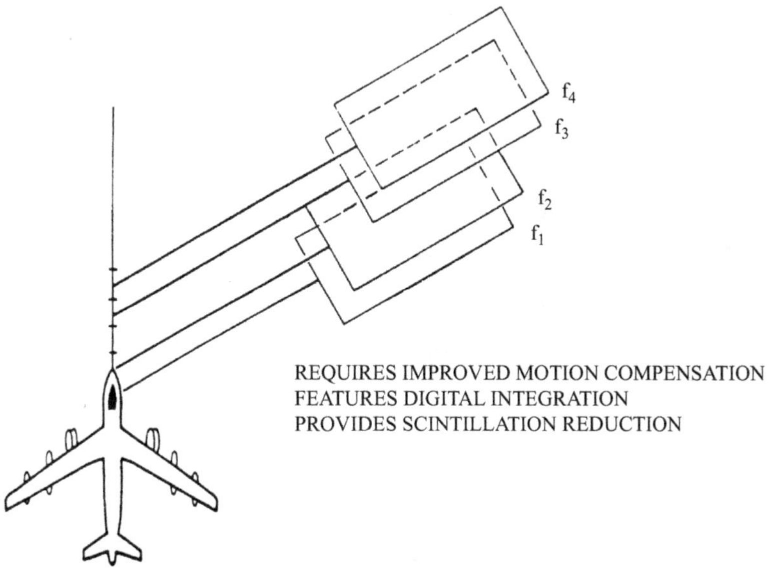


*Figure 7.72 Multifrequency SAR image overlay [3]*

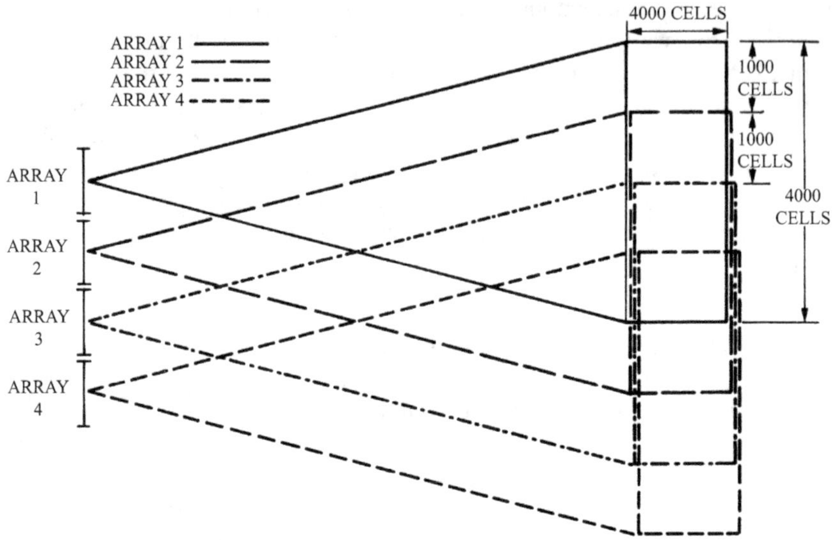


*Figure 7.73 Constant frequency SAR overlay example [1]*

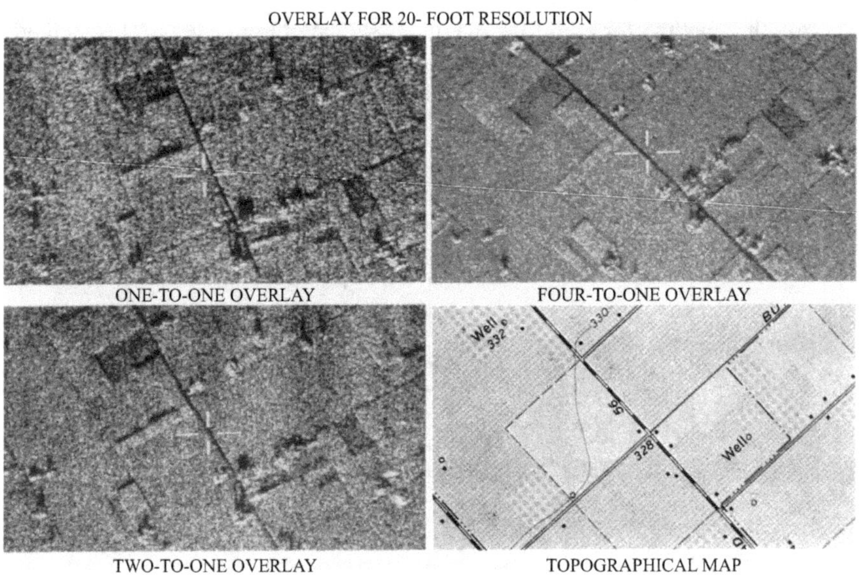


*Figure 7.74 Comparison of multiple SAR map overlays [3]*

to traverse the mapped patch. Then a second map is formed in which 3/4 of the pixels overlap with the first map. A third map is formed with an overlap of 1/2 pixels and so on until four partial overlaps are processed. The corresponding magnitude detected range-Doppler pixels are summed. If this process is carried out continuously, a strip map is formed with four overlays everywhere. Although from the exaggerated figure one might think that no angle change occurs, in fact, there is enough angle change to produce scintillation array-to-array. Magnitude addition of all overlapping maps produces more continuous tone maps with less "speckle."

Typical rotation angles map-to-map of a few degrees is usually adequate to provide angle diversity equivalent to frequency diversity of frequency offsets. Figure 7.74 is a typical example.

## *7.12.1 Polar format*

Polar format has been alluded to several times earlier in the chapter. Although the SAR notion has been discussed as though the raw data is in a rectangular array, in fact the raw data is gathered in a spherical coordinate system (range, azimuth, elevation) as suggested earlier by Figures 7.2 and 7.9. For coarse resolution relative to wavelength and limited range-angle coverage, treating the raw data as though it represents a rectangular array provides more than adequate performance. However, as resolution goes below $10\lambda$ or angular coverage goes above 0.1 radians, this curvature must be taken into account. One way to account for this curvature is polar format. Polar format is most advantageous when the transmit waveform is a chirp. Figure 7.75 shows the polar format geometry.

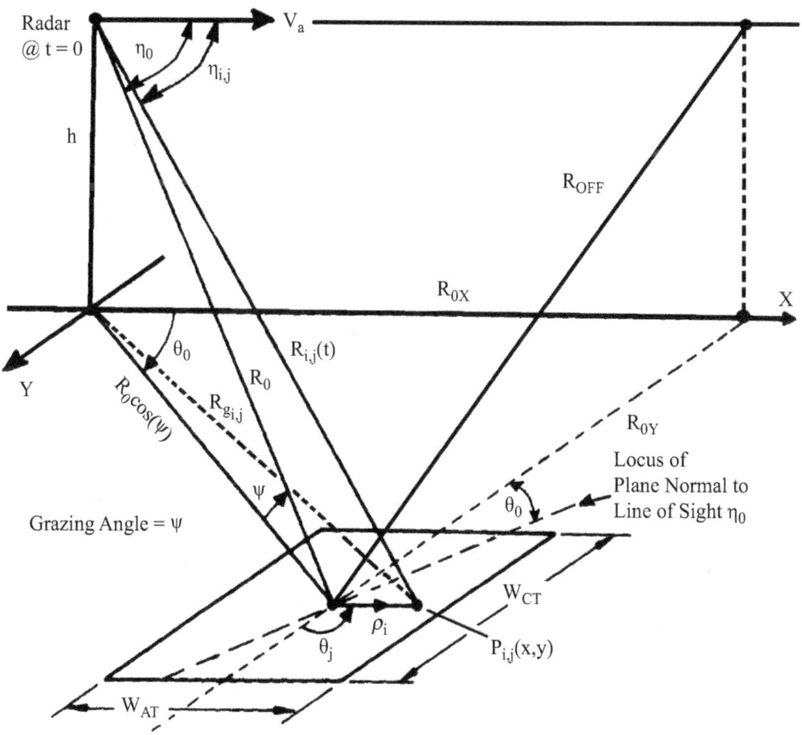


*Figure 7.75 Polar format geometry [1]*

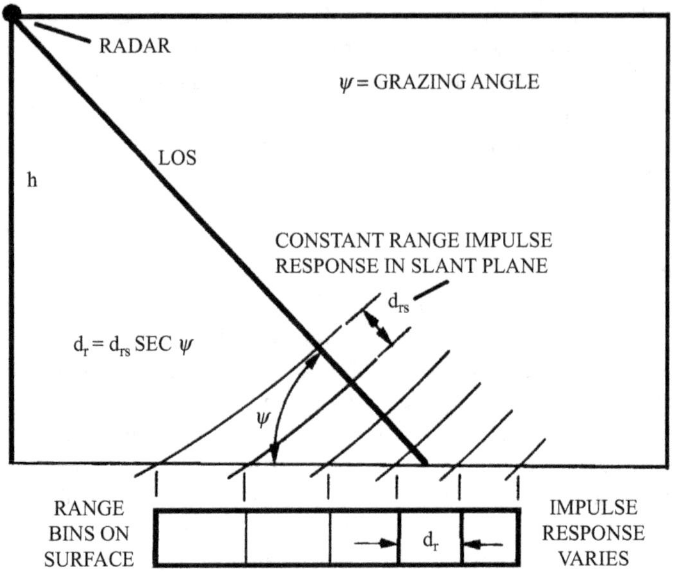


*Figure 7.76 Range resolution variation with grazing angle*

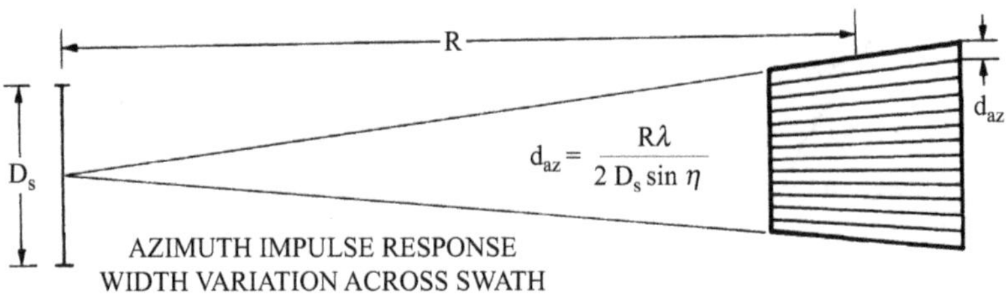


*Figure 7.77 Azimuth resolution variation with range [1]*

First, the range resolution varies with grazing angle so that a fully processed and rectified output map will have uniform range pixel resolution for minimum geometric distortion. This issue is shown in Figure 7.76. At shallow grazing angles from an aircraft platform, this ground range resolution variation can be small but, if the range is long, the Earth's curvature comes into play. For a spacecraft platform, ground range resolution variation is almost always an issue. The data mapping must take the range resolution variation into account.

Second, azimuth resolution varies with range and it must be compensated as well so that azimuth pixels are rectified and uniform. Figure 7.77 and (7.1) repeated in the figure shows the azimuth variation with range. Similarly, the data mapping must compensate for azimuth variation to provide a rectified image after processing.

Figure 7.78 shows the typical variation in range and azimuth resolution for a low-resolution SAR spacecraft in LEO. One will note that the range resolution

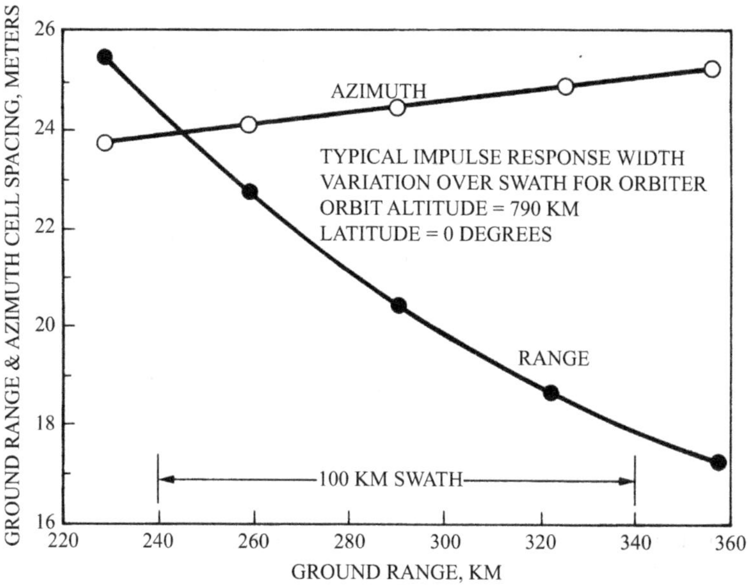


*Figure 7.78 Typical spacecraft low-resolution SAR variation [1]*

variation is much larger than the azimuth resolution variation. The elevation depression angle from the spacecraft (not grazing angle) to beam center is approximately 20° in the figure.

Equations (7.18) provide an approximation to the phase returned from each scatterer (*i, j*) in a map patch based on the geometry of Figure 7.75. It uses a different approximation than presented earlier but is the one most commonly used with polar format. Again, without loss of generality the velocity vector, $V_{ax}$, is parallel to the x axis at an altitude, h. There is a fixed offset in y with slant range of $R_{OFF}$ at the point of closest approach. The matched filter is usually done in two steps. The first is called bulk filtering matched to the midpoint of the map, which is typically at $i_{max}/2$ and $j_{max}/2$. The midpoint will be perfectly focused but every other point will have some residual phase with time, which must be compensated. (See Figure 7.36 and RangeChangevsTime007 in Appendix A.7.)

Define some intermediate variables:

$$Rx_0 = R_0 \cdot \cos(\psi) \cdot \cos(\theta_0) \text{ , } Ry_0 = R_0 \cdot \cos(\psi) \cdot \sin(\theta_0)$$

$$px_{i,j}(0) = \rho_i \cdot \sin(\theta_j) \text{ , } py_{i,j}(0) = \rho_i \cdot \cos(\theta_j)$$

$$Rx_{i,j}(0) = Rx_0 + px_{i,j}(0) \text{ , } Ry_{i,j}(0) = Ry_0 + py_{i,j}(0) \text{ , } R_{OFF} = \left[ Ry_{i,j}(0)^2 + h^2 \right]^{1/2}$$

Range to each scatterer: $R_{i,j}(t) = \left[ \left( Rx_{i,j}(0) - V_{ax} \cdot t \right)^2 + Ry_{i,j}(0)^2 + h^2 \right]^{1/2}$

(7.18.1)

$$R_{i,j}(t) = R_{OFF} \cdot \left[1 + \frac{\left(Rx_{i,j}(0) - V_{ax} \cdot t\right)^2}{R_{OFF}^{\ 2}}\right]^{1/2}, \text{ Let: } Z(\mathrm{t}) = \frac{\left(Rx_{i,j}(0) - V_{ax} \cdot t\right)^2}{R_{OFF}^{\ 2}}$$

Observe that: $(1+Z)^{1/2} = 1 + \frac{Z}{2} - \frac{Z^2}{8} + \frac{Z^3}{16} - \frac{5 \cdot Z^4}{128} \ldots$

Then: $R_{i,j}(t) \cong \left[R_{OFF} - \frac{Rx_{i,j}(0)^4}{8 \cdot R_{OFF}^{\ 3}} + \frac{Rx_{i,j}(0)^2}{2 \cdot R_{OFF}}\right] + \left[\frac{V_{ax}^{\ 2} \cdot t^2 - 2 \cdot V_{ax} \cdot t \cdot Rx_{i,j}(0)}{2 \cdot R_{OFF}}\right]$

The first bracketed term is a constant and can be ignored when match filtering.

Phase to each scatterer with time: $\varphi_{i,j}(t) \cong \frac{4 \cdot \pi}{\lambda} \cdot \left[\frac{V_{ax}^{\ 2} \cdot t^2 - 2 \cdot V_{ax} \cdot t \cdot Rx_{i,j}(0)}{2 \cdot R_{OFF}}\right]$

(7.18.2)

There are basically two ways to account for this sample curvature. One way is to match filter the data in each sample bin using the exact 2D matched filter, which is usually done by transforming to the spatial frequency domain, 2D filter multiplication and then inverse transform, followed by spatial rectification to rectangular coordinates. The second way is to map the raw data into a rectangular grid, dechirp and 2D FFT.

This is the method shown in Figure 7.79. The second way may or may not reduce the total number smashing but isn't as simple as it sounds. Much of the improvement, if any, is very dependent on the signal processing hardware architecture because the raw data mapping requires irregular memory addressing and interpolation. The basic idea of the second method data mapping is shown in Figure 7.79. The first mapping interpolates into x and azimuth coordinates. The second data mapping interpolates into x and y coordinates. The figure shows an obvious drawback for this method. The final resolution will be poorer than the waveform and array length would imply because the output data array is smaller.

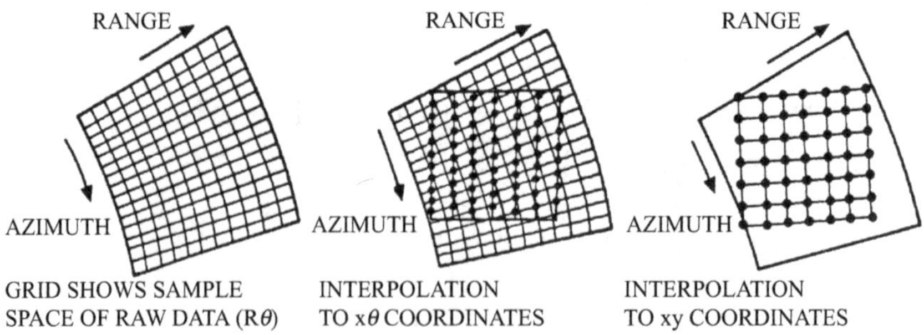


*Figure 7.79 Polar format data mapping notion [1]*

Clearly, as resolution increases interpolation is required to provide rectified images. The issue, when the input and output sequence have different sampling intervals, is that the distance between samples is continuously sliding. Figure 7.80 shows that the delta, $\delta_n$ , between the input and output steadily changes no matter what the initial offset alpha, $\alpha$, is. In addition, it is known that the input samples don't represent constant resolution, whereas the desired output resolution should be matched and uniform in surface x and y.

Equations (7.19) provide an example of a typical interpolation function. The example is a combination of raised cosine weighting and the sins function. The first function in (7.19) to be calculated well before data gathering is the normalized weighting function table. Usually to save memory space, the weighting table is limited to a specific quantized set, $j_{max}$ in number, of offsets between input and output data. This quantization does lead to some error but it can be made small by making $j_{max} = 4\text{-}8$ as shown in Figures 7.81 and 7.82.

Interpolation is nothing but lowpass filtering to calculate intermediate values between samples. A rectangular lowpass filter in the frequency domain is a sin(x)/x convolution (commonly abbreviated sinc(x)) in the time domain. Parenthetically, one must observe that since both inputs and outputs of the interpolation are sampled data, the true function implemented is sin(Nx)/(Nsin(x)) (commonly abbreviated sins(Nx)) which is folded around the sample rate. Usually to reduce passband ripple, the sinc function is multiplied by a weighting function of the type mentioned in Chapter 2, Table 2.2 and Chapter 4, Table 4.1. A weighted sinc is shown in Figure 7.81 and note that the locations of the desired interpolation weights for a

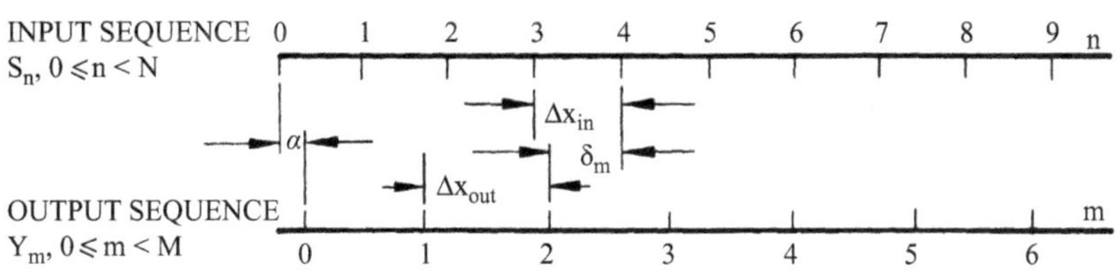


*Figure 7.80 Input interpolation samples slide relative to output*

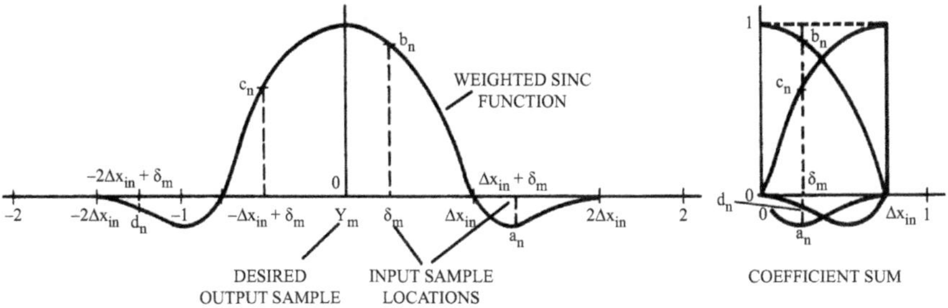


*Figure 7.81 Typical interpolation function [1]*

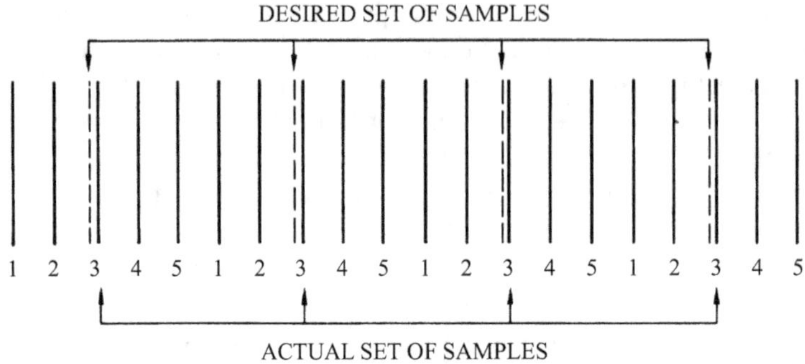


*Figure 7.82 Picking quantized rather than desired sample locations [1]*

single output $Y_m$ are a function of a specific offset $\delta_m$ as shown in Figure 7.81. If the input sequence is long, almost all values of the interpolation function could be used. For every output $Y_m$, $\delta_m$ must first be calculated and the appropriate weighting values retrieved from a table in memory, multiplied by the corresponding input sequence $S_n$ and the result summed.

Calculate weighting coefficients for every approximate offset between $n$ and $m$:

$$i_{max}=N_k\ ,\ i=\frac{-i_{max}}{2},1+\frac{-i_{max}}{2}\ldots(i_{max}-1)+\frac{-i_{max}}{2}\ ,\ j_m=0,1\ldots j_{max}\ ,\ \delta s=\frac{\Delta x_{in}}{j_{max}}$$

$$WU_{i,j}=\left(a+(1-a)\cos\left(2\cdot\pi\cdot\frac{i-j_m\cdot\delta s}{N_k}\right)\right)\cdot\frac{\sin(\pi\cdot i\cdot N_k\cdot\Delta x_{in}-j_m\cdot\delta s)}{\pi\cdot i\cdot N_k\cdot\Delta x_{in}-j_m\cdot\delta s}$$

Calculate normalization factors for each approximate offset:

$$WN_{j_m}=\sum_i WU_{i,j_m}\ ,\ W_{i,j_m}=\frac{WU_{i,j_m}}{WN_{j_m}}$$ Store normalized weighting coefficients in a table.

Calculate an approximate offset for a given $n$ and $m$: $j_m=\frac{j_{max}}{\delta s}\cdot\mathrm{Trunc}\left(\frac{\delta_m}{j_{max}}\right)$

For each $m$ conditioned on $n$, retrieve coefficients for a specific $j_m$

and calculate: $Y_{m|n}=\sum_i S_{n+i}\cdot W_{n+i,j_m}$

(7.19)

Where $N_k$ is the length of the weighting kernel, $\Delta x_{in}$ is the spacing between input samples, *WU* is the not normalized weighting coefficients, *WN* is the normalizing factor for each set of offset coefficients, $\delta_s$ is the minimum offset step size, $\delta_m$ is the actual offset between the specific *n* and *m* for the current interpolation calculation. An example interpolation is the Appendix A.7 DVD-R or internet in Mathcad and PDF format titled Interpolation001, 002.

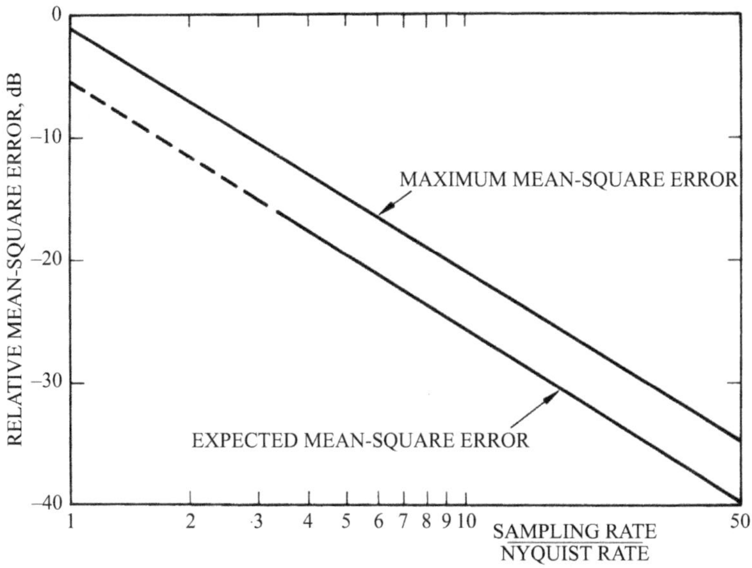


*Figure 7.83 Errors caused by quantized sample locations*

After calculating the table, the next step is to calculate a specific $j_m$ based on the actual calculated offset between input and output. That value is then used to access the specific coefficients in the table for the output sample. Finally, the sum of the products of weighting function and input samples is calculated. Since both the input range and azimuth samples may be nonlinear in rectilinear space, the kernel length and input spacing will change over the whole array. Those changes may require a new table every few hundred samples.

It would be fortuitous if the raw data samples lay exactly where the math would result in the least error. Usually samples closest to the desired locations are selected as shown in Figure 7.82. This gives rise to small errors in a map.

Figure 7.83 shows the errors created by quantizing sample locations. It is typical in interpolation to significantly oversample to simplify processing. Usually, a $-20$ dB relative mean square error is considered adequate, which implies about a 4:1 oversampling rate. Figure 7.84 shows an actual example in which the range interpolation and the azimuth presummation and interpolation are both oversampled at the filter outputs, which allows the nearest quantized data sample to be selected as the sample to be transformed into a rectilinear output image.

Another strategy for simplification of polar format processing is to allow significantly different memory read in and readout addressing. This only works well if the image processing bulk memory has random access. An example scheme is illustrated in Figure 7.85. The input is written along radial lines corresponding to the angle off the velocity vector while the readout is read along a rectilinear grid

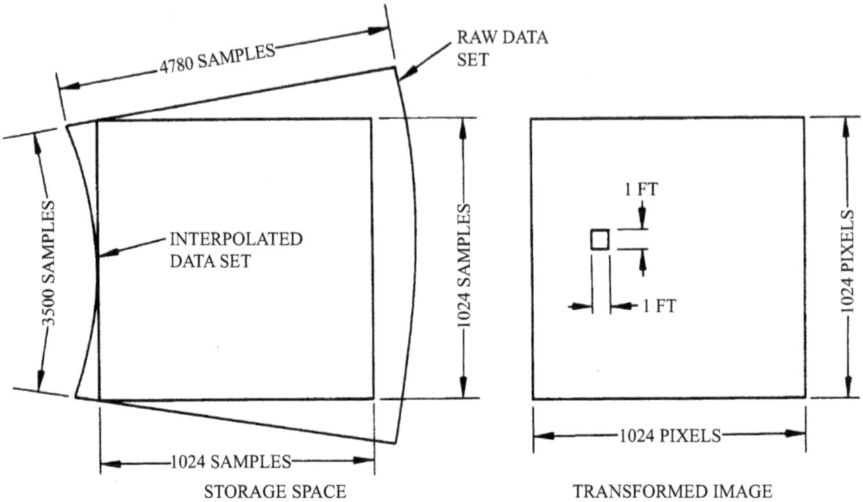


*Figure 7.84 Oversampling to allow nearest sample selection [1]*

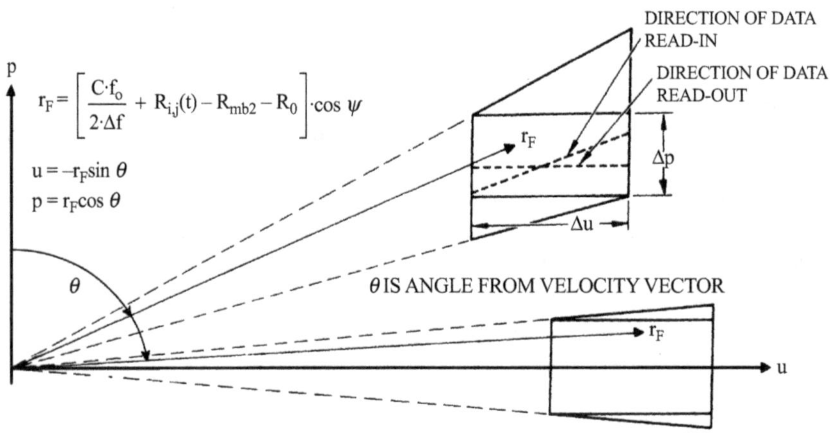


*Figure 7.85 Polar memory write in, rectangular readout notion [20]*

line. The idea is to create a new coordinate space for the SAR samples in u and p. The actual memory used is only Δp by Δu words.

The variables in Figure 7.85 are $r_F$ = the radial distance to an input data point in u, p space, c = velocity of light, $R_{i,j}(t)$ = range to a specific data point in patch, $f_0$ = transmit center frequency in cycles/s, $\Delta f$ = chirp slope in frequency change per unit time, $\theta$ = angle from velocity vector to patch center, $t$ = time, $R_{mb2}$ = range to center of chirp main bang, $R_0$ = range to patch center reference point, $\psi$ = grazing angle (see Figure 7.75).

## 7.13 Slow-moving target indication by change detection

### *7.13.1 Non-coherent change detection*

Although CCD has been much discussed in the literature, often various versions of NCD are very effective. The reason is the time and aspect changes between images may make phase coherence of limited use. Many times area normalized NCD is dramatically better. One of the biggest challenges for SAR-based change detection is image localization, registration and geometric rectification. The SAR images may have been taken at different times, geometrical aspects and even frequency bands. Resolving the geometrical aspect requires mapping the radar pixels into a common coordinate frame as well as correcting for propagation and motion-induced distortion. Some of this can be done from known trajectory and aspect data taken as part of the data gathering but some of the potential offsets and distortions must be done by correlating and warping the images to overlay properly on each other. This often requires interpolation (Interpolation001, 002 in Appendix A.7). Often the resolution of the change map will be poorer than the original images due to the required warping and overlay. (See CCD003, -003a in Appendix A.7.) When map data are taken rapidly, both NCD and CCD can be employed together to obtain best mover detection. Equation (7.20) gives a typical NCD method.

Normalized difference magnitude between map1 & map2 each pixel:

$$g1_{n,j} = \frac{64 \cdot \left(\left|F1_{n,j}\right| - \left|F2_{n,j}\right|\right)}{\left(\left|F1_{n,j}\right| + \left|F2_{n,j}\right|\right) + 1}, \text{ Find magnitude: } g1M_{n,j} = \left|g1_{n,j}\right|$$

Average magnitude across azimuth bins each range bin:

$$g1MA_j = \frac{1}{(n_{\max} + 1)} \left( \sum_{n=0}^{n_{\max}} g1M_{n,j} \right)$$

Threshold each pixel using average magnitude & threshold multiplier:

$$NCDT1_{n,j} = \text{if}\left(g1M_{n,j} \geq TM \cdot g1MA_j, 250, 1\right)$$

Combine map with non coherent detections for display:

$$NFC_{n,j} = \left|F1_{n,j}\right| + NCDT1_{n,j}$$

Where: $F1_{n,j}, F2_{n,j}$ = focused map IQ data, $n$ = azimuth bin index,

$j$ = range bin index, $TM$ = CFAR threshold multiplier. (7.20)

Figure 7.86(a and b) shows shadow and Doppler detections using NCD. Resolution was approximately 1.5 m and the car-sized simulated targets were ~2 × 6 radar pixels. Figure 7.87 shows both coherent and non-coherent detection in a single-range bin. The NCD are much noisier and may require a higher threshold, typically 20–22 dB. The successive SAR maps were 1 and 8 s apart in Figure 7.86(a and b). A combination of coherent and non-coherent detection should be used for tactical targets.

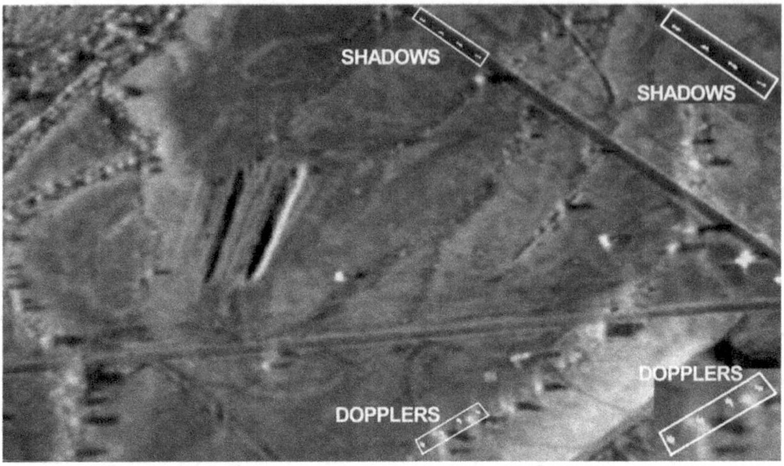


*Figure 7.86 (a) Non-coherent change detection 8 s between maps w/ enlargements, opening target @ 3.6 m/s (CCD003)*

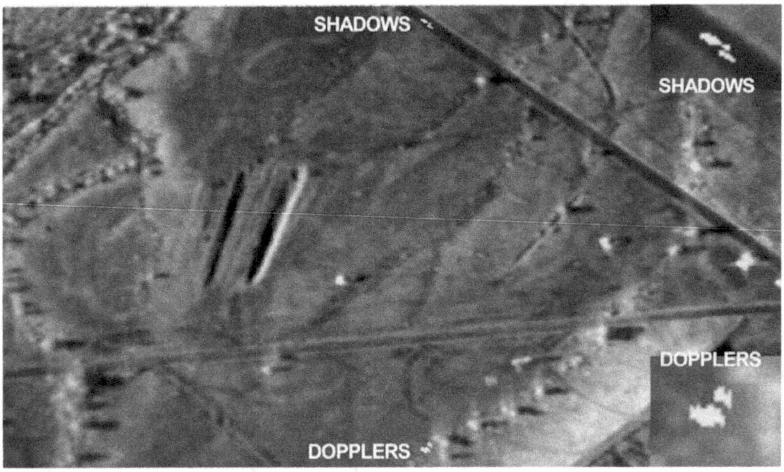


*Figure 7.86 (b) Non-coherent change detection 1 s between maps w/enlargements, opening target @ 3.6 m/s (CCD003a)*

One simple way to match maps is to adjust FFT zero fill so that the same number of filters spans each map even though the Doppler spread will be different map to map (can also be done in range).

### *7.13.2 Coherent change detection*

Suppose that one calculated the coherence between two SAR maps by calculating the normalized cross-coherence of the complex signals at each radar pixel as shown

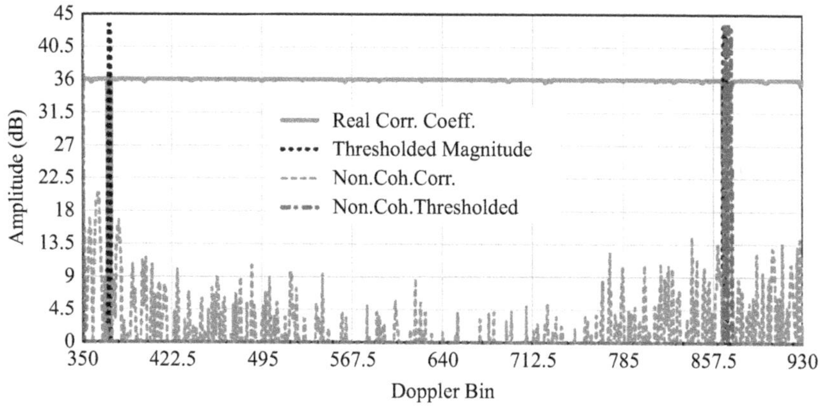


*Figure 7.87 Range bin 556 showing coherent and non-coherent change detections (CCD003a)*

in (7.21). If the coherence was almost perfect in a radar pixel, then no change has occurred between images in that pixel. One could then substitute the original map at that pixel. Less than perfect coherence would be marked by a lower amplitude or dramatically changed phase shift while a radar pixel with coherence less than ½, which has obviously undergone a significant change could be represented by lowest or highest amplitude (black or white?).

$$\gamma(n,j) = \frac{\mathrm{E}\left(s1_{n,j}\cdot\overline{s2_{n,j}}\right)}{\left(\mathrm{E}\left(s1_{n,j}\cdot\overline{s1_{i,j}}\right)\cdot\mathrm{E}\left(s2_{n,j}\cdot\overline{s2_{n,j}}\right)\right)^{0.5}} = \left|\gamma(n,j)\right|\exp\left(j\cdot\arg\left[\gamma(n,j)\right]\right)$$

Where: $s1_{n,j}$ is the complex amplitude in SAR map 1 radar pixel $n, j$,

$s2_{n,j}$ is the complex amplitude in SAR map 2 radar pixel $n, j$,

$\left|\gamma(n,j)\right|$ is magnitude of coherence and $\arg\left[\gamma(n,j)\right]$ is phase in radians.

(7.21)

Several authors recommend using NCD to screen coherence calculations to emphasize low coherence (large changes in a radar pixel). CCD depends primarily on recognizing a large phase change relative to the local background. Figure 7.88 shows a change detection image for vehicle sized objects (10 m$^2$ corner reflectors) moved after the first image and before the second along highway 99 (diagonal multilane road in upper part of image) in California with resolution of 20 ft. As one can see there are many reflectors, which show some decorrelation image-to-image (e.g., vegetation and fence movement) but still do not represent a real change such as might be expected from a tactical vehicle. The 8 dB threshold change detection image on the right in Figure 7.88 shows the real and significant changes. The large changes are circled in the 0 dB threshold image, which are obvious in the 8 dB thresholded detection image. This SAR map image was taken in a controlled test.

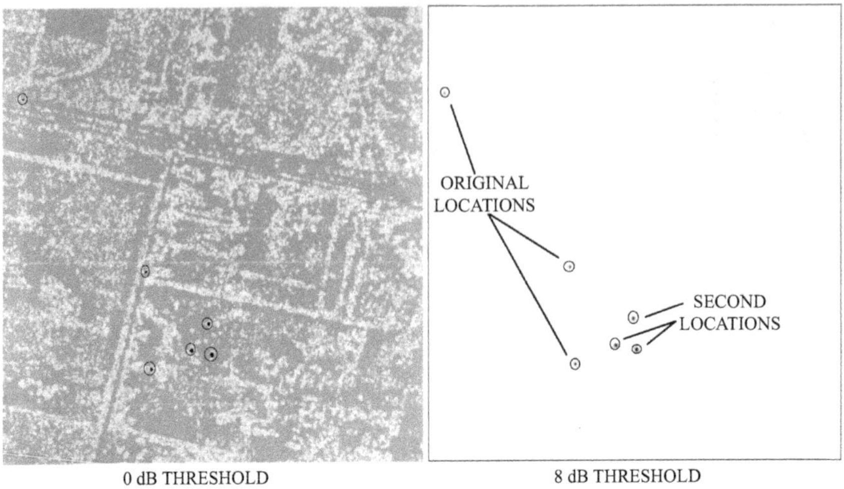


*Figure 7.88 CCD detection of vehicle-sized targets [3]*

There were three artificial targets that were moved from the locations showing the fainter changes to locations near one another whose detected changes are significantly brighter.

## 7.14 Video SAR and shadow detection MTI

The idea behind video SAR is to have a high enough SAR map rate to detect movement of objects between images. This usually requires high enough rates that there is a significant overlap in the data used in each image. You might call this the first of the high-speed change detection methods. Short-term CCD and NCD may be used to separate stationary targets from slow-moving endo-clutter targets. Short-term CCD is a method in which two or more coherent SAR maps taken within a short time of one another at the same frequency are registered and coherent change detected radar pixel by pixel. A common SAR-MTI display may be presented to the operator showing coarse resolution MTI targets superimposed on a higher-resolution SAR map as shown in Figure 7.89. In many cases, this is much better than slow MTI based on STAP.

Shadow detection MTI is the second of the high-speed change detection methods in which the primary emphasis is the detection of shadows of objects that have moved fast enough not to focus but the shadow will focus and show the true location of the object. By using change detection image-to-image, the trajectory of the object can be accurately determined. Of course, the moving target Doppler will cause the return from the target to be displaced from its true location. A mover

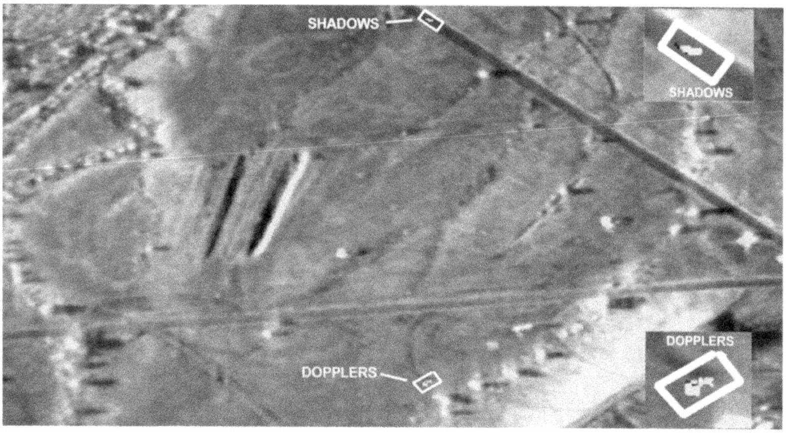


*Figure 7.89 Slow-moving targets by CCD – 1 s between maps w/enlargements (CCD003a)*

depending on vehicle speed and radar processor presum ratio may or may not appear on the SAR map at all.

Since multiple maps will be taken and processed, they each will be at a slightly different Doppler and frequency spread. This means that the number of radar pixels will span different numbers of Doppler bins. As mentioned previously, one way to get samples from different maps to have matching sample spaces is to add zeros to the FFT so that the same number of samples span the spectrum in each map. No more resolution is achieved but the correlation is more accurate. Another important factor is the accumulated phase change between each map and the next. As a certain radar pixel is processed one map to the next, there will be a phase change associated with the opening or closing range to that pixel. For squinted SAR off the normal to the velocity vector, the phase change map to map might be as much as ~4589 radians/s as it is in Figure 7.89. The center phase of each map to be processed can be matched but the phase differences in each radar pixel away from the center will be different even with no change in the pixel movement map to map. For CCD this must be compensated by using some form of sliding window average to obtain a phase to be subtracted from each pixel. Then, any additional phase change map to map could represent a moved or moving target. The sliding window (see Section 4.6) must have a large enough blanked center so that movers, which will span multiple range-Doppler cells don't affect phase compensation or thresholding (SlidingWindow002 in Appendix A.4). Since the phase of each azimuth bin naturally will exhibit a linear phase delay with frequency, a sliding window that favors range averages is required.

Equations (7.22) show a sliding window usable for regular phase advance Doppler bin-to-bin. It shows a 6-5-6 sliding window to be used to compensate for the argument of $\gamma$ as the phase advances across the image.

Let: $\Phi_{n,j} = \arg\left[\gamma(n,j)\right]$

Then the early average is: $THE_{n,(j1+3)} = \left(\dfrac{1}{j2_{\max}+1}\right)\cdot\sum_{j2}\Phi_{n,(j1-j2)}$

Late average is: $THL_{n,j1} = THE_{n,(j1+2\cdot j2_{\max}+2)}$

Compensate for data array edge effects:

$THE_{n,(j2+1)} + THE_{n,(j2_{\max}+3)}$ , $THE_{n,(j1_{\max}+4+j2)} = THE_{n,(j2_{\max}+3)}$,

$THE_{n,(j1_{\max}+5+j2+j2_{\max})} = THE_{n,(j2_{\max}+3)}$

Combine early and late averages: $TH1_{n,j1} = 0.5\cdot\left(THE_{n,j1} + THL_{n,j1}\right)$

Phase compensate coherence: $\gamma C_{n,j} = \gamma_{n,j}\cdot\exp\left(-\mathrm{j}\cdot TH1_{n,j}\right)$

Find phase discriminant: $TH_{n,j} = \left|\gamma C_{n,j}\cdot\mathrm{Im}\left[\gamma(n,j)\right]\right|$ (7.22.1)

Where: $\mathrm{Im}\left[\gamma(n,j)\right]$ = imaginary part of $\gamma$; $j2 = 0,1\cdots j2_{\max}$

$j2$ = number samples averaged in early or late windows,

$j1 = j2_{\max},\left(j2_{\max}+1\right),\cdots j1_{\max};\ j1_{\max} = j_{\max};\ j = 0,1\cdots j_{\max}$ (7.22.2)

---

**Example 7.9** For example, consider the radar parameters of Table 7.13 Example 7.9 used to generate Figures 7.89 and 7.90.

The images in Figures 7.89 and 7.90 show four SAR maps cross-correlated each to the next at intervals of 1 and 8 s apart. The maps were generated by SoCalSAR030a(6, 7, 8, 9) from raw I/Q data SoCalSum006, 007, 008, 009 provided in Appendix A.7 DVD-R or internet. The maps were processed using the equations in (ex 7.3) and in (7.21) and (7.22). Each map pair must be thresholded separately. In the example, only the maps that were adjacent in time were processed for coherence. In general, all maps should have cross-coherence calculated. A 2D sliding window was used on the CCD discriminant to minimize false alarms while providing detections. Range threshold multipliers were about 2.7 and azimuth threshold multipliers were about 4.2. In this case, the range and azimuth sliding

*Table 7.13 Example 7.9 video SAR*

| Parameter | Value | Parameter | Value |
|---|---|---|---|
| Radar altitude, km | 10 | PRF, Hz | 3598 |
| Radar velocity, m/s | 225 | Presum sample rate, Hz | 514 |
| Antenna width, m | 1 | Array time, s | 2.24 |
| Antenna height, m | 0.5 | Initial look angle, ° | 60 |
| Antenna orientation | Sidelook | Wavelength, m | 0.03 |
| Max. ground range, km | 50 | Raw IQ samples | 1024 × 1152 |

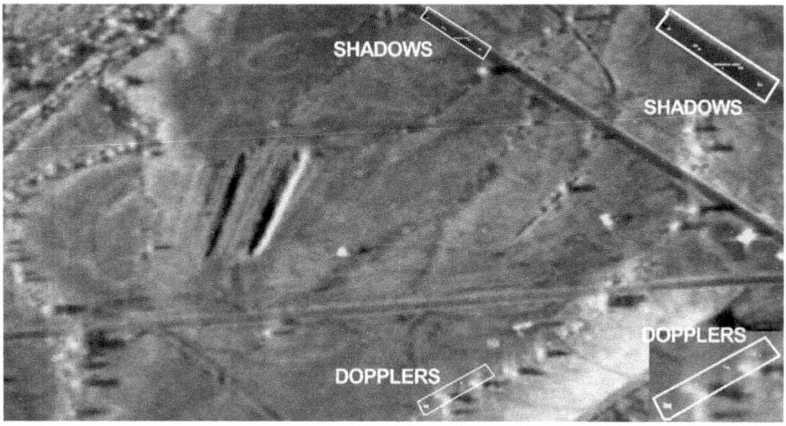


*Figure 7.90 Slow-moving targets by CCD – 8 s between maps w/enlargements (CCD003)*

windows were additive rather than multiplicative as used in other examples in Section 4.6 and Chapters 6 and 7. Thus the composite threshold multiplier is 2.7 + 4.2 = 6.9. The thresholded CCD outputs similar to Figure 7.86 are designated by full brightness pixels superimposed on one of the SAR maps. The detections are boxed in the images Figures 7.89 and 7.90 (see enlargements at right in images). These detections are from a simulated moving target on the road in the upper part of the image. The simulated mover is the size of a car, 2 × 6 m, traveling at 3.6 m/s (3 m/s radial) on the road away from the radar. The radar is illuminating the patch from the left of the image. In this simulation, there are many mover velocities, which are Doppler blind but the shadows always show up at the true location. There are some radar pixels that are range and Doppler blind to movers using the discriminant in (7.22). This results due to artifact induced by the argument of $\gamma$ going from $\pi$ to $-\pi$. Since those phase transitions are unique to the two maps being compared, those blind spots go away if each map is compared to all others.

## 7.15 Inverse SAR

A far more reliable method of vehicle, ship, aircraft and spacecraft recognition is ISAR [4,21,22]. The basic notion is that the motion of a rigid object can be resolved into a translation and rotation with respect to the LOS to the target. The translation is usually removed by conventional range and angle tracking of the entire object to be imaged. So the object appears to be quasi-stationary in the imaging window with coordinates, $x'$, $y'$, $z'$, centered at the center of the target rotation. Figure 7.91 gives the basic idea. A rotating rigid body has different parts moving at different speeds depending on the distance, $\rho_i$, from the center of rotation. Also the apparent speed along a LOS will vary sinusoidally if observed long

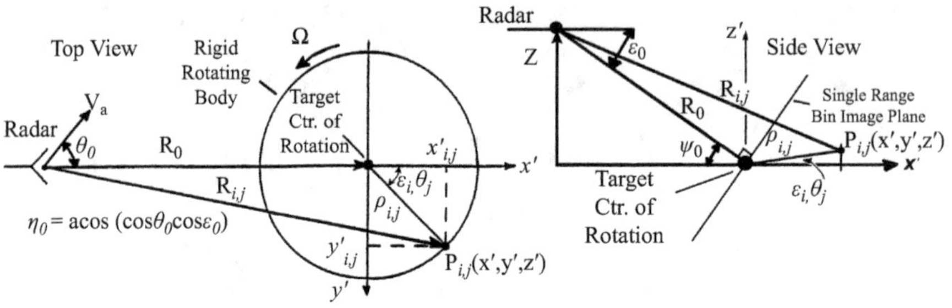


*Figure 7.91 Inverse SAR notion*

enough. The rotation gives rise to a differential rate of phase change across the object. The phase history differences can be match filtered to resolve individual scatterers in a range cell. Conceptually, such a matched filter is no different than a filter used to match a phase coded pulse compression waveform. This is the basis of all SAR, RCS range imaging, observed geometric target acceleration, turntable imaging and ISAR. For turntable and RCS range imaging the rotation rate is known with considerable accuracy and exact matched filters are used radar pixel by pixel.

For general ISAR of ground vehicles, ships, aircraft and spacecraft rotation rates must be guessed or estimated and then an appropriate filter bank is applied. Typically, the cross-range dimension is resolved using an FFT Doppler filter bank matched to a guessed rotation period. The 2D ISAR picture will have LOS range as one axis of the image and the second axis will be normal to the LOS. One complicating factor is the velocity of the sensor platform will cause the image to shift, rotate and change the cross-range scale factor depending on the ratio between the sensor platform velocity and the apparent speed of the scatterer, $\mathbf{V_t}$. The distortion will be on the order of $\mathbf{V_t/(V_a\ R)}$.

Figure 7.92 provides a more detailed geometry for an object rotating near the earth's surface observed by an airborne or spaceborne sensor. There will be a patch on the surface that has an along track, vertical and cross-track area containing the rotating object desired to be imaged. At the onset of the imaging (t = 0), each scatterer will have an angle, $\varepsilon_i$, $\theta_j$ , and a radius, $\rho_{i,j}$ , from the center of rotation as shown in Figure 7.91. As time passes, this point will undergo partial sinusoidal motion, which will give rise to a Doppler that allows cross-range separation of individual scatterers. The geometry shown in Figure 7.92 is especially useful for recognition of ground vehicles and aircraft in turns as well as air-to-air diagnostics. In straight-line travel, there often is not enough angle change for ISAR on ground vehicles to allow recognition. However, in turns it is possible to form an ISAR image of tactical ground vehicles. A ground vehicle can easily turn through 90° in 3 s or 30°/s. Earlier we saw that a turn of 3° can yield 0.3 m cross-range resolution (7.1).

Figure 7.93 shows a more detailed geometry for a tactical vehicle. The grid on the vehicle has spacing of 0.3 × 0.3 m. There probably will be 350 radar pixels of 0.1 $m^2$ area on a typical light truck. Depending on aspect roughly 200 radar pixels

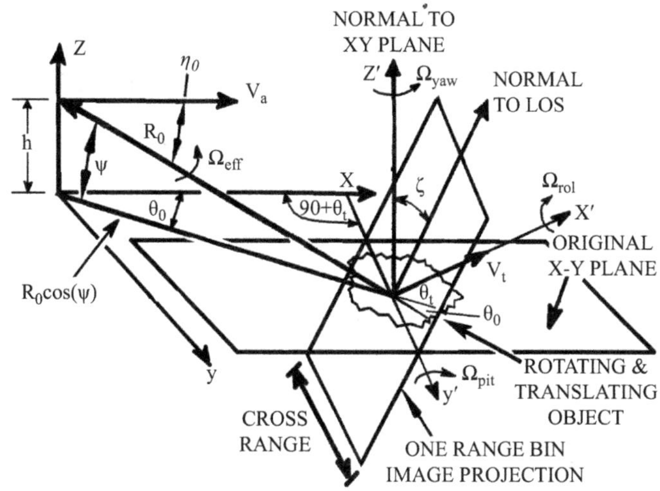


*Figure 7.92 ISAR detailed situation geometry surface tactical vehicles*

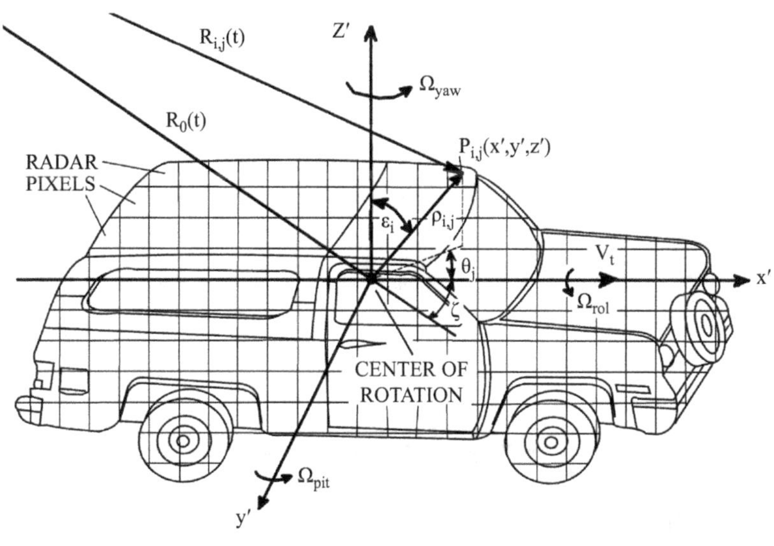


*Figure 7.93 Rotation axes tactical vehicle*

will be visible. Each point $P_{i,j}(x',y',z')$ represents a radar pixel in a grid on the scattering surface of the vehicle, which will undergo rotation during the observation time in roll ($\Omega_{rol}$), pitch ($\Omega_{pit}$) and yaw ($\Omega_{yaw}$). This will give rise to a change in range (phase) for each pixel with time and hence Doppler. Assuming the target vehicle is in track, the relative range change will be just $\Delta R_{i,j}(t) = R_0(t) - R_{i,j}(t)$.

Even though the gross range and counterpart Doppler to the target center of rotation will have been removed by tracking, there will still be translation motion by the observer and target, which will give rise to translation induced yaw. The grid shown on the pickup truck with camper shell has 0.3 × 0.3 m spacing will provide adequate resolution for recognition. It will not provide the technical definition of identification. Although Figure 7.93 seems to show the center of rotation high in the vehicle, the actual center of rotation will be in a plane close to the axles of the wheels and in the center of the front wheel track or about 1.6 m below where it is shown in Figure 7.93.

The equations governing the geometry of Figures 7.92 and 7.93 are shown in (7.23). One can see from (7.23) that the ownship velocity has the effect of changing the apparent velocity of the rotating object (in the example case of Figure 7.91 only in the x dimension since the platform velocity is assumed along the x axis). Unfortunately, the platform velocity and target aspect may not be favorable for most encounters. In addition, for ships there may be motion with different periods for roll, pitch and yaw so that the apparent velocity is continually changing. The trick is to form coarse short-term ISAR subarrays, perhaps as many as 16, then shift the observed filter outputs so that each output covers the same set of pixels as well as accounting for the time delay between subarrays. Then, when these subarrays are coherently combined across the entire ISAR imaging interval, the output image is dramatically better.

Define some intermediate variables:

$$\mathbf{P}_{i,j}(x',y',z') = Px'_{i,j}\cdot\mathbf{i} + Py'_{i,j}\cdot\mathbf{j} + Pz'_{i,j}\cdot\mathbf{k}$$

Where: **i, j, k** are unit vectors in the x', y', z' directions and

$$Px'_{i,j} = \rho_{i,j}\cdot\sin\varepsilon_i\cdot\cos\theta_j,\ Py'_{i,j} = \rho_{i,j}\cdot\sin\varepsilon_i\cdot\sin\theta_j,\ Pz'_{i,j} = \rho_{i,j}\cdot\cos\varepsilon_i$$

$$\zeta = \operatorname{acos}\left(\cos(\psi)\cdot\cos(\theta_0+\theta_t)\right),\ \eta_0 = \operatorname{acos}\left(\cos(\varepsilon_0)\cdot\cos(\theta_0)\right)$$

Object total rotation is translation induced yaw plus attitude roll, pitch and yaw.

$$\mathbf{\Omega}_\Sigma(t) = \mathbf{\Omega}_\mathbf{T}(t) + \mathbf{\Omega}_\mathbf{a}(t)$$

$$\mathbf{\Omega}_\mathbf{T}(t) = \frac{R_0\cdot V_a\cdot\sin\eta\cdot\mathbf{k}}{R_0^2+V_a^2\cdot t^2+2\cdot R_0\cdot V_a\cdot t\cdot\cos\eta} + \frac{R_0\cdot V_t\cdot\sin\zeta\cdot\mathbf{k}}{R_0^2+V_t^2\cdot t^2+2\cdot R_0\cdot V_t\cdot t\cdot\cos\zeta}$$

$$\mathbf{\Omega}_\mathbf{T}(t) \cong \frac{V_a\cdot\sin\eta+V_t\cdot\sin\zeta}{R_0}\cdot\mathbf{k} \text{ and } \mathbf{\Omega}_\mathbf{a}(t) = \Omega_{rol}(t)\cdot\mathbf{i} + \Omega_{pit}(t)\cdot\mathbf{j} + \Omega_{yaw}(t)\cdot\mathbf{k}$$

$$\text{Where: } \Omega_{rol}(t) = \omega_{r0}\cdot\cos\left(2\cdot\pi\cdot {t}/{T_r}+\theta_{r0}\right),\ \Omega_{pit}(t) = \omega_{p0}\cdot\cos\left(2\cdot\pi\cdot {t}/{T_p}+\theta_{p0}\right),$$

$$\Omega_{yaw}(t) = \omega_{y0}\cdot\cos\left(2\cdot\pi\cdot {t}/{T_y}+\theta_{y0}\right),\ \omega_{r0} = {2\cdot\pi\cdot A_r}/{T_r},\ \omega_{p0} = {2\cdot\pi\cdot A_p}/{T_p},\ \omega_{y0} = {2\cdot\pi\cdot A_y}/{T_y}$$

$A_r, A_p, A_y$ = angular amplitudes, $T_r, T_p, T_y$ = angular periods, $\theta_{r0}, \theta_{p0}, \theta_{y0}$ = initial angle

Range to each scatterer with time: $\Delta\mathbf{R}_{i,j}(t) = \mathbf{\Omega}_\Sigma(t)\times\mathbf{P}_{i,j}(x',y',z')$

Phase and frequency with time: $\mathbf{\Phi}_{i,j}(t) = \frac{4\cdot\pi}{\lambda}\cdot\Delta\mathbf{R}_{i,j}(t),\ \mathbf{F}_{i,j}(t) = \frac{2}{\lambda}\cdot\frac{d\Delta\mathbf{R}_{i,j}(t)}{dt}$

(7.23)

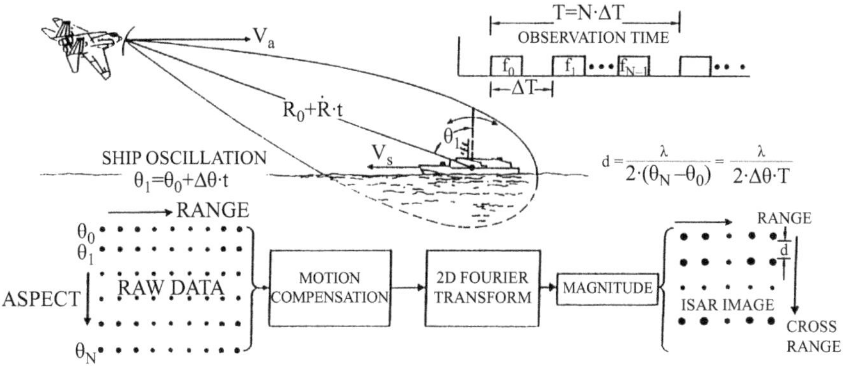


*Figure 7.94 ISAR ship processing [2]*

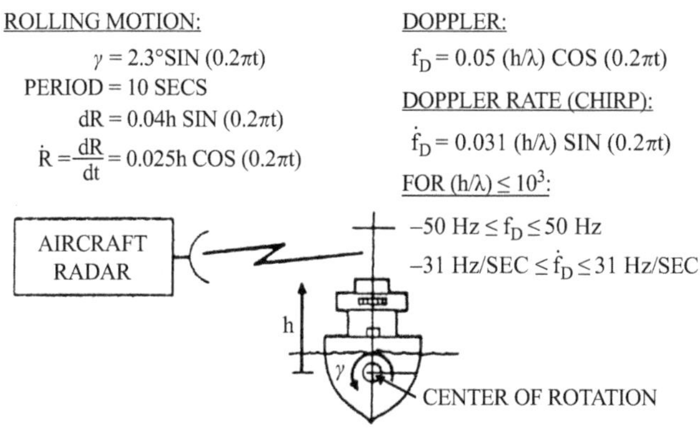


*Figure 7.95 ISAR ship example [2]*

Figure 7.94 shows typical processing of a stepped chirp transmitted waveform common to many ship recognition ISAR systems. During data gathering, motion compensation on data subarrays is performed, followed by a full resolution transform and magnitude detection.

Recall from Section 7.1 equations (7.1) that depending on wavelength the amount of angle change necessary to achieve a given resolution is relatively small. At X band, total angle change of under 3° allows for 0.3 m resolution in most cases. For ships depending on aspect, this may require from 0.5 to 30 s observation time (see Table 3.5). Subarray FFTs are used to estimate position, velocity and acceleration on a short-term basis. Then successive coarse images are compared to correct estimated rotation rate. Each subarray of I/Q data after rotation is then compensated for phase shift between subarrays and integrated into a larger discrete Fourier transform.

A ship in open water exhibits roll, pitch and yaw motions about its center of rotation (c.r.). For example, Figure 7.95 shows a rolling motion of ±2.3° that might

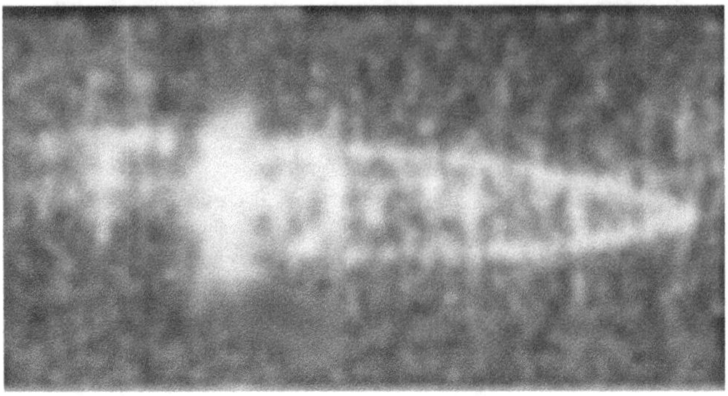

*Figure 7.96 Single ISAR ship image [23]*

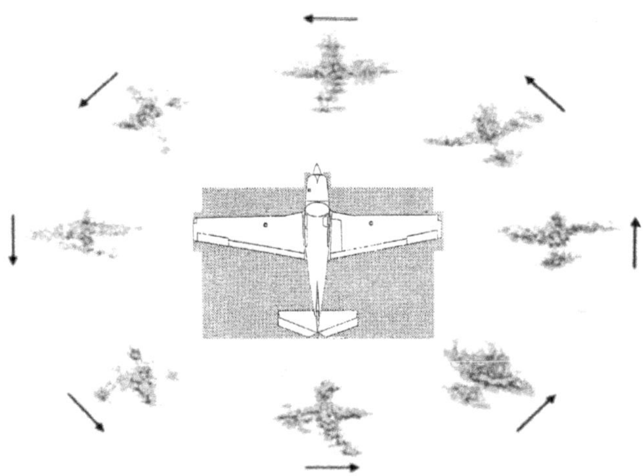

*Figure 7.97 Unrectified surface to air ISAR [21]*

be exhibited by a ship in calm seas. The roll motion, $\gamma$ , might have a period of 10 s (not the same as $\gamma$ in (7.21) and (7.22)). The motion of almost all the scatterers on a large combatant are moving in arcs of circles projected as segments of ellipses to a radar observer. For a radar observer the change in range, dR, associated with a roll movement is a function of the height (radius), h, of the scatterer above the center of rotation, which may be close to the center of gravity. The approximate range rate for each scatterer in rolling (or pitch or yaw) motion at a height, h, is the time derivative of R shown in Figure 7.95. For a given desired cross-range resolution with reasonable sidelobes, $d_{az}$ , $\gamma$ must be equal to $\lambda/d_{az}$. For the example, 0.375 m cross-range resolution is obtainable with a 10 s observation time. The corresponding Doppler and Doppler rates (chirp) are also given in Figure 7.95.

For a ship whose principal scatterers are less than 85 ft above the center of rotation, the Doppler's will be in the range of $\pm 50$ Hz at X band with a rate of change of up to $\pm 31$ Hz/s. As long as the image resolution is not too great, each

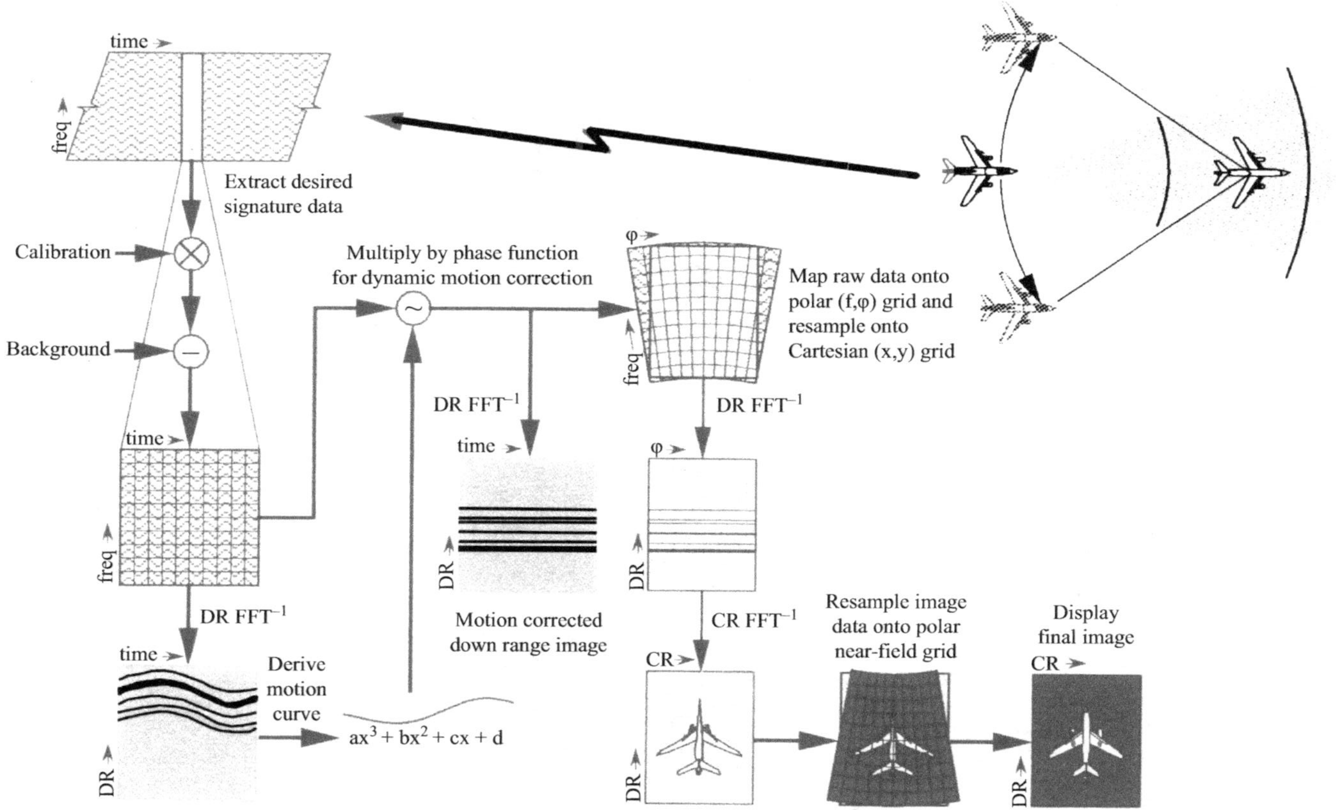


*Figure 7.98 Air-to-air ISAR processing [24]*

range-Doppler bin can be match filtered using the hypothesized motion for each scatterer and an image can be formed on the ship. Each range bin may contain multiple scatterers from the ship in a given roll plane and they may be distinguished by their differing phase history. However, scatterers in the pitch axis at the same range and roll height could not be separated. Although pitch and yaw motions are much slower, they also exist and allow separation in other similar planes.

Reasonably good images coupled with experienced radar operators allow recognition of most surface combatants. Recognition aids using prestored ship profiles allow identification to hull number in many cases. An example of a single ISAR image of a landing assault ship is given in Figure 7.96. The radar in this case is illuminating the ship from the bow at 30 km and 6° grazing. The bright scatterers exhibit cross-range sidelobes, which can be partially reduced by sensing large returns, then applying amplitude weighting and display compression as has been done in this image. Integration of multiple ISAR images also dramatically improves quality.

As previously mentioned, (7.23) also apply to surface to air ISAR. An example of raw unrectified aircraft imaging is shown in Figure 7.97. It shows a sequence of ISAR images of a Mooney light aircraft during real-time turning flight. It is easy to estimate the wingspan and fuselage length from these images. One can also see the geometric distortions caused by uncompensated aircraft velocity even though the aircraft is in track.

Air-to-air ISAR can be used for aircraft or spacecraft recognition as well as RCS diagnostics of stealth performance. Figure 7.98 shows the processing sequence for air-to-air ISAR. The raw data is not much different than any high-resolution airborne or spaceborne radar. Obviously, the data must be calibrated to account for hardware deficiencies. Since high resolution is required, background interference must be cancelled just like any radar. Next a down range (DR) coarse FFT is performed to identify relative motion between the aircrafts. That output is used to derive a third-order motion compensation curve, which is applied to the data. A second coarse DR FFT may be formed to verify that motion has been corrected. The motion-compensated data is entered into a polar format as described in the polar format subsection. Then as shown in Figure 7.98, a DR FFT and CR FFT are performed as in any SAR or ISAR system. One other wrinkle that may be required is compensation for near-field distortion depending on distance between the aircraft. The end result is a rectified image of the target aircraft.

## References

[1] Tylka, J., J.M. Swiger, J. Pearson, D. Lynch, "Synthetic Aperture Radar", *ETS Short Course Notes*, 1977.

[2] Lynch, D., J.O. Pearson and E. Shamash, "Principles of modern radar", *Educational Technology Inc., Short Course Notes*, June 1988.

[3] Hughes Aircraft Co., "Forward Looking Advanced Multimode Radar (FLAMR) Report P78-09R", 1977, Declassified 12/31/87.

[4] Lacomme, P., J-P. Hardange, J-C. Marchais, E. Normant, *Air and Spaceborne Radar Systems: An Introduction*, William Andrew Publishing, Norwich, NY (2001), pp. 329–335, 371–425, 171–176, 469.

[5] Lynch, D., *Introduction to RF Stealth*, SciTech Publishing, Raleigh, NC (2004), pp. 467–470, 516.

[6] Pearson, J., "FLAMR Signal To Noise Experiments", *Hughes Aircraft Report No.* P74-501, December 1974, declassified 12/31/1987.

[7] Craig, D., Hershberger, M., "FLAMR Operator Target/OAP Recognition Study", *Hughes Aircraft Report No.* P74-524, January 1975, Declassified 12/31/1987.

[8] Lynch, D., "Signal Processor for Synthetic Aperture Radar", *SPIE Technical Symposium East 1979*, paper no. 180-35.

[9] Hovanessian, S., *Introduction to Synthetic Array and Imaging Radars*, Artech House, Norwood, MA (1980), Ch. 5.

[10] Curlander, J., R. McDonough, *Synthetic Aperture Radar Systems and Signal Processing*, Wiley & Sons, New York, NY (1991), pp. 99–124, 427–535.

[11] Kovaly, J., *Synthetic Aperture Radar*, Artech House, Norwood, MA (1976), pp. 72–79, 118–123, 249–271.

[12] Lewis, B., *Aspects of Radar Signal Processing*, Artech House, Norwood, MA (1986).

[13] Schlolter, R., "Digital Realtime SAR Processor for C & X band Applications", *IGARSS 1986*, Zurich, Vol. 3, p. 1419.

[14] Fabrizio, R., "A High Speed Digital Processor for Realtime SAR Imaging", *IGARSS 1987*, Ann Arbor, Vol. 2, p. 1323.

[15] Hughes Aircraft Co., "AFCD Final Report", 1982.

[16] Hughes Aircraft Co., Pave Mover TAWDS Final Report, 1984.

[17] General Atomics, Interim Reports 2009, 2011.

[18] Lynch, D., P. Bogdanovic, J.O. Pearson, G. Graham, "Radar Data Processing", *Evolving Technology Institute Course Notes*, 1986.

[19] harris, f.; "Observer Adaptive PRF for SAR Imaging", *IEEE Conference on Acoustics, Speech and Signal Processing*, 1984.

[20] Lynch, D., J.M. Swiger, J. Pearson, R. Thone, "Synthetic Aperture Radar", *ETS Short Course Notes*, 1978.

[21] Jain, A., I. Patel, C. Boerman, W. Kessler, "Imaging and RCS Diagnostics of Aircraft Using a Ground-To-Air Measurement System (GTAMS)", *37th Tri-Service Symposium*, 1991.

[22] Sullivan, R.J., *Radar Foundations for Imaging and Advanced Concepts*, Scitech Publishing, Raleigh, NC (2000) pp. 267–286.

[23] Skolnik, M. Ed., D. Lynch, C Kopp; *Radar Handbook, Ch. 5*; McGraw Hill, New York, NY (2008).

[24] Whelan, D., Original work private communication.

# Appendices

*Disclaimer*: The following software appendices can be obtained online from the IET Digital Library at http://digital-library.theiet.org. From the homepage search for this book by title and once on the book's homepage click on the "supplementary material" tab. For more information, email books@theiet.org. Alternatively, the reader may purchase the complete appendices on a DVD-R from the author by e-mail request at: davidlynchjr@ieee.org. The cost is USD 15.00 to cover mailing. The software appendices are for information only. There are counterpart Adobe Acrobat files of most of the programs with output for a user to read first.

The programs are written in M/S Excel 2000, M/S Quick Basic and Mathcad 11 and 15. If the user has any Mathcad host between 11 and 15, these application programs should run. Files ending in .prn are space-delimited raw data. Files ending in .bmp are amplitude-detected input or output images. The software worked on four different computers with both AMD and Intel CPUs in M/S Windows 7 and XP. But as anyone familiar with these host applications knows, they may not work on your machine and operating environment without some modifications. If the application listed has no file type extension, the extension is .mcd or .xmcd. There is no warranty! The following items of software are in Chapters 1–7 Appendices on the DVD-R or internet.

## A.1 Appendix Chapter 1

1. Cellphone13 – This program calculates the search rate and location accuracy for single cellphone emitters from a LEO spacecraft with a large antenna.
2. Cellphone14d2 – This program calculates the search rate and location accuracy of cellphone emitters from an aircraft. It applies a maximum likelihood estimate of location using a series of angle detections and position estimates.
3. cornoise2 – This program calculates simulated experimental statistics for a correlation-type intercept receiver. The typical configuration multiplies the left and right channels of a monopulse antenna–receiver system together. It follows this with a square law detector and integrator over the look time. For typical monopulse feed spacing, the two channels will be adequately close in TOA for any emitter in the mainbeam so that only a few dB will be lost in decorrelation for a broadband phase code. There will be recurrent ambiguous sidelobe peaks that will also correlate sidelobe interferers. X and Y are the input channels that are assumed Gaussian. The simulation assumes homodyne operation but the statistics are the same for IF processing. The experimental statistics are plotted along with other density functions to

provide an intuition of the relative performance. Moments are calculated. Output mean and variance are estimated. The inputs and comparison statistics have variances of 1.

4. cornoise3 – Similar to cornoise2 but with different parameters.
5. cornoise4 – Similar to the above but with different parameters.
6. cornoise5 – Similar to the above but with different parameters.
7. GMTIDwell – This Excel spreadsheet calculates required dwell times for a hypothetical spacecraft for a series of orbital locations and look angles. Be sure to turn on macros.
8. GMTITotals – This Excel spreadsheet summarizes and compares performance for a series of spacecraft configurations based on "beginning of life on orbit mass." Be sure to turn on macros.
9. footprnt81 – This Mathcad program calculates footprints and power at an intercept receiver using the more accurate model including antenna pattern, weather losses and flat earth range geometry.
10. ftprt3 – This Mathcad program calculates cookie cutter footprints.
11. ftprt4 – This Mathcad program calculates cookie cutter footprints.
12. HA6GMTISAR – This Excel spreadsheet calculates GMTI and SAR performance for a hypothetical 6 ton spacecraft used in the overall evaluation provided in GMTITotals above. Be sure to turn on macros.
13. Monopulsel – This program calculates a simple 1D monopulse antenna pattern using parabolic weighting, which is easy to realize and has a linear difference pattern.
14. PassiveDF6a – This Mathcad program calculates the idealized Gaussian patterns for broadband cavity backed spirals used for amplitude comparison passive direction finding (amplitude AOA). The channel match is assumed 1.25 dB. Quadrant beamwidths chosen to limit discriminant range to less than 20 dB.
15. PassiveDF7a – This Mathcad program calculates the idealized Gaussian patterns for broadband cavity backed spirals used for amplitude comparison passive direction finding. The SNRs are held constant but noise voltage is assumed to be 10 times lower at the log detectors (it makes a large difference and is one of the problems with log detectors). The channel match is assumed 1.5 dB.
16. PassiveDF8 – This Mathcad program calculates the total error for phase comparison passive direction finding (phase AOA) as a function of SNR as well as antenna and processing mismatch.
17. PassiveDF9 – This Mathcad program calculates the total error for phase comparison passive direction finding (phase AOA) as a function of SNR as well as antenna and processing mismatch with somewhat different parameters than PassiveDF8.
18. PassiveDF15 – This Mathcad program calculates the total error for range estimation using ground-based interceptors and phase comparison passive direction finding (phase AOA) as a function of various ranges and SNR as well as antenna and processing mismatch.
19. PassiveDFl6 – This Mathcad program calculates the total error for range estimation using ground-based interceptors and phase comparison passive

direction finding (phase AOA) as a function of a different set of ranges and SNR as well as antenna and processing mismatch.

20. PassiveDF17 – This Mathcad program calculates the total error for range estimation using ground-based interceptors and phase comparison passive direction finding (phase AOA) as a function of a different set of ranges and SNR as well as antenna and processing mismatch.
21. Radiometer2 – This Mathcad program calculates the SNR for an intercept receiver operating in a radiometer mode. The signal is assumed to integrate according to Swerling 1 and grows approximately to the 1.5 power since the input SNR begins at less than 1. (See pages 81 and 82 in Nathanson, *Radar Design Principles*, 1st ed., 1969.) It assumes that there is no interference.
22. rdrcls52 – This is a very simple program to calculate the number of emitters visible to an observer in masking terrain as a function of altitude and ground range to the horizon.
23. Tempapprox1 – This Mathcad program is a temperature approximation for use with radiometer interceptors.

## A.2 Appendix Chapter 2

Please see the disclaimer in the beginning of the Appendices. All of this software is offered to help understanding. Every effort has been made to improve usability and operation. But there is no warranty!

24. AESAWeighting002 – This program calculates an ideal AESA beam pattern in azimuth and elevation using raised cosine (cosine squared) amplitude weighting with $M \times N$ elements.
25. FIRFiltersl – This program calculates a finite impulse filter with a cascade of two filters. One to suppress far sidelobes and other to form the central band lowpass filter.
26. FIRFilters2 – This program calculates two bandpass FIR filters: one for negative frequencies and the other for positive frequencies.
27. FIRFilters3 – This program calculates two FIR filters that might be used in a synthetic monopulse.
28. FIRFilters3c – This program calculates two FIR filters with a linear difference pattern.
29. GMTIPower006d – This program calculates the required transmit power for a given set of parameters. The power profiles with range for target, clutter, noise and jammer are calculated.
30. ISLR2a – This program calculates the ISLR for unweighted, triangular weighted and Parzen weighted FFT filters. This is primarily oriented toward chirp stretch processing. **C** is the multiplier of the 3 dB down point to be used in the ISLR calculation. $N_{rb}$ is the number of equivalent range bins, $T_E$ is the effective pulse width and $\Delta f_{IF}$ is the input bandwidth to the A/D.
31. Islr3 – Antenna sidelobe and ISLR calculation with element pattern and raised cosine weighting, square linear array of $50 \times 50$ elements at 1/2 wavelength spacing.

32. Islr4 – Antenna sidelobe and ISLR calculation with element pattern and Hamming cosine weighting, square linear array of 50 × 50 elements at 1/2 wavelength spacing.
33. Islr5 – This program calculates the ISLR and sidelobes for cosine weighted with edge value 0.25 for a square linear array of $n \times n$ elements at 1/2 wavelength.
34. Islr7 – This program calculates the ISLR and antenna sidelobes with element pattern and segmented parabola weighting, square linear array of 50 × 50 elements at 1/2 wavelength spacing.
35. Islr8 – This program calculates the ISLR and antenna sidelobe with element pattern and cubic segment – Parzen weighting, square linear array of 50 × 50 elements at 1/2 wavelength spacing.
36. Islr9 – This program calculates the ISLR and antenna sidelobe calculation with element pattern and cosine cubed weighting, square linear array of 50 × 50 elements at 1/2 wavelength spacing.
37. Monopulse3b – This program calculates a monopulse pattern for two Hanning weighted subapertures and the corresponding sum and difference patterns.
38. Monopulse4 – This program calculates a monopulse pattern for two parabolic weighted subapertures and the corresponding sum and difference patterns. It also calculates a null in a mainbeam jammer direction.
39. Monopulse7 – This program calculates a monopulse pattern for two Hamming weighted subapertures for sum and Taylor weighted difference patterns.
40. Monopulse8x – This program calculates a monopulse pattern for two Hamming weighted subapertures for sum and Taylor weighted difference patterns. The difference pattern is orthogonal to the sum pattern.
41. pointerr – This Mathcad program calculates the allowable RMS per element phase error for an array as a function of required angular accuracy or cross-beam dimensional accuracy. The approach is taken from *Antenna Engineering Handbook* by H. Jasik ed., pp. 2-36 through 2-41 and *Reference Data for Radio Engineers*, 6th ed., pp. 27-34 through 27-36.
42. Probdet1a – This Mathcad program calculates the probability of detection versus SNR with probability of false alarm as a parameter using the Rice model for linear detection with steady target in Gaussian noise.
43. Probdetxa – This Mathcad program calculates the probability of detection versus SNR with probability of false alarm as a parameter using the Rice model for linear detection with steady target in Gaussian noise. It also calculates a power series linear fit approximation for $P_{FA}$ of $10^{-6}$.
44. R90RCS – This Mathcad program is an analysis of the R90 acquisition range versus radar cross section for several different radars.
45. RadarLookAngle001 – This Mathcad program calculates a radar look angle from ownship latitude, longitude and altitude and a corresponding target latitude, longitude and altitude.
46. rdrclasla – This Mathcad program calculates the antenna sidelobes of a linear array with element pattern and periodic error.

47. rdrclas7a – This Mathcad program calculates antenna sidelobes for a 50-element linear array with raised cosine weighting where the paired echoes are three beamwidths apart instead of four beamwidths. This is a much easier weighting function to realize in a real antenna but it has poorer sidelobes.
48. rdrclas8a – This Mathcad program is an antenna sidelobe calculation with element pattern, Hamming cosine weighting for a linear array of *N* elements of d spacing at 1 wavelength.
49. RDRCLAS19 – This Mathcad program calculates patterns of multibeam electronically scanned linear array antennas with Hamming weighting and 100 MHz beam step frequency. Array size, element spacing, center frequency and frequency step size are easily changeable.
50. Soninel – This Mathcad program is the calculation of the Sonine distribution and sidelobes for circular and rectangular antennas. Integrated sidelobe ratios are also calculated.
51. SynMonoAIQ – This is a space-delimited output file of complex impulse response coefficients for the A side of a synthetic monopulse spatial filter.
52. SynMonoBIQ – This is a space-delimited output file of complex impulse response coefficients for the B side of a synthetic monopulse spatial filter.

## A.3 Appendix Chapter 3

Please see the disclaimer in the beginning of the Appendices. All of this software is offered to help understanding. Every effort has been made to improve usability and operation. But there is no warranty!

53. 5 Scatter RCS1 – This program calculates the radar cross section with angle of a synthetic target made up of five spherical scatterers, one in the middle and four each on moment arms 2 m long and 90° apart.
54. ClutterSpectralSpread – This program calculates clutter exponential Doppler spectral spread based on wind velocity.
55. CluttervsGrazingl – This program calculates relative clutter cross section versus grazing angle for three different models.
56. GMTIDopplersl – This program calculates the probability that a certain Doppler and RCS will be observed for surface targets assuming uniform random velocity and direction.
57. LogNormal1 – This program calculates lognormal distributed clutter probability density.
58. Probden1 – This program calculates one side of several different probability density functions useful in approximating terrain statistics including the Cauchy density function.
59. Probden2 – This program calculates one side of several different probability density functions useful in approximating terrain statistics including the gamma and Weibull density functions.

60. Probden3 – This program calculates one side of several different probability density functions useful in approximating terrain statistics including the Gaussian and chi-squared density functions.
61. Probden4 – This program calculates one side of several different probability density functions useful in approximating terrain statistics including a Cauchy-like density and chi-squared density functions.
62. SlantRange3a – This program calculates slant range and elevation footprint as a function of desired ground range and radar altitude.
63. ClutterSpectralSpreadl – This program calculates clutter velocity spread as a qualitative function of wind velocity.

## A.4 Appendix Chapter 4

Please see the disclaimer in the beginning of the Appendices. All of this software is offered to help understanding. Every effort has been made to improve usability and operation. But there is no warranty!

64. BarkDopp2a – This program calculates ISLR degradation as a function of Doppler or phase shift across a Barker encoded pulse.
65. ChanSepla – This program calculates the output spectrum and adjacent channel leakage for *n* pole filters on transmit and receive using a 13-chip Barker code.
66. ChanSep2a – This program compares the output spectrum and adjacent channel leakage for *n* pole filters on transmit and receive using a 13-chip Barker code and a 169-chip Barker-Barker code.
67. Frankl6 – This program calculates ISLR degradation as a function of Doppler or phase shift across a 16-chip Frank encoded pulse. There are several pulses in the range swath.
68. GMTIStLouis001 – This program calculates conventional monopulse GMTI using real clutter data from a synthetic aperture radar illuminating downtown St. Louis. The original SAR data was presummed before recording and so most of the moving targets were removed. Ten simulated moving targets in each of three range bins are inserted in the data for this program.
69. HilbertTransform002 – This program calculates a typical Hilbert transform half band filter with a 21-point impulse response.
70. ISATPcode4 – This program calculates ISLR degradation as a function of Doppler or phase shift across a 256-chip P4 encoded pulse with sidelobe suppression.
71. ISATPcode4d – This program calculates ISLR degradation as a function of Doppler or phase shift across a 1024-chip P4 encoded pulse with sidelobe suppression.
72. Kalman014 – This program calculates an example Kalman filter tracking a surface target with a large change in target direction.

73. LinFMAmb2 – This Mathcad program calculates the matched filter output for a linear FM in the presence of Doppler. It is related to the ambiguity function of the waveform. It doesn't appear to be stable for large compression ratios somewhere between 100 and 1000.
74. noise004 – This program simulates time domain Gaussian noise through a receiver IF filter, heterodyned to baseband and then applied to a Hilbert half band filter. The Hilbert output is shifted to be symmetrical around DC.
75. SlidingWindow001b – This program calculates a 2D sliding window and its impulse response.
76. SlidingWindow002 – This program calculates a 1D sliding window and its impulse response used for CCD phase compensation.
77. ZeroFill001 – This program shows the relative advantage gained by zero filling the data record and calculating a larger FFT with closer spaced filters.
78. PCOUT9.xls – This Excel spreadsheet hosts a visual basic for applications (VBA) program 'compcode()', which calculates simple complementary codes and plots the resulting outputs. Be sure to turn on macros.
79. Compcdz2.bas – COMPCDZ2 is a Quickbasic program that generates interleaved biphase complementary codes, applies a video filter and timing mismatches, compresses the code, and sums the complements for final output. The generated code is written to a file called PCCODEZ2. The compressed output is written to a file called PCOUTZ2, the maximum code length supported by the program is L chips.
80. Compcodg.bas – This is a Quickbasic program which generates polyphase and biphase complementary codes, applies a series of video filter and phase/timing mismatches, compresses the code and sums complements for final output. It calculates the peak sidelobe ratio, integrated sidelobe ratio and the RMS sidelobe ratio for a single code and the sum of the complements. The generated code is written to a file called PCCODEG. The compressed output is written to a file called PCOUTG, the maximum code length supported by the program is 512 chips.
81. Compcodz.bas – This is a simpler Quickbasic program than Compcodg.bas but still generates interleaved biphase complementary codes, applies a video filter and timing mismatches, compresses the code and sums complements for final output. It calculates the peak sidelobe ratio, integrated sidelobe ratio and the RMS sidelobe ratio for a single code and the sum of the complements. The generated code is written to a file called PCCODEZ. The compressed output is written to a file called PCOUTZ, the maximum code length supported by the program is L chips.
82. Polycoda.bas – This is a very simple Quickbasic program that generates type II polyphase complementary codes, applies a video filter and timing mismatches, compresses the code and sums complements for final output. The generated code is written to a file called Polycode. The compressed output is written to a file called Polyout, the maximum code length supported by the program is 512 chips.

83. Compcodb.bas – This is a very simple Quickbasic program to generate and compress biphase complementary codes. The code resides in A and the complement in B. The maximum code length supported is 1024 chips.
84. STLSARDataDif016.prn – This is a space-delimited file of complex IQ values for a difference channel map of downtown St. Louis. It is used in GMTIStLouis001.
85. STLSARDataGuard016.prn – This is a space-delimited file of complex IQ values for a guard channel map of downtown St. Louis. It is used in GMTIStLouis001.
86. STLSARDataSum016.prn – This is a space-delimited file of complex IQ values for a sum channel map of downtown St. Louis. It is used in GMTIStLouis001.
87. StLouis42ft512x1024.bmp – This is an output amplitude SAR map of downtown St. Louis.
88. Stretch001 – This program performs the stretch pulse compression process on a chirp waveform for both simulated targets and real clutter data.

## A.5 Appendix Chapter 5

Please see the disclaimer in the beginning of the Appendices. All of this software is offered to help understanding. Every effort has been made to improve usability and operation. But there is no warranty!

89. ADNoise – This program calculates noise created by an A/D converter from quantization and saturation.
90. ADNoise002 – This program calculates overall noise performance of A/Ds including drive amplifier, thermal noise, jitter, quantization and saturation. It then calculates the equivalent number of bits compared to an ideal A/D.
91. AESANoiseSidelobes001 – This program calculates sidelobe degradation in an $M \times N$ element AESA caused by phase errors in each TR channel.
92. GMTIPower006 – This program calculates required power from a phased array for a given set of parameters. It also calculates the power received for maximum calculated transmitted power for a target, clutter, jammer and noise at the input to the receiver front end. This version is just at the input to a single T/R module.
93. GMTIPower006a – Same as GMTIPower006 but at the output of the AESA signal distribution manifold.
94. GMTIPower006b – Same as GMTIPower006a but after the A/D converter.
95. GMTIPower006c – Same as GMTIPower006b but after pulse compression.
96. GMTIPower006d – Same as GMTIPower006c but after Doppler filtering.
97. IQImbalance001 – This program calculates the effects of in-phase and quadrature unbalance when shifting a single signal from IF to I/Q channels. The unbalance leads to “ghost” images and targets in phase and amplitude.

98. NoiseSidebandFold – This program calculates antenna sidelobe degradation and ISLR for T/R channel phase error including element pattern and Hamming cosine weighting for a rectangular linear array of $N \times M$ elements of d spacing at l wavelength.

## A.6 Appendix Chapter 6

Please see the disclaimer in the beginning of the Appendices. All of this software is offered to help understanding. Every effort has been made to improve usability and operation. But there is no warranty!

99. AircraftRdr001 – This analysis estimates generic GMTI and SAR radar performance for an aircraft surveilling near earth targets including radar internal spurious signals ("birdies"). The parameters are antenna size, 6 m, platform speed, 250 m/s, range resolution, 1 m and operating wavelength, 0.03 m. The program is very flexible as to parameters.
100. AnApSARDataDif017.prn – This print file is a space-delimited file of partially processed difference channel SAR IQ data. It is used in GMTIAnApSynMono006hl.
101. AnApSARDataSum017.prn – This print file is a space-delimited file of partially processed sum channel SAR IQ data. It is used in GMTIAnApSynMono006hl.
102. AndalusiaAP512x1024.bmp – This is a SAR map of Andalusia Airport, AL. It is used in GMTIAnApSynMono006hl.
103. CancellersB – This program calculates the frequency response for several different clutter canceller filters including ones shifted to the PRF/2. One of the cancellers is a birdie killer.
104. GMTIAnApSynMono006hl – This program takes SAR I/Q data from Andalusia Airport, AL, and uses it to create a synthetic monopulse for MTI clutter cancellation. In some ways this is like adaptive displaced phase center clutter cancellation. It puts 10 MTI targets in each of three range bins. Two samples each are used to synthesize the monopulse beams to create the discriminant
105. GMTIMontAISTAP005 – This program takes reconstructed I/Q data from Montgomery, AL, and uses it to provide STAP-based MTI in mainlobe and sidelobe clutter. Ten moving targets are inserted into each of three range bins. It uses a 2D sliding window for false alarm control followed by a threshold. It includes tilt logic in the sliding window. The data contains a known low-frequency birdie.
106. GMTIStLouis001 – This is the same program as in Appendix Chapter 4. The program calculates conventional monopulse GMTI using real clutter data from a synthetic aperture radar illuminating downtown St. Louis. The original SAR data was presummed before recording and so most of the moving

targets were removed. Ten simulated moving targets in each of three range bins are inserted in the data for this program.

107. GMTIStLouisMono005 – This program uses real radar IQ data taken from a SAR mapping flight of downtown St. Louis, MO. The data was presummed before recording so that there are almost no moving targets in the record. Ten simulated moving targets are inserted in each of three range bins. Monopulse clutter cancellation is performed. The output is run through a 2D sliding window with tilt logic and thresholded. There is a known low-frequency birdie in the data.
108. GMTIStLouisMono006b – This program takes three phase center clutter data for a part of St. Louis metropolitan area and forms sum and difference beams, which it uses to cancel clutter for GMTI detection. The GMTI targets are synthesized in multiple range bins. Three phase centers allow clutter cancellation and target angle location.
109. GMTIStLouisSTAP005x2 – This program performs clutter cancellation using monopulse STAP. The monopulse has both phase and amplitude offsets as most real AESAs have. The clutter data is downtown St. Louis and environs with 42 ft range resolution. Range pulse compression took place before this data was recorded (169:1 Barker-Barker with range sidelobe suppression).
110. GMTIStLouisSTAP010xd – This program uses monopulse with three phase centers to allow angle of arrival estimation as well as clutter cancellation. The antenna face is sidelooking to the velocity vector.
111. GMTIStLouisSTAP020x2g – This program performs clutter cancellation using monopulse STAP. The monopulse has both phase and amplitude offsets as most real AESAs have. The clutter data is downtown St. Louis and environs with 42 ft range resolution. Range pulse compression took place before this data was recorded (169:1 Barker-Barker with range sidelobe suppression).
112. Monopulse22 – This program calculates a monopulse pattern for two Hanning weighted (raised cosine) subapertures and uses them to perform clutter cancellation based on angle of arrival and Doppler.
113. Monopulse28f – This program calculates a monopulse pattern for two Hanning weighted (raised cosine) subapertures and uses them to perform clutter cancellation based on angle of arrival and Doppler.
114. MontAl59ft512x1024.bmp – This is a SAR map of Montgomery, AL. It is used in the GMTI/SAR program GMTIMontAISTAP005.
115. MontAlSARDataDif005.prn – This file is a space-delimited partially processed difference channel IQ data for use in other programs.
116. MontAlSARDataDif018.prn – This file is a space-delimited partially processed difference channel IQ data file for use in GMTIMontAISTAP005 program.
117. MontAlSARDataGuard018.prn – This file is a space-delimited partially processed guard channel IQ data for use in GMTIMontAISTAP005 program.

118. MontAlSARDataSum005.prn – This file is a space-delimited partially processed sum channel IQ data for use in GMTIMontAISTAP005 program.
119. MontAlSARDataSum018.prn – This print file is a space-delimited partially processed sum channel IQ data for use in other programs.
120. MontgomeryAl512.bmp – This is another SAR map of Montgomery, AL. It is used in MontAISARModel001d and MontAISARModel005b programs.
121. MontAISARModel001d – This program takes real magnitude detected SAR map data from Montgomery, AL, and calculates the equivalent I/Q data that would have been used to create the map based on the original antenna size, aircraft speed and altitude, antenna look angle, number of samples and antenna amplitude weighting. The antenna was a monopulse with two overlapping phase centers in azimuth.
122. MontAISARModel005b – This program is similar to MontAISARModel001d. It provides I/Q data output for further processing examples in other programs.
123. RangeChangevsTime007d – This program calculates the change of range with time for a given maximum range, look angle, wavelength, altitude and velocity. It centers the range swath and the phase to zero. It assumes that the velocity is along the $x$ axis. It also compares an approximation to the phase and Doppler.
124. SpaceRdr001 – This analysis estimates generic GMTI and SAR radar performance for a satellite surveilling near earth targets including radar internal spurious signals ("birdies"). The program supports a wide variety of parameters.
125. StLouis42ft512x1024.bmp – This is a SAR map of downtown St. Louis used in GMTIStLouis001, Mono005, STAP005x2 programs.
126. STLSARDATAA022.prn – This file is a space-delimited partially processed (pulse compressed and presummed) A channel IQ data for use in other programs that process three phase centers GMTIStLouisMono006b and GMTIStLouisSTAP010xd.
127. STLSARDATAB022.prn – This file is a space-delimited partially processed B channel IQ data for use in other programs that process three phase centers GMTIStLouisMono006b and GMTIStLouisSTAP010xd.
128. STLSARDATAC022.prn – This file is a space-delimited partially processed C channel IQ data for use in other programs that process three phase centers GMTIStLouisMono006b and GMTIStLouisSTAP010xd.
129. STLSARDataDif016.prn – This file is a space-delimited partially processed (pulse compressed and presummed) difference channel IQ data for use in GMTIStLouis001, -Mono005, -STAP005x2, -STAP20x2g programs.
130. STLSARDataGuard016.prn – This file is a space-delimited partially processed (pulse compressed and presummed) guard channel IQ data for use in GMTIStLouis001, -Mono005, -STAP005x2, -STAP20x2g programs.
131. STLSARDataSum016.prn – This file is a space-delimited partially processed (pulse compressed and presummed) sum channel IQ data for use in GMTIStLouis001, -Mono005, -STAP005x2, -STAP20x2g programs.

## A.7 Appendix Chapter 7

Please see the disclaimer in the beginning of the Appendices. All of this software is offered to help understanding. Every effort has been made to improve usability and operation. But there is no warranty!

132. AircraftRdr001 – This is the same program as in Appendix Chapter 6. The analysis estimates GMTI and SAR radar performance for an aircraft surveilling near earth targets including radar internal spurious signals ("birdies").
133. AircraftRdr002 – This is documentation of the same program as in Appendix Chapter 7, 132 except antenna size, 3 m, platform speed, 125 m/s, range resolution, 0.3 m and operating wavelength, 0.02 m, are different.
134. CCD003 – This program takes four sets of SAR IQ data and cross-correlates them to detect movement, that is, coherent change detection. In this case, the SAR maps are taken 8 s apart with an array time of about 2.25 s. The moving target is a car-sized vehicle traveling at 3 m/s radial velocity on a road in rural California. The SAR resolution is approximately 1.5 m. The moving target has two effects: a Doppler offset, which places the target response at the incorrect spot in the map, and a shadow cast by the vehicle at its true location. Those decorrelation changes can be detected and then an artificial bright spot is placed at each location.
135. CCD003a – This program takes four sets of SAR IQ data and cross-correlates them to detect movement. It is similar to CCD003. In this case, the SAR maps are taken 1 s apart with an array time of about 2.25 s. The moving target is a car-sized vehicle traveling at 3 m/s radial on a road in rural California. The SAR resolution is approximately 1.5 m.
136. ComIndSARIQ010.prn – This print file is a space-delimited partially processed southern California SAR IQ output data from SoCalSAR030a10 program.
137. CommercialIndustrial01ft.bmp – This output SAR map is used in the SoCalSAR030a10 program.
138. Interpolation001 – This program calculates a Hanning weighted eight-sample interpolation function.
139. Interpolation002 – This program calculates a Hamming weighted eight-sample interpolation function.
140. ParkedCars5 – This program calculates the SAR contrast ratio for cars parked in a lot. The contrast ratio for the whole scene as well as for a single car is calculated. The car chosen has for the shape contrast ratio is approximately 6.75 ft by 14.75 ft. Several other statistical measures are also calculated.
141. ParkedCars5.bmp – This output SAR map is used in the ParkedCars5 program.
142. RangeChangevsTime006 – This program calculates the change of range with time for a given maximum range, look angle, wavelength, altitude and

velocity. It centers the range swath and the phase to zero. It assumes that the velocity is along the $x$ axis. It also compares an approximation to the phase and Doppler.

143. RangeChangevsTime007d, -e, -g, -h, -i – These programs calculate the change of range with time for a given maximum range, look angle, wavelength, altitude and velocity with different parameters. It centers the range swath and the phase to zero. It assumes that the velocity is along the $x$ axis. It also compares an approximation to the phase and Doppler.
144. RangeHistory002 – This program calculates the change of range with time for a given maximum range, look angle, wavelength, altitude and velocity. It centers the range swath and the phase to zero. It assumes that the velocity is along the $x$ axis. It also compares an approximation to the phase and Doppler.
145. SoCal1mres002.bmp – This is a SAR map of rural southern California used as input in the CCD003, -003a and SoCalSAR030a6, -6a, -7a, -7c, -8, -8a, -9 programs.
146. SoCalSAR006.bmp – This is a SAR map output from SoCalSAR030a6 of rural southern California used in other SAR or NCD programs.
147. SoCalSAR006a.bmp – This is a SAR map output from SoCalSAR030a6a of rural southern California used in other SAR or NCD programs.
148. SoCalSAR007a.bmp – This is a SAR map output from SoCalSAR030a7a of rural southern California used in other SAR or NCD programs.
149. SoCalSAR007c.bmp – This is a SAR map output from SoCalSAR030a7c of rural southern California used in other SAR or NCD programs.
150. SoCalSAR008.bmp – This is a SAR map output from SoCalSAR030a8 of rural southern California used in other SAR or NCD programs.
151. SoCalSAR008a.bmp – This is a SAR map output from SoCalSAR030a8a of rural southern California used in other SAR or NCD programs.
152. SoCalSAR009.bmp – This is a SAR map output from SoCalSAR030a9 of rural southern California used in other SAR or NCD programs.
153. SoCalSARIQ006.prn – This print file is a space-delimited partially processed (pulse compressed and presummed) southern California SAR IQ output data usable in change detection program CCD003.
154. SoCalSARIQ006a.prn – This print file is a space-delimited partially processed southern California SAR IQ output data usable in change detection program CCD003a.
155. SoCalSARIQ007a.prn – This print file is a space-delimited partially processed southern California SAR IQ output data usable in change detection program CCD003a.
156. SoCalSARIQ007c.prn – This print file is a space-delimited partially processed southern California SAR IQ output data usable in change detection program CCD003.
157. SoCalSARIQ008.prn – This print file is a space-delimited partially processed southern California SAR IQ output data usable in change detection program CCD003.

158. SoCalSARIQ008a.prn – This print file is a space-delimited partially processed southern California SAR IQ output data usable in change detection program CCD003a.
159. SoCalSARIQ009.prn – This print file is a space-delimited partially processed southern California SAR IQ output data usable in change detection programs CCD003, -a.
160. SoCalSARIQ010.prn – This print file is a space-delimited partially processed southern California SAR IQ input data usable in SoCalSAR030a10 program.
161. SoCalSum.prn – This print file is a space-delimited partially processed southern California SAR IQ data input usable in SoCalSAR030 programs.
162. SoCalSum002.prn – This print file is a space-delimited partially processed southern California SAR IQ data input usable in SoCalSAR030 programs.
163. SoCalSum003.prn – This print file is a space-delimited partially processed southern California SAR IQ data input usable in SoCalSAR030 programs.
164. SoCalSum004.prn – This print file is a space-delimited partially processed southern California SAR IQ data input usable in SoCalSAR030 programs.
165. SoCalSum005.prn – This print file is a space-delimited partially processed southern California SAR IQ data input usable in SoCalSAR030 programs.
166. SoCalSum006.prn – This print file is a space-delimited partially processed southern California SAR IQ data input usable in SoCalSAR030 programs.
167. SoCalSum006a.prn – This print file is a space-delimited partially processed southern California SAR IQ data input usable in SoCalSAR030 programs.
168. SoCalSum007.prn – This print file is a space-delimited partially processed southern California SAR IQ data input usable in SoCalSAR030 programs.
169. SoCalSum007a.prn – This print file is a space-delimited partially processed southern California SAR IQ data input usable in SoCalSAR030 programs.
170. SoCalSum008.prn – This print file is a space-delimited partially processed southern California SAR IQ data input usable in SoCalSAR030 programs.
171. SoCalSum008a.prn – This print file is a space-delimited partially processed southern California SAR IQ data input usable in SoCalSAR030 programs.
172. SoCalSAR030a6 – This program takes SoCal Sum I/Q data and inserts a car-sized moving target on a rural road in the scene. It then forms the SAR map including the mover. It also inserts a shadow at the true mover position. It is the second in a four-map sequence 8 s from the starting position.
173. SoCalSAR030a6a – This program takes SoCal Sum I/Q data and inserts a car-sized moving target on a rural road in the scene. It then forms the SAR map including the mover. It also inserts a shadow at the true mover position. It is the second in a four-map sequence 1 s from the starting position.
174. SoCalSAR030a7c – This program takes SoCal Sum I/Q data and inserts a car-sized moving target on a rural road in the scene. It then forms the SAR map including the mover. It also inserts a shadow at the true mover position. It is the third in a four-map sequence 16 s from the starting position.
175. SoCalSAR030a7a – This program takes SoCal Sum I/Q data and inserts a car-sized moving target on a rural road in the scene. It then forms the SAR

map including the mover. It also inserts a shadow at the true mover position. It is the third in a four-map sequence 2 s from the starting position.

176. SoCalSAR030a8 – This program takes SoCal Sum I/Q data and inserts a car-sized moving target on a rural road in the scene. It then forms the SAR map including the mover. It also inserts a shadow at the true mover position. It is the fourth in a four-map sequence 24 s from the starting position.
177. SoCalSAR030a8a – This program takes SoCal Sum I/Q data and inserts a car-sized moving target on a rural road in the scene. It is the fourth in a four-map sequence 3 s from the starting position.
178. SoCalSAR030a9 – This program takes SoCal Sum I/Q data and inserts a car-sized moving target on a rural road in the scene. It then forms the SAR map including the mover. It also inserts a shadow at the true mover position. It is the start of two four SAR map sequences.
179. SoCalSAR030a10 – This program takes in raw IQ SAR data and match filters it. The antenna is a sidelooking ku-band ESA 1 m in length by 0.8 m in height. Angles are measured from the velocity vector. The SAR raw data is after pulse compression and presummation. Platform velocity is 100 m/s, range is 20 km, altitude 4.5 km, post-presum sample rate is 398.5/s. The output is compressed SAR IQ data and a SAR bitmap.
180. Space Radar Assumptions.pdf – This Adobe Acrobat file contains figures describing the geometries assumed in SpaceRdr001.
181. SpaceRdr001 – This is the same program as in Appendix Chapter 6, 124. This analysis estimates generic GMTI and SAR radar performance for a satellite surveilling near earth targets including radar internal spurious signals ("birdies"). The program supports a wide variety of parameters.

# Glossary

| | |
|---|---|
| 1D | One dimensional |
| 2D | Two Dimensional |
| 3D | Three dimensional |
| 4D | Three dimensions in space and one in time |
| A | Acceleration of platform |
| a | Amplitude weighting coefficient |
| $\alpha$ | Cone angle, secondary phase angle, SAR turn angle or surface roughness factor |
| $\alpha_p$ | Platform heading angle |
| A beam | One of a set of antenna beams combined to form an angle discriminant |
| A/D | Analog-to-digital converter |
| $a_{bias}$ | INS accelerometer bias |
| ADR | Automatic data reduction |
| ADRC | Automatic data reduction controller |
| AESA | Active electronic scan antenna |
| AM | Amplitude Modulation |
| AMTI | Airborne moving target indication. |
| AN | Analyst |
| AOA | Angle of arrival |
| AOR | Area of regard |
| AR | Arithmetic rate required for radar signal processor |
| ARM | Antiradiation missiles |
| ASARS | Advance strategic airborne radar system |
| ASIC | Application-specific integrated circuit, sometimes called a gate array |
| $a_{rms}$ | INS RMS accelerometer error |
| $a_{scale\ factor}$ | INS accelerometer scale factor |
| $a_T$ | Amplitude of target return |
| b | A/D quantization (bits) |
| $\beta$ | Angle, often measured from the center of the Earth |

| | |
|---|---|
| $\beta_p$ | Platform pitch angle |
| $\beta_{(\,)}$ | Secondary phase angle for subscripted item |
| $B_{(\,)}$ | Bandwidth for subscripted item |
| $B_{A/D}$ | A/D sampling rate |
| B beam | One of a set of antenna beams combined to form an angle discriminant |
| bandpass | Passing only those frequencies in the normal emitter band of operation |
| $B_{amlc}$ | Azimuth mainlobe clutter bandwidth |
| $B_{antDop}$ | Doppler bandwidth passed by antenna |
| BDA | Battlefield damage assessment |
| $B_{DL}$ | Data rate of data link |
| $B_{dop}$ | Mapping Doppler bandwidth |
| $B_{emlc}$ | Bandwidth of mainlobe clutter in elevation |
| *BER* | Bit error rate |
| $B_{IF}$ | IF bandwidth ($= B_RF$ for strip mapping modes) |
| BIT | Built-in test |
| BM | Bulk memory |
| BRF | Mapping RF bandwidth |
| $BW_{(\,)}$ | Beamwidth for subscripted item |
| $BW_{Exp}$ | Broadening ratio |
| c | Speed of light in free space, $2.9979 \times 10^8$ m/s |
| C | Clutter backscatter power |
| CAGC | Coarse automatic gain control |
| CCD | Coherent Change Detection |
| CFAR | Constant false alarm rate |
| chip | single time slot or symbol in a coded waveform |
| clutter | radar backscatter from Earth's surface; Unwanted signals, echoes, or images that appear on face of display screen, interfering with observation of desired signals. |
| CDPE | Change detector processing element |
| CDR | Clutter-to-distortion (or disturbance) power ratio |
| CMT | Covariance matrix taper |
| CNR | Clutter-to-thermal noise power ratio |
| Comint | Communications intelligence systems. |
| CoR | Center of rotation |
| CPI | Coherent processing interval |
| CPU | Central processing unit |
| CR | Contrast ratio |

| | |
|---|---|
| $CR_H$ | Contrast required for hole detection |
| crosstalk | Unwanted signals in communication channel caused by transference of energy from another circuit |
| CW | Continuous wave |
| $d_{( )}$ | Element spacing or incremental resolution for subscripted item |
| $\Delta G$ | Elevation beam gain loss |
| $\Delta$ | Difference antenna beam in angle space |
| $\Delta f$ | Change of frequency with time (chirp) |
| $\Delta\Delta f$ | Acceleration of frequency with time |
| $D_x$ | antenna horizontal length |
| $D_y$ | antenna vertical length |
| D/L | Data link |
| DARPA | Defense Advanced Research Projects Agency |
| data link | Radio used to transmit digital or analog information in machine-readable format. Used to transmit radar, sonar, imagery and other sensor data, as well as weapons guidance command, fire control information and missile telemetry |
| $d_{az}$ | Azimuth resolution |
| dB | Decibels, a logarithmic relative ratio, that is, 10 $\log_{10}$ (Power1/Power2) |
| DE | Data entry |
| deinterleave | To separate multiple mixed pulse trains from one another to allow measurement of each |
| DFT | Discrete Fourier transform |
| $d_m$ | Actual mapping resolution (multilook) |
| DME | Distance measurement equipment |
| DMS | Data management system |
| DN | Data network |
| $d_o$ | Potential mapping resolution (single look) |
| DOA | Direction of arrival systems |
| $d_p$ | Azimuth resolution projected on cross-track axis |
| DPCA | Displaced phase center antenna, clutter cancellation method |
| $d_r$ | Range resolution |
| $D_\phi$ | Distance between phase centers |
| $D_{PRF}$ | Distance between PRIs |
| $D_s$ | Synthetic array length |
| DSU | Disk storage unit |

| | |
|---|---|
| $\Delta v$ | INS velocity error |
| $E$ | Measured voltage |
| **E** | vector electric field |
| $\varepsilon_0$ | Permittivity of free space, $8.85 \times 10^{-12}$ F/m |
| $\varepsilon$ | Depression angle or total error standard deviation |
| $\varepsilon_r$ | Elevation or depression angle in radar coordinates |
| $\varepsilon_s$ | Elevation or depression angle in stabilized coordinates |
| $\varepsilon_{el}$ | Antenna elevation beamwidth |
| $\varepsilon_u$ | Depression angle to upper edge of swath |
| $E_{( )}$ | Electric field component for subscripted item |
| EC | Exploitation controller |
| ECCM | Electronic counter countermeasures |
| ECM | Electronic countermeasures |
| ELINT | electronic intelligence systems |
| emission | Electromagnetic waves radiated by any electronics or physical process; substances (exhausts) discharged into the air |
| EO | Electro-optical, system for visible or near-visible imaging |
| EOB | Electronic order of battle. Deployment and utilization of communications, radar, electronic countermeasures, and radar intercept equipment in (adversary) military doctrine and tactics. *See also* order of battle |
| EOC | Exploitation controller |
| ERP | Effective radiated power (gain × average power) |
| ERPP | Effective radiated peak power |
| ERTOS | Embedded real-time operation system |
| ESA | Electronically scanned array |
| ESM | Electronic support measures |
| $\phi$ | Electrical phase angle |
| $\varphi$ | Space angle around the $z$ axis measured from the $x$–$z$ plane |
| $\phi_q$ | Quadratic phase error, center-to-end of synthetic array |
| $\phi_{qa}$ | Quadratic phase error from INS accelerometer bias |
| $\phi_{qv}$ | Quadratic phase error from INS velocity error |
| $f$ | Frequency |
| $f_{( )}$ | Frequency for subscripted item |
| FAGC | Fine automatic gain control |
| FEBA | Forward edge of battle area |
| FEXT | Far-end crosstalk |

| | |
|---|---|
| FFT | Fast Fourier transform |
| FIR | Finite impulse response (filter). |
| FLAMR | Forward-looking airborne multimode radar (Hughes/AFAL) |
| FM | Frequency Modulation |
| $f_s$ | Sampling frequency, often the PRF |
| FSS | Frequency selective surface |
| $\gamma$ | Ground surface backscatter coefficient or coherent change correlation or roll angle |
| FT | Fourier Transform |
| $\gamma_{\mathbf{p}}$ | Platform roll angle |
| $\Gamma_{(\ )}$ | Complex reflection factor for subscripted item |
| $\Gamma_{\mathrm{EU}}$ | Incomplete Euler gamma function |
| $G$ | Antenna gain (one-way) |
| $G_{(\ )}$ | Antenna or processing gain for subscripted item |
| GERP | Antenna gain × effective radiated power |
| GHz | Gigahertz, $10^9$ Hz |
| GMTD | Ground-moving target detection |
| GMTI | Ground-moving target indication |
| GMTT | Ground-moving target tracking |
| GO | Geometric optics |
| GPP | General purpose processor |
| GPS | Global positioning system |
| $H(\omega)$ | Fourier transform of the impulse response |
| $\mathbf{H}$ | Vector magnetic field |
| $h(t)$ | Circuit impulse response |
| h | Altitude of platform |
| H | Hole backscatter power |
| $h_{(\ )}$ | Altitude for subscripted item |
| $H_{(\ )}$ | Magnetic field component for subscripted item |
| $\mathrm{h_{ant}}$ | Antenna height |
| Hz | Hertz (cycles per second) |
| $\eta$ | Compound angle made up of $\varepsilon$ and $\theta$ |
| $i$ | Index, usually representing angle in this text |
| $\mathbf{I}$ | $N \times N$ identity matrix, that is, ones along the diagonal, zeros everywhere else |
| I/O | Input–output circuitry specialized for application |
| $\mathrm{I_0}$ | Modified Bessel function of the first kind, order 0 |
| $\mathrm{I_a}$ | Amplitude of interference |

| | |
|---|---|
| IDP | Image display processor |
| IF | Intermediate Frequency, a convenient frequency for amplifying, gain adjustment and band limiting of a received signal |
| IMU | Inertial measuring unit (of INS) |
| INS | Inertial navigation system |
| interleave | To mix multiple independent codes or functions into a single time line |
| IOP | Input–output processor |
| IR | System for infrared imaging |
| ISAR | Inverse synthetic aperture radar, that is, Doppler processing to image a rotating body |
| ISLR | Integrated sidelobe-to-mainlobe power ratio (total) |
| $ISLR_a$ | Component of ISLR from INS RMS accelerator scale factor error |
| $ISLR_{A/D}$ | Component of ISLR from A/D distortion |
| $ISLR_{Amb}$ | Component of ISLR from ambiguities |
| $ISLR_{M/C}$ | Component of ISLR from motion compensation errors |
| $ISLR_V$ | Component of ISLR from INS velocity quantization |
| $ISLR_W$ | Component of ISLR from atmospheric dispersion (weather) |
| $j$ | Index, usually representing range in this text |
| j | Square root of –1 |
| $J_0$ | Bessel function of the first kind, order 0 |
| $J_1$ | Bessel function of the first kind, order 1 |
| JTIDS | Joint Tactical Information Distribution System, sometimes called Link 16 |
| $k$ | Index |
| $K_a$ | Attenuation caused by atmosphere |
| $k_B$ | Boltzmann's constant |
| $K_{cl}$ | Attenuation caused by clouds |
| $\mathbf{K}_k$ | Kalman filter gain matrix |
| $K_0$ | Mid elevations backscatter coefficient |
| $K_r$ | Attenuation caused by rain |
| Ku-band | 12–18 GHz |
| KFC | Ku-band fire control radar |
| $k_w$ | sidelobe weighting factor |
| kW | Kilowatts |
| $\ell$ | spacing between antenna elements |
| $l$ | Incremental length or index |
| $\lambda$ | Operating wavelength, typically 1 foot to 1/4 in. |

| | |
|---|---|
| $L_{( )}$ | Loss associated with subscripted item |
| LAPSIM | Low-altitude penetrator simulation (software program) |
| Lat. | Latitude |
| la | Latitude |
| Lon. | Longitude |
| lo | Longitude |
| LO | Local Oscillator; Low observable system designed to make object difficult to detect. *See also* signature, passive |
| LORAN | Long-range radio navigation |
| LOS | Line of sight |
| lossy | Causing weakening or dispersion of electrical energy |
| LR/SO | Long range/standoff |
| LREW | Long-range early warning radar |
| LVPS | Low-voltage power supply |
| $m$ | Index |
| m | Data rate multiplier for presuming, abbreviation for meters |
| $\mu$s | Microsecond, $10^{-6}$ s |
| $\mu_0$ | Permeability of free space, $4\pi \times 10^7$H/m |
| MAG. TAPE | Instrumentation magnetic tape recorder |
| MASINT | Measurement and signal intelligence |
| mB | Millibars |
| MCU | Microprocessor control unit, very limited programmable computer |
| MDL | Missile data link |
| MDV | Minimum detectable radial velocity of a moving target |
| MHz | Megahertz |
| MIMD | Multiple instruction, multiple data |
| MIMO | radar with Multiple Inputs and Multiple Outputs |
| MLE | Maximum likelihood estimate |
| Monte Carlo simulation | Random input variable stimulation to a system or model and output characterization; technique used to obtain approximate solution to mathematical or physical problem |
| MO | Master Oscillator, a low noise oscillator which controls most of the frequencies and timing in an interceptor or radar |
| MPPS | Million pulses per second |
| MR | Recording capacity |
| MTAE | Multiple time around echoes |
| MTI | Moving target indication or detection |
| MTT | Moving target tracking |
| $N$ | Noise power or if in a subscript a noise term |

| | |
|---|---|
| $n$ | Index, usually representing time or frequency in this text |
| $\mathbf{n}$ | Vector surface normal |
| $N_{( )}$ | Total number for subscripted item |
| NCD | Non-coherent Change Detection |
| NED | North East Down coordinate system |
| NEXT | Near-end crosstalk |
| $NF_{( )}$ | Noise figure for subscripted item |
| NRL | Naval Research Laboratories |
| $\omega$ | Angular frequency |
| $\Omega$ | Solid angle, rotation angle |
| $\Omega_{roll}$ | Roll rotation angle |
| $\Omega_{pit}$ | Pitch rotation angle |
| $\Omega_{\mathbf{T}}$ | total rotation |
| $\Omega_{yaw}$ | Yaw rotation angle |
| O.R. | PRF oversample ratio (relative to Doppler bandwidth of antenna) |
| order of battle | strength, command structure and disposition of the personnel, units and equipment of a military force. *See also* electronic order of battle |
| $\mathbf{\Phi}$ | state transition matrix |
| $P$ | Probability |
| $P_k$ | error covariance matrix |
| $P_N$ | noise power |
| $p$ | Index or incremental probability |
| P( ) | Position of point in coordinate units |
| P | Position of platform |
| $P_{( )}$ | Power for subscripted item |
| $P_{( )}$ | Probability for subscripted item |
| $P_{ave}$ | Average transmitted power |
| PCR | Pulse compression ratio |
| $P_d$ | Probability of target detection |
| PDI | Post detection integration |
| $P_{fa}$ | Probability of false alarm in target detection |
| $P_{hd}$ | Probability of hole detection |
| $P_{hfa}$ | Probability of false alarms in hole detection |
| $\mathbf{P}_{i,s}$ | Vector incident or reflected power |
| planform | Contour of object as seen from above |
| platform | A weapon system such as an aircraft, ship, tank, mobile shelter, spacecraft, etc. |
| PLSS | Precision Location Strike System |
| PO | Physical optics |

| | |
|---|---|
| $P_{pk}$ | Peak transmitted power |
| Presum Ratio | Factor relating input PRF to output sample rate of presum filter |
| PRESUM | filter before main signal processing |
| *PRF* | Pulse repetition frequency |
| *PRI* | Pulse repetition interval (= 1/PRF) |
| PSLR | Peak sidelobe ratio |
| PSP | Programmable signal processor |
| PUL($t,\tau$) | Rectangular pulse function, zero everywhere except during a pulse of height 1 and duration associated to $\tau$, starting at $t=0$ |
| θ | Angle, usually in azimuth or squint angle (measured from velocity vector) |
| $\theta_{\mathrm{az}}$ | Antenna beamwidth in azimuth |
| $\theta_r$ | Azimuth pointing angle in radar coordinates |
| $\theta_s$ | Azimuth pointing angle in stabilized coordinates |
| θ | Space angle measured from *z* axis in spherical coordinates |
| $Q_k$ | process noise covariance matrix |
| $R_{\mathrm{k}}$ | measurement noise covariance matrix |
| $R$ | Slant range from radar to target |
| $\rho$ | Radius from a central point in a map |
| $R_E$ | Range to an emitter or to a point |
| $\mathrm{R_e}$ | Radius of the earth, commonly approximated as 6,378,137 m |
| $R_g$ | Ground range |
| $R_0$ | Initial range to a point or to the center of a patch |
| $R_{slant}$ | Along antenna beam slant range |
| radar | (*Ra*dio *d*etecting *a*nd *r*anging) a system for locating objects by means of electromagnetic waves reflected from object for purpose of determining object location and direction of travel |
| radiometer | Instrument for measuring thermal, electromagnetic or acoustic total radiated energy |
| radome | Housing to protect antenna assembly of a communications, intercept receiver or radar set |
| RAM/RAM(s) | Radar-absorbing material(s) |
| range | Maximum distance at which a radio or radar transmission can be effectively received |
| RAR | Real aperture radar |
| RCS | Radar cross section; (1) ratio of the scattered power to the incident power in the direction of an observer at infinity, (2) image produced by radar signals reflected off target surface |

| | |
|---|---|
| REC | Instrumentation magnetic tape recorder |
| RF | Radio frequency, typically 0.5 MHz to 230 GHz |
| $RG_{max}$ | Maximum ground range |
| $RG_{min}$ | Minimum ground range |
| RHAW | Radar homing and warning receiver |
| rms | Root-mean-square |
| RMSLR | Root-mean-square sidelobe ratio |
| RSO | Radar systems operator |
| RSP | Radar signal processor |
| RSS | Root sum square |
| RWR | Radar warning receiver |
| S | Swath length in range |
| $\sigma_0$ | Clutter backscatter coefficient per unit ground area |
| $\sigma_{0\mathrm{sea}}$ | sea backscatter |
| $\sigma_{1\mathrm{rain}}$ | rain backscatter |
| $\sigma_{\mathrm{ave}}$ | Clutter backscatter coefficient for average clutter |
| $\sigma_{\mathrm{c}}$ | Clutter crosssection |
| $\sigma_{\mathrm{hole}}$ | Clutter backscatter coefficient for hole area |
| $\sigma_{\mathrm{T}}$ | Average radar crosssection for target |
| $\sigma_{\mathrm{v}}$ | clutter velocity standard deviation |
| $\sigma_{(\,)}$ | Radar cross section or standard deviation for subscripted item |
| $\Sigma$ | Sum beam in angle space |
| S&TA | Surveillance and Target Acquisition radar |
| $S_a$ | Desired ambiguity suppression |
| SAR | Synthetic aperture radar |
| SCR | Signal to clutter ratio |
| SDB | Sensor data buffer |
| Sh | Surface area of hole |
| signature, active | Observable emissions from platform: acoustic, chemical (soot and contrails), communications, radar, IFF, IR, laser and UV |
| signature, passive | Platform observables, requiring external illumination: magnetic and gravitational anomalies; reflection of sunlight and cold outer space; reflection of acoustic, radar and laser illumination; reflection of ambient RF |
| sinc( ) | Symbol for sin x/x function |
| sins( ) | Symbol for sin(Nx)/(N sin x) function |
| SLAR | Side-looking airborne radar |
| SLW | Sliding window threshold process |

| | |
|---|---|
| SMTD | Slow-moving target detection |
| SNR | Signal to noise ratio, peak-power-signal-to-mean-noise power |
| SINR | Signal to interference and noise ratio |
| SOG | System operations controller |
| $S_t$ | Spot size for track mode |
| STAP | Space-Time Adaptive Processing |
| STC | Sensitivity Time Control, receiver gain adjusted during the return from a radar swath |
| Stealth | Popular term for "low observables" that is, technologies, materials and designs intended to reduce the radar, infrared, acoustic and visual signatures of aircraft, missiles and ships |
| $t$ | Time or turbulence |
| T | Synthetic array time, interpulse period |
| $T_D$ | Interceptor dwell time |
| $T_e$ | effective temperature |
| $T_E$ | Effective pulse length |
| $T_0$ | Standard temperature, 290 K |
| TO | Transmit Oscillator, locked to the LO by an offset frequency equal to the IF frequency |
| $T_{sw}$ | Range swath time |
| $\tau$ | Pulsewidth or time delay |
| TA | Terrain avoidance |
| TACAN | Tactical air navigation |
| TACINT | Tactical Interface |
| TCR | Target-to-clutter power ratio |
| TDMA | Time division multiple access, a communications multiplex method |
| TDR | Target-to-disturbance power ratio |
| TF | Terrain following |
| *Th* | Threshold |
| *THF* | Frequency dimension threshold |
| *THR* | Range dimension threshold |
| $T_I$ | Interceptor search time |
| $t_L$ | Time per look in a single cell |
| *TM* | Threshold multiplier, factor chosen to minimize false alarms |
| *TMF* | Frequency dimension threshold multiplier |
| *TMR* | Range dimension threshold multiplier |
| TOA | Time of Arrival |

| | |
|---|---|
| $T_{OT}$ | Time on target |
| $\mathrm{T_R}$ | Recording time |
| $T_{rb}$ | time for a single range bin |
| TWT | Traveling wave tube (type of transmitter) |
| $u$ | sin θ cos φ, $x$ direction cosine |
| UCIF | Universal console interface; standard PC keyboard, mouse and display interface |
| UHR | Ultra-high resolution (0.1–2 ft) |
| *USAF* | U.S. Air Force |
| $v$ | sin θ sin φ, $y$ direction cosine |
| v, V, Va | Velocity of radar platform |
| VCO | Voltage-controlled oscillator |
| VFO | Variable frequency oscillator |
| $\mathrm{V_q}$ | Velocity quantization |
| $W_{AT}$ | Swath width along radar track |
| $W_{CT}$ | Swath width cross (orthogonal) radar track |
| $w$ | cos θ, $z$ direction cosine |
| $w_r$ | Amplitude weighting associated with a rectangular antenna |
| $w_y$ | Amplitude weighting in $y$ dimension |
| $w_x$ | Amplitude weighting in $x$ dimension |
| $\mathbf{W}\odot\mathbf{W}$ | Hadamard product, that is, each element in first matrix multiplied by counterpart element in second matrix |
| $\mathbf{W}\otimes\mathbf{W}$ | Kroneker product, that is, each element in first matrix multiplied by all elements in second matrix |
| $\mathbf{W}(n,i,j)^{-1}$ | Matrix inversion function |
| $\mathbf{W}(n,i,j)^{\mathrm{T}}$ | Transpose matrix function, that is, exchange rows for columns |
| $\mathbf{W}(n,i,j)^{\mathrm{H}}$ | Hermitian matrix function, that is, matrix conjugate transpose |
| WAHTD | Wide area hard target (stationary) detection |
| WAMTD | Wide area moving target detection |
| $\mathrm{W_c}$ | Water content of clouds (g/m$^3$) |
| $\zeta$ | Angle between line of sight and target velocity vector |
| $\psi$ | Grazing angle, that is, angle between the line of sight and Earth's surface |
| $\psi_{\mathrm{heel}}$ | Grazing angle to lower (closer) edge of range swath |
| $\psi_{\mathrm{toe}}$ | Grazing angle to upper (farther) edge of range swath |
| X-band | 8–12.5 GHz |
| XFC | X-band fire control radar |
| $\mathbf{Z}$ | Raw data vector, usually in Kalman filters or STAP |

# Index

www.ingramcontent.com/pod-product-compliance
Lightning Source LLC
LaVergne TN
LVHW020055220926
848484LV00002BB/484